DISCARD

Water Science and Application 1

Lake Champlain in Transition
From Research Toward Restoration

Thomas O. Manley
Patricia L. Manley
Editors

American Geophysical Union
Washington, DC

Published under the aegis of the AGU Books Board
Andrew Dessler, Chairman; John E. Costa, Jeffrey M. Forbes, W. Rockwell Geyer, Rebecca Lange, Douglas S. Luther, Walter H. F. Smith, Darrell Strobel, and R. Eugene Turner, members.

F
127
.C6
L34
1999
c.3

Lake Champlain in Transition: From Research Toward Restoration
Water Science and Application 1
ISBN 0-87590-350-9

Cover: A view over the Champlain Basin toward the Adirondack Mountains of New York (taken from Mount Philo State Park, Charlotte, Vermont). In this image, much of the land is agricultural and forest. The area of Lake Champlain that is viewed is that of the southern portion of the Main Lake where it narrows significantly between Thompsons Point and Split Rock. The inset on the back cover represents land-use coverage within Lake Champlain Basin; this image was created from Landsat Thematic Mapper satellite imagery and generated using GIS software. Although there are 19 separate land-use types, in general, the yellows, tans, blues, and greens represent urban, agricultural, wetlands and forest lands, respectively.

Photograph courtesy of Douglas Facey, St. Michael's College, Colchester, Vermont.

GIS imagery courtesy of Northern Cartographic, Burlington, Vermont.

Copyright 1999 by the American Geophysical Union, 2000 Florida Ave., NW, Washington, DC 20009, USA.

Figures, tables, and short excerpts may be reprinted in scientific books and journals if the source is properly cited.

Authorization to photocopy items for internal or personal use, or the internal or personal use of specific clients, is granted by the American Geophysical Union for libraries and other users registered with the Copyright Clearance Center (CCC) Transactional Reporting Service, provided that the base fee of $1.50 per copy plus $0.35 per page is paid directly to CCC, 222 Rosewood Dr., Danvers, MA 01923. 0065-8448/99/$01.50+0.35.

This consent does not extend to other kinds of copying, such as copying for creating new collective works or for resale. The reproduction of multiple copies and the use of full articles or the use of extracts, including figures and tables, for commercial purposes requires permission from AGU.

Printed in the United States of America

CONTENTS

Foreword
Senator Patrick Leahy — v

Preface
Thomas O. Manley and Patricia L. Manley — vii

ATMOSPHERICS

Current Knowledge of Air Pollution and Air Resource Issues in the Lake Champlain Basin
Timothy D. Scherbatskoy, Richard L. Poirot, Barbara J. B. Stunder, and Richard S. Artz — 1

Air Trajectory Pollution Climatology for the Lake Champlain Basin
Richard L. Poirot, Paul Wishinski, Bret Schichtel, and Phil Girton — 25

BASIN HYDROLOGY

The Hydrology of the Lake Champlain Basin
James B. Shanley and Jon C. Denner — 41

LAKE HYDRODYNAMICS

Aspects of Summertime and Wintertime Hydrodynamics of Lake Champlain
T. O. Manley, K. L. Hunkins, J. H. Saylor, G. S. Miller, and P. L. Manley — 67

Numerical Hydrodynamic Models of Lake Champlain
Kenneth Hunkins, Daniel Mendelsohn, and Tatsu Isaji — 117

Gravity Currents and Internal Bores in Lake Champlain
James H. Saylor, Gerald Miller, Kenneth Hunkins, Thomas O. Manley, and Patricia Manley — 135

Sediment Deposition and Resuspension in Lake Champlain
Patricia L. Manley, Thomas O. Manley, James H. Saylor, and Kenneth L. Hunkins — 157

CHEMISTRY

Response of St. Albans Bay, Lake Champlain, to a Reduction in Point Source Phosphorus Loading
Scott C. Martin, Richard J. Ciotola, Prashant Malla, and Subramanyaraje N. G. Urs — 183

PHOSPHORUS LOADING

Importance of Instream Nutrient Storage to P Export From a Rural, Eutrophic River in Vermont, USA
Deane Wang, Suzanne N. Levine, Donald W. Meals, Jr., James P. Hoffmann, John C. Drake, and E. Alan Cassell — 205

Slurry Sidedressing and Topdressing Can Improve Soil and Water Quality in the Lake Champlain Basin
Denis Côté, Aubert Michaud, Thi Sen Tran, and Claude Bernard — 225

TOXICS

Toxic Substances in Lake Champlain: An Overview
Alan McIntosh, Mary Watzin, and John King — 239

Ecological Effects of Sediment-Associated Contaminants in Inner Burlington Harbor, Lake Champlain
J. M. Diamond, A. L. Richardson, and C. Daley 261

Mercury Cycling and Transport in the Lake Champlain Basin
James B. Shanley, Andrea F. Donlon, Timothy Scherbatskoy, and Gerald J. Keeler 277

AQUATIC BIOLOGY

Lower Trophic Level Interactions in Pelagic Lake Champlain
S. N. Levine, M. A. Borchardt, A. D. Shambaugh, and M. Braner 301

A Survey of Lake Champlain's Plankton
Angela Shambaugh, Alan Duchovnay, and Alan McIntosh 323

Analysis of Fish DNA Integrity as an Indicator of Environmental Stress
Glenn A. Bauer, Brian T. Dwyer, Brennan J. Leddy, Ryan P. Maynard, and Leah M. Moyer 341

Physiological Indicators of Stress Among Fishes From Contaminated Areas of Lake Champlain
Douglas E. Facey, Cynthia Leclerc, Diana Dunbar, Denise Arruda, Lori Pyzocha, and Vicki Blazer 349

NONAQUATIC WILDLIFE

High Rates of Brown-headed Cowbird Occurrence in Champlain Valley Forests: Conservation Implications for Migratory Songbirds
Steven D. Faccio and Christopher C. Rimmer 361

Research Management of the Common Tern on Lake Champlain, 1987-1997: A Case Study
Mark S. LaBarr and Christopher C. Rimmer 371

CULTURAL AND SOCIAL RESOURCES

Lake Champlain Basin Education and Outreach Programs
Thomas R. Hudspeth and Patricia Straughan 381

Lake Champlain Cultural and Social Resource Management in the 1990s: You Can't Get Where You're Going Until You Know Where You've Been
Susan Bulmer, Art Cohn, and Ann Cousins 389

ECONOMICS

Economic Analysis of Lake Champlain Protection and Restoration
Timothy Holmes, Anthony Artuso, and Douglas Thomas 397

MANAGEMENT

Watershed Management at a Crossroads: Lessons Learned and New Challenges Following Seven Years of Cooperation Through the Lake Champlain Basin Program
Lee Steppacher and Eric Perkins 419

Phosphorus Management in Lake Champlain
Eric Smeltzer 435

List of Contributors 453

Foreword

Senator Patrick Leahy

For the people of Vermont, New York and Quebec, there is little doubt about the importance of Lake Champlain to their quality of life. The lake is the source of drinking water for thousands; a recreational jewel; critical habitat for fish, wildlife, and migratory birds; a direct and indirect economic engine; and an archeological/historical treasury of the Champlain Valley from the glacial age, through pre-European and Colonial times to the present day. Its importance cannot be overestimated.

In 1990, as a response to the growing interest in the Lake, we took legislative action. In 1991, shortly after the Lake Champlain Special Designation Act became law, the Senators from Vermont and New York set out our legislative intent to help guide the newly appointed members of the Lake Champlain Management Conference.

We urged the conferees to think and act comprehensively, to look beyond water quality to the broad scope of issues affecting Lake Champlain. We asked that they be inclusive and consider all aspects of human and ecological life within the Basin. Our legislation envisioned a broad geographic scope as well, including the Champlain Basin as a whole–from the Adirondacks to the Green Mountains, from the Richelieu to the Champlain Canal.

We saw our Lake at a crossroads, with many challenges similar to the Great Lakes, but also with the opportunity to prevent many of the worst problems found in other waters. With the Special Designation Act and the subsequent appropriation of significant federal dollars, we committed ourselves to protect this resource.

We tried to include as many interested sectors as possible. Although the diverse make up of the Management Conference itself would at times slow the planning process, we hoped this would lead to a stronger result with broad public support. "Opportunities for Action: An Evolving Plan for the Future of the Lake Champlain Basin" meets the intent of the Act, which called for the development of a comprehensive pollution prevention, control and restoration plan. Signed by Governors Dean and Pataki in 1996, it provides a comprehensive yet specific blueprint for us and established a Steering Committee to oversee plan implementation. Composed of strong state, federal, citizen and scientific representation, the Steering Committee will publish annual "State of the Lake" assessments to help us gauge our progress and make adjustments as needed.

We tried to work with the institutions already in place. Our bill embraced the existing Vermont-New York-Quebec Memorandum of Understanding and the Citizens Advisory Committees (CAC's), hoping to strengthen them and enhance their role, not replace them. With the Management Conference's work complete, the CAC's will play an ever more important role in both public outreach and guidance for the Basin Program.

A sincere effort was made to build on the work which had already been done. We wanted to gather what we knew about the Lake and the Basin, identify what we did not know and set about filling those gaps in our knowledge. So the Special Designation Act also charged the Management Conference with the establishment of a multi-disciplinary monitoring and research program, encouraging the identification and use of existing research and data as it set research needs. The Lake Champlain Research Consortium was specifically mentioned in this regard.

It was and remains of particular importance for research priorities to support the needs of resource management in making public policy decisions. In fact, we sought federal funds to support the phosphorus diagnostic work a year before the legislation was passed and earmarked appropriations for toxic assessment research the same year the Act passed. Perhaps that was arbitrary, but we felt there was sufficient consensus to move ahead in those areas to save precious time. Beyond those initial Congressional actions, however, we have been confident to defer to the Basin Program partners, particularly the Research Consortium and the Technical Advisory Committee, in the research arena.

To complement this effort, we have tried to expand the resources available to the research community and to draw in additional agencies that might contribute both fiscally and intellectually to our work. A modest but important example is the funding support provided by the National Oceanic and Atmospheric Administration (NOAA) over the past few years for atmospheric and hydrodynamic research.

A far more significant example is our successful amendment to include Lake Champlain in the Sea Grant Program. When my original amendment passed Congress, I never expected such a firestorm would erupt. But, I have to admit that the "Great Lake debate" has been a wonderful opportunity for Vermont and New York to showcase Lake Champlain, our partnerships to protect it and the research we are conducting to maintain its greatness ecologically, historically and economically. Now that the University of Vermont and Research Consortium members can compete for Sea Grant College status and research funds, we will be better prepared to solve the problems facing Lake Champlain. I am confident that UVM, the Research Consortium and the Sea Grant Program will form a successful partnership to bring Sea Grant to Vermont.

While the Sea Grant/Great Lake debate garnered headlines, the successful inclusion of Lake Champlain in the National Invasive Species Act, or NISA, was certainly significant–if not deemed as newsworthy by some. As we all know, zebra mussels are infesting Lake Champlain. Last year, the U.S. Geological Survey noted that Lake Champlain has one of the highest rates of expansion for zebra mussels anywhere in the country. By adding Lake Champlain to NISA, we now have access to federal funds and also gained a seat on the National Aquatic Nuisance Task Force. The Task Force advises all the federal agencies on how to implement NISA and, with a voice at the table, we will ensure that our lake and the needs of the Northeast are not ignored in battling invasive species.

Another threat to Lake Champlain, Vermont and other Northeastern states is the pollution carried into our states and our waters every day from coal-fired power plants in the Midwest. This spring, I introduced legislation to eliminate mercury, one of the most toxic pollutants. Although my legislation is focused on eliminating mercury at its source, I also included provisions for increased research and monitoring of mercury in areas like the Champlain Basin.

Another exciting development in federal research is the new $600 million research program launched in the Agriculture Research bill passed by the Senate in May 1998. This new research program includes a small-state provision that will give smaller schools like ours an advantage in competition for research funds.

As our work continues to enhance the research capacity in the Champlain Basin, we have additional reasons for optimism. Construction is underway on the Rubenstein Ecosystem Sciences Laboratory at the Science Center on the Burlington waterfront. The Lake Champlain Basin Program is, for the first time, included in the President's budget request to Congress, helping make it permanent. Just as importantly, presidents of member colleges in the Lake Champlain Research Consortium have expressed a renewed commitment of support for multidisciplinary research within the basin.

I believe that this monograph represents both a symbol of our commitment to research of the Lake Champlain Basin and a valuable summary of our evolving knowledge of this ecosystem.

As we further our understanding of the resources of the Lake Champlain Basin--human, cultural and natural, we must keep in mind that they are precious commodities we hold in trust for future generations. It is only with sound scientific knowledge upon which to base public policy that we can honor that trust.

Patrick Leahy

PREFACE

Lake Champlain is the largest body of fresh water in the United States after the Great Lakes. At 200 km long, a maximum breadth of nearly 21 km, 1124 square kilometers of surface area, shoreline nearing 800 km, and drainage basin of nearly 20,000 square kilometers, the lake also has the distinction of being shared by the states of Vermont and New York, with a small part in the Canadian province of Quebec. To ensure that the lake can continue to provide for the needs of its many constituencies, and to avoid some of the pitfalls that have affected the Great Lakes, major research initiatives have been undertaken as support for informed management and policy decisions.

This volume, an outcome of the 1998 Lake Champlain Research Consortium's (LCRC) Spring Conference, reflects our current understanding of the Lake Champlain ecosystem after seven years of dedicated research. The purpose is to provide a synthesis of past as well as recent research within the broad categories of chemistry, biology, atmospherics, hydrodynamics, hydrology, social economics, cultural studies, economics, land use and management. Included are not only papers from major research programs, but just as importantly, individual contributions pertaining to the lake's ecosystem. Environmental planners and managers as well as limnologists and hydrologists will find the volume of interest.

Considerable research was initiated on the lake and its basin following the enactment of the 1991 federal Lake Champlain Special Designation Act. Under this mandate, the LCRC, the Lake Champlain Management Conference (LCMC) and the Lake Champlain Basin Program (LCBP) were created to direct and oversee research that would lead to the first management plan devoted to preserving the lake ecosystem for generations to come. In 1996 the LCMC concluded its responsibilities and transferred its management plan to the reorganized Lake Champlain Steering Committee (LCSC). Through the LCSC's leadership of the LCBP, the management plan is being implemented while further research within the basin is continuing.

In the years following the Special Designation Act, better linkages within and between the lake's ecosystem and cultural, social and economic pressures have been well defined. The roles of the atmosphere in controlling the deposition of mercury in the lake and its basin along with its movement from the watershed into the rivers via episodic events are presented. There is a comprehensive summary of lake sediment toxicity and initial results of its effect on benthic and aquatic species. Trophic levels were studied, from the impacts of nitrogen and phosphorus on phytoplankton to multiple "trophic cascades" and management implications. Phosphorus loading and subsequent eutrophication has received a great deal of attention by looking at comprehensive loading budgets, a whole-lake mass-balance model and subsequent management schemes. The complex and remarkably unique hydrodynamics of the lake, during the summer months, generates a large-amplitude (up to 40 m) internal seiche which dominates the Main Lake's circulation. High-speed surges and gravity currents have also been observed and numerically modeled. Surface outcroppings of the thermocline are postulated; as is the movement of nutrients, biomass, and other chemical constituents caused by the internal seiche. Attention is also focused on the more sluggish flow during wintertime. Lake bottom sediment erosion at depths of up to 90 m as well as lake-level variations, associated lake flooding and various aspects of both watershed and submerged cultural resource management are also woven into the overarching issue of lake management. It is our intent that this book will, in some way, demonstrate the research efforts required as well as highlight the cooperation that made possible such large strides towards the more effective management of this unique resource--Lake Champlain.

We would like to acknowledge the LCRC, LCBP/EPA, NOAA, Lintilhac Foundation, Middlebury College, and the Vermont Water Resources and Lake Studies Center for supporting the 1998 Spring Conference. In particular, the American Geophysical Union, LCRC, NOAA, Lintilhac Foundation, and Middlebury College gave continued support for the publication of this monograph, as well as the distribution of this book to the institutional libraries within the basin. We thank the reviewers who not only gave up many hours of their summer to help keep the production of this book on schedule, but in many instances, who unselfishly donated their review 'honorarium' to support further student research in the basin. Finally, our thanks to Trish Dougherty for her attention to details and coordination of reviewers and authors.

Thomas O. Manley
Middlebury College

Patricia L. Manley
Middlebury College

Current Knowledge of Air Pollution and Air Resource Issues in the Lake Champlain Basin

Timothy D. Scherbatskoy, Richard L. Poirot, Barbara J. B. Stunder, and Richard S. Artz

ABSTRACT

The Lake Champlain basin, although predominately rural, is exposed to a variety of atmospheric pollutants and related environmental stresses, largely originating outside the basin. These include acid rain, dry deposition of sulfur and nitrogen compounds, organic and inorganic toxic substances in gaseous and particulate forms, tropospheric ozone, ultraviolet radiation, and climate change. The relatively large land area of the basin (18:1 land:lake area) is a sink for airborne pollutants which can be captured and transferred in the basin, producing direct and indirect effects on the lake and surrounding ecosystems. Because of the growing human population in this region, complex and diverse ecosystems, and multiple ecosystem management issues, it is important to assess the status and impacts of atmospheric contaminants in the basin. This report summarizes the relevant information on regional air quality during the past two decades, collected at several monitoring sites in northern Vermont, New York and Quebec. The data show patterns and trends for concentration or deposition of acidifying compounds, mercury, toxic compounds, fine particulates and ozone. Exposure to many of these contaminants continues to threaten human and ecosystem health in the basin, despite increased regulation of air pollutants. Analysis of regional meteorology and application of air transport models to these data show the probabilistic extent of the airshed affecting the basin and the contaminant source regions. Finally, information gaps, research needs and management issues are examined to provide guidance for future work to improve understanding and control of air pollution issues in the basin.

INTRODUCTION

> "The smoke of coal pits covered the land. The trees were swept away as if some gigantic scythe bearer had mowed them over. One may travel now for miles in that region and not find a tree large enough to make a respectable fish pole."
>
> - L.E. Chittenden, around 1890

As this quote demonstrates, air pollution in the Lake Champlain basin is not a new problem, but certainly the form and extent of pollution has changed substantially in the past 100 years. Today, pollution and its effects in the basin are likely to be comparatively subtle and largely invisible to the average observer. Despite the absence of gross water or air pollution, however, the public has become increasingly aware of and concerned about the degradation of Lake Champlain ecosystems by pollutants being transported into the basin by the atmosphere. Concern about mercury contamination of fish in Lake Champlain and the possible effects of acid rain on forests in the basin are but two examples of this awareness.

The Lake Champlain basin (Figure 1) is a region that is particularly exposed to air pollution in the eastern United States by virtue of its location "downwind" of many pollutant sources and its relatively large catchment area. The basin has a large (18:1) ratio of watershed to lake surface area, so the capture and processing of atmospheric pollutants by the terrestrial portions of the basin may be important in this region. Approximately 89% of the basin is forested or agricultural, with the lake itself representing only 5% of the total basin area (Table 1). This is in sharp contrast to the Great Lakes, for example, which themselves represent a large proportion of their watersheds. These characteristics and the relatively mountainous, high elevation terrain favor the capture and accumulation of air pollutants in the basin by processes including wet and dry deposition, foliar wash-off, leaf litter-fall, spring snow-melt and runoff, and stream transport, as well as direct deposition to surface waters.

Historical Development of Air Pollution Management in the Basin

Air pollution management and measurements in the Lake Champlain basin were initiated on a "routine" basis in the late 1960's and early 1970's, primarily in response to the 1963 Clean Air Act and its 1970 amendments. The federal 1963 legislation was motivated to a large extent by a growing public perception that air pollution was a serious "local" problem in many large urban and industrial areas. Pollution control efforts focused on the largest, most poorly controlled sources or source categories. In the Lake Champlain basin these included open burning at dumps, asphalt plants, wood burning boilers and a few large industrial sources. Early ambient measurements were conducted predominantly in urban areas or in the vicinity of large sources, with primitive methods including lead sulfation plates (for sulfur dioxide), folded rubber strips (for ozone) and dust-fall buckets (for particulate matter). The 1970 amendments required establishment of national ambient air quality standards for a number of common "criteria" pollutants (carbon monoxide, nitrogen dioxide, sulfur dioxide, particulate matter, hydrocarbons and photochemical oxidants).

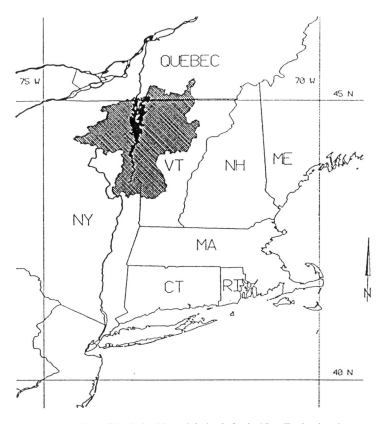

Figure 1. Map of the Lake Champlain basin in the New England region.

Improved measurement methods were developed, and routine monitoring was required to determine compliance with the ambient standards. Given the effectiveness of early control efforts at reducing the most obvious "smoking" sources, and the relatively small size of urban areas and industrial sources in the Champlain Basin, few exceedances of national standards were recorded, and the general public perception was of pristine air quality - rivaling the clarity of Lake Champlain's sparkling waters. During this period similar awareness and regulatory attention were given to surface waters in the U.S., although only recently has Lake Champlain water quality been recognized to be at risk.

TABLE 1. Area and percent of total basin area of major land use classes in the Lake Champlain basin, based on aerial photography in 1973 (Budd and Meals 1994).

Class:	Forest & Wetland	Agriculture	Urban	Surface Waters	Lake Champlain	Total Non-Lake	Total Basin
Hectares:	1,322,506	567,262	55,840	75,111	113,000	2,020,719	2,133,719
Percent:	62%	27%	3%	4%	5%	--	--

The "discovery" of "acid rain" by the popular press in the late 1970's, dramatically changed the public perception of air pollution in the region. It introduced the concepts of "long-range transport", and "effects of air pollutants in other environmental media" - like rainfall, surface waters and forest soils. Significant responses by the scientific community ensued, and the upland Adirondack and Green Mountain sections of the Champlain Basin became centers of major monitoring and research efforts to address questions of acid rain effects. The first air pollution monitoring effort to address basin-wide questions began in 1982 with the establishment of an acid rain monitoring station at the University of Vermont Proctor Maple Research Center in Underhill Center, Vermont. During the next decade a dozen more air quality monitoring studies were added at this site. Also during this period, acid rain monitoring and related air quality studies were upgraded at Whiteface, NY (where a MAP3S precipitation chemistry station had been established in 1976) and Sutton, Quebec (where the CHEF Program had been established in 1985). Added to these sites, air quality monitoring for criteria pollutants and assorted research measurements at several locations in Vermont, New York and Quebec, plus the network of National Weather Service (NWS) primary and cooperative stations were deployed to provide the basis for a modern monitoring infrastructure in the basin (Figure 2).

During the 1980s, new methods were developed and implemented to sample the chemistry of precipitation and cloud water [Galloway et al. 1978, Scherbatskoy and Bliss 1984]. Routine measurements of criteria pollutants were extended to rural areas, where surprisingly, the region's highest ozone concentrations were observed on the remote top of Whiteface Mountain. Accumulation of heavy metals was observed in forest soils and lake sediments in the "pristine" Adirondacks and Green Mountains [Friedland et al. 1984]. During this period, inter-governmental agreements were developed for more efficient management of shared air resources, and by the late 1980's, New York, Vermont and Quebec had each committed to reducing and maintaining emissions of acidifying sulfur emissions. In 1988, these three jurisdictions also signed a Memorandum of Understanding on environmental cooperation on the management of Lake Champlain. In concert with the rising awareness of the ecological interconnectedness among air, land and water resources, inter-disciplinary approaches were developed and applied to environmental measurement programs, leading to the formation in 1990 of new coordinating institutions such as the Lake Champlain Research Consortium and the Vermont Monitoring Cooperative.

The 1990 Clean Air Act Amendments included requirements for phased reductions in acidifying sulfur emissions from electric utility sources in areas upwind of the Lake Champlain basin (with similar emissions reduction commitments from Canadian sources). Also under the 1990 Amendments, New York and Vermont were included in the multi-state Ozone Transport Region - composed of 11 "Northeast Corridor" States from Virginia to Maine. Lake Champlain was specifically included, along with the Great Lakes and Chesapeake Bay, as one of the "Great Waters", for which additional monitoring and research was required to understand the extent, effects and causes of atmospheric deposition of hazardous air pollutants to these waters. In 1990, the Lake Champlain Special Designation Act provided a range of new inter-governmental management structures, and included direct support for needed monitoring and research programs. State-sponsored "air toxics" monitoring programs were initiated in the Vermont, New York and Quebec sections of the Lake Champlain Basin, and monitoring and research programs to assess concentrations, deposition, sources and ecological processing of airborne mercury were also

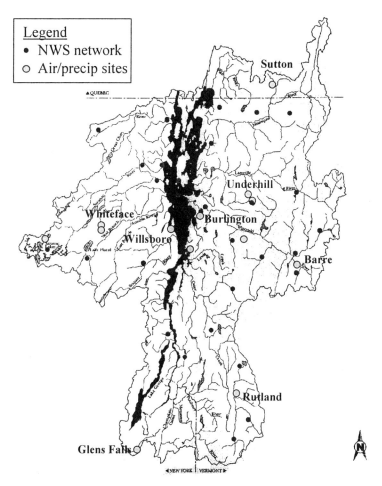

Figure 2. Map of the Lake Champlain basin showing locations of air and precipitation monitoring stations and weather stations.

established. Measurements of ultraviolet radiation, and additional monitoring and research of local meteorology and climate conditions were initiated, reflecting public and scientific concerns over emerging global air pollution issues.

Current Air Resource Issues in the Basin

Currently, media attention, public hearings and our experience suggest that public concern about air pollution issues in the Lake Champlain basin appears to be focused on mercury contamination of fish, global climate change, and acid rain, in that order of importance. Although not so widely perceived by the public, exposures to air toxics, fine particulates and tropospheric ozone are currently receiving scientific attention as significant contributors to environmental quality in the basin. As discussed further at the end of this report, public perception of immediate and obvious impacts on human health and economics often drives decisions about resource management and research priorities, sometimes independently of their scientific importance. Mercury contamination of waters in the basin captures public attention in this way. Questions about the effects of acid rain on the health of forest ecosystems in the basin have never been fully answered and recent findings of nutrient depletion [Likens *et al.* 1996], episodic acidification and nitrogen saturation [U.S. EPA 1995] in northeastern forests have stimulated renewed attention on additional acid rain controls. Although ozone concentrations have not exceeded federal standards in the basin in the past several years, there continues to be persistent exposures at levels of concern to forest ecologists. The recently revised, stricter ozone standards may increase public perception of this issue.

In light of this background, the purpose of this report is to discuss the recent history of air pollution monitoring in the Lake Champlain basin and to document the long term trends and current status of the major air pollutants affecting the lake and basin, including mercury, acid deposition, tropospheric ozone, air toxics and fine particulates. In addition, analysis of air transport patterns for these pollutants is used to show the probable geographic extent of the airshed of the basin.

AIR POLLUTANTS IN THE LAKE CHAMPLAIN BASIN

Acid Deposition

Acidic deposition in precipitation has been monitored at three sites in the basin since 1984 under the National Atmospheric Deposition Program/National Trends Network (NADP/NTN) at the Vermont Monitoring Cooperative monitoring station in Underhill, VT (400 m elevation), at Whiteface Mountain, NY (620 m) and at Sutton, Quebec (244 m). Weekly data for the U.S. sites are available at the NADP Internet Web site [NADP 1998] including meta-data and wet-only precipitation amount, pH, and concentration and deposition of major cations and anions. Dry deposition of acidifying compounds has been monitored in the basin since 1992 at the VMC site in Underhill, VT and since 1984 at the Whiteface site. Using weekly filter-pack samplers, air concentrations of sulfur dioxide, sulfate aerosol, nitric acid vapor and nitrate aerosol are measured and the corresponding deposition of sulfur and nitrogen is calculated by inferential methods [Hicks *et al.* 1991].

Nationally, SO_2 emissions have been declining slowly since the early 1970s, with a sharp drop in 1994-95 in electric utility emissions due to compliance with Title IV of the Clean Air Act Amendments. These trends are apparent in the SO_4 concentration in precipitation for the Lake Champlain basin as well as many other northeastern locations. Figure 3 shows pH and volume-weighted seasonally averaged concentrations for sulfate and nitrate in precipitation at Underhill and Whiteface. The long-term downward trend for sulfate and absence of trend for nitrate are obvious in these data. In addition, pH appears to be increasing only slightly, sulfate concentration is typically highest and nitrate the lowest in

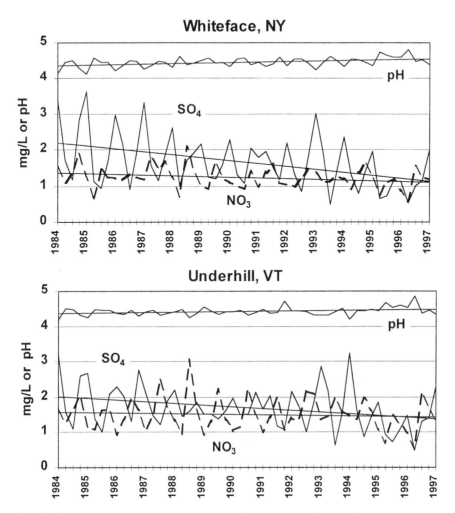

Figure 3. Trends in seasonal (quarterly) volume weighted average pH (upper line) and concentration of sulfate (solid line) and nitrate (dashed line) in precipitation in the Lake Champlain basin at Whiteface Mountain, NY and Underhill, VT, from NADP weekly samples in 1984 - 1997, with linear regression trend lines.

the summer period, and sulfate concentrations tend to be slightly higher at Whiteface, while nitrate concentrations are slightly higher at Underhill. During the summer seasons of both 1993 and 1994, sulfate concentration was noticeably elevated at both sites due to very high sulfate concentrations in June and August. Another interesting observation is that calcium concentration in precipitation at both sites has been declining at a rate of about 0.3% annually (data not shown). Reasons for this are obscure, but concern about the effect of this trend on forest ecosystem health has been expressed in other studies [Likens *et al.* 1996].

Figure 4 shows the concentration patterns of the compounds in the basin that dominate dry deposition of sulfur and nitrogen as gasses (SO_2 and HNO_3) and particulates (SO_4^{2-} and

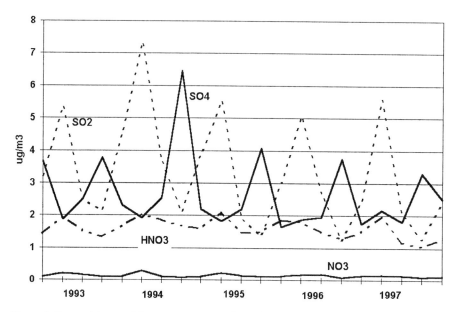

Figure 4. Seasonal (quarterly) average concentrations of ambient vapor and particulate sulfur and nitrogen compounds in the Lake Champlain basin at Underhill, VT, from NOAA AIRMoN weekly samples in 1992 - 1997.

NO_3^-). Note that the concentration of reduced (SO_2) and oxidized (SO_4^{2-}) sulfur compounds alternates seasonally, with sulfur dioxide concentration being greatest in the winter months and sulfate being highest in the warmer months (as for precipitation). Elevated sulfate concentrations during summer 1994 are also apparent, suggesting a possible relationship between particulate and precipitation sulfur.

Annual wet and dry deposition rates of sulfur and nitrogen compounds at Underhill and Whiteface are shown in Table 2. Over this period, annual wet deposition rates at Underhill averaged 32% higher for nitrogen and 10% higher for sulfur than at Whiteface, suggesting possible reduced influence of nitrogen sources at Whiteface. These data indicate that annual deposition of sulfur and nitrogen in the basin occasionally exceeds the annual critical loading rate of approximately 8 kg S/ha and 8.4 kg N/ha, levels at which changes in vegetation composition (due to to nitrogen) or soil aluminum mobilization (due to sulfur) can be expected in this region [DeVries 1993].

The NOAA Air Resources Laboratory (ARL) measures daily precipitation chemistry at the Underhill, VT site as part of the AIRMoN Network [AIRMoN 1998]. This site began operation in 1992, shrtly after the termination of the Whiteface, NY MAP3S station, and is the only long-term daily monitoring station in New England. Daily measurements allow a much clearer evaluation of meteorological conditions than is possible with the weekly NADP samples. ARL also maintains meteorological data and data analysis service via its "READY"(Real-time Environmental Analysis and Display sYstem) Internet site [NOAA 1998]. Among other things, the READY site provides access to the HY-SPLIT (HYbrid Single Particle Lagrangian Integrated Trajectories) trajectory model [Draxler and Hess 1997]. The HY-Split trajectories can be calculated backward (or forward) in time to track

TABLE 2. Annual wet and dry deposition as kg/ha of total sulfur (SO_4^{2-} S and SO_2 S) and total nitrogen (NO_3^- N, NH_4^+ N and HNO_3 N) at Underhill, VT and Whiteface, NY.

Year	S, wet		S, dry		S, total		N, wet		N, dry		N, total	
	VT	NY	VT	NY	VT	NY	VT	NY	VT	NY	VT	NY
1985	7.34	7.34		2.03		9.37	6.30	5.21		1.69		6.90
1986	7.90	7.68		1.45		9.13	6.32	4.77		1.58		6.35
1987	7.04	7.24		1.46		8.70	6.76	5.62		1.50		7.12
1988	6.53	5.73		1.83		7.56	5.49	3.85		1.96		5.81
1989	6.15	7.09		1.94		9.03	6.85	5.77		2.29		8.06
1990	8.49	7.03		1.96		8.99	7.72	5.70		2.97		8.67
1991	6.52	6.24		2.10		8.34	5.94	4.53		3.55		8.08
1992	4.64	4.89		1.42		6.31	4.78	3.57		2.42		5.99
1993	7.87	7.49	2.05	1.44	9.92	8.93	7.64	5.18	1.82	2.55	9.46	7.73
1994	7.70	5.50	2.34	2.23	10.00	7.73	7.31	4.64	2.08	3.25	9.39	7.89
1995	5.27	4.00	1.53	1.00	6.81	5.00	5.70	4.43	1.75	2.28	7.45	6.71
1996	4.27	3.89	1.40	0.89	5.67	4.78	5.30	4.48	1.79	2.10	7.09	6.58
1997	5.79	4.86	1.32	1.41	7.11	6.27	5.79	4.60	1.31	3.45	7.10	8.05
Mean:	6.58	6.08	1.73	1.63	7.90	7.70	6.30	4.80	1.75	2.43	8.10	7.23

the path of airmass motion for air arriving at (or leaving from) a specified receptor location. Trajectory Cluster Analysis [Stunder and Artz1996] is a powerful statistical technique which allows large numbers of trajectories to be grouped into a much smaller number of trajectory "clusters", for which trajectories have common spatial characteristics which are similar within each cluster and different from those in other clusters.

In a recent analysis of the Underhill site AIRMoN rain events for 1995, backward HY-SPLIT 36-hour trajectories were calculated for 88 of 99 precipitation events for which chemical analysis was available. For 11 cases trajectories could not be computed because of missing meteorological data. This analysis resulted in 5 clusters (Figure 5) representing the general paths of air masses resulting in the precipitation events occurring at Underhill in 1995. Using the AIRMoN data, the chemical characteristics associated with the trajectories in each cluster can be examined (Table 3). According to this analysis, precipitation events associated with the Southwest Long cluster (cluster #1), accounted for about 24% of the total 1995 precipitation volume at Underhill, but contributed about a third of the total sulfate and hydrogen ion deposition, and over 40% of the nitrate ion deposition. Together, the three West-Southwest clusters (clusters # 1, 2 and 4) accounted for 69% of the precipitation events, 77% of the precipitation volume and more than 85% of the 1995 deposition of sulfate, nitrate and hydrogen ion. Although the northwest sector is typically the source of fair weather and clean air in this region, the concentration of sulfate and hydrogen ion in precipitation from this sector in 1995 was nearly as great as any other. Because this sector contributed only a small quantity of the total precipitation, the total wet deposition of pollutants from these storms was small (5%). This analysis underscores the fact that precipitation events arriving in the Champlain Basin are both frequent and polluted. Fortunately, future reductions in sulfur and nitrogen emissions from Midwestern electric

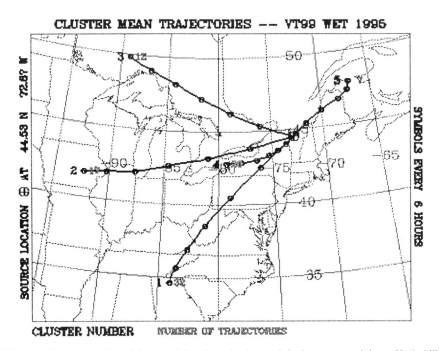

Figure 5. Mean trajectories of clusters of air trajectories for precipitation events arriving at Underhill, VT during 1995. Dots along trajectories represent 6-hour increments. Numbers in larger type at the end of each trajectory is the trajectory cluster label (see Table 3), and numbers in smaller type is the number of trajectories in the cluster.

utilities are anticipated from the 1990 Clean Air Act Amendments, and from EPA's recently proposed nitrogen oxide reductions based on recommendations from the Ozone Transport Assessment Group [OTAG 1998].

Ozone

Tropospheric ozone is a federally regulated air pollutant, harmful to all animal respiratory systems and plant life. It is routinely monitored at three locations in the Lake Champlain basin, at 394 m elevation in Underhill, VT, at 620 m and 1142 m at Whiteface Mt., NY, and at several elevations in Sutton, Quebec. Ozone is monitored by continuous photometric analysis and is reported as hourly average concentrations.

Ozone pollution is of concern for potential effects on human health, forest ecosystems and agricultural crops in the Lake Champlain basin. The health-based federal standard for ozone (formerly based on maximum 1-hour concentration of 125 parts per billion [ppb]) has recently been changed to protect against exposures at a lower threshold (85 ppb) over a longer averaging time (8 hours). Hourly data from Whiteface - which typically experiences among the highest concentrations in the Basin - indicate no exceedances of the

TABLE 3. Contributions of each trajectory cluster shown in Figure 5 to the precipitation chemistry at Underhill, VT in 1995.

Cluster #	1	2	3	4	5
Cluster Name	SW Long	W	NW	SW Short	NE
# of Trajectories per Cluster	32	17	12	20	7
Total Precipitation (cm)	12.2	9.2	2.4	17.9	9.8
VWM[a] SO_4 Concentration (μeq/L)	31.9	29.3	31.3	20.0	10.5
% of Total SO_4 Deposition	33	23	5	30	9
VWM NO_3 Concentration (μeq/L)	29.7	21.2	18.4	10.6	5.1
% of Total NO_3 Deposition	43	23	5	22	6
VWM H^+ Concentration (μeq/L)	44.3	32.7	38.0	24.6	13.4
% of Total H^+ Deposition	36	20	6	29	9

[a] Volume Weighted Mean

1-hour standard in the past five years (Figure 6, top panel), although this threshold is frequently approached, and a number of exceedances were recorded there in the late 1980's and early 1990's. The annual pattern of ozone concentration seen at Whiteface during the past five years is fairly representative of ozone behavior at the other monitoring sites in the basin. Analysis of historical data indicates that ozone concentrations in the basin would have been close to exceeding the new ozone standard on several occasions, and the new standard will be more stringent (more easily exceeded) than the 1-hour standard.

Forest health and agricultural scientists are more concerned with exposures exceeding 40 to 60 ppb, which occur quite frequently at all sites in the basin. As indicated in Figure 6 (top panel), regional ozone exposures clearly peak during the summer growing season and also exhibit distinct diurnal patterns and elevational gradients (Figure 6, bottom panel). At lower elevation sites like Bennington, VT (just south of the Lake Champlain basin), summer ozone levels typically fall off dramatically at night, as ozone formation requires sunlight and ozone destruction by chemical reaction and physical deposition takes place predominantly at or near the earth's surface. At higher elevations, the nocturnal destruction is minimal, such that mountain forests in the basin are chronically exposed to higher ozone exposures than low elevation counterparts [Poirot 1993]

Fine Particles and Air Toxics

Particulate matter pollution is of current concern for its potential effects on human health and visibility, and for its contribution to the deposition of acidifying compounds and toxic metals and organic contaminants in the Basin. The federal standards for particulate matter, based on particles of less than 10 micrometer (μm) diameter (PM-10), are currently being revised to focus on smaller particles, less than 2.5 μm (PM-2.5). These smaller particles penetrate and reside at further depths of the respiratory tract and tend to be composed of inherently injurious substances, including acidic sulfates, and potentially toxic organic compounds and trace metals. Analysis of historical data suggests that ambient fine particle

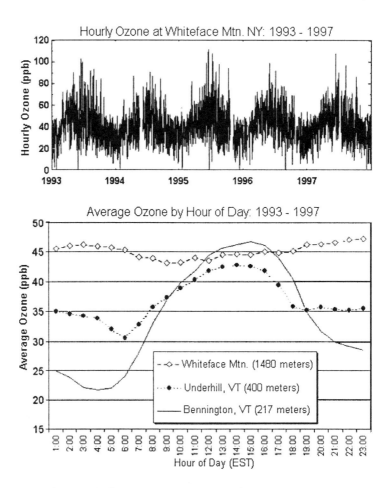

Figure 6. Trend in tropospheric ozone concentration at Whiteface Mountain, NY (top) and diurnal average hourly ozone concentration patterns for three sites in and near the Lake Champlain basin (bottom), based on hourly average ozone concentrations monitored in 1993 - 1997.

concentrations in the basin may be relatively close to the new PM-2.5 standard of an annual average not to exceed 15 micrograms per cubic meter ($\mu g/m^3$).

Fine particles and air toxics have been monitored in the basin at various urban and rural sites, including Burlington, Barre, Rutland and Underhill in Vermont, Glens Falls, Whiteface and Willsboro in New York, and Sutton, Quebec. Methods for collection of fine particles typically involve the use of a size-selective inlet to remove large particles and weekly or bi-monthly collection of a 24-hour air sample on a filter which is then subjected to various analysis by spectrometry, fluorescence or other techniques. Gaseous air toxics are typically sampled in canisters or absorbent cartridges with subsequent analysis by gas or liquid chromatography.

PM-2.5 data from 1993 in the Burlington, VT urban area and at two nearby, remote

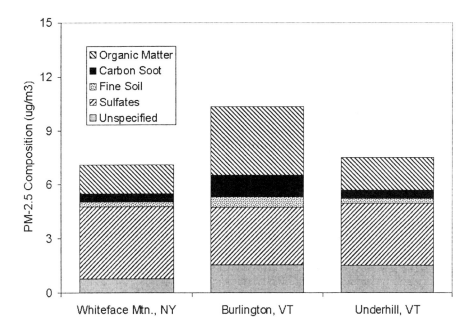

Figure 7. Fine particle (2.5 μm) composition and concentration at three sites in the Lake Champlain basin, calculated from every-sixth-day 24-hour samples in 1993.

"background" sites at Whiteface Mtn., NY and Underhill, VT are reported in Figure 7. These data indicate that average background fine particle mass concentrations are about half the level of the proposed standard, with about half of the background concentration composed of sulfate compounds. As these sulfate particles result primarily from distant sources, their concentration does not increase at the urban Burlington site. Organic compounds and elemental carbon (soot) are substantially higher at the urban site (adding about 3 µg/m^3 to the background levels) indicating significant local source contributions for these fine particle species.

Local urban sources are also important (predominant) contributors to elevated local concentrations of several toxic gaseous organic compounds, including formaldehyde, benzene and 1,3-butadiene. As indicated in Figure 8, these compounds slightly exceed Vermont's health-based toxic standards at the remote background site in Underhill, and increase by a factor of 2 to 4 in nearby suburban and urban areas. The Vermont toxic standards for these 3 contaminants are based on an estimated cancer risk of 10^{-6} (1 cancer death per million population) such that an ambient concentration of 10 times greater than the standard would represent an estimated risk of 10^{-5}. Local emissions of these contaminants result primarily from mobile sources (gasoline and diesel vehicles and associated refueling), and a reduction in the concentration of these toxic compounds will require some combination of cleaner (or alternative) transportation and/or major lifestyle changes for the basin's inhabitants.

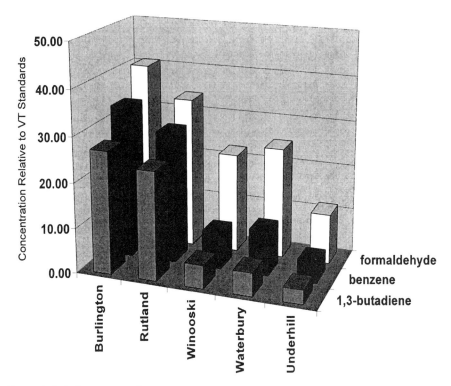

Figure 8. Relative concentrations of three toxic air contaminants at five sites in the Lake Champlain basin, calculated from 24-hour samples collected every 12 days in 1994.

Mercury

Mercury (Hg) contamination of fresh waters and their biota has become a widespread and serious problem in many parts of the world, including northeastern North America. Health advisories restricting consumption of some mercury contaminated fish in Lake Champlain and other waters in the basin and throughout the Northeast have focused public and scientific attention on this problem. There is concern that atmospheric sources of mercury are responsible for increasing Hg burdens in Lake Champlain [Scherbatskoy *et al.* 1997, Watzin 1992, Vasu and McCullough 1994], but mercury sources, transport mechanisms and accumulation are not yet well understood. There is growing evidence that forested ecosystems have an important and complex role in the transport of atmospheric Hg to aquatic systems [Bishop *et al.* 1995]. Recent studies have identified several processes affecting Hg movement in ecosystems, including organic complexation in soils and water, accumulation in plant tissues, foliar capture of atmospheric Hg, and a large watershed:lake area ratio [Schuster 1991, Rea *et al.* 1996, Swain *et al.* 1992]. The Lake Champlain basin is susceptible to mercury pollution because of its large proportion of forested lands where these processes occur and its proximity to relatively large Hg emissions in the region [NESCAUM 1998].

TABLE 4. Descriptive statistics for atmospheric Hg concentrations in Underhill, VT for the period December 1992 through December 1997.

	Mean	Std. Dev.	n	Minimum	Maximum	Median
Vapor Hg (ng/m^3)[a]	1.67	0.49	324	1.01	7.31	1.59
Particulate Hg (pg/m^3)[a]	9.24	6.32	356	0	42.0	7.79
Precipitation Hg (ng/L)[b]	8.23	8.48	553	1.15	89.8	8.03

[a] Corrected to Standard Temperature (25 C) and Pressure (1013.3 mb)
[b] Volume Weighted Mean

Atmospheric mercury has been studied at two locations in the basin: for precipitation, vapor and particulate forms of Hg since 1992 in Underhill, VT [Scherbatskoy et al. 1998] and for vapor and particulate forms in 1992-94 in Willsboro, NY [Olmez et al. 1996]. The Underhill site is the longest running continuous monitoring program for all three forms of atmospheric Hg in the world, and provides an important long-term and regional reference for other studies that have recently begun in other parts of the Northeast. The Underhill Hg research program is a collaborative effort between the Vermont Monitoring Cooperative and the University of Michigan Air Quality Laboratory. Precipitation for Hg and trace metal analysis is collected at this site on an event basis in a modified wet-only precipitation collector, and vapor and fine particulate (2.5 μm) air samples are collected for 24 hours on gold-coated sand vapor traps and glass-fiber filters, respectively. Sampling and analysis procedures are described in detail by Burke et al. [1995]. At Willsboro similar methods were used for Hg vapor and particulate sampling, using charcoal vapor traps and Teflon filters, respectively [Olmez et al. 1996].

The long-term patterns of Hg concentration and deposition in precipitation, based on the data of Keeler et al. [this volume], are shown in Figure 9. The magnitude and pattern of seasonal increase in Hg concentration during the summer is similar to that seen at other sites in the northern U.S. Because of greater concentration and precipitation in the summer months, Hg deposition in the basin is greatest during this period. Annual wet deposition of Hg at this site averages 87 mg/ha. Dry deposition of Hg vapor to the forest ecosystem is large during the warmer months, but because of large uncertainties about the re-emission of Hg from forests, it is difficult at this time to accurately fix the total wet plus dry net deposition. Recent analysis by Rea [1998] indicates that 50-120% of dry deposited Hg may be re-emitted to the atmosphere from forests. Nonetheless, according to Shanley et al. [this volume] about 92% of the measured wet and estimated dry Hg deposition is retained in the terrestrial ecosystem (with some unknown amount returning to the atmosphere by re-emission), and only 8% of this deposition is transported out of the ecosystem in stream water.

Descriptive statistics for concentrations of precipitation, vapor and particulate forms of Hg at Underhill are presented in Table 4. These values are generally similar to those for other sites in the northern U.S. There do not appear to be any significant trends in concentration at this site, except perhaps for Hg vapor which has declined from around 2 ng/m^3 in 1993 to around 1.6 ng/m^3 recently (Figure 10). Comparable values for Hg vapor were measured by Olmez et al. [1996] at Willsboro, NY during 1992-1994, except for two periods (winters of 1992 and 1993) when concentrations were approximately twice as high as Underhill.

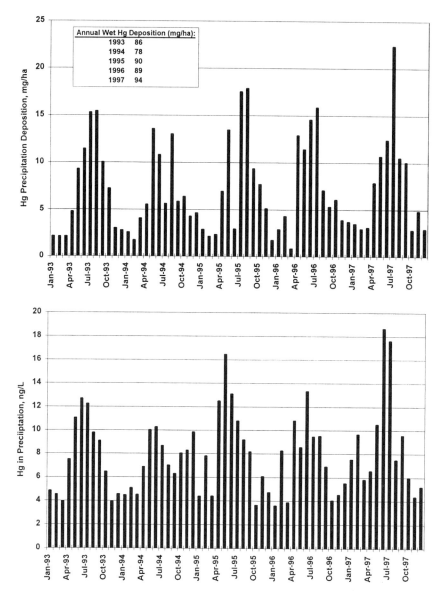

Figure 9. Monthly mercury deposition (top) and volume weighted monthly mean mercury in precipitation at Underhill, VT in the Lake Champlain basin, based on event samples in 1993 - 1997.

Meteorology and Ultraviolet Radiation

Two other important issues related to atmospheric resources in the Lake Champlain basin are global climate change and increasing solar ultraviolet radiation (UV-B). While these are both potentially important environmental factors in the Lake Champlain basin and

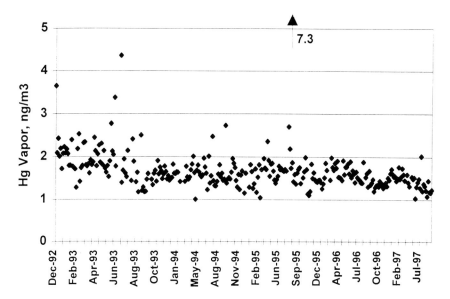

Figure 10. Ambient mercury vapor concentration at Underhill, VT in the Lake Champlain basin, based on every-sixth-day 24-hour samples in 1993 - 1997.

elsewhere, our data and understanding of their effects are very limited. Meteorological monitoring in the basin has been enhanced in recent years by the addition of automated weather monitoring stations located at regional airports, and by research weather stations operated by the Vermont Monitoring Cooperative at Colchester Reef on Lake Champlain and at several locations on Mount Mansfield. Data from these stations, plus data from the National Weather Service cooperative weather stations throughout Vermont are available through the Vermont Monitoring Cooperative Data Library. In addition, an assessment of precipitation patterns in the Lake Champlain basin has been recently conducted by A.R.D. Inc. [B. Hegman, pers. comm. 1998) for use in watershed pollutant loading studies. These data are valuable for interpreting recent and historical climate and ecological processes.

Ultraviolet radiation has been monitored in the basin since 1993 at the Vermont Monitoring Cooperative's forest research site in Underhill, VT using a broad-band pyranometer (Yankee Environmental Systems UVB-1). In 1996, the Underhill site became part of the U.S. Department of Agriculture UV-B Monitoring Program, a climatological network of 25 sites nationwide using several broadband instruments. Information about and data from this network are available from the program's Web site [USDA 1998]. Because of the limited data for this site, no trend or effects information is available at this time, but the availability of these data for the basin is important for future environmental research.

THE LAKE CHAMPLAIN AIRSHED

While many pollutants affecting the Lake Champlain basin are of predominantly local origin (see Figures 7, 8), others are of more distant origin and subjected to long-range transport. Toxic metals, for example, are a current concern for their potential effects on human or ecological health in the basin. Certain metals (like mercury) in our ambient air and precipitation have significant local source contributions such as commercial incinerators, while others (nickel, arsenic, selenium) may result predominantly from more distant sources. It is valuable to identify source regions for these contaminants (as for acid deposition, above), and to determine the extent of the airshed, the geographic region with significant sources potentially affecting the basin. Poirot *et al.* [this volume] examined the concept of the synoptic-scale "Lake Champlain airshed", employing 8 years of (12 per day) backward "air mass histories" to derive a long-term "probabalistic" definition of "areas most likely to be upwind" of the Lake Champlain basin. They found this probabilistic airshed varied in size and shape as a function of season, meteorological condition, and pollutant of concern. For high local concentrations of fine particle arsenic, the basin's air was likely to have previously resided to the northwest over a Canadian region of large primary smelter emissions. For high nickel concentrations, transport from East Coast metropolitan areas was indicated, while high selenium concentrations were associated with transport from the Ohio River Valley.

Similar methods are employed and shown here in Plate 1 to explore the "airshed" for 3 Lake Champlain basin monitoring sites on an "everyday" basis, and for high concentrations of ozone, PM-2.5 mass and PM-2.5 sulfur. The top row in Plate 1 displays "everyday upwind probability fields" for sites at Whiteface Mtn., NY, Underhill, VT and Lye Brook, VT (outside but near the basin boundary in southeastern Vermont). These are based on all available trajectories for these sites over an 8-year (1989-96) period. The shaded areas depict the smallest areas accounting for 20%, 40% and 60% of the trajectory "residence time hours" on the map. The patterns are similar for all 3 sites, and are generally oriented in a westerly (or WNW) direction from the receptors. The second row displays similar upwind probability fields, but in this case, only trajectories on days with high 8-hour ozone concentrations at the receptors were included. While there is a strong southwesterly orientation on high ozone days at all sites, note there is also a strong southerly influence (East Coast urban corridor) on the Lye Brook site. Lye Brook also experiences somewhat more southerly influence than the northern sites for fine particle mass and sulfur - which generally show a dominant southwesterly influence at all the sites. Similar patterns for fine particle mass and sulfur are not surprising, since half or more of the fine particle mass at these rural sites is sulfate). Sulfur in the fine particle phase can range from pure sulfuric acid to fully neutralized ammonium sulfate. All of these sulfate compounds efficiently absorb water from the atmosphere, which increases their light scattering efficiency and makes them the most important causes (accounting for 60-70%) of regional haze in the Champlain Basin. The results here for fine particle sulfur are quite consistent with the NOAA ARL cluster analysis results for Underhill precipitation sulfate (Figure 5). Hence, a common upwind source region is a major contributor to acid deposition, fine particle pollution, regional haze, and ozone pollution in the Basin, and is a key component of the Lake Champlain Airshed.

AIR RESOURCE ISSUES IN THE FUTURE

Perception, Policy and Management

While our understanding of air pollution, and of the priority air resource issues in the Lake Champlain basin is evolving and far from complete, much has been learned over the past

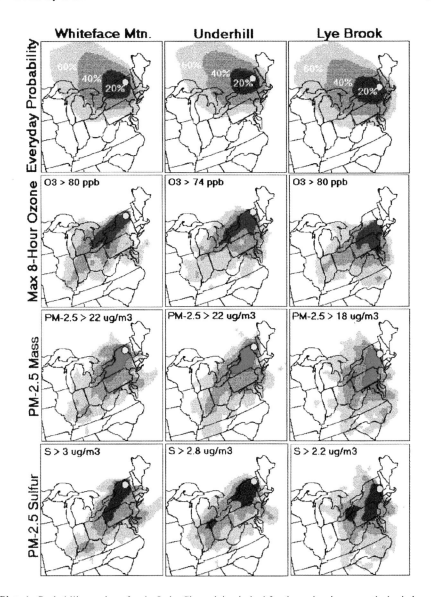

Plate 1. Probability regions for the Lake Champlain airshed for three sites in or near the basin based on meteorological data for 1989 - 1996 for days of elevated sulfur, fine particle, and ozone, and for all days ("everyday"). For each receptor site, the shading shows the area of probability that the air mass originated in the region with a probability of 20% (black), 40% (dark gray), or 60% (light gray).

30 years since routine measurement programs were initiated. The management of environmental resources can be viewed as a policy response to public perception of environmental issues of concern. Environmental measurement, through routine monitoring and more focused research efforts, can provide critical scientific guidance for effective management decisions by clarifying the nature, severity and causes of environmental problems. Over time, environmental measurements can lead to changes in public perception, and provide the means to assess environmental management decisions.

Specific issues facing air resource managers and the public in the Lake Champlain basin include acid deposition, mercury, fine particles, air toxics, greenhouse gasses, UV-B and climate change. There are numerous complex possible impacts of these stressors on human and ecosystem health, which may be caused by direct single-factor and multiple exposures, and by indirect processes. Effective responses to and management of these problems must be guided by good quality information. Access to data such as those presented in this report, however, faces several challenges. There have been several proposed cuts and actual lapses in funding for several monitoring programs described in this report, even during the period when air pollution levels were expected to respond to new regulations in the Clean Air Act. Only through continued monitoring of acid rain constituents were we able to see that indeed these regulations were effective and that additional controls were still needed. In addition, data access itself can be a problem for managers and policy-makers. In order to improve the relationship between air pollution science and management, easy and rapid access to monitoring data, is helpful. Finally, it is important to recognize that our perceptions of environmental problems change, and today's issues in the basin will evolve and new issues arise in the future. Good communication among stakeholders, scientists, managers and policy makers is necessary to ensure dynamic and accurate responses to perceived environmental problems. Thus, support for long-term monitoring, good data access, and integrated communications are critical components of effective air resource management in the Lake Champlain basin in the years ahead.

Air Resource Management Needs and Recommendations

Although we have good information about some aspects of air resource issues in the basin, there are a number of areas where improvements and additional data are needed. Broadly, there are three type of processes to pay attention to when addressing air resource issues in the basin. *Atmospheric exposure and deposition* of acidifying substances, toxic contaminants, and ozone are important atmospheric environmental stressors in the basin. *Ecosystem processes,* including element and energy cycling, population dynamics, and changing landscape structure, underlay the interactions between these stressors and ecosystems. Finally, *climate change processes* are probably having significant direct and indirect effects on the ecosystems of the basin and their responses to atmospheric stressors and other factors.

At a recent workshop of the Lake Champlain Research Consortium [McIntosh 1998], two broad areas of research and monitoring were identified to improve our understanding of air resource issues in the basin. These were to assess *source-receptor relationships* for atmospheric contaminants entering the basin, and to clarify the *transport mechanisms* for hazardous air pollutants within the basin. In addition, a number of strategic recommendations were made, including increased attention to air issues in the basin, under the clean air regulations of the state and the Clean Air Act Amendments (section 112m).

While many of the air pollutants and environmental stressors affecting the basin originate outside its borders, there are important local sources of some contaminants such as mercury and some volatile organic compounds. A better inventory of these emissions and an analysis of their contribution to pollutant exposures in the basin are needed. This represents a significant gap in our knowledge and an impediment to effective management of air resources in the basin.

Transport and deposition of regional and local air pollutants are strongly affected by regional and local weather patterns. As seen in this paper, regional and synoptic analysis of meteorology and air pollution data allows assessment of transport patterns and some understanding of source-receptor relationships. Local-scale processes, however, are poorly understood, particularly as they affect deposition patterns and movement of locally generated pollutants. Most of the modeling of deposition has been conducted at coarse scale resolutions (>100 km), while meso-scale modeling (<10 km resolution) is needed to assess deposition processes and source-receptor relationships within the basin. The NOAA Air Resources Laboratory, as well as several other federal and university research groups, have the capability to model wind-fields and pollutant deposition at this finer scale. Transferring this technology to the Lake Champlain basin would benefit our understanding and management of air resource issues in the basin.

Although this report documents much useful information about air contaminants in the basin, there remain a number of concerns that need specific attention. To achieve effective air resource management in the basin in the foreseeable future, the following additional needs must be addressed. *1. Support long-term monitoring.* Only with continuous, high-quality monitoring at the appropriate frequency will we be able to provide reliable information capable of detecting patterns and trends in spatial and temporal environmental data. *2. Assess source-receptor relationships.* Further analysis of air transport and high resolution chemical deposition data is needed to better identify regional and local sources of air contaminants in the basin. *3. Conduct elemental speciation of fine particulates.* Although new regulations set standards for fine particulate (2.5 μm) matter, there is currently inadequate information on the concentration of toxic elements in this material to evaluate specific health risks and identify sources. *4. Clarify watershed transport and cycling processes.* There are still important gaps in our understanding of how forested and agricultural watersheds affect the movement of air contaminants in the basin, particularly with respect to the cycling of mercury and land use effects on nutrients and contaminants.

5. Assess environmental and ecological effects of climate change. Although this is complex, collection of environmental data and development of ecological models are needed that help illuminate the direct, indirect, and interactive effects of climate change in the ecosystems of the basin.

Acknowledgments. This work described here has been funded by numerous grants over many years, and we wish to acknowledge continuing support of the National Oceanic and Atmospheric Administration, the Environmental Protection Agency, the Vermont Agency of Natural Resources, the Vermont Monitoring Cooperative, and the Cooperative Institute for Limnology and Ecosystems Research. Special appreciation also goes to Carl Waite, Andrea Donlon, Jessica Orrego, Krista Reinhart and Sean Lawson for their assistance in preparing this report, and to the many field personnel who dependably operate monitoring stations throughout the region.

REFERENCES

AIRMoN 1998. The Atmospheric Integrated Research Monitoring Network (AIRMoN): http://nadp.sws.uiuc.edu/airmon/.

Bishop, K., Y.H. Lee, C. Patterson and B. Allard. 1995. Methylmercury output from the Svartberget catchment in northern Sweden during spring flood. *Water Air Soil Pollut.* 80:445-454.

Burke, J., M. Hoyer, G. Keeler and T. Scherbatskoy. 1995. Wet deposition of mercury and ambient mercury concentrations at a site in the Lake Champlain Basin. *Water Air Soil Pollut.* 80:353-362.

DeVries, W. 1993. Average critical loads for nitrogen and sulfur and its use in acidification abatement policy in the Netherlands. *Water Air Soil Pollut.* 68:399-434.

Draxler, R.R. and G.D. Hess, 1997, Description of the Hysplit-4 modeling system NOAA Technical Memorandum ERL ARL-224: http://www.arl.noaa.gov/ss/models/hysplit.html.

Friedland, A.J., B.W. Craig, E.K. Miller, G.T. Herrick, T.G. Siccama and A.H. Johnson. 1992. Decreasing lead levels in the forest floor of the northeastern USA. *Ambio* 21:400-403.

Galloway, J.N., E.B. Cowling, E. Gorham, W.T. McFee, eds. 1978. *A National Program For Assessing The Problem Of Atmospheric Deposition (Acid Rain), A Report To The Council On Environmental Quality.* National Atmospheric Deposition Program. Natural Resource Ecology Laboratory, Colo. State University; Ft. Collins, CO. 97 pp.

Hicks, B.B., R.P. Hosker, Jr., T.P. Meyers and J.D. Womack. 1991. Dry deposition inferential measurement techniques. I. Design and tests of a prototype meteorological and chemical system for determining dry deposition. *Atmospheric Environ.* 25A:2345-2359.

Keeler, G., E. Malcolm, J. Burke, S. DeBoe and T. Scherbatskoy. Atmospheric mercury transport and deposition in the Lake Champlain Basin, 1992-1997. This volume.

Likens, G.E., C.T. Driscoll and D.C. Buso. 1996. Long-term effects of acid rain: response and recovery of a forest ecosystem. *Science* 272:244-246.

McIntosh, 1998. *Proceeding of a workshop on research and monitoring priorities in the Lake Champlain basin.* Lake Champlain Research Consortium, Burlington, VT. 12 pp.

NADP. 1998. National Atmospheric Deposition Program. http://nadp.sws.uiuc.edu.

NESCAUM. 1998. *Northeast states and eastern Canadian provinces mercury study; a framework for action.* M. Tatsutani, ed. Northeast States for Coordinated Air Use Management. Boston.

NOAA 1998. Real-time Environmental Applications and Display sYstem (READY): http://www.arl.noaa.gov/ready.html.

Olmez, I., M.R. Ames, G. Gullu, J. Che and J.K. Gone. 1996. *Upstate New York Trace Metals Program. Volume 1. Mercury.* Massachusetts Institute of Technology Nuclear Reactor Laboratory, Cambridge, MA. MIT Report MITNRL-064.

OTAG 1998. Ozone Transport Assessment Group (OTAG): http://www.epa.gov/ttn/otag/.

Poirot, R.L. 1993. 1992 Regional Ozone Concentrations in the Northeastern United States. Report for the Data Management and Ambient Monitoring and Assessment Committees of the Northeast States for Coordinated Air Use Management (NESCAUM). http://capita.wustl.edu/NEARDAT/REPORTS/TechnicalReports/CF/P92cf.html.

Poirot, R.L., P. Wisinski, B. Schichtel and P. Girton. Air trajectory pollution climatology for the Lake Champlain basin. This volume.

Rea, A.W. 1998. The processing of mercury in forested ecosystems. PhD Dissertation, University of Michigan.

Rea, A.W., G.J. Keller and T. Scherbatskoy. 1996. The deposition of mercury in throughfall and litterfall in the Lake Champlain watershed: a short-term study. *Atmos. Environ.* 30:3257-3263.

Scherbatskoy, T. and J.B. Shanley. 1998. Factors controlling mercury transport in an upland forested catchment. *Water, Air, and Soil Pollut.* 105:427-438.

Scherbatskoy, T., J.M. Burke, A.W. Rea and G.J. Keeler. 1997. Atmospheric mercury deposition and cycling in the Lake Champlain Basin of Vermont. In: J.E. Baker (ed.) *Atmospheric deposition of contaminants to the Great Lakes and coastal waters.* Soc. Envion. Toxicol. and Chemistry. Pensacola, FL.

Scherbatskoy, T. and M. Bliss. 1984. Occurrence of acidic rain and cloud water in high elevation ecosystems in the Green Mountains of Vermont. pp. 449-463 in P.J. Samson (ed.), *The Meteorology of Acid Deposition, Transactions of an APCA Specialty Conference, Hartford, CT. Oct. 16-19, 1983.* Air Pollution Control Association, Pittsburgh.

Schuster 1991. The behavior of mercury in the soil with special emphasis on complexation and adsorption processes - a review of the literature. *Water, Air, Soil Pollut.* 56:667-680.

Shanley, J.B., A.F. Donlon, T. Scherbatskoy and G.J. Keeler. Mercury cycling and transport in the Lake Champlain basin. This volume.

Stunder, B.J.B. and R.S. Artz. 1996. A comparison of 1993 and 1995.AIRMoN precipitation chemistry measurements using HYSPLIT trajectories. 1996 NADP Technical Committee Meeting, October 21-24, 1996, Williamsburg, VA.
http://www.arl.noaa.gov/ss/transport/cluster.html

Swain, E.B., D.R. Engstrom, M.E. Brigham, T.A. Henning and P.L. Brezonik. 1992. Increasing rates of atmospheric mercury deposition in midcontinental North America. *Science* 257:784-787.

U.S.D.A.1998. The USDA UVB Radiation Monitoring Program.
http://uvb.nrel.colostate.edu/UVB/home_page.html.

U.S. EPA. 1995. *Acid Deposition Standard Feasibility Study Report to Congress.* EPA-430-R-95-001A. http://www.epa.gov/acidrain/effects/execsum.html.

Vasu, A. B. and M. L. McCullough. 1994. *First report to Congress on deposition of air pollutants to the Great Waters.* EPA-453/R-93-055. U.S. EPA Office of Air Quality Planning and Standards, Research Triangle Park, North Carolina.

Watzin, M. C. 1992. *A research and monitoring agenda for Lake Champlain: proceedings of a workshop, December 17-19, Burlington, VT.* Lake Champlain Basin Program Technical Report No. 1. US EPA, Boston.

24

Air Trajectory Pollution Climatology for the Lake Champlain Basin

Rich Poirot, Paul Wishinski, Bret Schichtel and Phil Girton

ABSTRACT

Multiple air mass history calculations from the CAPITA Monte Carlo model are aggregated and sorted by residence-time analysis to investigate upwind, synoptic-scale meteorological conditions and associated ambient pollutant concentrations for several remote monitoring sites in (or near) the Lake Champlain Basin. The CAPITA model, with meteorological input from the National Climatic Center's Nested Grid Model (NGM) was run to calculate 5-day backward air mass histories, every 2 hours, for the years 1989 through 1996 for receptor sites surrounding the Lake Champlain watershed in Sutton, Que; Whiteface Mtn., NY; Underhill, VT and Lye Brook, VT. The results presented here begin to provide a probabilistic description of the Lake Champlain "airshed" (areas most likely to be upwind). This airshed has no fixed geographical boundaries, but varies with time (season, year, etc.), upwind meteorological conditions (pressure, wind speed, precipitation, etc.) and the characteristics (emissions patterns, atmospheric chemistry and phase functions) of the airborne pollutant(s) of interest. Over the long-term (8-year) period of record, areas which are persistently upwind prior to high concentrations of specific air pollutants in the Champlain Basin sites are interpreted as predominant source regions for these pollutants.

INTRODUCTION

Lake Champlain's watershed is relatively undeveloped (89% forest and agricultural), large relative to the Lake's surface area (by a factor of 18:1), and is clearly defined by fixed geographical boundaries (mountain ridges). Contaminants discharged to ground or surface waters anywhere within this large, rural watershed can potentially result in downstream consequences, including impacts in the Lake itself. In a similar way, contaminants discharged into a Lake Champlain "airshed" can result in downwind exposure to, or deposition of these contaminants to the Lake's surface or to upstream areas throughout the watershed. If airshed is taken to mean a geographic region which encompasses all emissions that occasionally have some (any) influence at a specified receptor area, then the airshed for any receptor may well extend to hemispheric or global scales. However, the likelihood or probability of impact from an upwind source is greatest near the receptor and decreases with distance, as modified by prevailing wind direction and other upwind meteorological conditions. For individual pollutants, the relative strength and spatial distribution of emissions sources can also modify the size and shape of the "effective" airshed. For reactive pollutants with short atmospheric lifetimes, local urban centers like Burlington and Plattsburgh may represent the predominant source areas for the Basin. Given the small sizes of these local urban areas and the relative absence of large industrial air pollution sources in the Lake Champlain watershed, the airshed for longer-lived pollutants may extend hundreds to thousands of kilometers beyond the watershed boundaries. For example, Dennis (1995) calculated an effective "airshed" of 900,000 km^2 to account for 70% of the nitrogen deposition in the Chesapeake Bay watershed.

When multiple pollutants are considered, the spatial and temporal variations of emission sources and meteorological conditions, make it virtually impossible to develop a fixed definition of airshed. However, by combining multi-year sets of regional-scale meteorological data and local ambient pollution monitoring data a long-term or "climatological" description of airshed can be developed in "probabilistic" terms. This paper begins the process of describing the Lake Champlain airshed by providing several example illustrations of upwind meteorological conditions, including areas most likely to be upwind when concentrations of selected trace elements are elevated in the Basin.

METHODS

A backward air trajectory estimates the path of airmass motion prior to arrival at a specified "receptor" location at a specified arrival time. The authors have previously reported on ensemble backward air trajectory techniques - referred to as "residence-time analysis" - to explore predominant source regions of regional haze, particulate matter, trace elements and ozone for various receptor locations and time periods [Poirot and Wishinski, 1986, 1998; Wishinski and Poirot, 1986, 1998]. The general approach [after Ashbaugh, 1983] involves calculating large numbers of backward trajectories from a receptor site where concurrent ambient air monitoring data are available. A grid is used to track the multiple trajectory locations, and the gridded trajectories are sorted and/or aggregated as a function of the resultant ambient concentration at the receptor. The automated calculation of multiple trajectories for many arrival times requires a computerized trajectory model (such as the NOAA HY-SPLIT model [Draxler, 1992] or the CAPITA Monte Carlo model [Patterson et al., 1981; Schichtel and Husar, 1996]) and a multi-day, 3-dimentional meteorological database (such as from NGM [Rolph, 1996] or RAMS [McQueen et al., 1997]).

In a currently ongoing analysis for the US Forest Service, the authors are developing an "Air Mass History Pollution Climatology for Northeastern Forests and Parks" [Poirot et al., 1998]. The objective is to compare long-term (8-year), upwind meteorological characteristics with resultant air pollutant concentration or deposition characteristics for a number of rural, regionally representative ambient monitoring sites in the Northeastern US and Eastern Canada. Four of these sites: Sutton, Quebec; Lye Brook, VT; Whiteface Mtn., NY and Underhill, VT (Figure 1), are located within (or just outside) the Lake Champlain watershed, and are positioned near the north, south, east and west watershed boundaries. Measurements at these sites include ozone, precipitation chemistry, and aerosol chemistry covering much (or all) of the 8-year period 1989 through 1996 for which archived meteorological data suitable for calculation of airmass histories are also available.

The term "airmass history" as applied here includes the spatial information provided in a backward trajectory, but also incorporates other meteorological information along the trajectory path. For example, during the trajectory calculation, information on trajectory height, pressure, atmospheric mixing height, temperature, humidity, precipitation, wind speed, etc. can be calculated and retained for each location along the trajectory. The airmass histories for this study are calculated by the CAPITA Monte Carlo model and NGM meteorological fields. The Nested Grid Model (NGM) is routinely run by the National Meteorological Center to support meteorological forecasts on a variety of grid scales, including global. Input includes: radiosonde, satellite observations of temperature and cloud motion, and surface observations. A subset of the NGM data covering most of North America has been routinely archived since 1989 by the NOAA Air Resource Laboratory [Rolph, 1992]. These archived NGM data include estimates of about 20 meteorological variables with a 2-hour time step, horizontal grid resolution of about 180 km, and 10 vertical (sigma) layers ranging from about 150 m to 7000 m above the surface terrain. The routine archival of NGM data was discontinued in early 1997 (replaced by more sophisticated ETA model results). Hence the period 1989-1996 represents an ideal period for "long-term" analysis based on consistent methodologies.

The CAPITA Monte Carlo model acts upon the Eulerian NGM data to calculate a Lagrangian airmass history, running (in this case) backward in time from a fixed receptor location and arrival time. A current version of the CAPITA model [Schichtel and Husar, 1995, 1996] is highly automated for PC operation, with options to run sequentially starting every 2 hours, retaining any or all of the NGM meteorological variables, at airmass positions every 2 hours backward in time for up to 5 days, for multiple receptor locations. The CAPITA model incorporates an estimate of vertical mixing by releasing 10 particles and allowing the vertical position of each to move randomly within the mixed layer - such that for each release time, 10 separate air mass histories are calculated for each starting location. No vertical motion is imparted to particles which are positioned above the mixed layer. In addition to the lat./long. (map) position of trajectory endpoints, various other aspects of the air mass histories are also retained - including trajectory pressure height, precipitation, mixing height, relative humidity, specific humidity, temperature, velocity, and age. So for each air mass history for each particle, values of about 12 meteorological variables are calculated and retained every 2 hours for 5 days backward in time. For each starting location, 120 air mass histories are calculated each day for 8 years. The starting locations and heights for the airmass histories are based on the locations and elevations of the ambient monitoring sites. For several of these sites, (including Sutton and Lye Brook) the ambient measurements were conducted at significantly different elevations, and two sets of airmass histories were calculated (see Figure 1). At night, and especially during the winter, higher elevation sites are often above the mixing height, and experience different meteorological flow conditions than lower elevation sites.

Example illustrations of some of the air-mass history information retained for individual release times for Acadia, NP are displayed in Plate 1. Note the influence of vertical mixing on the widely dispersed trajectories of particles arriving at the receptor on 8/4/95 at 8 AM, in contrast to the similar (horizontal and vertical) pathways traversed by all particles arriving on 8/1/95 at 10 AM. Note also the presence of precipitation along all trajectory paths just before arrival at Acadia on 8/4/95, and along several other locations further back along the particle pathways.

The individual air mass history information is tracked on a grid of 1,440 80x80 km squares, covering an approximately 10 million km^2 area of Eastern North America. To reduce data volume, and to facilitate direct comparison with ambient pollutant data (much of which is only available on a 24 hour basis), the air mass history data for each receptor site are converted to a 24 hour basis by aggregating all (120) air mass history results for each calendar day for each grid square. This aggregation assumes that the air mass history information for each of 10 particles released at each of 12 arrival times during a 24-hour period is equally representative of the upwind meteorological characteristics associated with pollutant concentrations measured during that 24-hour period. Rather than a single trajectory line, this approach identifies an upwind area (and associated meteorological parameters) as an upwind region of potential influence for each 24-hour measurement period. Examples of daily aggregate air mass history results for a site near the PA/NJ/NY border on 7/3/94 and 7/4/94 are displayed in Plate 2. In these examples, the spatial characteristics of all 120 airmass histories for each day are displayed as " residence-time probabilities", with 25% of the day's upwind residence-time hours included in each separately shaded area. Note that on 7/3/94, the upwind residence-time probability is relatively tightly constrained along a narrow upwind path - reflecting strong, persistent flow from the northwest of the receptor. On the following day, the upwind probabilities are more broadly distributed over nearby areas to the east and northeast of the receptor.

RESULTS

Plate 3 displays "Everyday" upwind probability fields for the Sutton and Lye Brook sites. These are based on the daily airmass history data (for the higher elevation starting heights at these receptors) for every available day, 1989 through 1996. The tracking grid is employed to aggregate the "residence-time" hours that each airmass has spent over each grid square in its path en route to the specified receptor. The hours are summed for all dates, arrival times and grid squares, and the probability for an individual square is expressed as the fraction of total hours for that square divided by the total hours in all squares. Each separately shaded area in Plate 3 contains 20% of the residence-time probability for the total (approximately 10 million square kilometer) tracking grid, with isopleths bounding the smallest areas accounting for 20%, 40%, 60% and 80% of the total probability within the grid. The least probable 20% of upwind locations are included in the large, lightly shaded, outermost area - which represents about 70% of the total grid area, but represents a very low upwind probability for any specific location.

Similar everyday upwind probability fields result for the Whiteface Mtn., and Underhill sites. In Plate 4, the everyday probability fields from all 4 Champlain Basin sites are combined (averaged by upwind grid square) - to provide a general estimate of the Lake Champlain "airshed" in probabilistic terms. A cutoff at the 80% probability contour is arbitrary - the "total airshed" extends well beyond this (and beyond our tracking grid too), but the upwind probability for individual locations beyond the 80 % contour is relatively small. For example, an individual 80x80 km grid square on the outer edge of the 80% contour line is upwind of the Basin only 0.07% of the time - which is coincidentally equal to the percentage that would result if the probabilities were distributed equally across all

Figure 1. Champlain Basin Ambient Monitoring and Airmass History Sites

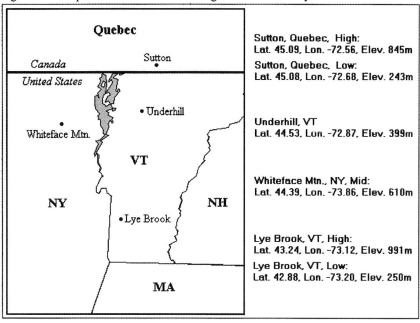

Plate 1. Example Airmass Histories for Acadia NP on 8/1/95 10 AM (left) and 8/4/95 8 AM (right)

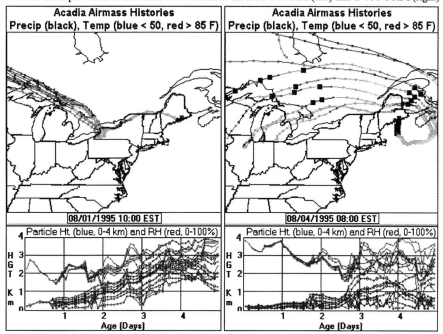

Plate 2. Example 24-hr. Aggregation of Airmass Histories for PANJNY site

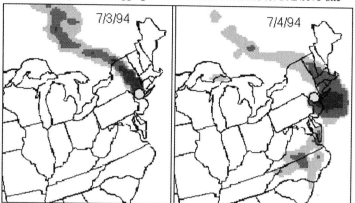

Plate 3. Everyday Upwind Probabilities for Sutton, Que. and Underhill, VT

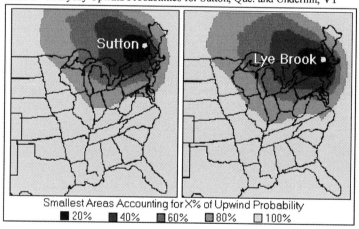

Plate 4. Probabilistic Lake Champlain Airshed and Upwind Source Regions

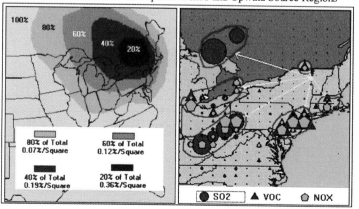

Plate 5. Seasonal Variations in Probabilistic Lake Champlain Airshed

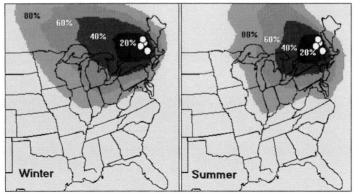

Plate 6. Average Pressure and Wind Speed Upwind of Underhill, VT

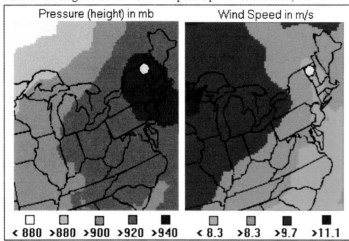

Plate 7. Seasonal Variation in Wind Speed Upwind of Underhill, VT

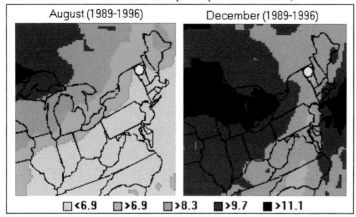

Plate 8. Upwind Probabilities for High Aerosol Se, Ni, As and Mn at Champlain Basin Sites

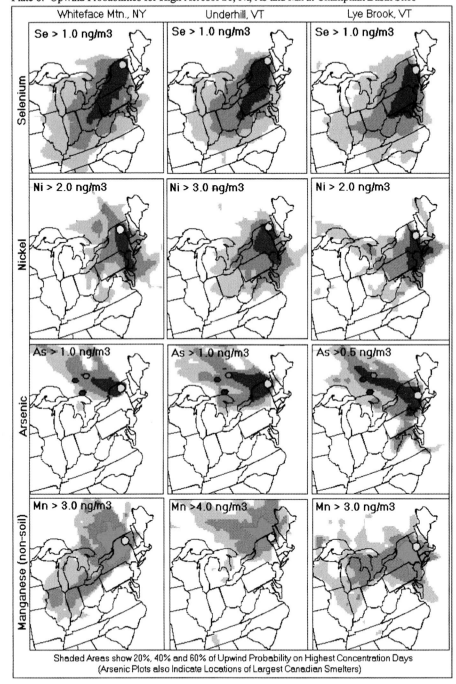

1440 grid squares (1/1440 = 0.07%). The 80% contour also approximately corresponds to a transport time of 2 to 3 days and contains a number of large source regions which are upwind on a relatively frequent basis.

These source regions, and certain aspects of their emissions characteristics are also displayed in Plate 4. The emissions of sulfur dioxide (SO_2), nitrogen oxide (NO_x) and volatile organic compounds (VOC) are taken from the 1985 NAPAP inventory, and are illustrative of predominant source types. They are plotted here with symbols sized relative to emission rates - with the exception that the SO_2 symbols are reduced by 50% for clarity (also, SO_2 emissions from both US and Canadian sources have been substantially reduced since 1985). The major source regions which are relatively close to, and likely to be periodically upwind of the Lake Champlain Basin include:

- Montreal & Boston Urban Areas (VOC and NO_x from area and mobile sources),
- East Coast Urban Corridor (VOC and NO_x from area and mobile sources),
- Ohio River Valley (SO_2 and NO_x from large utilities and industrial sources),
- Lower Great Lakes (mixed pollutants from urban & industrial sources),
- Canadian Smelters (SO_2 from a small number of large industrial sources).

These generalized source regions include a variety of other sources and pollutant emissions (trace elements, for example). As they represent large emissions densities, are frequently upwind, and nearly surround the (low emissions density) Champlain basin, the air quality and deposition in the Basin may be influenced by one or more of these source regions on a day to day and long-term basis. As they are located in many different directions from the Champlain Basin, and are characterized by different mixes of predominant source types, it should be possible to discern their impacts in the Basin through combinations of the airmass history and ambient pollutant data. If, for example, emissions from within the Champlain Basin and nearby Montreal and Boston are considered part of local "New England" sources, then the above-mentioned 5 source regions are approximately equivalent to the 5 regional "elemental fingerprints" identified by Rahn and Lowenthal [1984, 1985] in aerosol samples at Underhill, VT during the early 1980's. Those regional fingerprints were described as: NENG (New England and parts of southern Canada excluding smelter sources), SONT (southern Ontario including smelter sources), ECE (East Coast urban corridor), UMW (upper Midwest) and LMW (lower Midwest).

Seasonal differences in the upwind probability fields, from aggregation of the winter (DJF) and summer (JJA) quarters for the 4 Champlain Basin sites, are displayed in Plate 5. The airshed has a stronger northwesterly orientation in the winter, while areas to the southwest and south are relatively more likely to be upwind in the summer months.

Several different features of the Champlain airshed are displayed in Plate 6. In this case, the average pressure (left) and wind speed (right) are calculated as a function of prior upwind location for airmasses arriving at Underhill, VT. This plotting methodology differs from Plates 3-5 in that the frequency with which locations are upwind is not considered. Rather, an average value is calculated for each upwind grid square for all aimass histories which have passed over that square en route to the receptor. If the air passed over square X, then the average pressure (or speed) was Y. The pressure (representing the average pressure for all airmasses (particles) at each location upwind of Underhill) relates directly to the airmass height (above mean sea level) by the approximate relation:

$$\text{Height (meters above MSL)} = 10 \times [1{,}016 - \text{Pressure (mb)}]$$

These average upwind pressure (heights), particularly those at greatest distance from the receptors, include many particle trajectories which have passed well above the mixed layer, and could have no possible influence from emissions sources in the regions they traverse. They provide however a relative indication of the general height characteristics of airmasses approaching the Champlain Basin from different regions. Flows from the northwest tend to be subsiding, having previously resided at higher altitude, whereas air near the Basin in all directions and for some distance from the south/southeast of the Basin tends to have been at relatively low altitude. This pattern has a degree of coincidental correspondence to the general nature of the emissions sources in these upwind regions. Emissions within the Basin and at relatively nearby Montreal, Boston and East Coast urban areas tend to be released primarily from ground-level sources; Midwestern utility stacks are moderate to tall, and the Canadian smelter stacks are among the tallest in North America.

Average wind speed for areas upwind of Underhill is displayed on the right side of Plate 6. As with the pressure heights, these average upwind wind speeds include influence of some very high (and fast) trajectories, and should be considered in relative rather than absolute terms. Lowest wind speeds are typically associated with areas immediately surrounding the Basin and for long distances to the south; while highest wind speeds (on average) are typically associated with areas to the west. As with the pressure (altitude), there is some coincidental association with the emissions characteristics of upwind source regions. Air approaching the Basin from nearby urban areas (with large numbers of small, broadly distributed, ground level sources) tends to have moved slowly (stagnated) over these areas and along the relatively short transport route to the Basin. Opportunities for pollutant removal with the slow transport speeds are offset by the relatively short transport distance. Transport speeds from the more distant regions (characterized by smaller numbers of larger sources) tend to be faster, offsetting the potential for pollutant dispersion and deposition over the relatively long transport route to the Basin.

A strong seasonal variation in wind speed is indicated in the Plate 7 comparison of average August (left) and December (right) wind speed as a function of upwind location. Upwind summer wind speeds are generally lower than winter, especially over large sections of the Eastern US - south of the lower Great Lakes. Slower wind speeds are conducive to the formation and build-up of secondary pollutants (formed in the atmosphere) like sulfates and ozone. This secondary formation process is enhanced by temperature and solar radiation, and requires time - all of which are more abundant during the summer. Evaporative VOC emissions from urban areas (an important precursor to secondary ozone formation) are also greatest in summer over (East Coast and Midwestern) source regions with greatest summer stagnation. During the winter, the relatively lowest wind speeds are over areas to the south and north of the Basin. The large nearby population centers in these (winter stagnating) directions experience a considerable seasonal increase in emissions associated with residential space heating. Smelter emissions tend to exhibit relatively little seasonal variation and are located in areas which tend to experience relatively high wind speeds in all seasons. This is not conducive to buildup and transport of secondary sulfate species from the northwest (and smelter emissions are relatively unimportant for ozone formation), but strong, persistent winds do represent an efficient mechanism for efficient delivery of primary emissions (emitted directly to the atmosphere) such as trace metals with minimal dispersion/deposition en route from a small number of tall stacks. Thus, the Lake Champlain airshed not only includes a variety of large source regions which are frequently upwind, but is also characterized by natural climatological features which seem conducive to maximizing the potential impact from several of these upwind regions.

For three of the Lake Champlain Basin air monitoring sites - Whiteface, Underhill and Lye Brook - relatively long-term records of 24-hour fine particle (< 2.5 micron) concentration and composition (including trace elements) are available during the 1989-1996 period of available airmass histories, through the IMPROVE [Eldred et al., 1988] and NESCAUM [Flocchini et al., 1990] monitoring networks. The Lye Brook IMPROVE site collects fine particle samples 2 days/week, started operation in September, 1991 and is still operating (total 450 samples during 1989-96 period). The Whiteface and Underhill NESCAUM sites commenced operation in September, 1988, collecting 3 samples/week. The Whiteface NESCAUM site was discontinued in November, 1993 (total 676 samples) and the Underhill NESCAUM site was discontinued in May, 1995 (total 815 samples). Elemental analysis for both IMPROVE and NESCAUM samples has been conducted by Proton-Induced X-ray Analysis (PIXE) and X-Ray Fluorescence (XRF) at the Crocker Nuclear Laboratory, U. California at Davis [Eldred et al., 1988].

Plate 8 displays the upwind probabilities for airmasses which have resulted in "high" concentrations of the trace elements - selenium, nickel, arsenic and manganese - at the Lake Champlain sites. Each of these plots may be thought of as a probabilistic airshed, comparable to Plates 3-5, but constrained to the condition of high concentration of the individual pollutants. The "high" concentrations selected here are above a specified threshold or cut-point, for which we attempted to chose round numbers which represented approximately similar sample sizes and distribution percentiles across the 3 sites and 4 pollutants. In this case, the high subsets include 20 to 40 of the highest daily concentrations representing the highest 3 to 5% of the sample days for each pollutant at each site. Concentrations at or above these threshold levels might be expected to occur on roughly 10 to 20 days per year.

The upwind probability plots for high selenium in the top row of Plate 8 show a clear southwesterly orientation at all 3 receptor sites, pointing directly toward Midwestern (coal-burning) sources along the Ohio River Valley. While Se has not been identified as a contaminant of concern in Lake Champlain sediments or at ambient aerosol concentrations in the Basin, it does appear to be an excellent tracer for influence from coal combustion [Rahn and Lowenthal, 1984], which in turn, is a predominant source of sulfate pollution in the Basin [Rahn and Lowenthal, 1984, 1985; Poirot and Wishinski, 1986; Wishinski and Poirot, 1986]. Aerosol sulfate compounds contribute about half of the regional fine particle mass concentrations and 60% of the regional haze in the Basin [Poirot et al., 1992], while sulfate deposition is a predominant cause of acidification in upland areas in the Basin [Scherbatskoy et al., 1998]. Coal combustion is also an important contributing source for deposition of other assorted trace metals, including mercury, of concern in the Champlain Basin [Scherbatskoy et al., 1998].

The upwind probability plots for high nickel in the second row of Plate 8 show a strong southerly orientation at all 3 sites, pointing directly toward the East Coast urban corridor. Atmospheric emissions of Ni are dominated by oil combustion - both residual oil (a common utility and industrial fuel in the East Coast urban corridor) and distillate oil (commonly used for residential and commercial space heating in the northeast corridor). While ambient air concentrations of Ni at Champlain Basin sites are relatively low compared to larger eastern urban areas, and are not currently considered to pose a direct health threat in the Basin, they do appear to be a useful tracer for influence in the Basin from larger urban centers to the south. Elevated Ni concentrations have also been noted in Lake Champlain sediments in Outer Malletts Bay [McIntosh and Watzin, 1998], although the causes of these sediment concentrations are unclear.

The upwind probability plots for high arsenic in the third row of Plate 8 have a strong northwesterly orientation at all 3 sites, pointing directly toward the Canadian smelter region. Locations of several large Canadian smelters are also identified in these high As plots, with the Noranda smelter - identified as a green dot - appearing to be the most likely contributor. Previous "elemental fingerprint" analysis of aerosol samples at Underhill, VT during the early 1980's [Rahn and Lowenthal, 1985] apportioned approximately equal fractions (40%) of the ambient As to Canadian smelters and Midwestern (coal burning) sources. While there is no indication in Plate 8 of a Midwestern influence, the apparent discrepancy with these earlier studies may be due in part to the focus here on only the very highest As levels. Rahn and Lowenthal's analysis apportioned all the As (including days with low and moderate concentrations) and was also based on analytical methods (neutron activation) with lower detection limits for As than the PIXE and XRF methods employed in the IMPROVE and NESCAUM networks. In any event, high arsenic levels in the current data set appear to be excellent tracers for influence in the Basin from the Canadian smelter region. Elevated levels of As have been observed in sediments of outer Malletts Bay [McIntosh and Watzin, 1998], although the origin is unclear. Ambient air As concentrations in the Basin have declined markedly since 1989 (perhaps in association with acid rain-related improvements to smelter emissions control systems). Current concentrations remain of some concern in the Basin, as they remain close to the level of Vermont's Hazardous Ambient Air Standards (0.23 ng/m^3 - intended to prevent a cancer risk of 10^{-6}) even at rural background sites. The fine particle "non-soil" manganese concentrations used to determine "high" concentrations in the upwind probability plots in the bottom row of Plate 8 were adjusted to remove the fraction of manganese assumed to result from soil (after Rahn and Lowenthal, 1984, 1985). This adjustment was based on concurrent measurements of fine particle silicon, and an assumed Si:Mn ratio in soil of 277:1 (based on crustal composition estimates from Mason [1966]). Since there is typically only a small amount of soil or "crustal material" in the fine particle fraction, the effect of this "non-soil" adjustment was typically small (less than 10% on average). An exception was for the Underhill, VT site where a short period of local construction resulted in several very high concentrations of "crustal" elements (including Mn), which are eliminated as "high Mn days" by the non-soil adjustment. The upwind probability plots for non-soil Mn are less directionally distinct than for the other 3 trace elements, and are less similar between 3 receptor sites. The 20% isopleths of highest probability are more uniformly distributed around each receptor. Lye Brook and Whiteface indicate areas of relatively high probability extending west along the US Canadian border. Whiteface also indicates an area of high probability to the north. For Underhill, where peak Mn concentrations are higher than for the other 2 sites, the highest upwind probability is predominantly to the north. Extraordinarily high manganese levels have been observed in sediments in outer Malletts Bay, and have been associated with acute biological toxicity in sediment pore water tests [McIntosh and Watzin, 1998]. However the origin or cause of these high concentrations is unclear, and may be predominantly related to changes in redox chemistry related to periodically anaerobic conditions in deeper sections of the Bay. Ambient air Mn concentrations in the Basin are well below levels of concern from a human health perspective. However, potentially phytotoxic levels of Mn have been observed in Balsam Fir foliage on Roundtop Mtn. in Sutton, Quebec, where the patterns of foliar and soil Mn concentration suggest that foliar uptake from atmospheric exposures may be an important factor [Lin et al., 1995].

Atmospheric sources of Mn are not entirely clear. Methylcyclopentadienyl manganese tricarbonyl (MMT) was developed in the 1950s as an octane-enhancing fuel additive in leaded gasoline, but was banned by US EPA for use in unleaded gasoline in 1978. A similar ban was not imposed on MMT additive in Canada, where it has been used in unleaded gasoline since 1976. Wallace and Slonecker [1997] noted historical statistical associations between fine particle Mn and Pb at US monitoring sites that suggested leaded gasoline as a predominant source of airborne Mn prior to the elimination of leaded gasoline in the early 1990s. They also noted substantially higher Mn concentrations in more recent fine particle measurements throughout Canada compared to measurements at similar sites in the US - which they attributed to the influence of the MMT additive [Wallace and Slonecker, 1997]. The spatial patterns of upwind probabilities for high Mn concentrations at the Champlain Basin sites - showing highest probabilities to the north (Montreal) and west (along the US/Canada border) are consistent with an MMT source. However, Mn is also emitted by various industrial processes, including steel production (prevalent in the lower Great Lakes region on both sides of the border). A large Mn alloy production plant in Beauharnois, Quebec (about 20 miles southwest of Montreal) ceased operations in 1991, but may well have contributed to the northerly influence evident at the Whiteface and Underhill sites. Wood combustion has also been identified as an important Mn source in a recent US national toxics emission inventory [US EPA, 1997]. Additional analysis of the ambient data and airmass histories before and after the 1991 Beauharnois plant closing may shed additional light on these potential source influences. Similar analysis of future data may also prove illuminating, as a recent court ruling in the US [US Court of Appeals, 1995] has essentially allowed the use of MMT in US unleaded gasoline; while more recent Canadian legislation prohibiting the importation or intra-provincial sale of MMT may lead to substantial reduction of MMT use in Canada.

CONCLUSIONS

The above results represent preliminary examples from an ongoing investigation of airmass histories and associated pollutant concentrations in the Northeastern US. Several of the selected data sets are directly relevant to exploration of regional-scale atmospheric impacts in the lake Champlain watershed. The Lake Champlain airshed is an illusive concept, which can't be defined in precise geographical terms, but which can be approximately described in probabilistic terms. A number of relatively large air pollution source regions, including nearby urban centers to the north and east, the East Coast urban corridor to the south, Ohio River Valley to the southwest, urban/industrial areas to the west, and Canadian smelters to the northwest are all periodically upwind of the Lake Champlain watershed on a relatively frequent basis, and within a transport time of 1 to 3 days. The potential impacts from these upwind source areas is in several cases enhanced by natural meteorological characteristics of airmasses passing over these source regions en route to the Champlain Basin. Airmass history analysis of selected trace elements suggests that distinctly different upwind source regions are associated with high concentrations of selenium, nickel, arsenic and manganese in the Champlain Basin. Several of these trace elements exhibit (perhaps coincidentally) elevated concentrations in Lake Champlain sediments, and/or represent ambient air exposures of concern from a human health or ecological health perspective.

REFERENCES

Ashbaugh L. L., A statistical trajectory technique for determining air pollution source regions, *J. Air Pollut. Control Assoc.* **33**: 1096-1098, 1983.

Draxler, R. R., Hybrid Single-Particle Lagrangian Integrated Trajectories (HY-SPLIT): Version 3.0 -- User's Guide and Model Description, NOAA Technical Memorandum ERL ARL-195, 1992.

Dennis, R. L., Using the regional acid deposition model to determine the nitrogen deposition airshed of the Chesapeake Bay Watershed, in: *Atmospheric Deposition to the Great Lakes and Coastal Waters*, J. Baker ed., Soc. of Env Tox. and Chem., 1995.

Eldred, R. A., T.A. Cahill, L. K. Wilkinson, P. J. Feeney and W. C. Malm, Measurement of Fine Particles and their Chemical Components in the IMPROVE/NPS Networks, in: Proceedings of the 1988 Annual Meetings of the Air Pollution Control Assoc., 1988.

Flocchini, R. G., T. A. Cahill, R. A. Eldred, and P. J. Feeney, Particulate Sampling in the Northeast: a Description of the Northeast States for Coordinated Air Use Management (NESCAUM) Network, AWMA/EPA Int. Spec. Conf. on *Visibility and Fine Particles*, C. V. Mathai, ed., 1990.

Husar, R. B. and D. E. Patterson, Monte Carlo simulation of daily regional sulfur distribution: Comparison with SURE sulfate data and visual range observations during August 1977, *J. of Applied Meteorology* **20**: 404-420, 1981.

Lin, Z. Q., P. H. Schuepp, R. S. Schemenauer and G. G. Kennedy, Trace Metal Contamination In and On Balsam Fir (Abies Balsamea (L) Mill.) Foliage in Southern Quebec, Canada, *Water, Soil and Air Pollution*, **81**: 175-191, 1995.

Mason, B., *Principles of Geochemistry*, 3rd Edition, Wiley & Sons, Inc., NY, NY, 1966.

McIntosh, A. and M. Watzin, Toxic Substances in Lake Champlain: a Review of Research and Monitoring Efforts, this volume, AGU, 1999.

McQueen, J. T., R. R. Draxler, B.J.B. Stunder and G.D. Rolph, An Overview of the Regional Atmospheric Modeling System (RAMS) as applied at the NOAA/Air Resources Laboratory, NOAA Technical Memorandum ERL ARL-220, 1997.

Patterson, D.E., Husar R.B., Wilson W. E. and Smith L. F., Monte Carlo simulation of daily regional sulfur distribution: comparison with SURE data and visibility observations during August 1977. *J. Appl. Meteor.* **20**: 404 - 420, 1981.

Poirot, R. L. and Wishinski, P. R., Visibility, Sulfate and Air Mass History Associated with the Summertime Aerosol in Northern Vermont, *Atmos. Environ.*, **20**: 1457-1469, 1986.

Poirot, R. L., P. R. Wishinski,, P. Galvin, R. G. Flocchini, G. J. Keeler, R. S. Artz, R. B. Husar, B. A. Schichtel, and A. VanArsdale, Transboundary Implications of Visibility-Impairing Aerosols in the Northeastern U. S. (Part I), Spring 1992 AGU/CGU Meetings, Montreal, Canada, 1992.

Poirot, R. L. and P. R. Wishinski, Long-Term Ozone Trajectory Climatology for the Eastern US, Part II: Results, 98-A615, Air and Waste Manag. Assn. Annual Meetings, San Diego, CA, 1998.

Poirot, R. L., P. R. Wishinski, B.A. Schichtel and P. Girton, Air Mass History Pollution Climatology for Northeastern Forests and Parks, Forest Service Status Report, 1998, http://capita.wustl.edu/NEARDAT/Reports/TechnicalReports/ForestSer_TrajProp/fstrjsum.htm

Rahn, K. A. and D. H. Lowenthal, Elemental Tracers of Distant Regional Pollution Aerosol, *Science*, **223**: 132-139, 1984.

Rahn, K. A. and D. H. Lowenthal, Pollution Aerosol in the Northeast: Northeastern-Midwestern Sources, *Science* **228**: 275-284, 1985.

Rolph, G. D. (1996) NGM Archive TD-6140, January 1991 - June 1996, Prepared for National Climatic Data Center (NCDC), July 1996.

Scherbatskoy, T., R. Poirot and R. Artz, Current Knowledge of Air Pollution and Air Resources Issues in the Lake Champlain Basin, this volume, AGU, 1999.

Schichtel, B. A.; Husar, R. B. Regional Simulation of Atmospheric Pollutants with the CAPITA Monte Carlo Model, *J. of Air & Waste Manage. Assoc.*, **47**: 331-343, 1996.

Schichtel, B.A.; Husar, R.B. The Monte Carlo Model: PC- Implementation, CAPITA, http://capita.wustl.edu/CAPITA/CapitaReports/MonteCarloDescr/mc_pcim0.html#monte , 1995.

U.S. Court of Appeals, *Ethyl Corp. vs EPA*, No. 94-1516, decided Oct. 20, 1995.

US EPA, 1990 Emissions Inventory of Forty Section 112(k) pollutants - Supporting Data for EPA's Proposed Section 112(k) Regulatory Strategy, External Review Draft, 1997.

Wallace, L. and T. Slonecker, Ambient Air Concentrations of Fine (PM-2.5) Manganese in the US National Parks and in California and Canadian Cities: The Possible Impact of Adding MMT to Unleaded Gasoline, *J. Air & Waste Manage. Assoc.* **47**: 642-652, 1997.

Wishinski, P.R. and Poirot, R.L., Source/receptor relationships for a number of factors contributing to summertime variation in light extinction in northern Vermont, APCA Spec. Conf. on Visibility Protection: Research and Policy Aspects, P.S. Bhardwaja, Ed., September 1986, 807-822, 1986.

Wishinski, P. R. and R. L. Poirot, Long-Term Ozone Trajectory Climatology for the Eastern US, Part I: Methods, 98-A613, Air and Waste Manage. Assn. Annual Meetings, San Diego, CA, 1998.

The Hydrology of the Lake Champlain Basin

James B. Shanley and Jon C. Denner

ABSTRACT

Lake Champlain lies in a broad valley between the Adirondack Mountains of New York and the Green Mountains of Vermont. Lake waters discharge to the north via the Richelieu River to the St. Lawrence River at Sorel, Québec. The lake itself occupies only 5.4% of the 21,150-km^2 basin. Most of the inlets are high-gradient streams which peak within 24 hours in response to precipitation or snowmelt. In most of the mountainous basin, a high percentage of the winter precipitation is stored in the snowpack. Therefore, the dominant hydrologic event of the year is spring snowmelt, when nearly one-half of the annual streamflow typically occurs in a six to eight-week period. Because of the geometry of the Lake Champlain outlet, lake outflow cannot keep pace with stream inflow during snowmelt. As a result, the peak lake level lags the peak inflow by several days. In all but 3 of the 128 years of record, the peak annual lake level has occurred during spring. Although the highest historical water level of 31.12 m (102.1 ft) occurred in 1869, near-record levels occurred in 1993 and 1998; those events are analyzed in this paper. Alterations to the outlet channel have caused a general rise in the lake level of about 0.15 m (0.5 ft) since the 1960's.

INTRODUCTION

Interest in the hydrology of Lake Champlain dates to the earliest European explorers who were seeking a water passage from the St. Lawrence settlements to the Hudson River. A passage was not to be realized until two centuries later, when the Hudson-Champlain canal linked the south end of Lake Champlain to the Hudson in 1823, and the Chambly canal bypassed some unnavigable rapids between Lake Champlain and the St. Lawrence in 1843. By that time the widely fluctuating level of the lake had become a concern to commercial interests, and a lake level gaging station was established in Shelburne Bay in 1827 [Barranco, 1977]. As the Lake Champlain Basin became more populated, water-powered mills sprung up on the inlet streams, mainly for the pulp and paper industry, and to saw logs generated by the rapid conversion of land from forest to agriculture. The early dependence on water power dictated settlement patterns in flood plains where flooding has continually posed the risk of property damage and threat to human life. The largest flood of record in Vermont occurred in November 1927, and was followed by another large regional flood in the spring of 1938. The 1927 flood was the impetus for the establishment of many of the U.S. Geological Survey (USGS) stream gaging stations on the major inlets to Lake Champlain.

The Lake Champlain Basin experiences wet and dry climatic cycles on decadal time scales. Drought was an overriding concern in the Lake Champlain Basin in the 1960's; flooding has become a leading environmental issue in the wet decade of the 1990's. Two of the three highest water levels ever recorded in Lake Champlain occurred in 1993 and 1998. Localized areas within the basin were damaged by serious floods nearly every year in the decade. These events have heightened public awareness of hydrology in the basin and led to concerns over future flooding, especially in light of predictions of increased high-latitude winter precipitation in association with global climatic warming [Houghton, 1997]. The National Weather Service (NWS) has flood forecasting models for the Great Lakes, but does not have one for Lake Champlain [S. Hogan, NWS, personal communication, 1998].

In addition to flooding, other environmental issues have driven the need for better understanding of hydrology in the basin. Smeltzer and Quinn [1997] established a phosphorus input budget for the basin, for which they needed accurate hydrologic inflows. Recognizing the need for this type of information, part of a congressional appropriation that established the Lake Champlain Management Conference in 1990 was used to augment the USGS stream-gaging network in the basin. The number of stations was raised from 14 to 35, increasing the gaged area from 57 to 77% of the basin. Accurate inflows also are important in the investigation of mercury cycling in the basin [Scherbatskoy et al., 1998; Shanley et al., this volume].

In this paper we present a general overview of hydrology in the Lake Champlain Basin and characterize the effect of the physical features of the basin on its hydrology. We present a complete record of historical peak lake levels, and summarize some extensive analyses, conducted in the late 1970's and early 1980's but not made widely available, of the effects of human alterations to the lake outlet control on water levels on Lake Champlain. Finally, to demonstrate the hydrologic functioning of the basin, we analyze the extreme high water events of 1993 and 1998, taking advantage of data from the recently expanded USGS stream-gaging network.

LAKE CHAMPLAIN AND ITS BASIN

Geographic setting

The Lake Champlain Basin comprises 21,150 km² in Vermont, New York, and a small part of Québec (Figure 1). The lake itself occupies 1136 km² (excluding 133 km² of islands within the lake), or 5.4% of the basin. The basin is mountainous; the Adirondacks to the west of the lake in New York rise to 1629 m (5344 ft) above sea level at Mt. Marcy, and the Green Mountains to the east of the lake in Vermont rise to 1339 m (4393 ft) above sea level at Mt. Mansfield. Lake Champlain has a mean elevation of 29.5 m (96.5 ft). The

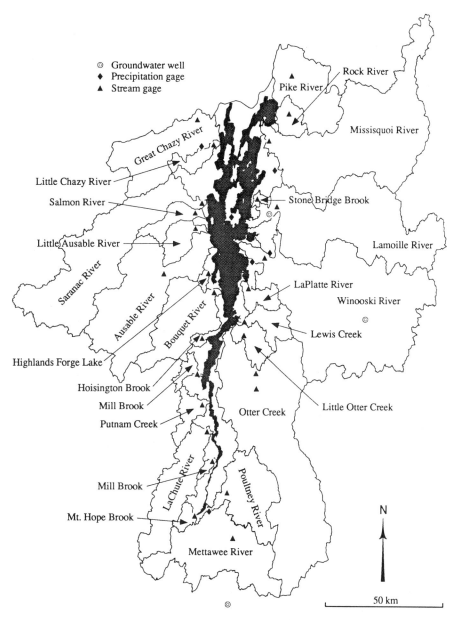

Figure 1. Map of Lake Champlain Basin showing major subbasins.

lake drains to the north via the Richelieu River to the St. Lawrence River at Sorel, Québec.

The basin is about 80% forested, predominantly second-growth northern hardwoods, with a transition to spruce/fir forest at higher elevations. Agricultural activity is concentrated near the lake shore and in the valleys of the tributaries. Dairy farming is the principal type of agriculture, thus much of the open land is pasture, hay-, and cornfields. The landscape is largely rural to unpopulated in the mountains, but the valleys have several urban/suburban centers including Burlington, Rutland, and Montpelier / Barre in Vermont, and Plattsburgh and part of Glens Falls in New York. The total population of the basin in 1990 was approximately 608,000.

Geology and geologic history

Lake Champlain occupies a graben-like depression within a long valley that extends from near Montreal in the St. Lawrence valley southward to New York City. The Adirondack Mountains to the west are highly metamorphosed Precambrian rocks overlain by Paleozoic sandstones and carbonate rocks [Myer and Gruendling, 1979]. The Green Mountains to the east consist of less metamorphosed Cambrian rocks [Myer and Gruendling, 1979]. Shales, carbonates, and sandstones underlie the eastern shore of the lake and probably the lake itself [Hunt et al., 1972]. Fractures in the bedrock yield groundwater in amounts adequate for homeowner wells in most areas, and occasionally enough for small-scale municipal supply.

Pleistocene glaciation left thick surficial deposits of lacustrine clays in the lowlands, dense silty glacial till in the uplands, and scattered sand and gravel outwash deposits in transitional areas. The thickness and permeability of the surficial deposits determine the quantity of water they transmit to surface streams. Generally, only the sand and gravel outwash deposits are significant aquifers in terms of water yield. The dense tills and lacustrine clays may hold a lot of water, but the water is released slowly and supplies base flow to streams. In the lowlands, the lacustrine deposits are areas of poor drainage which tend to support wetlands. Wetlands function as storage reservoirs which damp streamflow peaks.

An example of the heterogeneity of surficial deposits that one commonly encounters in this glaciated landscape is described in a hydrogeology study in the area of Bristol, Vermont [Mack, 1995]. Here, a glaciofluvial delta sequence of mainly coarse-grained sediments, but containing localized lenses of silt and clay, builds into the valley of a former glacial lake, whose floor contains a lacustrine clay layer overlying sand. Median grain size and hydraulic conductivities each span 2-3 orders of magnitude within a 1-km^2 area.

Lake characteristics

Lake Champlain extends nearly 200 km north to south and has a maximun width of 21 km (Figure 1). The average depth is 19.5 m and the maximum depth is 122 m. One-third of the lake water lies below mean sea level [Myer and Gruendling, 1979]. The mean residence time of water in the lake is 2.2 years. The lake is divided by islands and transportation causeways into several distinct parts. The South Lake, south of Crown Point, is long, narrow, and shallow, accounting for 40% of the length of the lake but less than 1% of its volume. The Main Lake extends from Crown Point to Rouses Point and contains 82% of the lake volume [Myer and Gruendling, 1979]. Some large islands in the north-central part of the lake separate the Main Lake from the relatively isolated northeast arm of the lake, which includes several bays. The elongated shape of the lake, in combination with the effects of wind and pressure gradients, gives rise to strong seiches and other dynamic transfers of water within the lake [Klein and Manley, this volume; Manley et al., this volume]. The lake generally stratifies in summer and winter, with overturn in spring and fall, but some bays remain mixed for most or all of the year [Myer and Gruendling, 1979].

Climate

The Lake Champlain Basin has a continental climate, sometimes modified by the Atlantic ocean about 300 km to the southeast. The average January daily minimum temperature is -16 C and the average July daily maximum is 24 C. The mean annual temperature for the basin is 7 C [Environment Canada, 1977]. The Adirondack Mountains to the west form a rain shadow which gives the lowlands of the Lake Champlain Basin one of the driest climates in the northeastern U.S.

HYDROLOGIC INPUTS, OUTPUTS, AND STORAGE

General hydrologic characterization

Rain and snow falling on the basin ultimately either run off as streamflow or return to the atmosphere by evapotranspiration (direct evaporation plus transpiration from vegetation). On an average annual basis, runoff and evapotranspiration in northern New York and New England are each about 50% of precipitation [Hornbeck et al., 1997]. Evapotranspiration is about the same each year, so precipitation in excess of the amount required for evapotranspiration runs off. During the spring snowmelt period, when the lake level usually peaks, evapotranspiration demand is low. Thus a high percentage of winter and spring precipitation runs off to surface water. Some fraction of the rain and meltwater infiltrates to recharge groundwater and restore soil moisture deficits. Newly recharged groundwater either contributes to evapotranspiration demand or discharges to surface water at some later time (not necessarily the same year).

Precipitation

Annual precipitation is approximately 800 mm at Plattsburgh and 850 mm at Burlington. As elevation increases, precipitation increases by orographic enhancement such that as much as 1500 mm per year falls on the mountain summits [Environment Canada, 1977]. Average monthly precipitation is fairly uniform throughout the year. The percentage of precipitation falling as snow ranges from less than 20% near the lake to as much as 40% in the mountains. Snow cover duration is intermittent near the lake but is continuous for 4 months or more in the mountains. Typical maximum depths are 1 to 2 m on the higher summits.

Variability of precipitation amount and type (snow vs. rain) in time and space has an important effect on the Lake Champlain Basin hydrology. The greater rainfall and snow accumulation at higher elevations delays and increases the magnitude of runoff peaks. Only part of the precipitation falling in inlet watersheds contributes to the lake as streamflow. Precipitation falling on the lake surface, however, contributes directly to lake level rise (minus any evaporation that may occur during the time period of interest). Condensation of atmospheric vapor directly on the lake surface when it is colder than ambient air constitutes a small additional water source.

Inlet rivers and streams

Twenty-five gaged inlet watersheds to Lake Champlain have drainage areas greater than 26 km^2 (10 mi^2) (Table 1; Figure 1). The 4 largest watersheds (Winooski, Otter, Missisquoi, and Lamoille) are in Vermont (part of the Missisquoi basin is in Québec), followed by the Saranac and Ausable in New York and the Mettawee, which drains parts of both states. These 7 largest watersheds (at least 1000 km^2) together represent 63% of the total basin area. Compared to most basins of comparable size in the U.S., the Lake Champlain Basin has a limited amount of artificial flow regulation such as lake outlet controls, interbasin water transfers, reservoirs, dams, and other flood control measures. Reservoirs and flood control measures do exist, particularly in the Winooski and Lamoille watersheds, but their effects are generally localized. A greater amount of regulation occurs on the LaChute River (Figure 1), the outlet to Lake George in the southern part of the basin. Spring runoff waters are retained in Lake George to maintain high water for summer recreational use.

The flow characteristics of individual inlet streams differ in relation to watershed size, distribution of precipitation, topography, surficial geology, and land use. Generally, streams in steep mountainous watersheds exhibit a "flashy" response to precipitation; that is, a short time to peak flow, a relatively high peak flow, and a quick recession. As watershed size increases and/or elevational gradients lessen, discharge peaks tend to become broader and more attenuated. Deforested landscapes tend to increase runoff due to lower evapotranspiration demand. Urbanization tends to increase runoff through the creation of impervious surfaces. Watersheds with large wetland areas, such as Otter Creek in Vermont, have the capacity to store a lot of runoff and thus have attenuated peak flows.

In the Lake Champlain Basin, the spring snowmelt is generally the most significant hydrologic event of the year. Most of the prior 3 to 5 months of precipitation, which has been stored in the snowpack, is released in a relatively short period. This input, coupled with spring rains and low evapotranspiration demand, combine to generate as much as half of the annual streamflow from mid-March to mid-May. In the larger watersheds, which span a large elevation range, the snowmelt period is extended by differential melt times; low-lying areas are often devoid of snow before significant melt begins in the uplands.

Although the snowmelt period nearly always causes the annual peak water level in Lake Champlain (discussed below), it does not necessarily cause the highest annual peak flow in the lake inlet streams. An interesting facet of the hydrology of this area is that the annual peak flow can occur at virtually any time of year, and for different reasons. Aside from the main snowmelt period, the following scenarios can cause high peak flows and flooding: (1) midwinter thaws, when there is often sufficient snowpack to contribute to runoff but not enough to absorb melt and rain water; (2) large post-snowmelt rainstorms falling on a landscape still nearly saturated from meltwater; (3) summer convective storms, when localized extreme rainfall amounts overcome summertime soil moisture deficits; and (4) fall frontal storms or hurricanes, which may bring high rainfall totals over a period of several days at a time when soil moisture deficits have recovered from the summer. For example, during a 5-year period on the Winooski River (Figure 2), the

TABLE 1. Inlet and outlet streams to Lake Champlain

• Inlet streams *Gaging station*	Area at lake km^2	Area at gage km^2	Area total/gage
Winooski River, VT	2828	2704	1.05
Otter Creek, VT[1]	2462		1.33
Otter Creek		1627	
New Haven River		298	
Missisquoi River, VT	2223	2202	1.01
Lamoille River, VT	1909	1777	1.07
Saranac River, NY	1575	1575	1.00
Ausable River, NY	1323	1160	1.14
Mettawee River, NY	1098	433	2.54
Great Chazy River, NY	769	640	1.20
Bouquet River, NY	712	712	1.00
Poultney River, VT	692	484	1.43
Lewis Creek, VT	209	200	1.05
Little Ausable River, NY	189	176	1.07
Little Otter Creek, VT	185	148	1.25
Salmon River, NY	175	160	1.09
Putnam Creek, NY	160	134	1.19
Little Chazy River, NY	139	137	1.01
Laplatte River, VT	137	116	1.18
Mill Brook (at Port Henry), NY	73	70	1.04
Stone Bridge Brook, VT	32	22	1.45
Mount Hope Brook, NY	30	30	1.00
Highlands Forge Lake, NY	30	28	1.07
Mill Brook (at Putnam), NY	27	27	1.00
Hoisington Brook, NY	28	17	1.65
Total area gaged streams	17005	14877	1.14
Other sites			
LaChute River, NY[2]	702		
Pike River, Québec[3]	517		
Rock River, Québec[3]	152		
Additional (ungaged) area	4902		
Total area ungaged/unavailable	6273		
• Outlet stream Richelieu River, PQ	21150[4]	22000	0.95
Total basin area	21150		
Watershed area (incl. islands)	20014		
Lake area	1136		

[1] Confluence of New Haven River with Otter Creek is downstream of Otter Creek gaging station, thus 2 gages are independent; ratio is for sum of 2 gaged areas.
[2] Outlet of Lake George; regulated.
[3] Streams have gaging stations; data not at hand.
[4] Recognized area of Lake Champlain basin; Richelieu River gaging station at Fryers Rapids (St. Jean) incorporates additional area.

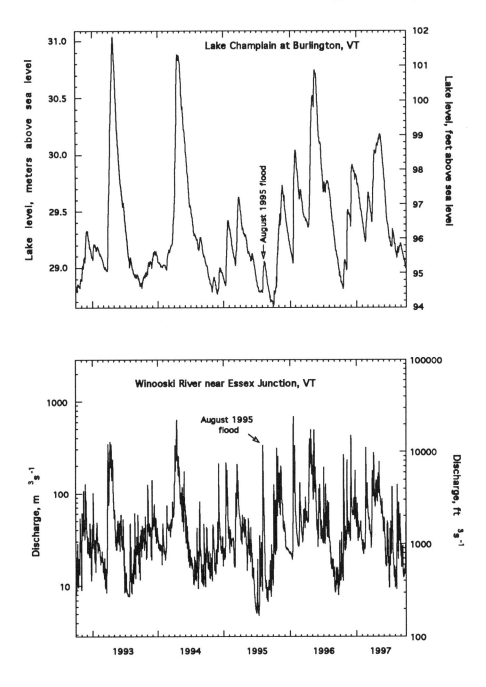

Figure 2. Daily mean water level on Lake Champlain at Burlington, Vermont and daily mean discharge at Winooski River near Essex Junction, Vermont, water years 1993-1997.

maximum daily average discharge for the water year resulted from snowmelt in 1993 and 1994, an August rainstorm in 1995, a January thaw in 1996, and a fall storm in 1997. Although the instantaneous peak flow may occur at different times of year, the largest runoff *volume* nearly always results from spring snowmelt. Note the broad character of the peaks associated with snowmelt (Figure 2). The broad peaks represent a large volume of water which, coupled with the storage characteristics of Lake Champlain, usually assures that the lake level peaks during the main snowmelt period (Figure 2).

Groundwater

Several factors combine to minimize the importance of groundwater to the hydrology of the Lake Champlain Basin. Foremost among these is the mountainous terrain, where high hydraulic gradients limit water residence time in aquifers, and the surficial deposits which are too thin in the mountains and too impermeable in the lowlands (lacustrine sediments) to support significant aquifers. For the lake itself, the high watershed:lake ratio assures that surface water inputs to the lake dominate over groundwater inputs.

Aquifers of localized extent occur in valley bottoms and in areas of sand and gravel outwash deposits, and may play an important role in the hydrology of individual tributaries. These aquifers usually do not discharge directly to the lake, but rather they discharge to inlet streams and their contribution to the lake is counted as surface water. Fractured bedrock aquifers provide water to most domestic and public supply wells in the basin. Groundwater discharging from the fractured rock aquifer likely makes a small but consistent contribution to the lake.

To varying degrees, water table fluctuations at various points in the Lake Champlain Basin mirror fluctuations in lake levels. At a well in Salem, New York, just outside the southern basin divide (Figure 1), groundwater fluctuates from 1 to 2.5 m below land surface (Figure 3). The small depth to water table causes groundwater to respond quickly to inputs; groundwater level peaks before the lake peaks (Figure 3a), leading to some scatter in the relation between groundwater level and lake level (Figure 3d).

As depth to water table increases, the response time to water inputs increases (Figure 3). The pattern of fluctuations at the Montpelier well (Figure 3b) (3.5 to 5.0 m depth to water) is quite synchronous with lake level fluctuations, giving rise to a very strong correlation of groundwater and lake levels (Figure 3e). The well at Milton, Vermont is close to the lake but has an upland setting where depth to water is 8 to 11 m. Its hydrograph lags that of the lake and has a prolonged recession from spring recharge which essentially continues until the following spring (Figure 3c). The two-stage recharge evident in the other 2 wells – the fall recharge and winter recession, followed by the more significant spring recharge and summer recession – collapses to a single annual cycle in the deeper Milton well. Accordingly, there is considerable scatter in the lake level - well level relation (Figure 3f).

Water withdrawals

Water is withdrawn from groundwater, streams and reservoirs within the basin, and from the lake itself for a variety of human uses (Table 2). Nearly half of the withdrawal is for domestic use. Industrial and commercial uses are each about half of the domestic requirement, and agricultural usage is minor. Approximately half of the water is supplied

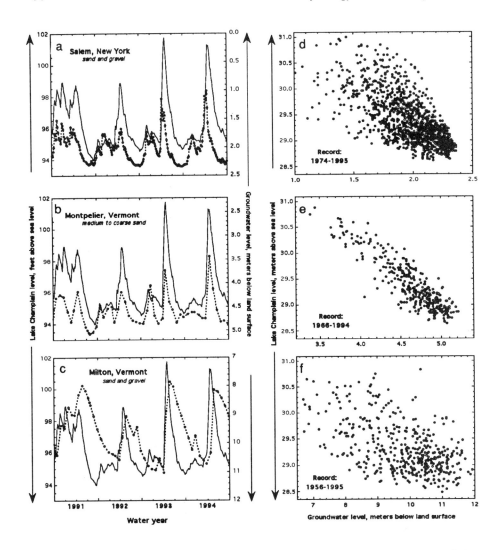

Figure 3. Time series of groundwater levels (dotted line) plotted with water levels on Lake Champlain at Burlington, Vermont (solid line) for water years 1991-1994 (a-c), and corresponding regression plot for entire period of groundwater record (d-f) for wells in Salem, New York (a,d); Montpelier, Vermont (b,e); and Milton, Vermont (c,f). The y-axes of all panels are identical and can be used to convert from ft to m.

by public suppliers. Most of the water withdrawn is returned to the hydrologic system as treated wastewater. About 20% of the water withdrawn is lost to additional evaporation and interbasin transfers. This is termed "consumptive use." Total withdrawals amount to less than 1% of the outflow from Lake Champlain; consumptive use is less than 0.2%.

TABLE 2. Water withdrawals in the Lake Champlain Basin. All values in hm yr^{-1} (10^6 m^3/yr). [Conversions: 1 hm yr^{-1} = 1.120 ft^3 sec^{-1} = 0.724 MGD]. Based on the period 1990-1995. Consumptive use is water not returned to surface- or groundwater.

Use	Ground-water	Surface water	Public supply[1]	Total withdrawn	Consump-tive use
Industrial/mining	3.65	16.93	17.79	38.37	5.65
Commercial	5.80	16.30	10.24	32.34	3.83
Domestic	21.55	0.55	39.82	61.92	8.19
Agriculture	5.94	5.73	0.00	11.67	10.51
Totals	36.93	39.52	67.85	144.30	28.18

[1] Public supply is derived 84% from surface water and 16% from groundwater.

Lake storage and outflow

The outlet control for Lake Champlain is a bedrock sill at St. Jean, Québec (Figure 4). The sill is overlain by silty moraine material. This elevational control limits the minimum possible lake level to 27.7 m (91 ft) [Downer, 1971]. The Richelieu River drops only 0.3 m (1.0 ft) in a distance of 35 km from Rouses Point to St. Jean, however the river is wide and offers little resistance to flow. The river narrows at St. Jean. This narrowing, combined with the rise in bed elevation at the sill, creates hydraulic flow resistance [Environment Canada, 1977]. In addition, human activity has restricted flow near St. Jean; the channel has been filled in to widen the adjacent Chambly canal, and riparian wetlands have been filled and dredged for residential settlement. All of these factors combine to create a "sensitive" control, in which a small increase in discharge causes a relatively large increase in stage.

During snowmelt, because of the restricted outflow, Richelieu River discharge cannot keep pace with inflows, even when inlet streams are well into recession. As a result, the lake level continues to rise and peaks several days after most of the inlet streams peak, and well after snow has disappeared from the lowlands along the lake. Because snowmelt is a basinwide event, and the volume of runoff during days to weeks of snowmelt tends to be large, the storage effect of the lake plays a key role in generating high lake levels.

At times of the year other than the snowmelt period, the lake level is generally much lower. The lake level recedes throughout the summer as evapotranspiration is at its maximum and inflows decline. Summer storms tend to be localized and have only a minor effect on the lake level. For example, the event of August 5, 1995, which caused extensive flooding in central Vermont, caused a rise in lake level of less than 0.3 m, from 28.77 m (94.4 ft) to 29.05 m (95.3 ft) (Figure 2). The great flood of November 1927, the largest flood in recorded history in much of Vermont, did not even induce the annual high lake level for that water year on Lake Champlain. For the water years 1993-1997, a generally wet period, the median lake level was 29.20 m (95.8 ft) (Figure 5). Fifty percent of the time the lake level fluctuated between 29.02 m (95.2 ft) and 29.63 m (97.2 ft), and 80% of the time it fluctuated between 28.86 m (94.7 ft) and 30.05 m (98.6 ft).

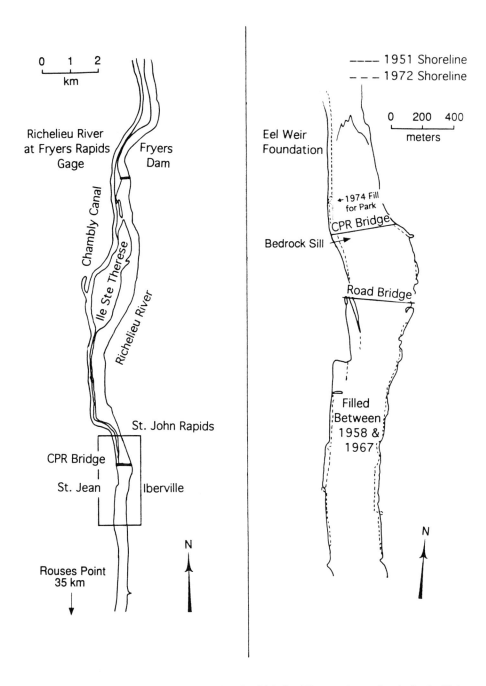

Figure 4. Map of area of St. Jean, Québec, showing Richelieu River gaging station, bedrock sill that controls Lake Champlain levels, channel modifications, and other natural and cultural features. Modified from Barnett [1978].

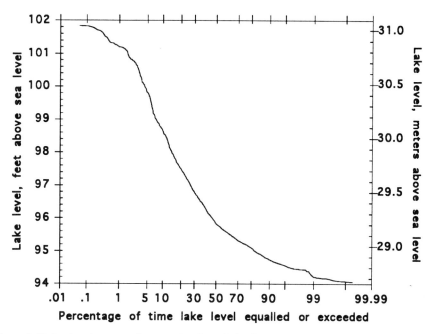

Figure 5. Daily duration curve for water levels on Lake Champlain at Burlington, Vermont, water years 1993-1997.

Hydrologic budget

A hydrologic budget was computed for Lake Champlain for water year 1993 (October 1992 through September 1993) (Figure 6). This year was chosen because most of the USGS streamgaging stations on inlet streams that were established or re-established in the late 1980's were in operation. Secondly, the lake level hydrograph in water year 1993 had a "classic" form, starting from the annual minimum at the beginning of the water year, rising to a small peak in autumn, a secondary peak from a January thaw, and a major peak during snowmelt, from which the lake level receded regularly to the annual minimum in late September. Most inlet streams on the New York side of the lake had above average flow for the year, whereas those on the Vermont side generally had less than average annual flow. The mean lake level for the year was above average, mainly from the spring peak which was the second highest ever recorded. The 1993 snowmelt event is analyzed later in this paper.

Annual discharges from the gaged inlet streams were summed. The average flow per unit area represented in this sum was applied to the ungaged area, including the Lake Champlain islands and the additional watershed area in Québec between the Rouses Point and Fryers Rapids gages. Measured flows from the gaged area and computed flows from the ungaged area were added to derive the total surface water input. The other measured input to Lake Champlain was direct precipitation on the lake surface. This was computed from the average of precipitation from 6 stations around the lake for the year (Figure 1). Groundwater input to the lake is unknown.

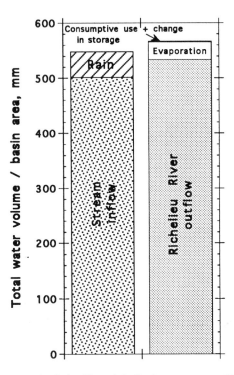

Figure 6. Hydrologic budget for Lake Champlain Basin, water year 1993. Total inflows plus precipitation directly on lake compared to outflow in Richelieu River. Discharge from ungaged area prorated assuming same discharge per unit area as gaged area.

Lake output was measured as flow in the Richelieu River at Fryers Rapids in St. Jean, Québec. Other outputs include evaporation from the lake surface, change in storage, and consumptive water use. Smeltzer and Quinn [1997], making appropriate adjustments to pan evaporation data from the NOAA station in Essex Junction, Vermont, computed the average annual evaporation from the lake in 1991-1992 as 650 mm ± 130 mm. This amount compares closely to the value of 610 mm stated in Environment Canada [1977]. We applied the 650 mm of annual evaporation to the 1136-km^2 lake surface to obtain a water loss, normalized to basin area, of 31.5 mm yr^{-1}. The lake was 30 mm higher at the end of water year 1993 relative to the end of water year 1992, yielding a "loss" to storage of 1.3 mm. Finally, the total consumptive water use in the basin averaged 1.3 mm yr^{-1} for the period 1990-1995. This may be regarded as an upper limit of water loss from the lake, because some of the consumptive use occurs within tributary subbasins and would already be accounted for by reduced inlet flows.

The hydrologic balance between inputs and outputs is reasonably close for water year 1993 (Figure 6). There is a 20-mm deficit in inputs which is probably within the compounded error of the measurements and assumptions. The deficit is on the proper side to account for groundwater inputs, which were not measured. Groundwater could also be lost from the basin, but the steep terrain suggests that the lake is a major regional groundwater discharge zone. The small difference in inputs and outputs in the balance is consistent with a generally minor role of groundwater contributions to the lake.

HISTORICAL LAKE LEVELS

History of gaging and elevation datum

Lake Champlain water level was recorded as early as 1827 at a site in Shelburne Bay, Vermont. Another measurement site was established at St. Jean, Québec in 1864. The U.S. Geological Survey established a gaging station at Rouses Point, NY in 1871, and another at Burlington, VT in 1907. The Rouses Point station was originally sited at Fort Montgomery but was moved to its present site near the Rutland Railroad bridge in 1923. Though moved a few times during their history, and subject to datum changes (discussed below), the Rouses Point and Burlington gaging stations have operated continuously since their establishment. A good discussion of the history of Lake Champlain gaging is given by Barranco [1977].

Published USGS water levels for Rouses Point prior to 1923 and for Burlington prior to 1939 are based on once-daily staff-gage readings. The staff gages were in open water, where wave action often led to considerable error in the readings. Consequently, these early data were of poor resolution and are published to the nearest .05 ft (15 mm). In contrast, lake level data collected after the late 1930's are of high resolution because stilling wells and continuous recording instrumentation were added to the gaging stations. These data are published as daily-mean lake levels to the nearest .01 ft (3 mm). Recorded data also include maximum and minimum instantaneous lake levels for each day. Data users should be aware of the difference between pre-1930's data (single observation) and more recent data (true daily averages).

During the history of gaging on Lake Champlain, stations occasionally have been moved, new surveys have revised benchmark elevations, and the sea level datum itself has been revised. These events have generated considerable confusion over conflicting datums in published lake-level records. Meaningful comparison of long-term lake-level records requires the maintenance of a single datum plane for the period of record. An arbitrary datum may be chosen for convenience but eventually the gage datum is adjusted to elevation above mean sea level. Maintaining the datum requires vigilance because gages are subject to settling, frost heaving, vandalism or other damage. If a gaging station is relocated, the datum must be carefully transferred to the new site.

Barranco [1977] investigated the history of Lake Champlain gage datums in detail. The study concluded that the gage datum at Rouses Point was transferred properly from Fort Montgomery. Arbitrary gage datums were maintained at both Rouses Point sites and at Burlington but confusion was generated when the gage datums were converted to mean sea level. In the late 1800's and early 1900's several surveys were done to establish benchmark elevations in the Lake Champlain region. Each survey resulted in different determinations of gage datum because of changes in mean-sea-level determinations. These changes do not detract from the validity of the lake-level data because the relative gage datums at Rouses Point and Burlington were consistent.

Lake Champlain water-level data from 1871 to present are based on the National Geodetic Vertical Datum of 1929 (NGVD 1929) mean-sea-level determination. NGVD 1929 is a geodetic datum derived from a general adjustment of first order levels in the United States and Canada [USGS, 1997]. Data published by the USGS for the period of record (1871 to present) are adjusted for this change in datum.

Lake levels at the active gaging stations at Rouses Point and Burlington are generally in close agreement. The internal seiches in the lake may cause differences in the levels of up

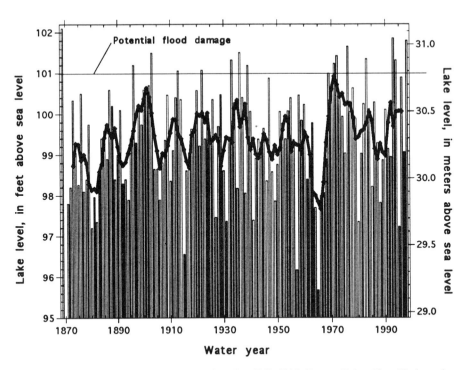

Figure 7. Lake Champlain annual lake level peaks, 1869-1998. Rouses Point, New York gaging station data for 1871-1906; Burlington, Vermont gaging station data for 1907-1998. Peak for 1869 from high-water mark as described in text; no data for 1870. Five year moving average of peak levels (heavy line) overlaid.

to 0.1 m (0.3 ft). The seiches, a phenomenon driven by wind in conjunction with internal lake dynamics, cause oscillations in the lake water surface at several different frequencies. The net difference in lake levels at the two sites averages to very near zero. The seiche phenomenon is discussed in other papers in this volume [Klein and Manley, this volume; Manley et al., this volume].

Annual extreme lake levels

The level of Lake Champlain varies considerably both within each year (Figure 2) and among years (Figure 7). At Rouses Point, the extreme levels of record are 31.12 m (102.1 ft) on May 4, 1869 and 28.09 m (92.17 ft) on October 23, 1941 [USGS, 1997]. The latter figure is contested by Barranco [1977] as unrealistically low relative to the average lake level reported for that day. He suggests that the minimum of record should be recognized as 28.16 m (92.40 ft) on Nov. 13, 1908. More than half of this all-time 2.96-m (9.7-ft) range is typically realized in any given year (Figure 2). The median annual high lake level, nearly always associated with spring snowmelt, is 30.30 m (99.4 ft) and the median annual low lake level is 28.65 m (94.0 ft). The high water extremes are of concern because of the proximity of residential dwellings, streets and highways, and railroad lines

TABLE 3. Lake Champlain peak levels exceeding 101 ft (30.78 m) above sea level. Values from Burlington, Vermont gaging station except where noted.

	Date	Year	Lake level (m)	Lake level (ft)
1	May 4	1869[1]	31.12	102.10
2	April 27	1993	31.05	101.86
3	April 5	1998	31.03	101.80
3	March 26	1903[1]	31.03	101.80
5	April 4	1976	30.99	101.66
6	March 27	1936	30.94	101.51
7	May 9	1972	30.92	101.43
8	May 12	1983	30.89	101.35
9	April 21	1933	30.89	101.33
10	April 21	1994	30.88	101.32
11	May 9	1971	30.86	101.24
12	April 21	1896[1]	30.85	101.20
13	May 1	1939	30.84	101.19
14	April 19	1922	30.81	101.08
15	April 7	1913	30.80	101.06
16	April 27	1969	30.78	101.00

[1] Level at Rouses Point, NY gaging station

to the lake shore. Property damage becomes likely when the lake level approaches 30.5 m (100 ft), especially if winds are active. The presence of ice blocks under conditions of wind and high water can further exacerbate the threat to property.

The maximum known lake level of 31.12 m (102.1 ft), in 1869, predates the Rouses Point gaging station. It was taken from a high-water mark on a railroad bridge and is believed to be reliable [Board of Engineers, 1900]. Other reports of higher historical levels, as listed in Downer [1971], are given little credence and may reflect improper datum adjustments. Based on available records from the Shelburne Bay gage in Vermont, the 1869 mark appears to be the all-time highest since at least 1827 [USGS, 1997]. The 1869 peak may have been artificially enhanced by an eel fishery, a massive stone structure placed in the Richelieu River at the head of St. John Rapids (Figure 4). The structure was short-lived, but its foundation remains and is still visible on aerial photos.

The second highest lake level of record occurred in April, 1993 at 31.05 m (101.86 ft) in Burlington (101.88 ft at Rouses Point, NY), followed by the third highest level in 1998 at 31.03 m (101.80 ft). In all, the lake level reached 30.78 m (101 ft) or greater in 16 years since 1869 (Table 3). There is little interrannual autocorrelation in the annual lake level peak (Figure 7). However, there does appear to be a clustering in the occurrence of unusually high annual peaks. Particularly noteworthy high lake levels occurred during the decades of the 1990's, 1970's, 1930's and 1895-1903. The drought of the 1960's is manifest in a series of lower than average peak annual levels. Barnett [1978] performed a power spectrum analysis on the time series of peak lake levels and found a moderately strong 10-year cycle and another cycle at 3-4 years. The 5-year moving average in peak heights (Figure 7) tends to highlight the 10-year cycles.

What causes the annual peak? Barnett [1978] performed several analyses examining the relation of peak height to rain and snow. He found that peak height was highly correlated ($r^2 = 0.64$) to the maximum snow water equivalent (SWE) measured at 18 sites in New

York and Vermont. The correlation improved to $r^2 = 0.85$ when 3-week precipitation following the time of maximum SWE is added to maximum SWE as the independent variable. The role of snowpack in the mountains to the generation of high lake levels is suggested by the comparatively lower correlation of annual lake level peak to February-to-April precipitation in Burlington of only $r^2 = 0.35$.

As discussed above, the hydrologic damping and restricted outflow of the lake nearly assure that the annual peak lake level occurs in association with spring snowmelt because it represents a basinwide contribution of a large volume of water. The annual peak lake level occurred during spring in all but 3 of the 128 years of record. Moreover, the storage characteristics of the lake tend to cause a delayed peak. The median date of peak is April 26 (Figure 8). The peak occurred on May 11 or later 17 times, including twice in June. Only 10 times did the peak occur before April 1.

Is the lake level changing?

At streamgaging stations, the relation between stage and discharge is established by a hydraulic control in the stream channel. Numerous discharge measurements are made to define this relation in a range of conditions from low to high flow. Because control conditions are dynamic, frequent rating checks are needed to either verify or modify the existing rating. Natural changes such as sediment scouring and filling can cause a rating shift, particularly after flood events. Seasonal rating changes can occur because of aquatic plant growth on the streambed or ice formation in the channel. Human-induced changes, such as channelization and construction of bridges or dams, can shift the rating. The accuracy of the streamflow record is not adversely affected by such changes if the rating curve is adjusted when necessary.

As mentioned previously, the principal control for Lake Champlain is a bedrock sill at St. Jean, about 35 km downstream of Rouses Point (Figure 4). The bedrock sill, in combination with frictional resistance of the channel, control the lake level throughout most of its range. However, it appears that channel modification from 1971-74 at Chambly has also affected the stage-discharge relation. The widening of the Chambly Canal into the river channel reduced the effective width of the main channel and thereby constricted flow. The constriction at Chambly Canal (Figure 4) has the effect of raising the level of Lake Champlain.

Barnett [1978] reviewed the relation between the level of Lake Champlain and the discharge of the Richelieu River at Fryers Rapids for the period 1937-1977. A shift in this relation was detected starting in the late 1960's. The shift increased sharply during 1971-74, the period of Chambly Canal modification. The magnitude of change determined was 0.15 m (0.5 ft) during high flow. For example, at a discharge of 1,000 $m^3 s^{-1}$ (35,000 $ft^3 s^{-1}$) Lake Champlain stage is 0.15 m (0.5 ft) higher than it was for the same discharge during the pre-shift period. Environment Canada [1977], using a physical model, computed that the channel modifications should result in a shift of 0.1 m (0.3 ft).

Another rating shift occurred starting in the mid-1960's, similar to but lower in magnitude than the shift associated with the 1970's canal work at Chambly [Barnett, 1978]. Reclamation of the Richelieu River in and around St. Jean during that period reduced channel width locally from 30 to 60 m, but surveying indicated no backwater effect in the reach. Increased erosion and sediment deposition in the Richelieu River, in part from the draining of riparian swamp forest, has been suggested as a possible cause of this earlier rating change. A vegetation-sediment mat may develop on the streambed during summer and persist after plant death in early winter [Myer and Gruendling, 1979].

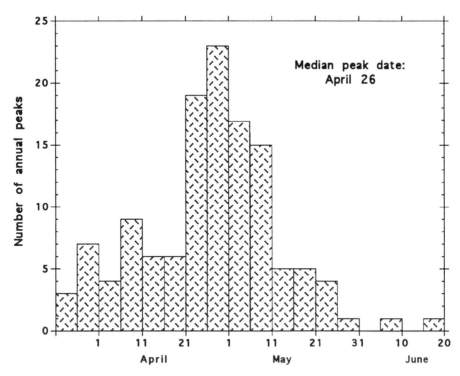

Figure 8. Frequency distribution of date of peak lake level during water years 1869-1998. Excludes the 3 years in which annual peak did not occur in spring, and 1870 (no data).

The stage-discharge relation during the 1993 flood was consistent with the relation found by Barnett [1978], i.e. a shift of 0.15 m (0.5 ft) during high discharge (Figure 9). This shift represents a reduction in discharge of 6 to 8% at high flow relative to the rating from the late 1930's before the channel alterations. The magnitude of the shift exceeds the standard 5% error limit commonly used in streamgaging, therefore the rating change is not the result of gaging error. One implication of this change is that a present-day flood causes a higher lake level than an identical flood (same discharge from the lake) prior to the 1960's.

Barnett [1978] also described an upward trend of minimum lake levels for Lake Champlain. Annual minimum lake levels increased markedly in the 1950's and even more significantly in the late 1960's. The rise in minimum lake levels was attributed to backwater from aquatic plants in the control section between Rouses Point and St Jean. This plant growth is attributed, in turn, to an increase in plant nutrients (nitrogen and phosphorous) in Lake Champlain and the Richelieu River since the 1930's. Unlike backwater created by a permanent obstruction, the effect of vegetation growth is variable over the growing season, and has the greatest influence from mid-June through mid-October [Myer and Gruendling, 1979]. Barnett [1978] concluded that the seasonal backwater effect caused summer lake levels to be about 0.2 m (0.7 ft) higher in the 1970's compared to the 1930's. For a given stage, the reduction in discharge is about 15% since the change.

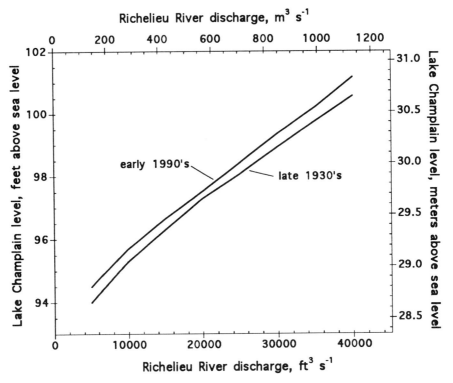

Figure 9. Stage-discharge rating curves for discharge of Richelieu River at St. Jean, Québec as function of lake level at Rouses Point, New York, before and after 1970's alterations to channel at St. Jean.

The Chambly Canal modification affects the lake level - discharge rating most significantly at high flows, whereas the backwater from aquarian plant growth is primarily a low-flow phenomenon. In the middle part of the range, the deviation between the modern rating and the pre-1950's rating is less than at the high and low ends (Figure 9). The rating appeared to stabilize in the late 1970's. For a given discharge, lake level is about 0.3 m (1.0 ft) higher at low flows and 0.15 m (0.5 ft) higher at high flows, compared to the historical rating (Figure 9). A marked shift is evident in annual peak lake levels, annual mean lake levels, and annual minimum lake levels during the 1970's (Figure 10). There does appear to be a slight long-term increasing trend in annual peak lake level. This trend is heavily influenced by the recent high peaks in the 1970's and 1990's, which in turn are affected by the rating change. It should be noted that Barnett [1978] found a *decreasing* trend from 1900 into the 1970's.

The increase in lake level from channel constriction, aquatic vegetation growth, and sedimentation may be countered by a downward trend in runoff. Barnett [1978] suggested that land use changes have reduced snowmelt runoff since the 1940's. In the Lake Champlain Basin at the end of the last century, deforestation and agricultural development had reached a maximum, with as much as 70% of the land cleared. Reforestation has been extensive since that time as Vermont is now 80% forested. Forest cover tends to reduce

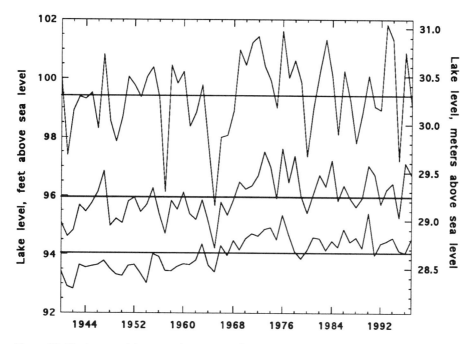

Figure 10. Maximum, minimum, and mean annual water levels on Lake Champlain at Burlington, Vermont, water years 1940-1997, showing shift during the mid-1960's to late-1970's. Horizontal lines through each curve denote means for the period.

runoff by increasing infiltration and evapotranspiration. The transition to forest cover is continuing today, a trend that should decrease the potential for spring lake flooding. However, the continued incidence of high lake levels and flooding suggests that the effect of the channel alterations has outweighed the effect of reforestation. Other factors that may promote high lake levels include increased urbanization (now at about 3% of the basin) and altered climatic patterns.

Climatic cycles within this century have brought periods of drought alternating with periods of excessive precipitation. These cycles tend to balance (no trend), but the contrast between the severe drought in the 1960's and the spring floods of the 1970's and 1990's illustrates the extremes. The changes to the stage-discharge relation for Lake Champlain have increased the susceptibility of the lake to flooding. A long-term change toward a wetter climate, or changes in the pattern of snow accumulation and melt, could further enhance the likelihood of flooding on Lake Champlain.

THE FLOOD OF 1993

On April 27, 1993, Lake Champlain at Burlington reached a level of 31.05 m (101.86 ft) (101.88 ft at Rouses Point), the highest level since at least 1869. The flood was unexpected, as the peak level occurred well after many of the major inlet streams had peaked (Figure 11), especially in the Vermont part of the basin. The magnitude of the flood was also surprising because the lake level at the end of the winter stood at 28.96 m

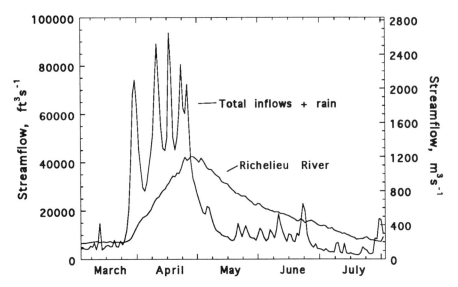

Figure 11. Sum of inlet flows (plus precipitation directly on lake) and outlet flow (Richelieu River) in Lake Champlain Basin during 1993 flood. Discharge from ungaged area prorated assuming same discharge per unit area as gaged area.

(95.0 ft), a relatively low level. A major factor in the near-record high level was the deep and unusually large areal extent of the snowpack in the basin. The "Blizzard of '93" had struck the region in mid-March with 500 mm or more of snow throughout most of the basin. The basinwide snowpack, coupled with deeper than average snowpack in the mountains and some large rain-on-snow events in April provided the hydrologic input that resulted in the near-record lake levels.

Prior to the onset of the main 1993 snowmelt period, outflow from the Richelieu River was in approximate balance with the inflows to the lake (Figure 11). During the initiation of the main snowmelt period in late March, stream inflows rapidly increased by an order of magnitude, while outflow in the Richelieu River merely doubled. Flow in the Richelieu River continued to increase monotonically for the entire month from late March to late April, despite large fluctuations in streamflow in response to individual snowmelt pulses and rain-on-snow events. Until the day of peak outflow, however, inflows consistently exceded outflows, frequently by a large margin. Inflows then receded rapidly through mid-May. The Richelieu River receded much more slowly as the stored water was gradually dissipated through the restrictive lake outlet. Flood damages from the high lake levels of late April 1993 were compounded by high winds and the presence of ice blocks in the lake.

The flow characteristics of the inlet streams to Lake Champlain differ considerably, as indicated by the hydrographs of the two largest inlets on the Vermont side and the largest inlet on the New York side during the 1993 snowmelt (Figure 12). The response of Otter Creek was quite damped in comparison to the Winooski River watershed to the north. The Otter Creek hydrograph has a similar form to that of the Richelieu River (Figure 11), indicating a similar degree of storage and attenuation in its watershed. The attenuation results from the great storage capacity within the riparian wetlands of the Otter Creek

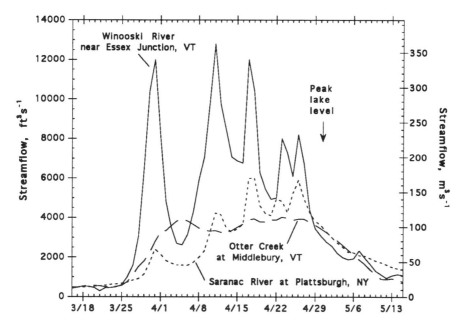

Figure 12. Daily discharge of three large tributaries to Lake Champlain during 1993 flood: Saranac River, NY, Winooski River, VT, and Otter Creek, VT.

watershed. The behavior of Otter Creek is somewhat anomalous for the Lake Champlain Basin, but it is noteworthy given that Otter Creek has the second largest inlet watershed in the Lake Champlain Basin, occupying 12% of the basin area. In contrast, the Winooski River, the largest inlet watershed with 14% of the basin area, has a hydrograph as flashy as the total inlet hydrograph (Figure 11), suggesting its behavior is more typical for the basin. The Saranac River had daily peak flows nearly coincident in timing with those of the Winooski River, but the largest daily peaks occurred later in the melt season, suggesting that the snowpack may melt more slowly on the west side of the lake.

A hydrologic budget for March 1 to July 31, 1993, indicated that the total inflow and outflow were in approximate balance for the period (Figure 13). Inputs were computed as the sum of (1) gaged flows, (2) an estimate of ungaged flow by proration based on ungaged area, and (3) rain falling directly on the lake surface. Output was taken as Richelieu River flow reduced by the percentage of additional watershed area incorporated between the lake outlet at Rouses Point and the gage at St. Jean (Table 1). There are two potentially significant terms excluded from this balance; inputs by groundwater and outputs by evaporation from the lake surface. Groundwater discharging to the lake is accounted for in part by the estimated contribution from ungaged area, as much of the ungaged area is area adjacent to the lake between inlets that is not drained by surface streams but rather by groundwater. There is probably a larger contribution from regional groundwater flow not accounted for in the balance. Evaporation from the lake surface would be considerable during late spring and summer, probably of similar magnitude to the precipitation input (Figure 13).

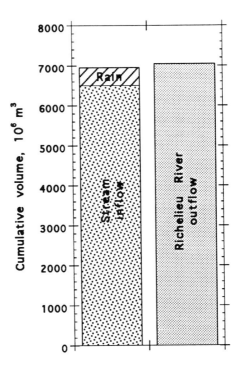

Figure 13. Hydrologic budget for Lake Champlain Basin during 1993 flood (March 1 through July 31). Sum of inlet flow volume (plus precipitation directly on lake) and outlet flow volume (Richelieu River). Discharge from ungaged area prorated assuming same discharge per unit area as gaged area.

THE FLOOD OF 1998

In 1998 the spring high water level in Lake Champlain reached 31.03 m (101.80 ft), very nearly matching the level of 1993 (101.86 ft). The 1998 flood was different in several respects, however. The great 1998 ice storm of January 11-12 fell as rain in much of the basin and pushed the lake level above 30.2 m (99 ft). The level never receded below 29.6 m (97 ft) before the onset of the main snowmelt, so the starting lake level was much higher than it was for the 1993 event (Figure 14). The lake level peak resulted from a single event at the end of March, a thermal melt from 3 days of record-setting warm temperatures over the totally snow-covered basin. The lake level rose rapidly from 29.9 m (98 ft) to the 31.03 m (101.80 ft) peak. Based on preliminary data, the peak inflow appeared to be at least 50% greater than it had been for the 1993 event, causing the more rapid response. Another factor in the more rapid response (compared to 1993) is that the ground was more saturated in 1998 from mid-winter events, causing the additional snowmelt input to run off more readily.

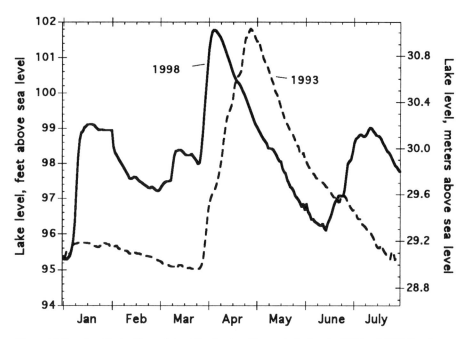

Figure 14. Levels of Lake Champlain at Burlington, Vermont, during the 1993 and 1998 floods.

SUMMARY

Lake Champlain hydrology is characterized by the mountainous terrain of the basin ("flashy" inlet streams), long snowy winters (snowmelt-dominated system), and a restrictive lake outlet (lagging and fluctuating water levels). These factors have led to periodic flooding on tributaries and to high water levels in the lake itself. Flooding on tributaries may occur at any time of year and is often localized. Flooding on the lake is exclusively a snowmelt-driven spring-time phenomenon. Although the record peak lake level occurred in 1869, recent high lake levels have raised concerns about flooding. Constrictions to the lake outlet channel in the 1970's appear to have raised lake levels by about 0.15 m (0.5 ft). There are additional concerns over the possible effects of increased precipitation associated with global warming, and the hydrologic effects of changing land use within the basin. No model currently exists to forecast lake levels, a shortcoming that should be rectified in light of the changes taking place in the Lake Champlain Basin.

Acknowledgements. Guy Morin of the Water Survey of Canada provided flow data from the Richelieu River. Tom Willard of Vermont Department of Environmental Conservation supplied valuable reference materials. Gardner Bent, USGS-Massachusetts, Rich Lumia, USGS-New York, and Laura Medalie, USGS-Vermont provided thorough and constructive reviews that improved the manuscript. Laura Medalie also provided information on water withdrawals and helped with figures, as did Kim Kendall. Thor Smith was a great help in the final preparation of text and figures.

REFERENCES

Barnett, S.G., Man induced changes in Lake Champlain hydrology, Regional Studies Report No. 7, Institute for Man and Environment, Plattsburgh, NY, 86 p., 1978.

Barranco, A.P,. Jr., Lake Champlain Water Levels, State of Vermont, Dept. Water Resourc., 23 p., 1977.

Downer, R.N., Extreme mean daily annual water levels of Lake Champlain, Vermont Water Resources Research Report 3, 18 p., 1971.

Environment Canada, The stage-discharge relationship of Lake Champlain-Richelieu River, 99 p., 1977.

Hornbeck, J.W., S.W.Bailey, D.C. Buso, and J.B. Shanley, Streamwater chemistry and nutrient budgets for forested watersheds in New England: variability and management implications, *Forest Ecology and Management.* 93, 73-89, 1997.

Houghton, J.T., *Global Warming: the Complete Briefing*, Cambridge University Press, Cambridge, 1997.

Hunt, A.S., E.B. Henson, and D.P. Bucke, Sedimentological and limnological studies of Lake Champlain, 64th ann. meeting, N.E. Intercollegiate Geol. Conf., Guidebook for field trips in Vermont, 1972.

Klein L. and T.O. Manley, Effects of the internal seiche in the south main lake of Lake Champlain, this volume.

Mack, T.J., Hydrogeology, Simulated groundwater flow, and groundwater quality at two landfills in Bristol, Vermont, USGS Water Resources Investigtion Rept. 94-4108, 111 p., 1995.

Manley, T.O., K. Hunkins, J. Saylor, G. Miller, and P. Manley, Aspects of summertime and wintertime hydrodynamics of Lake Champlain, this volume.

Myer, G.E. and G.K. Gruendling, Limnology of Lake Champlain, Lake Champlain Basin Study Report No. LCBS-30, 429 p., 1979.

Scherbatskoy, T., J.B. Shanley, and G. Keeler, Factors controlling mercury transport in an upland forested catchment, *Water Air and Soil Pollution*, 105, 427-438, 1998.

Shanley, J.B., A.F. Donlon, T. Scherbatskoy, and G. Keeler, Mercury cycling and transport in the Lake Champlain Basin, this volume.

Smeltzer, E. and S. Quinn, A phosphorus budget, model, and load reduction strategy for Lake Champlain, Lake Champlain Daignostic-Feasability Study final report, Vermont Dept. Environ. Cons., New York State Dept. Environ. Cons., 129 p., 1997.

U.S. Board of Engineers, Report of the Board of Engineers on Deep Waterways between the Great Lakes and the Atlantic Tide Waters, Part I, U.S. Govt. Printing Office, 1900.

U.S. Geological Survey, Water Resources Data, New York, Water Year 1996, Vol. 1, Eastern New York excluding Long Island, 1997.

Aspects of Summertime and Wintertime Hydrodynamics of Lake Champlain

T. O. Manley, K. L. Hunkins, J. H. Saylor, G. S. Miller
and P. L. Manley

ABSTRACT

Two and a half years of temperature and current observations obtained from longterm moorings within the central region of Lake Champlain were analyzed for wintertime and summertime circulation patterns. During wintertime periods, currents were consistently weaker than summertime observations. Water columns at individual sites were nearly isothermal with coldest temperatures approaching 0.5° C, however horizontal gradients were apparent. From early Spring to late Fall, the hydrodynamics of the Main Lake were dominated by the presence of a uninodal internal seiche. Spectral analysis showed the dominance of atmospherically controlled oscillations (7.1 and 10.7 days) and basin dominated modes (4.3, 2.7 and 1.8 days). A shifting of dominant periods inter- as well as intra-annually can be accounted for by varying conditions of the metalimnion as well as atmospheric forcing. The internal seiche also possessed a rotary component. Bottom currents tend to be preferentially aligned with bottom topography. Nonlinear aspects of the internal seiche in the form of internal surges and bores (gravity currents) were observed. Wave heights exceeding 10 m showed pronounced asymmetry of wave shape. Highly nonlinear events exceeding 60 m have been observed. Sudden increases in hypolimnetic temperature appear to be tied to these events. Whether these deep temperature shifts were a direct result of surface outcropping of the hypolimnion and/or significant internal mixing at the shoaling ends of the lake has yet to be determined. Near-bottom velocities of 30-50 cm/s have direct consequences for toxic sediment resuspension and dispersal.

I. INTRODUCTION

Lake Champlain, located between the Adirondack Mountains of New York and the Green Mountains of Vermont, is the sixth largest body of fresh water in the United States. From its most southern limit at Whitehall, NY, it extends roughly 170 km along the north-south trending Champlain Valley and into the Canadian province of Quebec (Figure 1). With an average depth and width of 19.2 m and 6.6 km, respectively, it comprises a total volume of approximately 26 km^3 [Myer and Gruendling, 1979]. Its maximum width of 20.2 km is located near Burlington, Vt., while the maximum depth of 121 m is found at the Thompsons Point - Split Rock gap (hereinafter referred as Thompsons Point). To the north and south of Thompsons Point, the lake bottom rises in a somewhat linear fashion along the thalweg to Rouses Point and the Crown Point Bridge, respectively (Figure 2). Approximately 91% of water influx into the lake is derived from river input over its 20,000 km^2 drainage basin; the remaining 9% obtained via direct precipitation and condensation over its 1130 km^2 surface [Henson and Potash, 1969]. The net northward flow of the lake exits via the Richelieu River; eventually to end up in the St. Lawrence. Gauging of the Richelieu River in Chambly, Quebec shows average daily discharge to be approximately 300 m^3/s; the minimum and maximum observed being 40 m^3/s and 1237 m^3/s [Myer and Gruendling, 1979].

Natural and man-made restrictions have divided Lake Champlain into three major regions. The first region, or South Lake, extends 53 km below the Crown Point Bridge. This section of the lake is typically viewed as a river environment, having an average width and depth of 1.1 km and 2.7m, respectively. North of the Crown Point Bridge lies the largest of the three regions; the Main Lake. This body of water covers the largest percentage of surface area (60%) and water volume (82%) of the three regions [Henson and Potash, 1969]. Numerous shoals, islands and shallow bays are found within its boundary. The large islands located in the north central part of Lake Champlain as well as several causeways and bridge structures have formed a somewhat isolated region of the lake that encompass Missisquoi Bay, Northeast Arm, and Mallettes Bay (hereinafter referred as the Restricted Arm). Narrow natural channels or bridge openings represent the only paths for interchange of water between the Restricted Arm and the Main Lake. Due to the limited flow to and from the Restricted Arm as well as the very narrow channel-like characteristics of the South Lake, only the hydrodynamics of the central and most significant component of Lake Champlain will be presented.

In 1979, virtually all information pertaining to the physical, chemical and biological characteristics of Lake Champlain from as early as 1902 (some of which are presently unobtainable) were compiled into a single manuscript by G. Myer and G. Gruendling of the State University of New York, Plattsburgh (much of the research being their own work). Although listed as a technical report, it remains as one of the most important reference documents for the lake to date. After 1979, little research in hydrodynamics was completed until 1991 when the first longterm mooring program began in the Main Lake (the subject of this paper). Therefore, it is from Myer and Gruendling [1979] that the fundamental hydrodynamics of Lake Champlain was obtained unless otherwise referenced.

Lake hydrodynamics are primarily driven by the combination of solar insolation and wind stress [Hutchinson, 1957]. Lake Champlain is no different in that wind driven circulation is prevalent throughout the Main Lake and Restricted Arm, however, to a lesser degree in the South Lake due to its river-like characteristics. Flow within shallow

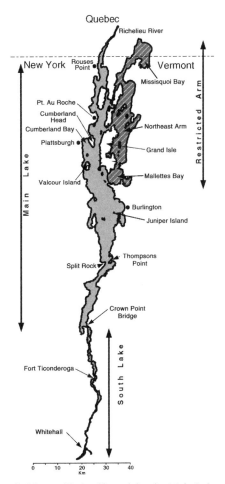

Figure 1. The three major divisions of Lake Champlain; the Main Lake, the South Lake, and the Restricted Arm. Specific locations of islands, cities, towns and bodies of water that will be mentioned in this article are defined. Observations that will be presented were obtained within the central portion of the Main Lake from Thompsons Point to Valcour Island.

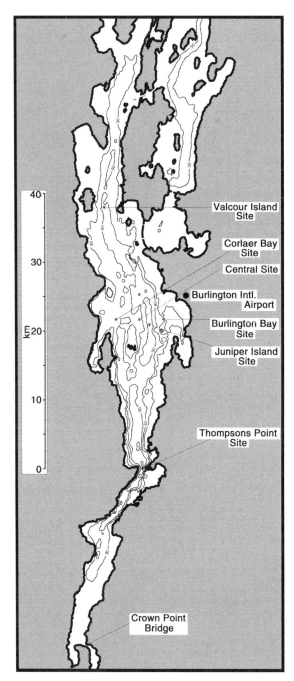

Figure 2. Bottom topography of the Main Lake with locations of all longterm mooring sites used to collect the 2.5 years of continuous hydrographic data.

and/or restricted bays have shown to be predominantly wind driven and typically accompanied by near-shore boundary currents as well as upwelling and downwelling events. The effect of the Earth's rotation on circulation has been verified in several instances via the observation of modified Ekman Spirals (when wind direction was steady for periods greater than 6 hours) and quasi-geostrophic currents within the more central portions of the lake. During the winter months, very few measurements were taken.

Associated with wind driven circulation were the observed seiches, both surface and internal. For excellent reviews on seiches, the reader is referred to Mortimer [1953, 1979], Hutchinson [1957], and Korgen [1995]. Information obtained from a suite of water level recorders located around the periphery of the lake provided the first, albeit only, view of the complex nature of surface seiches. Horizontal modes, up to quinquenodal, were classified in both the along-axis (north-south) and cross-axis (east-west) direction with the uninodal components having periods of 4.2 and 0.58 hours, respectively. The standard equation, often referred to as Merian's equation (Eqn. 1), for the determination of the surface seiche period (T_{sn}) of any mode (n) within a rectangular basin of length l and depth d (g being 9.8 m/s^2) predicts a shorter period of 3.5 hours for the Main Lake uninodal component (n=1; l=110 km; d=31 m). By including more representative structure.

$$T_{sn} = \left(\frac{2l}{n}\right) / \sqrt{gd} \qquad (1)$$

of the basin through the use of Defant's method, a better approximation to the observed periods, amplitudes and nodal positions were obtained. In general, results from this modeling (that extending no lower than the Crown Point Bridge) agreed well with observations, although many calculated periods were shorter than observed and some modes could not be specifically duplicated. For example, the Main Lake uninodal component was calculated to be 3.9 hours; still 0.3 hours off. More recent modeling of the surface seiche [Connell, 1994; Prigo et al., 1996], also using Defant's method, but with more detailed bottom topography, finer horizontal divisions and the inclusion of the South Lake, also showed good agreement with much of the previous modeling. Some modes, however, required different geographic ('leaky') boundaries to match the observed periods. The use of 'leaky' or 'near-closure' geographic boundaries such as those found at the Crown Point Bridge, Thompsons Point, and Cumberland Head were required by Myer and Gruendling [1979] to account for many of the observed periods of the surface seiche. Boundaries such as these act as weak, yet important, reflectors for the long wavelength seiche. The net effect being the creation of several sub-basin length scales possessing their own set of principle modes.

Very little information regarding the observed amplitudes of the surface seiche exist, although Myer and Gruendling [1979] did note that wind generated lake setup in the along-axis direction of the Main Lake is typical in the range of 1-5 cm, with the maximum reported at Rouses Point (10 cm). Cross-lake setups were typically less than 1 cm. A simplified predictive model for lake setup [Volker, 1949] showed that the observations were generally less than predicted although within close agreement [Myer and Gruendling, 1979]. By using this model for various wind speeds and sections of the lake, Myer and Gruendling [1979] concluded that for consistent 9 m/s winds, lake setup could be as high as 12 and 54 cm for the Main and South Lake, respectively. Accompanying surface seiche activity during periods of lake stratification is the internal seiche; a standing wave that occurs on the metalimnion (the density interface between the warmer epilimnion and colder hypolimnion). Simplified linear dynamics for the fundamental mode of an internal seiche within a narrow constant-depth lake with vertical walls show that the maximum horizontal and vertical water velocities would be found at the node and antinodes, respectively. The epilimnic and hypolimnetic layers would, however, move in opposite directions due to the constant adjustment of the internal pressure field; the layer of no motion being the metalimnion. As the complexity of bottom topography, lateral boundaries, depth of the metalimnion, wind forcing, and the stratification of a lake increases, so does the complexity of the internal seiche and hence, associated circulation patterns.

According to Myer and Gruendling [1979], along-lake modes up to binodal have been observed, although the most prevalent is that of the uninodal internal seiche. Unfortunately, much of the original observations and statistical information pertaining to the internal seiche from earlier observations is lacking or presently unobtainable. Although not explicitly stated, it was apparent that the period of the fundamental mode did vary from a maximum of 4.6 days (several weeks of observations in July 1974) to an inferred minimum of less than the noted average of 3.9 days (obtained from 4 months of observations in 1976 near Pt. Au Roche). The only time that the second mode (binodal) was weakly indicated was during the aforementioned survey of July 1974; its associated period being 3.1 days. Using similar terminology as in equation. 1, the period of any mode internal seiche (Tin) within a rectangular basin of uniform depth is provided in equation 2, where ρ_e and ρ_h and t_e and t_h are the densities and thicknesses of the epilimnion and hypolimnion, respectively [Hutchinson, 1957]. Due to the slight contrast in densities across the metalimnion,

$$T_{in} = \left(\frac{2l}{n}\right) / \left(\sqrt{\frac{g(\rho_h - \rho_e)}{\left(\frac{\rho_h}{t_h} + \frac{\rho e}{t_e}\right)}}\right) \quad (2)$$

compared to the air-water boundary, the internal seiche propagates (on average) nearly 25 times slower than its associated surface seiche. Using l=100 km, re = 0.998 g/cm^3, ρ_h = 1.0 g/cm^3, t_e = 20m and t_h =10 m, the estimated period of the along-axis uninodal internal seiche at 6.4 days is well off of the averaged observed period of 4.3 days. While this equation can only be used as an estimate because of the assumptions involved, a more intuitive estimate can be obtained by restricting the length of the basin to that defined by the average depth of the metalimnion (i.e., the thermocline shore). For Lake Champlain, the removal of the shallower ends reduces the length by 28 km. Changing l to 82 km and t_h to 15 m (since the average depth has increased) yields a closer estimated period of 4.6 days. Modeling predictions of the internal seiche by Defant's method (Connell, 1994; Prigo et al., 1996] with realistic bathymetry of the lake indicated the fundamental mode of the Main Lake to be at 4.0 days; closer to the minimum period observed by Myer and Gruendling [1979].

Cross-lake modes of the internal seiche were indicated by Myer and Gruendling [1979], yet appear to be linked to the along-lake fundamental mode. It should be noted that observations of the internal seiche south of Thompsons Point were and still are sparse. As a result, the activity of the internal seiche existing south of this 'boundary' to the Crown Point Bridge is speculative even though some models and observations suggest its presence. For example, Myer and Gruendling [1979] pointed out that the binodal internal seiche required a southern terminus of the Crown Point Bridge in order to account for its period. A more recent inference of maximum southern extent of the internal seiche was that of hypolimnetic water observed as a thin (~ 0.5 m) bottom layer of water possessing lower temperature and conductivity in the South Lake near Fort Ticonderoga, N.Y. [Roger Binkard, Aquatec Inc., personal communication].

The most diagnostic and easily mapped indicator of the internal seiche is that of the oscillatory motion of the metalimnion. Many local studies within the lake have shown that the height of the internal seiche can range from several to 10s of m, typically less than 20 m. Myer and Gruendling [1979] reported the largest depression of the metalimnion to be at the sediment-water interface (35 m) near Pt. Au Roche while in the more southern part of the Main Lake (north of Thompsons Point), hypolimnetic water was present at the surface in limited regions. More recently, the largest variation in the metalimnion in the northern sector of the lake was during an environmental survey for the installation of a Grand Isle Fish Hatchery near Valcour Island. During a single oscillation, the epilimnion spanned almost the entire water column, nearly 61 m. In the southern portion of the Main Lake, the deepest observation of the metalimnion was that by one of the authors in October of 1990. After strong north winds for nearly a week, a CTD station near Thompsons Point showed the top of the metalimnion to have reached a depth of 91 m; an increase of almost 70 m from its observation prior to the storm.

Based on the vertical range of the internal seiche, the effects of the internal seiche were readily observed throughout the Main Lake. Hypolimnic waters periodically flood onto and ebb off shallow bays based on accounts of maritime archeologists working on many of the submerged wrecks [Mr. Art Cohn, Dir. of the Lake Champlain Maritime Museum, personal communication]. A 14-day hydrographic survey of a shallow bay directly south of Thompsons Point showed that the internal seiche significantly modified the characteristics of the water column within the bay through the incursion and subsequent removal of deep hypolimnetic water [Thompson, 1994]. Within this survey region, which also included the deep water sector of Thompsons Point, the largest change in hypolimnetic waters over a single cycle of the seiche was calculated to be 7×10^7 m^3. During this time period the winds were light to moderate.

Internal seiches having periods above the local inertial period (17 hours for Lake Champlain) would exhibit a cyclonic rotation of the wave form around the perimeter of the basin [Hutchinson, 1957; Mortimer, 1963 and 1993]. Prior to 1991, the only reported indication of this rotational effect was coincident reversal of the E-W tilt of the metalimnion and internal-seiche driven N-S flow near Valcour Island [Myer and Gruendling, 1979]. Although the average length to width ratio of the Main Lake is approximately 20, indicating minimal effect of rotation [Rao, 1966 and 1977], irregularity of the basin width may suggest otherwise. The ratio of the internal Rossby radius of deformation (~5 km) to basin width (0.25 - 0.75), would however, suggest that the internal seiche should propagate as a Kelvin Wave [Stocker et al., 1987].

Few measurements of currents attributable to the internal seiche have been taken [Myer and Gruendling, 1979], and of these, all were taken over periods of a week or less and none synoptically. During summer conditions, currents within the Main Lake were clearly dominated by those resulting from the internal seiche (near-surface wind forced circulation neglected) and rarely exceeding 20 cm/s. Reversal of flow above and below the metalimnion was routinely encountered provided water depth was in excess of 30 m and the shoreline was no closer than 0.5 km. On a few occasions, it was noted that flow reversal was not centered along the metalimnion.

Indirect observations of currents near the lake bed (obtained from other researchers as noted in Myer and Gruendling [1979]) were based on the presence of sand and gravel on a clay substrate in portions of the deep thalweg of the lake. Other indirect evidence of deep erosion of sediments was from the Grand Isle Fish Hatchery located east of Cumberland Head. Occasionally, water from its deep intake at 55 m (2 m above the bottom) was found to contain very high loads of suspended sediment (silts and clays) that, after filtering, still made it difficult to see the fish in the raceways for more than a few feet [Dan Marchant (Manager of the Grand Isle Fish Hatchery), personal communication]. If caused by bottom resuspension, current speeds would have to have been in excess of 15 cm/s. Similar indirect support of the effect of strong bottom currents can be found in the thalweg of the lake south of the Grand Isle intake near the southern end of Valcour Island. Here, bottom furrows which are created by strong secondary circulation patterns near the sediment-water interface [Flood, 1981,1983; Viekman et al., 1989], have been mapped by side scan sonar [Manley, 1991; Leuke, 1995]. Work in the deep ocean and other lakes have shown that peak erosive conditions can occur in furrow fields with speeds of 30 cm/s [Flood and Hollister, 1980; Flood, 1989; Viekman et al., 1992].

Undoubtedly, the surface and internal seiches represent the longest waves residing in Lake Champlain. The surface seiche plays a very minor role in the circulation of the Main Lake, although it can affect local water levels. The internal seiche, on the other hand, dominates the circulation of the Main Lake during the period of summer stratification, yet there is little information about wintertime hydrodynamics and ice cover. Ice coverage has been documented to widely vary over space and time [Myer and Gruendling, 1979], however, there were no concurrent observations that could be used at the time of writing of this paper. The paucity of observations and their associated temporal and spatial scales, even though providing many important aspects of Main Lake circulation, still indicate that our understanding of its complex hydrodynamics is incomplete. In support of this conclusion are topics such as, 1) the extremely small amount of winter time observations such as currents, water temperature, and ice coverage, 2) the difficulty of determining mean lake circulation from an oscillatory controlled system during periods of stratification, 3) the added complexity of sub-circulation systems (gyres/eddies) that have been alluded to within the confines of the Main Lake [Myer and Gruendling, 1979], 4) a lack of bottom current observations to corroborate the indication of sediment erosion/resuspension, 5) the three-dimensional nature of the internal seiche when considering the effects of the Earth's rotation, 6) the free- versus forced-wave aspect of the internal seiche, 7) the controlling aspects of the high amplitude internal seiche as well as its associated dynamics, 8) and the effects of internal seiche activity on water intake and sewage effluent.

To address many of these issues, a 2.5 year field program consisting of several longterm moorings was designed. During any given year, three to five mooring sites were occupied in the central portion of the Main Lake extending from Valcour Island to Thompsons Point. This article will present an overview of wintertime and summertime hydrodynamics based on the collected data. Inclusion of detailed modeling and analysis of specific hydrodynamic processes found to exist within Lake Champlain is beyond the scope of this paper, however, research covering these topics is presented by Saylor et al., [this volume], Manley et al., [this volume], and Hunkins et al., [1998].

II. LONGTERM MOORINGS AND DATA ANALYSIS

The moorings that were placed in Lake Champlain consisted of several sub-surface taut-wire mooring designs depending on instrumentation. In common was the requirement to measure the vertical variability within the region of the summer metalimnion, currents at representative depths of the epilimnion and hypolimnion, and finally the sedimentation rate at three different levels of the water column. In most cases where water depths were greater than 50 m, all instrumentation was contained on a single mooring. In cases where water depth was shallower and interference between some instruments would have resulted or where the upper most device was well below the standard SCUBA diving limits, two tethered moorings placed within 100 m of each other were used.

Mooring structure consisted of a bottom anchor weighing up to 230 kg and 1/4 inch stainless steel wire rope extending to either one or two flotation spheres; the top sphere never coming closer to the lake surface than 6 m. No intermediate flotation was employed. The 6 m depth limitation was required to avoid all interference with boat traffic. To ensure proper depth of the upper flotation sphere below the surface, best possible positioning information, and minimum dynamic stress on the instrumentation, all moorings were deployed in an anchor-first mode.

Instrumentation employed on the mooring sites generally consisted of two Aanderaa RCM5 current meters (equipped with temperature, conductivity, pressure, speed and direction), a 40 m Aanderaa thermistor chain (T-chain) with 4 m sensor spacing, and three 41 cm long polycarbonate sediment traps (Figure 3). The near-surface RCM5 was placed directly beneath the deepest flotation sphere, while the T-chain spanned the distance from the bottom of the upper RCM5 to well below the average depth of the metalimnion. Directly below the T-chain was the second and last RCM5. Sediment traps were taped onto the line at depths equal to 12 m below the surface, 50% of the water column, and 5 m above the bottom. As the number of moorings increased over time, Acoustic Doppler Current Profilers (ADCPs) were used in place of the RCM5 current meters.

Many of the older style reel-to-reel tape RCM5s and T-chains failed to record properly and the recorded tapes had to be sent away to Bergen, Norway for special processing. Over 3/4 of these data tapes were successfully recovered while the remaining were partially or never recovered. Final reduction of the RCM5 current meter data was completed at Middlebury College while a majority of the ADCP and T-chain data were finalized at the NOAA Great Lakes Environmental Research Laboratory (GLERL, Ann Arbor, MI). While some instruments were set up to record at different intervals, all data reported herein represent hourly observations. Directions are reported relative to true North, and temperatures are in °C. Although accuracy of the various sensors used in this program varied, errors should not exceed factory specifications of 0.1 °C and +/-1 cm/s (Aanderaa T-chain and RCM5) and 0.1 °C and +/-1 cm/s (ADCP) for temperature and speed unless otherwise noted (e.g., bio-fouling or corrections during data processing).

Moorings were operated for a period of approximately six months. Starting in the summer of 1991, all moorings were recovered and subsequently redeployed over a 2 week period in October and May in an effort to provide continuous measurements over the first 2.5 years of the program. Subsequent to that, measurements were taken from May to October. Locations of the moorings also varied over time depending on the type of investigation performed during a specific field season. Two sites, Valcour Island and

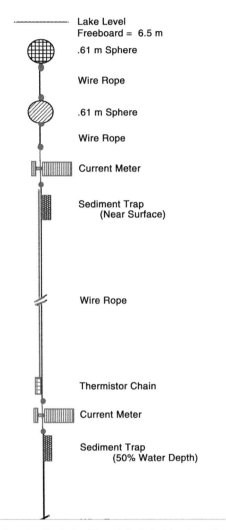

Figure 3. Typical taut-wire mooring plan for Lake Champlain that did not employ an ADCP. All moorings were sub-surface with a free board of approximately 7 m. Aanderaa RCM5 currents meters were placed above and below the T-chain. Sediment traps were located 12m below the surface, 50% of water depth and 5 m above the bottom.

Thompsons Point were always maintained in order to provide continuity of the data and an indication of inter-annual variability. Table 1 summarizes the instrumentation at each of the mooring sites from June 1991 to October 1993 as well as the amount of data recorded by each.

For the same time period, atmospheric data were also obtained from the National Weather Service for Burlington International Airport (BIA). At the time that most in-lake measurements were taken, BIA was the only reliable source of atmospheric information within the basin. Plattsburgh Airport was not used due to the delay required in the conversion of hand written logs as well as inherent errors and missing data. BIA data were reported as hourly observations based on Local Standard Time. As an aid for comparison and analysis requirements, wind directions were reversed so they have the same connotation as current direction (i.e., the direction to which flow is headed). Additionally, all times were converted to Julian Days (January 1, 1991 = day 1).

III. WINTERTIME OBSERVATIONS AND RESULTS

The longest thermal record was obtained at Thompsons Point, an area that represents both the deepest part of the lake as well as one of the natural restrictions of the Main Lake that limits some of the horizontal modes of the surface and internal seiche [Myer and Gruendling, 1979]. This record spans 864 days (June 17, 1991 to October 27, 1993) with brief gaps in the Fall and Spring when the mooring arrays were serviced. The range of observations varied from a minimum of 48 m (1991) to nearly 70 m (1992-1993) depending on the instrumentation used in addition to the standard T-chain. During the first year, temperature data from the upper and lower RCM5 current meters were available. In subsequent years, temperature data from the deeper upward-looking ADCP was used in place of the RCM5s. The composite 2.5 year scattered data set was subsampled to a three-hour interval, gridded using a minimum tension algorithm [Manley and Tallet, 1990] with 2 m and 0.2 day spacing in depth and time, respectively, and then contoured (Plate 1). Blanked out regions define periods of no data due to instrument servicing, no observations or sensor failure. The horizontal axis is in Julian Days referenced to January 1, 1991 being day 1. For reference, the first of the month for July, August, September, and October are days 182, 213, 244, and 274, respectively. The continual presence of the internal seiche during all summertime records was reflected in all other observational data within the central portion of the Main Lake. High amplitude internal seiches were found in each summertime record.

Intra- and inter-seasonal variations are visible over this record. During the time period from mid-November to mid-May (days 315-500 and 675-870), which will hereinafter be referred to as 'wintertime' to reflect the center weighting of observations, the lake remains in nearly isothermal conditions. Rapid transitions of temperature from 10 °C to 4 °C (within the upper 60 m of the water column) caused by rapid cooling and wind mixing are shown by the nearly isothermal structure. Weak stratification can be observe d after the fall turnover of the lake, although it is not strong enough to prevent the water column from obtaining uniformly cold temperatures near 0.5 °C due to continued cooling and wind mixing as evidenced in other deep lakes (e.g., Johnson [1964], Carmack and Farmer, 1982, Carmack and Weiss [1991]).

1. Horizontal Temperature Gradients Within the Lake

Detailed information regarding the variation in the velocity and temperature fields during December, 1992 at Thompsons Point and Valcour Island are presented in Plates 2 and 3, respectively. As in Plate 1, the thermal records at both sites show a consistent, gradual trend of decreasing temperature over time although closer inspection reveals that Valcour Island was consistently cooler than Thompsons Point. Vertically-averaged hourly temperatures along the lower 40 m of the T-chain confirmed that the water column at Valcour Island was, on average, 1.2 °C cooler than Thompsons Point for the month of

Table 1: Temporal and spatial coverage of hydrographic information during the first 2.5 years of observations at the various mooring sites within Lake Champlain. Time periods are grouped according to deployments and recoveries. Asterisks imply shorter than normal records. Dashes imply that the instrument was not used at the mooring site.

Mooring Site (Water Depth) Instrument	1991 June - Oct. Depth (m)	Duration (days)	91-92 Nov. - May Depth (m)	Duration (days)	1992 June - Oct. Depth (m)	Duration (days)	92-93 Nov. - May Depth (m)	Duration (days)	1993 June - Oct. Depth (m)	Duration (days)
Valcour Island 63 m										
T-Chain	13-53	129	13-53	94*	FAILED		14-54	203	11-51	137
RCM5-upper	11	129	12	201	-	-	-	-	-	-
RCM5-lower	57	129	58	129*	-	-	-	-	-	-
ADCP	-	-	-	-	9-55	116	10-57	199	9-53	135
Corlaer Bay 64 m										
T-Chain	-	-	-	-	12-52	120	12-52	204	FAILED	
RCM5-upper	-	-	-	-	11	120	10	205	10	135
RCM5-lower	-	-	-	-	57	120	57	205	57	135
Central Site 65 m										
T-Chain	-	-	-	-	FAILED		13-53	31*	13-53	136
RCM5-upper	-	-	-	-	12	120	12	49*	12	137
RCM5-lower	-	-	-	-	FAILED		59	205	59	137

Table 1 (continued): Temporal and spatial coverage of hydrographic information during the first 2.5 years of observations at the various mooring sites within Lake Champlain. Time periods are grouped according to deployments and recoveries. Asterisks imply shorter than normal records. Dashes imply that the instrument was not used at the mooring site.

Mooring Site (Water Depth) Instrument	1991 June - Oct. Depth (m)	Duration (days)	91-92 Nov. - May Depth (m)	Duration (days)	1992 June - Oct. Depth (m)	Duration (days)	92-93 Nov. - May Depth (m)	Duration (days)	1993 June - Oct. Depth (m)	Duration (days)
Juniper Island 93 m										
T-Chain	13-53	129	13-53	202	-	-	-	-	-	-
RCM5-upper	11	129	11	85*	-	-	-	-	-	-
RCM5-lower	57	129	58	201	-	-	-	-	-	-
Burlington Bay 47 m										
T-Chain	-	-	-	-	8-46	113	FAILED		7-45	137
RCM5-upper	-	-	-	-	11	121	10	142*	12	137
RCM5-lower	-	-	-	-	43	121	FAILED		44	137
Thompsons Pt. 121 m										
T-Chain	14-54	129	14-54	202	11-51	131	12-52	202	12-52	124
RCM5-upper	12	126	10	201	-	-	-	-	-	-
RCM5-lower	59	126	57	201	-	-	-	-	-	-
ADCP	-	-	-	-	9-67	118	15-73	199	11-70	134

December 1992. During the same time period, minimum and maximum water column differences between the two sites were 0.7 and 2.1 °C, respectively. Further inspection of all wintertime data showed that this difference was present in almost all observations. For the wintertime periods of 91/92 and 92/93, the average water column temperatures at Valcour Island were found to be 0.8 and 0.5 °C cooler than Thompsons Point (with standard deviations of 0.4 and 0.5 °C, respectively). Although this pattern of Valcour Island being cooler than Thompsons Point was seen to occasionally reverse with temperature differences reaching 0.6 and 0.8 °C during the two wintertime periods, the amount of time that these reversals occupied were relatively small (7%). During the 91/92 winter, T-chain data obtained at Juniper Island confirmed the northward cooling trend from Thompsons Point, however, the percentage of time that this site was cooler than Thompsons Point was less (77%; versus the 93% at Valcour Island).

Assuming that atmospheric cooling of the lake can be considered uniform over its surface area, it can be inferred that the variation of thermal properties along the lake is probably due to the thermal inertia associated with varying depth water columns along with the more sluggish wintertime flow. Since the Valcour Island site is nearly 61 m shallower than the Thompsons Point region, cooling would have to progress for a longer time at Thompsons Point for average water column temperatures to reach those observed at Valcour Island. Mid-way between these two terminal sites (both spatially and with depth), vertically integrated temperatures near Juniper island were similarly found to be intermediate. Dynamically, the result of such a consistent thermal imbalance between Thompsons Point and the northern end of the lake would be a weak internal pressure field driving flow to the north (supporting the mean northward flow of water in the Main Lake [Myer and Gruendling, 1979]). Similarly, southward flow from Thompsons Point to the Crown Point Bridge could also be created strictly due to decreasing water depth in this direction. Without the Coriolis effect considered, such an imbalance in the dynamic height along the lake should be redistributed at the speed of a long surface wave (several hours). If, however, geostrophic flow existed, a longer lived anticyclonic circulation 'gyre' could be created. As the water columns move below 4 °C, an opposite circulation pattern could emerge. In either case, cyclonic or anticyclonic flow patterns should co-exist with a net southward flow through Thompsons Point due to topographic constraints.

While speculative, as there are no measurements to confirm either a thermal difference or current direction in the southern portion of the Main Lake, velocity observations at Thompsons Point showed a fairly consistent southward flow during wintertime. Deep currents in the central region of the Main Lake, however, indicate the presence of weak cyclonic circulation.

2. Valcour Island - Thompsons Point ADCP Comparison

ADCP measurements at Valcour Island and Thompsons Point are also presented in Plates 2 and 3. These platees show hourly averages of the E-W, N-S and vertical components of water velocity obtained from measurements every 1 m in the vertical and away from all interference boundaries such as the sediment-water and air-water interfaces. On average, horizontal velocities at Thompsons Point were sluggish (few cm/s with speeds rarely exceeding 10 cm/s) compared to those at Valcour Island that showed a component of north-south oscillation superimposed on a general west southwest trend extending over the entire water column and approaching 10 cm/s. This is in keeping with

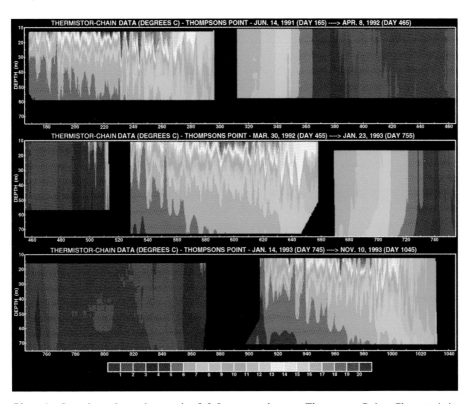

Plate 1. Complete thermal record of 2.5 years taken at Thompsons Point. Characteristic development of the epilimnion from spring to late fall and the lack of stratification during wintertime were well defined. See text for more details.

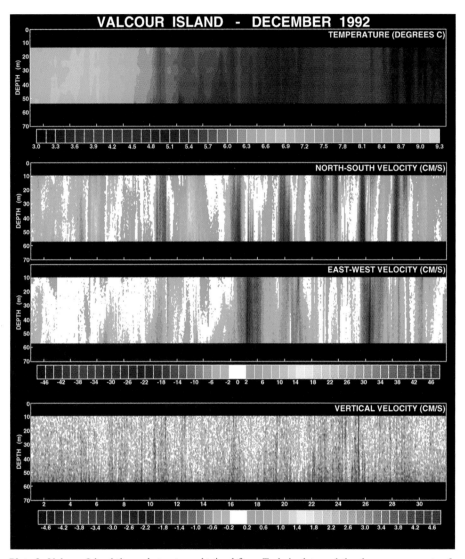

Plate 2. Valcour Island thermal structure obtained from T-chain data and the three components of velocity observed by the upward looking ADCP during the month of December, 1992. Temperature, horizontal and vertical velocity scales accompany the appropriate color panels.

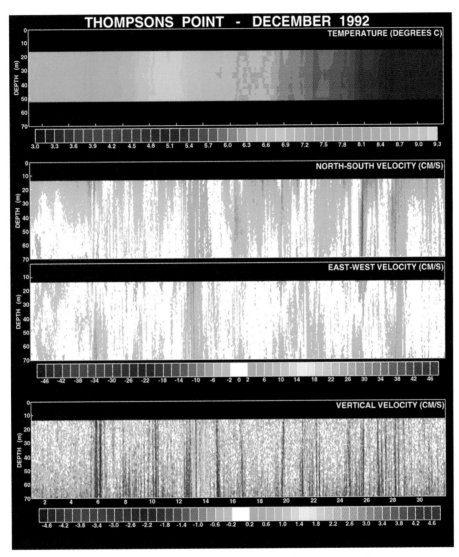

Plate 3. Thompsons Point thermal structure obtained from T-chain data and the three components of velocity observed by the upward looking ADCP during the month of December, 1992. All other aspects are the same as in Figure 5. Comparison between Valcour Island and Thompsons Point during this month are provided in the text.

general observations of weak currents observed in small and medium size lakes during the wintertime as compared to stratified periods where internal wave dynamics predominate (e.g., Lemin and Imboden [1987]). Spectral analysis of N-S winds at BIA and the N-S component of velocity at Valcour Island shows the predominant forcing period of the atmosphere to be near 3.3 days; with currents at Valcour Island lagging behind the wind by 1.5 days. Thompsons Point spectral results are similar in that the dominant in-lake period remains at 3.3 days, however, the energy and coherence levels are less (potentially due to the weaker currents). Phase information indicates that the N-S flow through Thompsons Point lags behind Valcour Island by approximately 4 hours. Unfortunately, the disparity between the magnitude of the flow at Thompsons Point and Valcour Island can not be explained; in part due to the lack of atmospheric observations along the length of the lake. It can only be suggested, therefore, that the typical trend of lower velocity currents at Thompsons Point might have been a combination of a consistent whole-lake wind stress acting on a larger (non-stratified) volume of water, varying wind conditions over the lake, and/or the presence of an opposing internal pressure field within this region.

3. Convection Dynamics

The vertical component of velocity at the Valcour Island and Thompsons Point sites frequently shows short term oscillations of upward and downward velocity events lasting several hours. These events represent either the propagation of convection cells past the sensor or short term events of vertical convection at a given location; most likely, a combination of both. While it is difficult to conclusively define the duration or structure of these convective cells since they may be propagating past the observation site, some characteristics of these features maybe obtained with the use of averaged observations and basic assumptions.

Using typical vertical velocities on the order of 1 cm/s for Valcour Island and nearly three times higher (and more defined) at Thompsons Point (possibly an effect of weaker horizontal currents or narrowness of the channel), the longest time needed for vertical convection to reach the bottom at both sites would be approximately 3.4 hours. The highest vertical velocity recorded during the month of December (4.6 cm/s at Thompsons Point) would take only 0.74 hours to convect to the bottom. Additionally, most convective events tend to last no longer than 3 hours, well within the range of values suggesting that convection may dissipate after the bottom was encountered.

Estimation of the vertical convection cell width is more speculative due to many of the assumptions that have to be made (e.g., convective cells must pass over the sensor along their diameters and they must be sufficiently long lived for horizontal propagation velocities to completely move them past the observation site). Nevertheless, by using a typical observational time scale of 2 hours and average horizontal motion of the water column to be 2 cm/s, an approximate cell diameter would be 150 m. When considering that the accuracies of the ADCP approach the observational values at Valcour as well as other assumptions, errors of 2-5 times the estimated 150 m cell diameter could also be realistic. What is interesting, however, is that Kelvin-Helmholtz and convective scaling parameters suggest horizontal cell sizes on the order of 100 m [Chu, 1991; Carmack and Weiss, 1991] which are in close agreement with our initial estimate of 150 m.

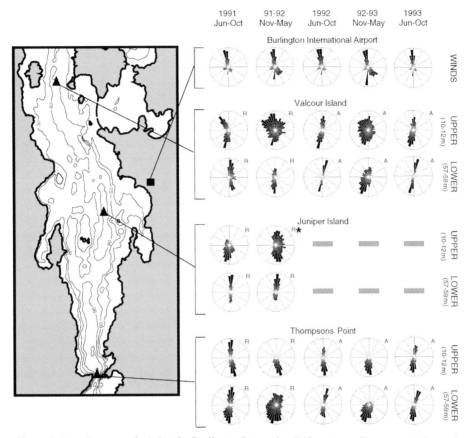

Figure 4. Rose diagrams of wind at the Burlington International Airport as well as currents observed in the epilimnion and hypolimnion at the Valcour Island, Juniper Island, and Thompsons Point mooring sites during the deployment periods defined at the top. Locations of these mooring positions are superimposed on more detailed bottom bathymetry of the central portion of the Main Lake. Note that all wind directions have been converted to reflect the same meaning as water currents in this paper. Superscripts of R and A define whether the diagram reflects RCM5 or ADCP data, respectively. An asterisks is used to denote a shorter than normal record. Horizontal bars indicate that the mooring site was not used or the instrument failed during that time period.

4. Main Lake Circulation Patterns & Statistics

The location of current meters within the central sector of the Main Lake varied during the first two years. From June 1991 to May 1992, three moorings were located along the thalweg of the lake and extended from Thompsons Point to Valcour Island; with a central site at Juniper Island (Figure 4). During this time period, only two RCM5 currents meters were used per site; each set at a representative depth of the epilimnion (10-12 m) and hypolimnion (57-59 m). In the second year (June 1992 to May 1993), the use of two ADCPs at Thompsons Point and Valcour Island and the acquisition of two additional T-chains permitted the expansion of the Juniper Island site into three E-W sites along a

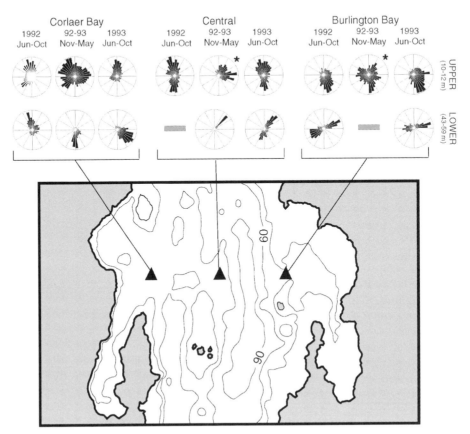

Figure 5. Similar diagram as Figure 7, however, the E-W aligned sites (Corlaer Bay, Central, and Burlington Bay) along the broadest part of the lake are presented.

similar latitude (Figure 5). The ADCPs were programmed to obtain current velocities at 1 m depth intervals. Depending on the depth of the water column, observations were obtained over a range from 9-73 m. In order to compare the ADCP measurements to those of the RCM5s, only the closest ADCP level to the RCM5 depth at that site was used.

Figures 4 and 5 provide polar histograms (rose diagrams) of current direction for each deployment period for the two representative depths of the epilimnic and hypolimnetic layers. Figure 4 also provides similar histograms for wind directions at BIA. It should be reiterated that all wind directions have been reversed so that they correspond with the same meaning as current direction. For statistical comparison, the minimum, mean, standard deviation, maximum, depth of observation, and record duration of temperature, speed, u (E-W), and v (N-S) for the two representative depths (corresponding to those in Figures 4 and 5) are provided in Table 2. Also provided in Table 2 are the same statistics for hourly observations of wind and air temperature at BIA.

As seen in Figures 4 and 5, wintertime flow patterns obtained from current meters having complete records were found to be dissimilar from those of corresponding

Table 2: Statistics from current meters and BIA during the first 2.5 years of observations at the various mooring sites within Lake Champlain. Time periods are grouped according to deployments. Depth of sensor and number of observational days are additionally shown. Statistics are shown as minimum, mean, standard deviation, and maximum. Trailing (A), (R), or (T) imply data from ADCP, RCM5, or T-chain, respectively. Closest T-chain information is provided for upper layer temperature since the ADCP can not remotely measure temperature away from its housing. Asteriks imply shorter than expected records. Dashes imply that the instrument was not used at the mooring site.

Mooring Site Location Sensor	1991 June - Oct. Depth - Duration Min Avg Dev Max Inst	91-92 Nov. - May Depth - Duration Min Avg Dev Max Inst	1992 June - Oct. Depth - Duration Min Avg Dev Max Inst	92-93 Nov. - May Depth - Duration Min Avg Dev Max Inst	1993 June - Oct. Depth - Duration Min Avg Dev Max Inst
Burlington Intl. Airport					
Air Temperature	-2.2 17.7 7.0 35.6	-27.2 -0.1 9.9 32.8	-0.6 17.9 5.6 31.7	-32.2 -0.4 9.6 28.3	-3.3 17.4 7.1 34.4
Wind Speed	0.0 4.4 2.3 14.0	0.0 4.7 2.5 15.5	0.0 4.3 2.3 13.5	0.0 4.2 2.6 14.5	0.0 4.0 2.3 12.4
Wind - N-S	-9.3 1.1 4.0 13.9	-10.3 0.4 4.4 15.0	-9.0 1.5 3.8 13.0	-9.4 0.5 4.0 14.4	-8.9 1.5 3.8 12.4
Wind - E-W	-8.5 0.1 2.6 8.5	-11.9 0.5 3.0 11.8	-7.6 0.1 2.6 9.3	-9.0 0.5 2.7 10.9	-5.7 0.2 2.3 8.5
Valcour Island					
UPPER	11 m - 129 days	12 m - 201 days	12 m - 116 days	12 m - 199 days	12 m - 135 days
RCM5/TCHN - Temp.	9.0 17.5 3.2 21.4 R	0.2 3.4 2.8 9.5 R	Failed T	0.1 3.2 3.0 10.8 T	7.3 16.1 3.7 22.1 T
RCM5/ADCP - Speed	0.4 7.3 5.8 32.5 R	1.5 2.3 0.9 7.5 R	0.0 7.6 5.2 33.8 A	0.0 3.9 3.4 24.9 A	0.1 6.8 5.0 31.4 A
RCM5/ADCP - N-S	-29.3 0.3 7.9 26.5 R	-5.9 0.1 1.8 5.8 R	-26.8 -0.4 8.7 33.2 A	-19.9 -0.6 3.8 11.7 A	-25.4 -0.6 7.7 31.0 A
RCM5/ADCP - E-W	-26.9 -2.5 4.2 10.6 R	-6.3 -0.1 1.6 5.5 R	-13.9 -0.7 3.0 10.4 A	-23.4 -1.0 3.3 12.6 A	-19.0 -0.8 3.1 12.3 A
LOWER	57 m - 129 days	58 m - 129 days*	55 m - 116 days	55 m - 199 days	53 m - 135 days
RCM5/ADCP - Temp.	4.7 6.4 1.2 9.5 R	0.6 3.9 2.6 9.3 R	4.7 6.4 0.6 8.1 A	0.3 3.2 2.5 9.6 A	3.9 6.4 1.7 14.2 A
RCM5/ADCP - Speed	0.0 7.5 6.3 44.5 R	1.5 2.3 1.0 7.4 R	0.1 8.3 7.0 49.0 A	0.0 4.5 4.4 30.2 A	0.1 8.6 7.7 54.9 A
RCM5/ADCP - N-S	-40.4 0.5 7.6 36.2 R	-7.4 -0.3 2.1 5.1 R	-48.6 0.0 10.6 28.6 A	-23.5 -0.8 5.0 14.0 A	-53.8 -0.6 10.9 29.2 A
RCM5/ADCP - E-W	-5.6 3.6 4.9 37.9 R	-6.0 -0.4 1.2 2.9 R	-11.0 -0.2 2.5 7.4 A	-29.9 -1.4 3.4 7.9 A	-11.1 -0.1 2.5 7.4 A

Table 2 (continued): Statistics from current meters and BIA during the first 2.5 years of observations at the various mooring sites within Lake Champlain. Time periods are grouped according to deployments. Depth of sensor and number of observational days are additionally shown. Statistics are shown as minimum, mean, standard deviation, and maximum. Trailing (A), (R), or (T) imply data from ADCP, RCM5, or T-chain, respectively. Closest T-chain information is provided for upper layer temperature since the ADCP can not remotely measure temperature away from its housing. Asteriks imply shorter than expected records. Dashes imply that the instrument was not used at the mooring site.

Mooring Site Location Sensor	1991 June - Oct. Depth - Duration Min Avg Dev Max Inst	91-92 Nov. - May Depth - Duration Min Avg Dev Max Inst	1992 June - Oct. Depth - Duration Min Avg Dev Max Inst	92-93 Nov. - May Depth - Duration Min Avg Dev Max Inst	1993 June - Oct. Depth - Duration Min Avg Dev Max Inst
Corlaer Bay					
UPPER			11 m - 120 days	10 m - 205 days	10 m - 135 days
RCM5/ADCP - Temp.	---	---	6.3 16.1 2.9 19.6 R	0.3 3.3 3.1 10.6 R	6.3 15.8 3.6 21.4 R
RCM5/ADCP - Speed	---	---	1.5 2.3 0.7 5.7 R	1.5 1.9 0.7 5.8 R	Failed
RCM5/ADCP - N-S	---	---	-4.0 0.3 1.8 5.3 R	-4.1 0.1 1.3 3.7 R	Failed
RCM5/ADCP - E-W	---	---	-4.7 -0.5 1.4 4.3 R	-3.8 0.1 1.5 5.8 R	Failed
LOWER			57 m - 120 days	57 m - 205 days	57 m - 135 days
RCM5/ADCP - Temp.	---	---	4.9 6.6 0.7 8.1 R	0.4 3.7 2.7 9.7 R	4.1 6.6 1.5 10.5 R
RCM5/ADCP - Speed	---	---	1.5 1.7 0.4 5.4 R	1.5 1.8 0.5 4.3 R	1.5 1.8 0.5 4.5 R
RCM5/ADCP - N-S	---	---	-3.8 0.4 1.3 4.8 R	-4.1 -0.6 1.3 4.0 R	-3.6 -0.4 1.2 4.2 R
RCM5/ADCP - E-W	---	---	-2.8 0.1 1.1 3.0 R	-3.6 0.3 1.1 3.8 R	-3.7 0.8 1.1 3.8 R
Juniper Island / Central	Juniper Island	Juniper Island	Central	Central	Central
UPPER	11 m - 129 days	11 m - 85 days*	12 m - 120 days	12 m - 49 days*	12 m - 137 days
RCM5/ADCP - Temp.	10.1 16.5 2.6 21.5 R	0.8 5.7 2.7 10.1 R	8.7 16.5 2.7 20.6 R	6.0 8.2 1.1 10.5 R	7.6 15.6 3.4 20.8 R
RCM5/ADCP - Speed	0.1 5.3 3.6 25.8 R	1.5 2.5 0.9 7.1 R	1.5 2.6 0.9 7.7 R	1.5 2.2 0.8 5.4 R	1.5 2.7 1.1 11.0 R
RCM5/ADCP - N-S	-24.2 -1.7 5.0 25.8 R	-4.7 -0.2 2.1 7.0 R	-6.2 0.1 2.3 6.7 R	-2.6 0.7 1.5 5.3 R	-6.8 0.3 2.4 10.3 R
RCM5/ADCP - E-W	-16.7 -0.5 3.6 14.7 R	-5.4 0.5 1.5 5.0 R	-5.0 -0.2 1.5 4.3 R	-3.7 1.0 1.5 4.7 R	-6.1 0.0 1.6 5.6 R
LOWER	57 m - 129 days	58 m - 201 days	59 m - 120 days	59 m - 205 days	59 m - 137 days
RCM5/ADCP - Temp.	4.9 6.4 1.1 10.5 R	0.2 3.4 2.7 10.0 R	Failed	0.6 3.6 2.8 10.0 R	4.0 6.5 1.5 10.2 R
RCM5/ADCP - Speed	0.4 4.9 3.8 26.1 R	1.5 2.1 0.8 8.3 R	Failed	1.5 1.8 0.7 9.4 R	1.5 2.3 1.2 10.7 R
RCM5/ADCP - N-S	-25.2 0.8 5.9 18.7 R	-5.0 0.4 2.0 7.7 R	Failed	-4.1 0.6 1.2 9.1 R	-9.8 0.1 2.1 6.9 R
RCM5/ADCP - E-W	-5.6 0.2 1.6 10.3 R	-6.1 -0.1 1.1 3.0 R	Failed	-5.7 0.4 1.4 4.6 R	-4.1 0.7 1.4 6.8 R

Table 2 (continued): Statistics from current meters and BIA during the first 2.5 years of observations at the various mooring sites within Lake Champlain. Time periods are grouped according to deployments. Depth of sensor and number of observational days are additionally shown. Statistics are shown as minimum, mean, standard deviation, and maximum. Trailing (A), (R), or (T) imply data from ADCP, RCM5, or T-chain, respectively. Closest T-chain information is provided for upper layer temperature since the ADCP can not remotely measure temperature away from its housing. Asteriks imply shorter than expected records. Dashes imply that the instrument was not used at the mooring site.

Mooring Site Location Sensor	1991 June - Oct. Depth - Duration Min Avg Dev Max Inst	91-92 Nov. - May Depth - Duration Min Avg Dev Max Inst	1992 June - Oct. Depth - Duration Min Avg Dev Max Inst	92-93 Nov. - May Depth - Duration Min Avg Dev Max Inst	1993 June - Oct. Depth - Duration Min Avg Dev Max Inst
Burlington Bay					
UPPER			11 m - 121 days	10 m - 142 days*	12 m - 137 days
RCM5/ADCP - Temp.	----	----	6.2 16.8 3.0 21.5 R	6.5 8.4 1.0 10.3 R	6.0 15.4 3.5 21.4 R
RCM5/ADCP - Speed	----	----	1.5 2.4 0.8 6.8 R	1.5 2.7 1.0 7.6 R	1.5 2.3 0.8 6.7 R
RCM5/ADCP - N-S	----	----	-6.1 -0.7 1.8 4.1 R	-7.1 -0.3 2.1 4.2 R	-6.4 -0.6 1.7 4.1 R
RCM5/ADCP - E-W	----	----	-3.8 0.5 1.5 4.7 R	-4.0 0.1 1.9 4.8 R	-4.0 0.7 1.5 4.9 R
LOWER			43 m - 121 days	44 m - 205 days	44 m - 137 days
RCM5/ADCP - Temp.	----	----	5.0 6.9 0.9 12.9 R	Failed	4.1 6.9 1.9 14.5 R
RCM5/ADCP - Speed	----	----	1.5 2.4 0.9 7.5 R	Failed	1.5 2.4 1.0 7.2 R
RCM5/ADCP - N-S	----	----	-4.4 0.0 1.4 3.1 R	Failed	-4.6 0.3 1.2 3.6 R
RCM5/ADCP - E-W	----	----	-4.6 -0.1 2.1 7.1 R	Failed	-5.1 -0.1 2.3 7.2 R
Thompsons Pt.					
UPPER	12 m - 126 days	10 m - 201 days	11 m - 118 days	12 m - 199 days	15 m - 134 days
RCM5/TCHN - Temp.	6.2 16.2 3.1 21.5 R	0.5 3.6 2.8 10.0 R	5.5 15.2 2.7 21.0 T	0.7 3.7 3.1 10.6 T	5.6 13.4 3.1 10.6 T
RCM5/ADCP - Speed	0.0 25.717.2 71.6 R	1.5 1.7 0.4 5.5 R	0.2 12.0 7.5 44.8 A	0.1 4.3 2.4 15.0 A	0.1 10.4 7.7 47.9 A
RCM5/ADCP - N-S	-67.2 -1.929.2 71.5 R	-4.0 -0.2 1.3 4.7 R	-42.6 -4.113.0 28.8 A	-14.9 -2.6 3.6 11.9 A	-47.2 -2.612.2 33.8 A
RCM5/ADCP - E-W	-43.0 2.7 9.6 50.8 R	-4.7 -0.2 1.1 4.0 R	-17.4 0.3 3.7 19.8 A	-9.4 -0.3 2.0 10.0 A	-15.0 -0.5 3.4 13.7 A
LOWER	59 m - 126 days	57 m - 201 days	59 m - 118 days	59 m - 199 days	59 m - 134 days
RCM5/ADCP - Temp.	4.9 6.5 1.1 9.7 R	0.3 3.6 3.0 10.0 R	4.7 6.1 0.7 7.9 A	0.6 3.4 2.8 9.7 A	4.1 6.0 1.2 9.5 A
RCM5/ADCP - Speed	0.3 5.1 4.4 41.4 R	1.5 2.3 0.9 8.1 R	0.0 6.0 4.8 34.4 A	0.0 2.5 1.8 13.2 A	0.0 6.0 4.2 28.8 A
RCM5/ADCP - N-S	-16.9 1.0 6.2 40.6 R	-8.1 -1.3 1.7 6.8 R	-16.5 0.6 7.3 33.7 A	-13.2 -1.2 -2.2 8.1 A	-18.1 -0.4 6.8 25.8 A
RCM5/ADCP - E-W	-12.5 -0.7 2.4 12.7 R	-6.8 0.1 1.3 5.3 R	-6.7 0.7 2.3 14.1 A	-9.1 -0.6 1.7 5.2 A	-10.7 0.2 2.6 13.0 A

summertime observations. As might be expected, upper layer flow showed the highest directional variability (e.g., Valcour Island, and Corlaer Bay), and even when records did not extend for the full deployment period (Juniper Island, Central Site and Burlington Bay) dissimilar flow patterns were still observable. While it is not surprising that Thompsons Point did not agree with this characteristic because of the alignment of flow to the narrow gap, it is important to note that wintertime flow was strongly unidirectional throughout the water column (with a slight variance in the lower layer of 1991-92) and not bi-directional as found in the summertime.

In contrast to the wintertime flow in the upper layer, lower layer flow shows a higher degree of directional stability that suggests control by bottom topography which has been observed in Lake Geneva over several years [Bohle-Carbonelle, 1986]. Mooring sites located within the thalweg of the Main Lake (Thompsons Point, Juniper Island, and Valcour Island) show a clear relationship between preferred alignment of flow and the N-S trend of the central axis, with slight variations between sites due to local bathymetric control. At the Burlington Bay site, the nearly E-W alignment of bottom currents and a similar trending buried river channel can be seen in Figure 5. West of the thalweg in the central part of the lake, bottom currents at the Central Site also showed some preferential alignment, however, they did not appear to be as strongly related to bottom contours. This may be explained by the fact that Lake Champlain bottom bathymetry has had no detailed mapping. Corlaer Bay represents the region with the strongest disparity of bottom flow over the last 1.5 years. Although there is preferential bottom alignment during each Corlaer Bay deployment, no consistent pattern from season to season emerged. When considering that all mooring sites were positionally maintained to within 100 m, it is difficult to explain the Corlaer Bay observations unless bottom topography is very rugged or as noted by Myer and Gruendling [1979], there may be independent circulation cells found within the lake.

Hodographs at approximately 10 and 60 m at all sites and all deployment periods in the Main Lake provided additional information on central lake circulation during wintertime. While the amount of data required to accurately define closed circulation cells are lacking, hodographs of deep currents (~60 m) during the winter of 92-93 at Thompsons Point, Valcour Island, Corlaer Bay and the Central sites suggest the presence of a weak, counter-clockwise circulation within the central lake (even though flow reversals were observed for brief periods of time). During the same time period, whole lake circulation patterns near the surface (~15 m) were less conclusive; possessing currents that both agreed (Thompsons Point and Valcour Island) or were only partially aligned with their associated deeper flow. During the previous winter, similar patterns were observed at Thompsons Point and Valcour Island, however Juniper Island displayed a weak initial southeast flow that was later replaced with a weaker northward flow (supporting a cyclonic circulation cell).

While largescale wintertime circulation patterns can only be considered speculative, it should be noted that the most unique and best defined shift in currents at Thompsons Point was from summertime bi-directional to that of a prevailing wintertime southward flow. When considering that the mean wintertime flow of water within this local region was previously inferred to be northward [Myer and Gruendling, 1979], a new picture of a systematic, seasonal change in circulation within the Main Lake becomes evident. Additionally, the fairly consistent southward flow throughout the upper 60 m of the water column at Thompsons Point is intriguing in that such a flow could not exist without the presence a compensating return flow to conserve mass in the local region. Whether this return flow is located at depth or along the periphery of the channel has yet to be determined.

Clearly, there are many aspects of wintertime circulation patterns within the Main Lake, (possibly including several smaller scale gyres as indicated by Myer and Gruendling [1979]), which are not well understood with the available information. What can be stated, however, is that significant variations in bathymetry which constrains deeper and sometime surface flow, the presence of nearly-homogeneous water columns of varying temperature causing internal pressure gradients, and longterm wind forcing appear to be the most significant forcing components of wintertime circulation patterns within the Main Lake.

IV. SUMMERTIME OBSERVATIONS AND RESULTS

1. The Internal Seiche

At the beginning of April (Plate 1; days 460 and 820), an increase in lake temperature rapidly progresses past spring turnover until the metalimnion is established. Subsequently, the less rapid alteration of the hypolimnetic thermal structure can be observed. While no observations were taken with the upper 10-12 m of the lake, maximum surface temperatures (which can typically be associated with maximum stratification of the metalimnion) can be inferred to have existed during the month of August (days 213-243, 579-609, and 944-974 from 1991 - 1993, respectively).

When near surface temperatures approach 5 °C, oscillations of the metalimnion are apparent even though only weak stratification exists. Having a period in the range of 4-6 days, these early oscillations represent the immediate onset of the internal seiche near mid-May (near days 500 and 870). The activity of the internal seiche remains unabated during the ensuing summer and into late fall. Being dependent on lake stratification, wind speed, wind direction and duration, it can be expected that the period and height of the internal seiche would vary both inter- as well as intra-annually. This variation in wave height and period is apparent in all three summertime periods. Ranging from a minimum of 7m (day 222) and 1.5 days (days 228-229) to a maximum in excess of 45 m (days 538 and 938 based on the excursion of the 5 °C isotherm) and 9 days (days 560-570), the internal seiche clearly represents the largest and highly variable wave within Lake Champlain.

Lateral and bottom damping of the internal seiche is primarily related to the shape, structure, and depth of the basin. Lake Zurich was found to have a relatively high frictional damping of approximately one period, yet Lake Lucerne appears to have one of the lowest with 5-7 cycles [VanSenden and Imboden, 1989]. Other work by Mortimer [1955], Thorpe [1971], and Saylor et al., [this volume] support an average damping of approximately three periods. Therefore, an average damping characteristic for Lake Champlain would equate to 12 days. From the thermal data depicted in Plate 1, there is little evidence of the internal seiche being completely damped into a quiescent state. Rather, the internal seiche is continually active from the onset of stratification in early spring to the disappearance of stratification in late fall.

2. Along-Axis Thermal Structure

Thermal records along the thalweg of the Main Lake at Valcour Island, Juniper Island and Thompsons Point during the summertime of 1991 (Plate 4) show the activity of the internal seiche, both laterally and vertically. As can be seen at any of the three sites (as well as in all other summertime records), the thermal structure at all depths was clearly affected by the internal seiche; vertical motion clearly in phase at any given site, yet out of phase when comparing opposing ends (e.g., days 232 and 245). Oscillating at the same period as the terminal sites, vertical motion at Juniper Island was comparably less. This is in agreement with the site's location being a few km south of the fundamental node position based on a Defant's model for Lake Champlain [Prigo et al., 1996].

With respect to the antinodes, Valcour Island and Thompsons Point will be used to represent close approximations of the actual terminal positions (which are estimated to be some 15 km farther north and south). As expected, Thompsons Point and Valcour Island were consistently out of phase during all oscillations of the internal seiche (Plate 4). Quantitatively, cross-spectral analysis of the T-chain data from these two sites confirmed high coherence and power at a period of 4.3 days, and a phase difference (~150°) which confirms their close positioning to the antinodes. High amplitude events such as those observed at days 231 and 245 can be used to demonstrate this nearly out-of-phase relationship observed at the two sites.

Current velocities observed at the various sites are typically saturated with oscillations in the range of 4 to 10 days. Taken from the lower RCM5 at Thompsons Point (located at 59 m), speed and direction for the summertime of 1991 are displayed in Figure 6. The consistent pattern of oscillation between the preferred along-channel directions of approximately 180° and 360° is highly visible. Transitions between the more stable, preferred directions were very rapid and did not last longer than a few hours. The readjustment of preferred alignment to less than a 180 degree differential after the beginning of September (day 244) is also apparent in Figure 6 and directly attributable to the weakening of the metalimnion and conversion to wintertime circulation. Consistent with the lower amplitudes of the earlier summertime internal seiche, speeds in the hypolimnion were less than 15 cm/s. At day 231, there was a pronounced peak of approximately 42 cm/s; subsequently remaining consistently higher until the last 10 days of the record. The high velocity peak at day 231 was associated with the onset of one of the largest amplitude internal seiches (Plate 4); the strong flow in the deep and shallow layers linked with the rapid depression of the metalimnion.

3. ADCP Measurements and the Metalimnion - Linear and Nonlinear Theory

The more complex structure in velocity and temperature at Thompsons Point and Valcour Island can be observed by combining the higher accuracy ADCP information with T-chain measurements for the month of September, 1993 (Plate 5). The velocity scale for both horizontal velocity components is shown at the center of the diagram. Superimposed on the velocity structure are the 10, 12, 14, and 16 °C isotherms that defined the average metalimnion during this time period. Areas of no data represent regions where near-surface or near-bottom acoustic interference prohibits accurate determination of speed. For Valcour Island, the range from 60-70 m represents bottom; the scale being kept the same as Thompsons Point (depth of 122 m) for intercomparison

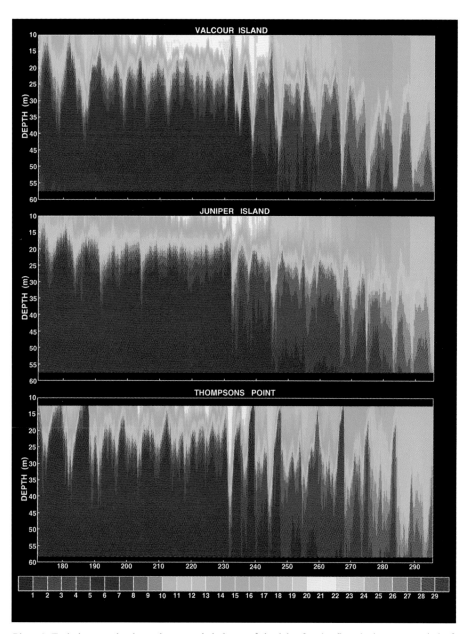

Plate 4. T-chain records along the central thalweg of the lake for the first deployment period of 1991. Temperature scale is provided at the bottom of the diagram. For reference, the first of the month for July, August, September, and October are days 182, 213, 244, and 274, respectively.

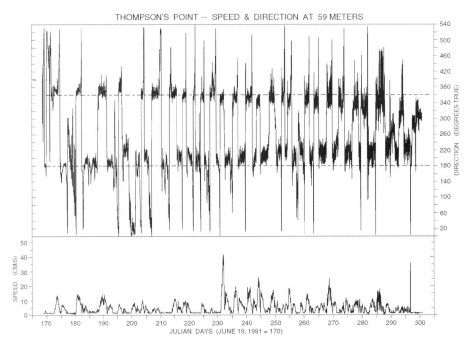

Figure 6. Speed and direction obtained from the deeper RCM5 current meter (59 m) at Thompsons Point during the summertime deployment period of 1991. Julian Days (as defined in Figure 9) are used in the horizontal axis. The alignment of the preferential flow along 180 and 360 degrees reflect the dominant N-S alignment of the narrow channel at Thompsons Point. During wintertime, velocities are weaker and southward directed flow, that is contrary to the expected averaged northward flow of water, is predominant over the entire water column.

purposes. The strong correlation between the metalimnion and velocity reversals at Valcour Island agree well with linear internal seiche theory, however, there were indications of a nonlinear component present at both Valcour Island and Thompsons Point. As might be expected from the narrow, almost N-S alignment of the channels at these two sites, E-W velocities were always the weaker of the two components and typically less than 10 cm/s. Velocity reversals due to internal seiche motion were typically associated with sharp transitional interfaces defining the zero velocity crossover 1) between upper and lower layer horizontal flow and 2) within each layer over time. According to 2-layer, constant-depth linear theory, the vertical separation interface (VSI), which represents the division between opposed flow in the epilimnion and hypolimnion, coincides with an infinitely thin metalimnion. The second, or temporal separation interface (TSI), coincides with motionless upper and lower layers at the time of flow reversal and coincides with the maximum elevation/depression of the metalimnion. At Valcour Island, both transitional interfaces (shown in white for speeds of -2 to 2 cm/s) were well defined for N-S flow, but indiscernible within the low velocity signatures of the E-W component. The vertically aligned TSIs can be observed within the epilimnion and hypolimnion at times when the metalimnion begins to reverse its previous trend at its maximum elevation or depression (e.g., 2 Sept. at 1200; 3 Sept. at 2000).

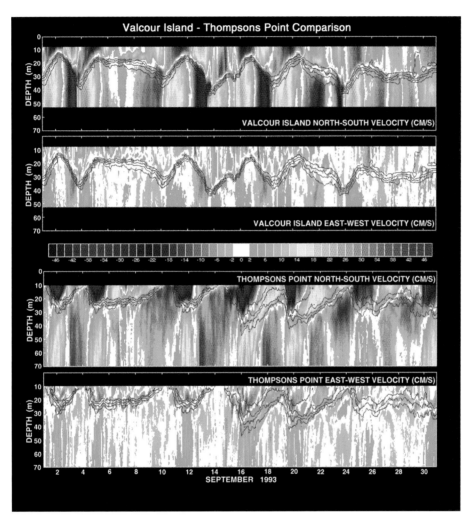

Plate 5. N-S and E-W velocity records obtained from ADCP located at Valcour Island and Thompsons Point during September of 1993. In contrast to Valcour Island, Thompsons Point defines a more complex set of dynamics. See text for more details.

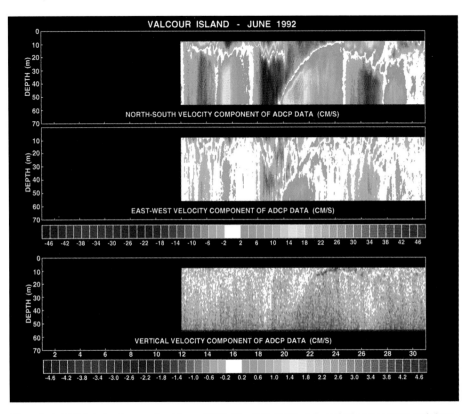

Plate 6. ADCP data for the month of June, 1992. The N-S, E-W, and vertical components and there associated scales are provided in descending order. The internal bore represents the extreme nonlinear mode of the internal seiche and has only been observed so far at Valcour Island.

Alignment of the N-S component VSI and the metalimnion were extremely close during time periods of epilimnic thinning (e.g., 1-2 and 4-5 Sept.), however, the departure away from linear theory was evident during many of the metalimnic depression phases observed in September. An example of this would be from mid-day on the 2nd to late on the 3rd of September, when the metalimnion falls from 12 m to a depth of nearly 35 m. During that time, the VSI failed to track the metalimnion past 22 m; thereby creating a case where flow above and below the metalimnion was in the same direction. Only after the occurrence of the following TSI (prior to 4 Sept.) did the VSI drop to the level of the metalimnion, after which both rose together in very close agreement. Comparison of this scenario with other oscillations in Plate 5 show that departures of the VSI and metalimnion occurred during the depression phase of the metalimnion while the associated degree of misalignment appears to be proportional to the strength of upper and lower layer flow fields during this phase. This can be seen in the decreased severity and occurrence of these departures over the month of September.

The reason for such departures away from simplified theory maybe explained by the fact that the internal seiche possessed a nonlinear component. Analytical and numerical modeling by Hunkins et al., [1998] indicated that division between linear and nonlinear modes of the internal seiche occurred at a wave height of roughly 10 m. From Plate 5, only 3 oscillations with wave heights less than 10 m were observed (6-8 and 26-30 Sept.) and in all of these, a higher degree of alignment between the VSI and metalimnion was observed. Similarly, the severity of the departure also increased as wave height increased past 10 m.

In contrast to the rather orderly pattern of events at Valcour Island, Thompsons Point showed a more complex system of currents and thermal variability. One of the most obvious differences was found in the variability of the metalimnion as depicted by the isotherms in Plate 5. Up until the 14th of September, the structure of the metalimnion (roughly defined by the vertical gradient of the displayed isotherms) was similar to that of Valcour Island. After the 14th, isotherms at Thompsons Point showed a significant departure from Valcour Island in the abnormally large divergence of the isotherms (vertical expansion of the metalimnion). Prior to the 14th, oscillatory motion at the two sites showed a consistent pattern of nearly out-of-phase metalimnic and nearly in-phase current (within each layer) oscillations that agree well with the longterm results of cross-spectral analysis at Valcour Island and Thompsons Point. Whether coincidence or not, this divergence of isotherms began at the same time that current and temperature observations at Thompsons Point failed to coincide with an abbreviated internal seiche oscillation at Valcour Island from the 13th (2000) to the 15th (1800). The TSI after the 16th of September also displayed a marked change from its typical vertical orientation to that of temporary inclined. This implies that the upper and lower layers were not moving as unified blocks of water that change direction instantaneously (as at Valcour Island), but rather reversing current direction from deeper to shallower layers over time (i.e., a vertically propagating interface). These observations may indicate the presence of a second internal vertical mode [LaZerte,1980; Munnich et al., 1992; Lemmin, 1987; Wiegand and Chamberland, 1987] that may be occasionally excited due to rapidly varying forcing events or tied to local basin configuration.

It is also possible that restriction of flow through Thompsons Point and/or interference with its own reflected wave (from Crown Point) may accentuate and/or distort the characteristics of the internal seiche and its various modes. One specific effect of this narrow but deep channel can be seen in the higher (lower) velocities of the epilimnion (hypolimnion) when compared to Valcour Island. Within the epilimnion, velocities at Thompsons Point are uniformly higher; particularly during times of southward flow. Exhibiting an opposite effect due to increased thickness (by a factor of 2) were the smaller velocities found in the hypolimnion. The depth of the metalimnion did not appear to be as strongly affected although it was apparent that the VSI at Valcour Island was more closely aligned with the upper part of the metalimnion (16 °C) but at Thompsons Point, it was closer to the lower portion of the metalimnion (10 °C). While strong departures of metalimnic structure along the length of a lake can be attributed to differential mixing of large amplitude internal seiches [Wiegand and Carmack, 1981], it is unclear whether or not channel restriction and/or wave-wave interference could have produced the observations at Thompsons Point.

4. The Rotary Nature of the Internal Seiche

The effect of the earth's rotation on long-period ($> 2\pi/f$ where $f = 1.454 \times 10^{-4}$ sin(latitude)) internal seiches is the creation of Kelvin waves that propagate cyclonically around the basin with the largest amplitude being found along the shore (exponentially decreasing away from shore) and to the right of the direction of travel [Mortimer, 1979]. Kelvin waves have been observed and modeled in many lakes [e.g., Mortimer, 1963; Defant (as reported by Mortimer, 1979)].

The first indication of a rotary nature of the internal seiche within Lake Champlain was noted by Myer and Gruendling [1979] to be the reversal of the metalimnion's E-W tilt in conjunction with the reversal of the N-S flow within the upper and lower layers at Valcour Island. To investigate this phenomena over a larger portion of the Main Lake, temperature data at 15 m from Corlaer Bay, Thompsons Point and Burlington Bay during September of 1992 were used (Figure 7). Valcour Island temperature data was not available during the same time period due to the failure of its T-chain. In Figure 7, three large temperature pulses lasting roughly a day and defining upward movement of the metalimnion were observed at Corlaer Bay on the 4th, 7th and 21st of September (days 613, 616, and 630). Lagging behind Corlaer Bay and of comparable magnitude were similar occurrences of these pulses at Thompsons Point. Timing of the passage of these events (calculated using the lowest temperature of the pulse's trailing edge) showed that Thompsons Point consistently lagged behind Corlaer Bay by an average of 22.3 hours. Burlington Bay, however, followed Thompsons Point with a larger time delay of some 38 hours (based on the first and last peaks). Phase information obtained from cross-spectral analysis of these three sites also confirms this cyclonic rotation of the internal seiche; clearly supporting the conclusion of Myer and Gruendling [1979].

The phase speed of the internal seiche can also be seen to vary along its path by using the 16 hour travel time difference needed for a 2-way transit time south of Thompsons Point. Using an average phase speed of 1 km/hr (determined from the Corlaer to Thompsons Point pulse) provides an estimate of the seiche's southern terminus to be located only 10 km south of Thompsons Point; roughly half of what was expected. Considering the close results obtained from numerical modeling by Connell [1994] and Hunkins et al., [1998] using the initial condition of an 82 km effective length of the basin and realistic bottom bathymetry, it would appear as though the propagation of the wave must be faster in the narrow channel south of Thompsons Point. Solving for the phase speed of the wave at Thompsons Point (Eqn. 2), shows that the simple increase of

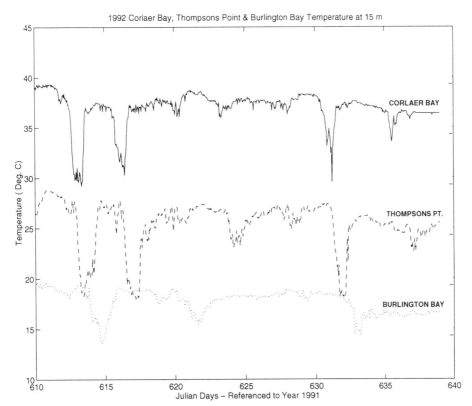

Figure 7. Temperature traces obtained from the T-chains at Corlaer Bay, Thompsons Point, and Burlington Bay during the month of September, 1992. Julian Days 610 and 640 correspond to the first of September and October. All observations were taken as close to 15 m as possible. Corlaer Bay, Thompsons Point and Burlington Bay temperatures are defined by the solid, dashed, and dotted lines, respectively. Burlington Bay temperature can be read directly off the vertical axis of the plot, however Thompsons Point and Corlaer Bay values have been offset by 10 and 20 °C, respectively. The rotary nature of the internal seiche can be observed in the cyclonic propagation of events around the southern sector of the Main Lake.

hypolimnetic depth to 100 m is sufficient to double the initially calculated phase speed. This, in turn, is sufficient to bring the estimate of the seiche's southern terminus into closer agreement with the existing thermocline shore, and therefore, model results.

The lack of correspondingly high amplitude pulses at Burlington Bay, when compared to those of Corlaer Bay and Thompsons Point, should also be noted. Unfortunately, there is not enough information to conclusively state that these reduced amplitude pulses were due to the restricted channel of Thompsons Point, wave-wave interference patterns, and/or the distance of the Burlington Bay site from shore. The similarity of the Corlaer Bay site with that of Burlington Bay in relative positioning from shore and weak indicators of other horizontal internal seiche modes suggests a higher likelihood of significant amounts of energy being lost from the internal seiche due to dynamic processes south of Thompsons Point.

If internal vertical modes exist within the vicinity of Thompsons Point it may be suggested that interactions between horizontal and vertical modes may provide an additional means by which energy can be effectively dissipated; particularly within a confined channel. The lack of data south of Thompsons Point offers very little latitude in the determination of the dominant dissipative processes within this region although increased lateral/bottom frictional effects, shear related turbulence, and internal breaking waves are highly probable.

5. Spectral Analysis of the Atmospheric and Internal Lake Modes

In order to provide a better understanding of the forcing and response modes within the Main Lake, spectral analysis was completed for each deployment over the entire 2.5 year observational program. The complexity of the system can be observed in the first five highest auto-spectra peaks of the principle components of winds, currents, and temperature for periods less than 14 days (Table 3). A summary of these wintertime and summertime forcing and response periods are provided in the histograms found in Figure 8. Due to the disparity of observations between seasons and instruments, data were normalized to provide a better means of intercomparison. Normalization of a dominate period was accomplished by summing all observations of that period over a specific time frame and dividing by the maximum possible number of observations; in effect creating a percentage of occurrence. As can be seen from wind information in Table 3 and Figure 8, dominant forcing periods were variable from summertime to wintertime, however, some periods were continually present (10.7, 7.1, 4.3, 3.9 and ~2.8 days).

The response of the Main Lake to such forcing was very different from summertime to wintertime. During the wintertime, there was a more uniform response of the lake with no clear preferred mode of oscillation (Figure 8). This reflects the observed sluggish circulation patterns of the lake during minimal or non-existent stratification, ice cover that can weaken atmospheric coupling to the water, and the predominance of NE directed winds (Table 2). Since the N-S and E-W modes of the Main Lake will be preferred over those forced along a different direction, it would be expected that the change of winds from predominantly along-axis summertime regime to that of approximately 45o to the main axis in the wintertime would further weaken the forcing of the preferred lake modes. Of the in-lake modes that did exceed the a 20% occurrence, there was a fairly even response at periods greater than 7.1 days which can be accounted for by the larger synoptic-scale winds. A more diverse response was observed at periods of 3.1, 2.4, 1.9 and 1.4 days. Surprisingly, these specific periods have no direct counterparts in atmospheric forcing and must therefore represent preferred lake modes. What physical boundaries these modes are related to still remains undefined.

In contrast to the nearly flat response of the lake during the wintertime, preferred modes during summertime were pronounced. As seen in Figure 8, there are both individual dominant periods and grouped response periods. The individual periods of 7.1 and 10.7 days displayed high occurrences (greater than 30%) and are typically separated from surrounding peaks. Grouped response, which spreads over a broader shaped peak with occurrences from 20-40%, can be observed within the ranges of 5.3-3.9, 3.3-2.5, and 2.3-1.8 days. Within these grouped responses is one predominant peak, about which the others are clustered. Within the two longest period groupings, the predominant peaks

Table 3: Dominant periods obtained from spectral analysis of velocity and temperature measurements. Only those periods of less than 14 days and possessing power spectra magnitudes greater than 1 are shown. Periods are arranged in descending order of spectral energy. Trailing A or R imply data from ADCP or RCM5, respectively. Asterisks denote shorter than expected records. Dashes imply that the instrument was not used at the mooring site or failed to record data.

Mooring Site Location Sensor	1991 June - Oct. Depth - Duration Dominant Periods (Days)	91-92 Nov. - May Depth - Duration Dominant Periods (Days)	1992 June - Oct. Depth - Duration Dominant Periods (Days)	92-93 Nov. - May Depth - Duration Dominant Periods (Days)	1993 June - Oct. Depth - Duration Dominant Periods (Days)
Burlington Intl. Airport					
Wind - N-S	4.3 6.1 10.7 3.3 2.5	2.8 2.5 2.1 1.5 10.7	4.3 7.1 10.7 2.7 2.0	3.3 10.7 3.9 2.5 6.1	2.7 4.7 7.1 3.6 1.8
Wind - E-W	7.1 1.0 3.1 4.3 2.5	10.7 2.5 2.8 4.7 3.9	1.0 7.1 4.7 2.8 2.4	7.1 3.3 3.9 2.3 2.8	1.0 2.7 5.3 3.3 3.9
Valcour Island					
T-CHAIN TEMPERATURE	15 m - 129 days 10.7 4.3 6.1 2.5 3.3	15 m - 95 days* 8.5 2.5	Failed	15 m - 204 days Very Weak	15 m - 137 days 7.1 5.3 3.9 3.3 2.8
UPPER	11 m - 125 days	12 m - 104 days*	12 m - 115 days	12 m - 198 days	12 m - 133 days
RCM5/ADCP - N-S	4.3 6.1 10.7 2.5 3.1 R	1.9 2.8 4.3 10.7 1.4 R	8.5 3.9 2.7 2.1 1.3 A	5.3 3.3 2.1 3.9 1.4 A	7.1 5.3 2.8 10.7 3.9 A
RCM5/ADCP - E-W	4.3 7.1 2.5 2.0 1.4 R	5.3 3.1 10.7 2.4 3.9 R	4.3 7.1 1.3 1.1 1.9 A	4.7 3.6 2.7 2.4 1.3 A	4.7 2.1 3.6 1.6 10.7 A
LOWER	57 m - 125 days	58 m - 104 days*	55 m - 115 days	55 m - 198 days	53 m - 133 days
RCM5/ADCP - N-S	3.9 7.1 2.5 1.9 1.6 R	10.7 3.9 3.3 1.9 5.3 R	7.1 4.3 2.7 2.1 1.6 A	5.3 7.1 3.3 3.9 2.1 A	2.7 5.3 3.3 7.1 3.9 A
RCM5/ADCP - E-W	7.1 2.0 3.9 2.5 1.5 R	3.3 5.3 2.3 4.3 2.0 R	7.1 3.9 2.7 5.3 2.1 A	8.5 4.7 3.1 2.5 1.9 A	5.3 2.7 3.3 7.1 10.7 A
Corlaer Bay					
T-CHAIN TEMPERATURE	---	---	15 m - 120 days 7.1 3.9 2.1 2.5 1.5	15 m - 205 days 8.5 (Weak)	Failed ---
UPPER	---	---	11 m - 115 days	10 m - 198 days	10 m - 133 days*
RCM5/ADCP - N-S	---	---	7.1 3.1 2.4 3.9 1.6 R	7.1 3.3 2.4 4.7 1.6 R	5.3 3.9 3.1 2.4 1.8 R
RCM5/ADCP - E-W	---	---	7.1 1.6 4.3 0.7 0.9 R	10.7 3.9 2.4 1.7 1.2 R	4.7 2.7 1.9 1.6 7.1 R
LOWER	---	---	57 m - 115 days	57 m - 198 days	57 m - 133 days
RCM5/ADCP - N-S	---	---	8.5 3.9 2.8 2.3 1.8 R	5.3 3.1 8.5 1.9 2.5 R	2.8 10.7 1.0 2.3 3.6 R
RCM5/ADCP - E-W	---	---	5.3 4.3 1.6 2.7 1.0 R	6.1 10.7 3.6 2.8 2.4 R	10.7 2.8 2.1 1.6 4.7 R

Table 3 (continued): Dominant periods obtained from spectral analysis of velocity and temperature measurements. Only those periods of less than 14 days and possessing power spectra magnitudes greater than 1 are shown. Periods are arranged in descending order of spectral energy. Trailing A or R imply data from ADCP or RCM5, respectively. Asterisks denote shorter than expected records. Dashes imply that the instrument was not used at the mooring site or failed to record data.

Mooring Site Location Sensor	1991 June - Oct. Depth - Duration Dominant Periods (Days)	91-92 Nov. - May Depth - Duration Dominant Periods (Days)	1992 June - Oct. Depth - Duration Dominant Periods (Days)	92-93 Nov. - May Depth - Duration Dominant Periods (Days)	1993 June - Oct. Depth - Duration Dominant Periods (Days)
Juniper IslandCentral	Juniper Island	Juniper Island	Central	Central	Central
T-CHAIN TEMPERATURE	15 m - 129 days 7.1 4.7 10.7 3.1 2.5	15 m - 202 days 10.7 (Weak)	Failed	15 m - 31 days* Very Weak	15 m - 137 days 7.1 3.3 3.9 2.8 2.0
UPPER RCM5/ADCP - N-S RCM5/ADCP - E-W	11 m - 125 days 4.3 6.1 10.7 2.5 2.1 R 4.3 7.1 2.5 2.0 1.4 R	11 m - 84 days* 1.9 2.8 10.7 4.3 1.4 R 5.3 3.1 10.7 2.4 3.9 R	12 m - 120 days 7.1 5.3 2.1 1.9 3.1 R 4.7 8.5 0.9 0.7 3.1 R	12 m - 49 days* 7.1 3.1 1.6 2.1 1.4 R 7.1 2.1 1.0 1.4 0.9 R	12 m - 133 days 7.1 4.7 2.8 1.0 1.9 R 7.1 10.7 2.4 2.7 3.9 R
LOWER RCM5/ADCP - N-S RCM5/ADCP - E-W	57 m - 125 days 4.3 6.1 2.5 2.1 1.6 R 10.7 4.3 6.1 2.5 1.8 R	58 m - 201 days 7.1 3.6 1.9 1.3 1.5 R 10.7 4.7 6.1 3.6 2.8 R	59 m - Failed	59 m - 198 days 4.7 2.8 1.8 2.0 1.5 R 3.3 7.1 2.0 4.7 1.4 R	59 m - 133 days 2.8 5.3 7.1 1.8 1.5 R 2.7 4.7 7.1 1.8 1.0 R
Burlington Bay					
T-CHAIN TEMPERATURE	---	---	15 m - 113 days 8.5 4.3 2.5 3.3 2.0	Failed	15 m - 137 days 7.1 3.9 5.3 3.3 2.8
UPPER RCM5/ADCP - N-S RCM5/ADCP - E-W	--- ---	--- ---	11 m - 115 days 7.1 2.4 0.7 0.9 4.7 R 7.1 2.7 4.3 0.7 1.6 R	10 m - 39 days* 10.7 3.1 1.4 1.0 0.8 R 4.3 2.1 7.1 1.4 1.0 R	12 m - 133 days 7.1 5.3 3.3 2.7 1.8 R 7.1 3.1 4.3 1.8 2.3 R
LOWER RCM5/ADCP - N-S RCM5/ADCP - E-W	--- ---	--- ---	43 m - 115 days 2.0 3.9 3.1 8.5 5.3 R 7.1 2.0 3.9 2.7 1.4 R	44 m - Failed -	44 m - 133 days 4.7 2.8 2.0 10.7 7.1 R 2.8 1.8 5.3 7.1 3.9 R

Table 3 (continued): Dominant periods obtained from spectral analysis of velocity and temperature measurements. Only those periods of less than 14 days and possessing power spectra magnitudes greater than 1 are shown. Periods are arranged in descending order of spectral energy. Trailing A or R imply data from ADCP or RCM5, respectively. Asterisks denote shorter than expected records. Dashes imply that the instrument was not used at the mooring site or failed to record data.

Mooring Site Location Sensor	1991 June - Oct. Depth - Duration Dominant Periods (Days)	91-92 Nov. - May Depth - Duration Dominant Periods (Days)	1992 June - Oct. Depth - Duration Dominant Periods (Days)	92-93 Nov. - May Depth - Duration Dominant Periods (Days)	1993 June - Oct. Depth - Duration Dominant Periods (Days)
Thompsons Pt.					
T-CHAIN					
TEMPERATURE	15 m - 126 days 10.7 4.3 2.1 3.1 1.4	15 m - 202 days 8.5 (Weak) 10 m - 201 days	15 m - 131 days 7.1 5.3 3.9 2.7 1.9 11 m - 115 days	15 m - 203 days 8.5 (Weak) 15 m - 198 days	15 m - 124 days 7.1 3.3 2.8 2.3 1.6 15 m - 133 days
UPPER					
RCM5/ADCP - N-S	12 m - 125 days 4.3 6.1 10.7 3.3 2.5 R	3.9 6.1 2.4 2.8 1.5 R	7.1 3.9 1.9 1.6 1.2 A	5.3 7.1 3.1 2.5 2.1 A	4.7 7.1 2.7 3.9 1.9 A
RCM5/ADCP - E-W	3.9 6.1 10.7 2.1 1.7 R	3.9 5.3 1.3 1.6 1.9 R	8.5 3.9 2.3 5.3 1.5 A	4.3 6.1 3.1 8.5 1.9 A	3.6 2.8 7.1 5.3 10.7 A
LOWER	59 m - 125 days	57 m - 201 days	59 m - 115 days	59 m - 198 days	59 m - 133 days
RCM5/ADCP - N-S	4.3 2.5 8.5 3.1 2.1 R	8.5 3.9 2.4 3.1 1.6 R	7.1 5.3 3.9 3.1 2.3 A	3.6 7.1 10.7 2.4 1.9 A	4.7 7.1 2.7 10.7 3.9 A
RCM5/ADCP - E-W	6.1 10.7 3.6 3.1 2.0 R	4.3 3.6 8.5 1.3 1.7 R	7.1 4.7 3.1 1.1 1.2 A	7.1 2.7 3.6 4.7 1.5 A	4.7 7.1 10.7 2.7 3.9 A

were 3.9, and 2.7 days. The shortest term grouping possessed a dominant peak of 2.1 days with a minimum occurrence value 20%.

Dominant in-lake periods of 10.7 and 7.1 days can be directly related to atmospheric forcing at the same periods since periods greater than 6 days do not appear to be related to any preferred mode of internal lake oscillation. The second largest in-lake mode of 3.9 days (within the broader group extending from 5.3 to 3.9 days) had weak atmospheric forcing at that period although one of the strongest atmospheric components (4.3 days) showed less of an in-lake response (at 4.3 days) than the 3.9 day period. Upon closer examination of Table 3, the predominance of atmospheric forcing at 4.3 days was from June to October, 1991 when both the N-S and E-W wind components possessed this period. During the same time, the lake response was dominated by oscillations at 4.3 days. Historical data [Myer and Gruendling, 1979] show that the most frequently observed along-axis internal seiche periods ranged from 3.9 - 4.6 days, while modeling of average summertime conditions by Connell [1994] and Prigo et al., [1996] predicted a resonant period of approximately 4.0 days. Although some variability of the fundamental internal mode does exist in the historical data, it is important to stress that the Main Lake represents a specialized case where a dominant period of atmospheric forcing is closely aligned with the fundamental period of the internal seiche. Such similarity between forcing and response modes suggests the presence of internal lake resonance which may, in turn, account for a portion of the high amplitude internal seiches observed in the 2.5 year record on Lake Champlain as well as internal surges and bores that have been observed in Loch Ness [Thorpe, 1971, Thorpe et al., 1972].

During the summertime of 1992, a decrease in the occurrence of the 4.3 day atmospheric peak (i.e., the E-W component of wind no longer showed this period although it was still the strongest in the N-S component) was similarly accompanied by a decreased occurrence of the in-lake 4.3 day period. Since the resonant mode of the lake should easily be set in motion even by light winds of a similar period, it is probable that the decreased activity of in-lake response was due to a shifting of the fundamental period during the summer of 1992. This alteration of the fundamental mode could be facilitated by changes in stratification, layer thicknesses, or the effective length of the basin based on the thermocline shore [Eq. 2]. Examination of Plate 1 shows that the depth of the metalimnion was deeper in 1992 than in 1991 and similarly, the metalimnion was more diffuse (thicker). Weaker stratification and a deeper metalimnion (which decreases the effective length of the basin) have the effect of opposing each other in modifying the internal resonant period, however, the alteration of the effective length has a linear effect compared to the square root of the density difference. The more frequent occurrence of the 3.9 day spectral period during the summertime of 1992 supports the premise that the net effect of the metalimnic variations was to decrease the effective length of the basin on the order of 9% (82 to 75 km). Therefore, the slight modification of the internal resonance period of the lake from 4.3 to 3.9 days was sufficient to weaken the effect of the peak atmospheric forcing period at 4.3 days.

Also expected is that variations of the metalimnion would be just as important over shorter time periods of weeks to months. As can be seen from Plate 1, the parametric representation of the metalimnion was constantly changing from early Spring to late Fall, and with it, the fundamental period of the internal seiche. Spectral analysis of a long term record (greater than a month) would therefore be expected to show a range of peak modes centered about the most dominant mode or as previously defined, a grouped response. The best example is that of the 5-3-3.9 day grouping. Over the three summertime records,

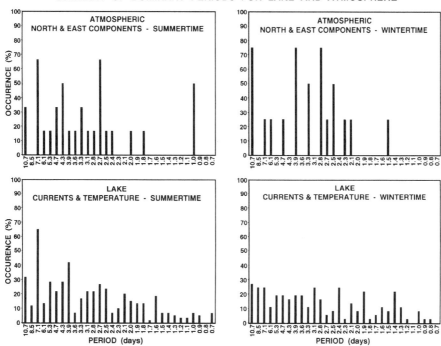

Figure 8. Frequency distribution plots of the first five dominant spectral peaks for the atmosphere and lake for all summertime and wintertime observations. See text for further details.

there was a consistent shift in the period of the fundamental internal mode from 4.3 to 3.9 days. The other peaks of 5.3 and 4.7 days that fall within the same grouping can be accounted for during times of weaker stratification. Calculations show that weakening the density contrast of the metalimnion from 0.002 g/cm^3 to 0.001 g/cm^3 (in the case of the 4.3 and 3.9 day modes) will force a readjustment of the fundamental period to 6.1 and 5.5 days, respectively. While it must be acknowledged that thickness of the various layers should be included in these simple calculations, a number of these permutations still supported the original concept of longer periods under conditions of weak stratification. While periods in excess of 7 days could be generated, these conditions can only be considered short lived, and therefore not a true indication of the more persistent in-lake modes that appear to begin at 5.3 days and decrease thereafter.

The relatively strong grouping centered about 2.7 days, however, deserves some additional comment. While it is apparent that atmospheric forcing appears to dominate the in-lake signal of the same period, the broad distribution of reasonably high occurrences from 3.1 to 2.5 days without strong atmospheric forcing at those periods suggests that there may be another principle mode within the lake or a principle sub-basin mode. Modeling by Connell [1994] concluded that the binodal internal seiche would be at 2.9 days based on average 1991 summertime conditions. It is possible that the 2.7 day peak may represent the preferred second mode of the lake. The second hypothesis is that of a fundamental mode for sub-basin within the Main Lake. By holding stratification and layer thicknesses constant as in the 4.3 day case, a length of approximately 50 km would be required to account for a 2.7 day uninodal period. Interestingly enough, this length corresponds to the distance between two natural restrictions within the Main Lake; Thompsons Point and Cumberland Head. While it is possible for the Main Lake to exhibit sub-basin modes [Myer and Gruendling, 1979], the agreement between the 2.7 period and the Thompsons Point-Cumberland Head sub-basin is speculative. Regardless of whether this mode represents the binodal component of the Main Lake or a fundamental sub-basin mode, it would still represent only one of the weaker modes within the lake.

The last grouping with a central peak of 2.1 days may provide an indication of a trinodal internal seiche since the spectral grouping does include the predicted value of 1.8 days [Connell, 1994]. Again, this can only be viewed as speculative and would require further modeling before confirmation. Estimates of cross-lake modes existing within the broadest part of the Main Lake show the fundamental period to be within 0.5-0.4 days for stronger stratification and no greater than 0.75 days for weak stratification. These modes were clearly not present in the most dominant energy peaks of current or temperature records and therefore suggest that their importance is negligible.

6. Nonlinear Aspects: Internal Surges and Bores

An indicative parameter of internal seiche structure is the time derivative of temperature at a given depth [$(dT/dz)_z$; hereinafter referred to as the temporal gradient]. While acknowledging that vertical gradients (dT/dz) vary with time according to factors such as insolation, cooling, and wind mixing, the structure of the internal seiche showed a consistent relationship with temporal gradients. Low-amplitude sinusoidal structure lasting several days (e.g., days 600-610) was associated with lower temporal gradients. Higher temporal gradients had the appearance of sharp fronts as seen at days 231, 246, and 928. During these three days, the higher amplitude propagating fronts were responsible for altering the temperature at 22 m at an average rate of 0.6-0.7 °C/hr for 13-16 hours with maximum hourly gradients of 2.4-3.0 °C/hr. Visually, a majority of oscillations possessed a more symmetric structure (i.e., the leading and trailing edges of the wave having a similar shape), however, the marked asymmetry associated with the passage of higher gradient fronts suggests that the internal seiche possesses a nonlinear component. Two-layer modeling of Lake Champlain by Hunkins et al., [1998] showed that for the waveform of the internal seiche to possess a symmetric structure, associated amplitudes would have to be less than 5 m for typical summer conditions. As amplitudes exceed this value, nonlinear effects become increasingly dominant until the wave interacts with the bottom topography or the surface. Under these conditions, a progressive nonlinear wave, known as a surge, will propagate away from the region [Horn et al., 1986]. The structure of a surge is highly asymmetric in that its leading and trailing edges become over steepened and more ramp-like, respectively. Surges have been observed in many lakes [Thorpe, 1971; Thorpe et al., 1972; Thorpe, 1974; Mortimer, 1955, Horn et al., 1986, Wiegand and Carmack, 1986], however, extreme examples of this over steepening of an internal wave face to nearly vertical, although not observed in Lake Champlain, would be that of internal solitons observed in Lake Seneca [Hunkins and Fliegel, 1973].

On occasions where high winds (typically in excess of 10 m/s) lasted for a period greater than a day, abnormally large modifications of the internal pressure field, caused by unusually large setups of the lake surface, were created. In such situations, the vertical displacement of the metalimnion would be large enough to intersect both the air-water and sediment-water interfaces. In response to this rapid modification of the metalimnion, a surge would propagate in an upwind direction. The response of the deep layer would be to form an internal bore that propagates in the opposite direction of the surge but along the sediment-water interface. Speeds of the internal bore (or gravity current) have been observed within the range of 27-30 cm/s and are dependent on stratification parameters [Saylor et al., 1994, Saylor et al., this volume]. Plate 6 depicts the passage of an internal bore through the Valcour Island site from 20-24 June, 1992. From 12-15 June, normally oscillating bi-directional flow associated with the internal seiche can be observed. On the 17th of June, winds at BIA switched from southward directed at 8 m/s to that of northward with peak speeds of 13 m/s. This wind event lasted for approximately 2.5 days, agreeing well with that of the strong reversal in currents (i.e., northward(southward) in the upper(lower) layers) over the same time period. A direct indication of the strength of the wind forcing can be seen in the speeds of the upper and lower layers that were in excess of 40 cm/s. Concurrently, the metalimnion was forced to abnormally deep (shallow) depths in the northern (southern) sector of the Main Lake as evidenced by the largest vertical excursion of the 8 °C isotherm over the entire summer at Thompsons Point. The elevation of this isotherm to less than 11 m suggests that further to the south, it may have outcropped at the surface although there are no measurements to confirm this. Five days later at Thompsons Point, the same isotherm was located at a depth of 57 m (Plate 1; days 535-540); a total excursion of greater than 46 m. Unfortunately, no concurrent information could be obtained on the thermal structure at Valcour Island since the T-chain failed during this deployment period. The readjustment of the deep metalimnion to the cessation of this long duration, high-speed wind event (which switched to southward flow at about 4 m/s on 20 June) was the internal bore as defined by high-velocity wedge-shaped northerly moving pulse observed from 20-25 June in Plate 6. The transitional interface between northerly and southerly flow, which is indicative of the location of the metalimnion and typically located in the upper 30 m of the water column, was found near the bottom at the onset of the return flow associated with the internal bore.

While several of these events have been observed in the Main Lake during the summertime, most appear to be centered around periods of weaker stratification found during the early summer and late fall. Since reduced stratification permits larger vertical oscillations of the metalimnion for any given wind forcing, the ability to force the metalimnion to abnormally large depths would be optimized during the spring and fall and not during peak summer conditions where winds are typically light and stratification the strongest. To this date, all internal bores in the Main Lake have been observed only in the northern sector, specifically in the region around Valcour Island. While there is no reason to exclude the presence of such features in the southern part of the Main Lake, there is a lack of observations south of Thompsons Point to support their presence. For more detailed observational analysis and theoretical aspects of the internal bore, the reader is directed to Saylor et al., [this volume] and Hunkins et al., [1998].

The extent to which surges and internal bores travel within the confines of the Main Lake and their effects on mean circulation, transfer of chemical and biological constituents, mixing, and sediment resuspension is largely unknown. Studies within other lakes suggest that the circulation and/or vertical mixing can be considerably altered prior to, during, and shortly after non-linear events. Horn et al., [1986] showed that shear induced vertical mixing during surge events could significantly enhance nutrient transfer from the hypolimnion to the epilimnion. Hypolimnic mixing could be enhanced by several processes ranging from Kelvin-Helmholtz instabilities [Thorpe et al., 1977; Thorpe, 1977] and downward propagating internal waves from a sheared interface as modeled by Stevens and Imberger [1994], to what appears to be catastrophic destruction of an internal surge, possibly due to the breaking of its internal wave field, as evidenced by the rapid and permanent increases of deep water temperature observed in Lake Seneca [Hunkins and Fliegel, 1973].

Internal seiche activity within other lakes has already been linked to near-bottom processes such as sediment resuspension [Baker et al., 1985; Halfman and Scholz, 1993; Gloor et al., 1994] as well as wetland environments and tributaries [Bedford, 1992]. Within Lake Champlain, bottom currents associated with nonlinear internal seiches (including surges and bores) clearly exceed those needed for resuspension of sediments. These erosional currents are not uniformly distributed throughout the basin, but rather within specific regions of the Main Lake [Manley et al., this volume]. Associated with many of these regions are bottom furrows that are believed to have been created by secondary circulation cells set up within the bottom boundary layer [Flood and Hollister, 1980; Flood, 1981, 1983; Flood and Johnson, 1984; Flood, 1989; Viekman et al., 1989, 1992]. Furrows in Lake Champlain therefore indicate the consistent presence of current patterns in these regions over periods of at least several years and in all likelihood, 10-100+ years. The disposition of resuspended sediment (with associated chemical and biological constituents) away from their original deposition site, however, is still a subject for research [Manley and Manley, 1993; Manley et al., 1996; Manley et al., this volume].

The effect that nonlinear events have on the mean circulation of the Main Lake is difficult to assess although the relative minor frequency of these events suggests that resulting changes may be small. Based on significant alterations of biomass [Levy et al., 1991; Gaedke and Schimmele, 1991] and nutrients [Lemin and Mortimer, 1986] from normal internal seiche activity, extrapolation would indicate that nonlinear events may have longer lasting effects in the areas of mixing, hypolimnetic ventilation, and sediment resuspension/redistribution. Unfortunately, the lack of observations relating to these specific topics forces speculation on nonlinear effects while at the same time underlining the importance of their future investigation.

7. Direct Atmospheric Ventilation of the Hypolimnion and Internal Mixing

By focusing on the thermal structure of the deep water in Plate 1, it can be seen that during the summertime periods, hypolimnetic temperature increases over time. While the warming of the hypolimnion can be accounted for by many inputs of heat such as sediment laden river input that sinks to deeper levels due to increased density, strong wind mixing events, diffusive processes, and groundwater [Ford, 1990], the rate of deep water warming suggests additional processes tied to the high amplitude nonlinear internal seiches. Specifically, extreme excursions of the metalimnion can ventilate the hypolimnion through direct access to the atmosphere. Concurrent high winds lead to strong mixing and therefore enhanced heat exchange of the deep hypolimnetic water. After relaxation of wind forcing, high internal velocities and associated shears would promote further mixing within the epilimnion and hypolimnion. If vertical shear (du/dz)

dominates over stability ($N^2 = gd\rho_* / \rho_* dz$ where g is acceleration due to gravity and ρ_* is the potential density of the medium) within the metalimnion (i.e., the gradient Richardson number, $Ri = N^2/(du/dz)^2$), another avenue of heat exchange to the hypolimnion would be available. With depths decreasing to the north and south of Thompsons Point, it is also possible that the internal breaking waves could be created, further facilitating the transfer of heat into the hypolimnion. In partial support of this was the conclusion of significant mixing occurring south of Thompsons Point based on lake chemistry over a period of 10 years [Potash and Henson, 1977].

Regardless of whether some processes described above are more significant than others or even if specific processes exist within Lake Champlain, the net effect of those that do facilitate hypolimnetic warming due to nonlinear internal seiche events should be evident in the time rate of change of the deep thermal structure. If observed, these changes in the heat content of the hypolimnion (which can be equated to the abrupt losses of colder isotherms) should be rapid and lag slightly behind the high amplitude internal seiches. For example, the 7 and 6 °C isotherms prior to day 230 were found at average depths of 30 and 50 m. After the large amplitude excursion of the metalimnion on day 231, temperatures colder than 6 °C were barely observed and the 7 °C isotherm was located some 20 m lower in the water column. By day 250 and two additional high amplitude events later, the 7 °C isotherm was almost completely removed from the thermal record. An exceptional example is found in the following year (1992), when the removal of all water cooler than 7 °C occurred directly after the first of 3 high amplitude events at day 633. Further investigation of the 2.5 year thermal record shows many of these rapid warming events of the hypolimnion occurred immediately following large amplitude internal seiches, even to the extent of accelerating the process of fall turnover (days 285-295, and 1010-1030) and was similar to that observed at Kamloops Lake by Carmack and Farmer [1982]. The result of modifying the thermal structure of the lake in discrete jumps over time could accordingly modify the internal pressure field and therefore the mean dynamics of the lake. To what degree is again a matter for further investigation.

V. CONCLUSIONS

Two and a half years of current and temperature data collected within the central portion of Lake Champlain have supported previous conclusions pertaining to the importance of the uninodal internal seiche within the Main Lake. Clearly, summertime hydrodynamics of Lake Champlain are dominated by the complex internal circulation patterns of the atmospherically forced internal seiche. The highest spectral power and coherence between atmospheric forcing and the currents/thermal structure observed in the lake existed with the N-S component of the wind. This was expected considering that the long axis of the Lake Champlain is aligned in the same direction. Both atmosphere and in-lake spectra possessed two dominant periods (10.7 and 7.1 days) that were well above the expected fundamental period of the basin and, therefore, reflected purely atmospherically controlled motion.

Estimates of the fundamental resonant period of the Main Lake's internal mode ranged from 4.6 to 4.0 days from equation 2 and modeling [Connell, 1994; Hunkins et al., 1998] using an effective length of the basin based on an average thermocline shore. By creating a frequency distribution chart of the five dominant spectral peaks observed in currents (E-W and N-S) and temperature at all sites over the 2.5 year data set, groupings around the dominate periods were observed. While the dominant peak represented the preferred period of a specific internal mode of the lake, the grouping of less dominant peaks around the preferred period could be explained by small departures of the average depth of the metalimnion and density differences between the epilimnion and hypolimnion. In other words, the period of the fundamental (and other) internal modes are continually being modified over time due to changing stratification and epilimnion/hypolimnion thicknesses. It is believed that periods up to the third internal mode were defined by the spectral groupings, however, sufficient variations in predicted and observed periods suggest that sub-basin fundamental modes may also be possible. Inter-annual shifting of dominate spectral peaks could also be observed due to varying atmospheric forcing from 1991 to 1993 as the dominant in-lake periods shifted from 4.3 to 7.1 and 3.9 days.

Rose diagrams describing the preferred alignment of currents over each deployment period at two depths that defined the epilimnion and hypolimnion provided some interesting insights to lake circulation. On average, upper layer summertime currents were strongly bimodal and tied to the oscillatory motion of the internal seiche. Lower layer currents also displayed a bimodal structure, but were strongly aligned with the local bathymetric contours. As an example, the bottom Burlington Bay current meter showed a predominant E-W alignment that agrees well with a deep channel oriented in the same direction. Wintertime current directions, on the other hand, showed very little preferential alignment except in the regions of confining channels or near bottom observations.

The internal seiche of the Main Lake displays linear as well as nonlinear characteristics. Modeling by Hunkins et al., [1998] agrees with observations indicating that nonlinear aspects of the wave become increasingly apparent after exceeding wave heights of 10 m. While high amplitudes can be directly caused by high, long duration winds, the Main Lake has a unique characteristic of possessing a fundamental period similar to that of wind forcing and therefore the potential for resonance. Nonlinear effects of the internal seiche can be observed in the asymmetry of its leading and trailing edges as well as the failure of the metalimnion and VSI to track each other (particularly during the depression phase of the metalimnion). During more extreme cases of metalimnic intersection of air and sediment interfaces, internal surges and internal bores (gravity currents) can be created. Both have been repeatedly observed in the northern sector of the lake at Valcour Island and were associated with bottom velocities as high as 40 cm/s; sufficient to resuspend sediments. Theoretical treatment of the surge and bore as well as other sediment resuspension characteristics associated are provided by Saylor et al., [this volume], Hunkins et al., [1998], and Manley et al., [this volume]. Nonlinear cases also lead to direct ventilation of the hypolimnion which is believed to be associated with rapid warming events of the deep water directly after the occurrence of a high amplitude internal seiche. Enhanced mixing related to high amplitude events though breaking of internal waves and shear instabilities would further facilitate warming of the hypolimnion.

Simplified linear calculations show that the volume of water that is cycled around the lake by the internal seiche is on the order of 1 km^3 per layer per half period within the Main Lake. During periods of higher-amplitude nonlinear events, the volume would increase accordingly. Although not previously discussed, one specific high amplitude event during 1993 was analyzed for volume displacement of the epilimnion and hypolimnion within the restricted boundaries from Thompsons Point to Valcour Island. At the time of maximum displacement of the metalimnion, three-dimensional calculations using the modeled thermal structure of the Main Lake bounded by actual bottom topography indicated volume displacements of 1.8 and 1.6 km^3 away from a static (level) metalimnion within the upper and lower layer, respectively. Inclusion of the remaining portions of the Main Lake (north of Valcour Island and south of Thompsons Point) would increase the volume displacements by an estimated 1 km^3 within each layer bringing the estimated maximum displacement in both layers to 5.4 km^3 during a half period. Volumetrically, this represents a transfer of 26% of water within the Main Lake in approximately 2 days.

The rotary nature of the internal seiche was also confirmed using temperature pulses related to specific high amplitude events at Corlaer Bay, Thompsons Point, and Burlington Bay. These well defined pulses were seen to propagate cyclonically around the basin at an average depth-integrated phase speed of 1 km/hr although calculations that include the deeper hypolimnion of Thompsons Point would define a phase speed of 2 km/hr. This higher phase speed agrees with estimates requiring the seiche to reach the southern reflection boundary defined by the effective length of the basin. While the amplitude of the pulses were similar at Corlaer Bay and Thompsons Point, the shape and amplitude of the related pulses observed at Burlington Bay were sharply reduced. This significant loss of energy south of Thompsons Point maybe linked to the narrow channel extending to the south of Thompsons Point, increased turbulence due to high shears, interaction with different internal modes and/or reflected wave interference.

The rotary nature of the internal seiche also poses additional problems relating to the disposition of pollutants and resuspended sediments and, therefore, environmental management of the lake. Newer information indicates that the region around Juniper Island (near the node of the fundamental internal seiche) may represent a sink for toxic metals that are easily attached to sediment [Dr. A. MacIntosh, Univ. of Vermont, School of Natural Resources, personal communication]. The presence of an internal Kelvin wave would support these observations in that the node would be a region of low velocities that would enhance deposition; in strong contrast to a longitudinal (N-S) internal seiche that possesses maximum velocities at the node and therefore minimal deposition.

Wintertime information supports a more sluggish circulation field within the Main Lake. With water temperatures uniformly approaching 0.5 °C during the maximum wintertime conditions and minimal to non-existent stratification during most of the time, currents were more uniform in the vertical compared to the summertime internal seiche dominated regime. Over the length of the lake, however, thermal structure was surprisingly different with the thicker water column around Thompsons Point continually lagging behind the cooler and shallower water found to the north. Reasonable assumptions suggest that a similar temperature gradient would occur towards the Champlain Bridge. Although speculative, it is possible that such internal pressure fields could create weak geostrophically balanced basin-scale cyclonic or anticyclonic gyre dependent on whether the lake temperature is above or below 4 °C. This would be in agreement with observations at Thompsons Point that showed a realignment of flow from the oscillatory N-S pattern of summer to that of a weak but consistent southward flow during the winter.

Ranging from sluggish wintertime flow that is, in part, hypothetically created by thermal gradients between the central deeper portions of the lake and its shallower ends, to that of an internal seiche dominated oscillatory flow capable of transporting several cubic kilometers of water around the basin every half-period (2-3 days) from early Spring to Late Fall, Lake Champlain represents a very complex hydrodynamic system. Two and a half years of observations within the central portion of the Main Lake and recent modeling efforts have provided several new insights of this system. Management issues of this lake are becoming more dominant each year, but it is clear that our present knowledge is insufficient to address questions related to transport, mixing, and dispersal processes related to a 3-dimensional internal Kelvin wave. This primarily reflects the thrust of past research to be directed, for the most part, in a nearly two-dimensional manner under the assumption of long, narrow-lake theory. While some aspects of lake circulation could be adequately addressed in this manner, it is reasonable to expect that the next phase of research will evolve about a better understanding of the three-dimensionality of the entire Main Lake over time, both observationally and via modeling.

Acknowledgments. This work was funded by the Cooperative Institute of Limnology & Echosystems Research (CILER) under cooperative agreements from the Environmental Research Laboratory (ERL), National Oceanic & Atmospheric Administration (NOAA), U.S. Departemnt of Commerce under Cooperative Agreement No. NA90RAH00079. The U.S. Government is authorized to produce and distribute reprints for governmental purposes notwithstanding any copyright notation that may appear hereon. This study was supported in part by the funds of U.S. Geological Survey, Grant No. 14-08-0001-G2050. Underwater equipment was purchased through a grant from the Lintilhac Foundation (Vermont) as well as from EPA Grant No. 50WCNR306021. We wish to acknowledge Captain Fred Fayette of the research vessel *Neptune* and we would like to further acknowledge the assistance of Captain Richard Furbush and the Univ. of Vermont research vessel *Melosira*, as well as our master SCUBA diver, Mr. Art Cohn for their continued support of this work under sometimes difficult conditions. Two and three-dimensional modeling as well as geographical display were accomplished using *EarthVision* (Dynamic Graphics, Alameda, CA).

REFERENCES

Baker, J. R., S. J. Wiseureich, T. C. Johnson and B. M. Halfam, 1985, Chlorinated hydrocarbon cycling in the benthic nepheloid layer of Lake Superior, *Env. Sci. and Tech., 19(9),* 854-861.

Bedford, K. W., 1992, The physical effects of the Great Lakes on tributaries and wetlands, *J. Great Lakes Res., 18 (4),* 571-589.

Bohle-Carbonell, M., 1986, Currents in Lake Geneva, *Limnol. Oceanogr., 31(6),* 1255-1266.

Carmack, E. C. and D. M. Farmer, 1982, Cooling Processes in deep temperate lakes: A review with examples from two lakes in British Columbia, *J. Mar. Res., 40 (suppl.),* 85-111.

Carmack, E.C., and R. F. Weiss, 1991, Convection in Lake Baikal: An example of thermobaric instability in deep convection and deep water formation in the oceans, in *Proceedings of the International Monterey Colloquium in Deep Convection and Deep Water Formation in the Oceans, Elsevier Oceanographic Series 57*, eds. P.C. Chu and J.C. Gascard, 215-228.

Chu., P.C., 1991. Geophysics of deep convection and deep water formation in oceans, in *Proceedings of the International Monterey Colloquium in Deep Convection and Deep Water Formation in the Oceans, Elsevier Oceanographic Series 57*, eds. P.C. Chu and J.C. Gascard, 3-16.

Connell, B., 1994, A study of numerical models of the surface and internal seiche of Lake Champlain, Middlebury College thesis for BA in Dept. of physics, May, 138 pp.

Flood, R. F. and C. D. Hollister, 1980, Submersible studies of deep-sea furrows and traverse ripples in cohesive sediments, *Marine Geol. (36)*, M1-M9.

Flood, R. D., 1981, Distribution, morphology, and origin of sedimentary furrows in cohesive sediments, South Hampton Water: *Sedimentology, 28*, 511-539.

Flood, R. D., 1983, Classification of sedimentary furrows and a model for furrow initiation and evolution, *Geol. Soc. Amer. Bull., 94*, 630-639.

Flood, R. F. and T. C. Johnson, 1984, Side-scan targets in Lake Superior-evidence for bedforms and sediment transport, *Sedimentology, 31*, 311-333.

Flood, R. D., 1989, Submersible Studies of Current-Modified Bottom Topography in Lake Superior, *J. Great Lakes Res., 15*, 3-14.

Ford, D. E., 1990, Reservoir Transport Processes, in *Reservoir Limnology; Ecological Perspectives* (K.W. Thornton, B.L. Kimmel and F.E. Payne, eds.), John Wiley & Sons, Inc., 15-42.

Gaedke, U. and M. Schimmele, 1991, Internal seiches in Lake Constance: influence on plankton abundance at a fixed sampling site, *J. Plankton Res., 13* (4), 743-754.

Gloor, M., A. Wuest and M. Munnich, 1994, Benthic boundary mixing and resuspension induced by internal seiches, *Hydrobiologica, 284*, 59-68.

Halfman, J. D. and C. A. Scholz, 1993, Suspended sediments in Lake Malawi, Africa: A reconnaissance study, *J. Great Lakes Res., 19(3)*, 499-511.

Henson, E.B. and M. Potash, 1969, Lake Champlain as related to regional water supply, *Proc. 12th Conf. of Great Lakes Res.*, Intl. Assoc. Great Lakes Res.

Horn, W., C. H. Mortimer, and D. J. Schwab, 1986, Wind-induced internal seiches in Lake Zurich observed and modeled, *Limnol. Oceanogr., 31(6)*, 1232-1254

Hunkins, K. L. and M. Fliegel, 1973, Internal undular surges in Seneca Lake: a natural occurrence of solitons, *J.Geophys. Res., 78(3)*, 539-548.

Hunkins, K.L., T.O. Manley, P. Manley, and J. Saylor, 1998, Numerical studies of the four-day oscillation in Lake Champlain, *J. Geophys. Res., 103(C9)*, 18425-18436.

Hutchinson, G. E., 1957, *A Treatise on Limnology, Vol. I*, John Wiley and Sons, Inc., 1015 pp.

Johnson, L., 1964, Temperature regime of deep lakes, *Science, 144*, 1336-1337.

Korgen, B.J., 1995, Seiches, *Amer. Sci., 83*, 330-341.

La Zerte, B. D., 1980, The dominating higher order vertical modes of the internal seiche in a small lake, *Limnol. Oceanogr., 25(5)*, 846-854.

Lemmin, U. and C. H. Mortimer, 1986, Tests of an extension to internal seiches of Defant's procedure for determination of surface seiche characteristics in real lakes, *Limnol. Oceanogr., 31(6)*, 1207-1221.

Lemmin, U. and D. M. Imboden, 1987, Dynamics of bottom currents in a small lake, *Limnol. Oceanogr., 32 (1)*, 62-75.

Lemmin, U., 1987, The structure and dynamics of internal waves in Baldeggersee, Limnol. *Oceanogr., 32 (1)*, 43-61.

Levy, D. A., R. L. Johnson and J. M. Hume, 1991, Shifts in fish vertical distribution in response to an internal seiche in a stratified lake, *Limnol. Oceanogr., 36 (1)*, 187-192.

Lueke, M., 1995, Sediment dynamics in a furrow field east of Valcour Island, Lake Champlain, Middlebury College thesis in Dept. of Geology, 83 pp.

Manley, P. L., 1991, Side-scan targets in Lake Champlain - Evidence of active sedimentary processes, *Green Mountain Geologist, 17*, 7-8.

Manley, P. L. and T. O. Manley, 1993, Sediment-current interactions at Valcour Island, Lake Champlain - A case of helical flow in the bottom boundary layer, Northeastern section GSA, *GSA Abstracts with Programs, 25(2)*, 36.

Manley P. L., Lueke, M. J. and T. O. Manley, 1996, Sediment dynamics in a furrow field, Lake Champlain, *GSA Abstracts with Programs, 28*, 78.

Manley, P. L., T. O. Manley, K. L. Hunkins and J. Saylor, Sediment deposition and resuspension in Lake Champlain, this volume.

Manley, T.O. and J. Tallet, 1990, Volumetric visualization: an effective use of GIS technology in the field of oceanography, *Oceanography, 3(1),* 23-29.

Mortimer, C. H., 1953, The resonant response of stratified lakes to wind, *Swiss J. of Hydrol, 15,* 94-151.

Mortimer, C. H., 1955, Some effects of the earth's rotation on water movements in stratified lakes, *Int. Ber. Theor. Agnew. Limnol. Verh., 12,* 66-77.

Mortimer, C. H., 1963, Frontiers in physical limnology with particular reference to long waves in rotating basins, Proc. 6th Conf. Great Lakes Res., *Great Lakes Res. Div., Pub. No. 10,* 9-42.

Mortimer, C. H., 1979, Strategies for compiling data collection and analysis with dynamic modelling of Lake motions, in *Hydrodynamics of Lakes,* eds. W.H. Graf, and C.H. Mortimer, Elsevier, Amsterdam, 183-221.

Mortimer, C. H., 1993, Long internal waves in lakes: review of a century of research. Center for Great Lakes Studies, Univ. of Wisconsin-Milwaukee, Special Report 42, 117 pp.

Munnich, M., A. Wuest and D. M. Imboden, 1992, Observations of the second vertical mode of the internal seiche in an alpine lake, *Limnol. Oceanogr., 37 (8),* 1705-1719.

Myer, G. E. and G. K. Gruendling, 1979, *Limnology of Lake Champlain,* Lake Champlain Basin Study Technical Report, New England River Basins Commission, Burlington, VT, 407 pp.

Potash, M. and E. B. Henson, 1977, Dilution in Lake Champlain from Whitehall to Rouses Pt., *Proc. of the Lake Champlain Basin Env. Conf.,* Chazy, NY, 235-246.

Prigo, B., T. O. Manley and B. Connell, 1996, Linear one-dimensional models of the surface and internal standing waves for a long narrow lake, *Am. J. Phys., 64(3),* 288-300.

Rao, D. B., 1966, Free gravitational oscillations in rotating rectangular basins, J. Fluid Mech., 25, 523-55.

Rao, D. B., 1977, Free internal Oscillations in a narrow, rotating rectangular basin, Mar. Sci. Directorate, Dept. Fish. Environ., Ottawa (Canada), MS-report 43, 391-398.

Saylor, J., G. Miller, T. O. Manley and P. L. Manley, 1994, Observations of high speed bottom currents in Lake Champlain, *EOS, Trans. Amer. Geophys. Union, 75,* 230.

Saylor, J., K. L. Hunkins, P. L. Manley, T.O. Manley and G. Miller, Gravity currents and internal bores in Lake Champlain, this volume.

Stevens, C. and J. Imberger, 1994, Downward propagation internal waves generated at the base of the surface layer of a stratified fluid, *Geophys. Res. Let, 21(5),* 361-364.

Stocker, K., K. Hutter, G. Salvade, J. Trosch, and F. Zamboni, 1987, Observations and analysis of internal seiches in the southern basin of lake of Lugano, Ann. Geophysicae, 5B(6), 553-568.

Thompson, S. D., 1994, Documenting the effects of the internal seiche in Lake Champlain on a shallow bay: Thompsons Point Bay, Lake Champlain, VT, Middlebury College thesis in Dept. of Geology, 71 pp.

Thorpe, S. A., 1971, Asymmetry of the internal surge in Loch Ness, *Nature, 231,* 306-308.

Thorpe, S.A., A. Hall, and I. Crofts, 1972, The internal surge in Loch Ness, *Nature, 237,* 306-308.

Thorpe, S. A., 1974, Near-resonant forcing in a shallow two-layer fluid: a model for the internal surge in Loch Ness, *J. Fluid Mech., 63,* 509-527.

Thorpe, S. A., A. J. Hull, C. Taylor, and T. Allen, 1977, Billows in Loch Ness, *Deep Sea Res., 24,* 371-379.

Thorpe, S. A., 1977, Turbulence and mixing in a Scottish loch, *Phil. Trans. Roy. Soc. London, A, 286,* 125-181.

Van Senden, D. C. and D. M. Imboden, 1989, Internal seiche pumping between sill separated basins, *Geophys., Astrophys. Fluid Dynamics, 48,* 135-150.

Viekman, B. E., R. D. Flood, M. Wimbush, M.Fagri, Y. Asako and J. C. Van Leer, 1992, Sedimentary furrows and organized flow structure: a study in Lake Superior, *Limnol. and Oceanog., 37(4),* 797-812.

Viekman, B. E., M. Wimbush and J. C. Van Leer, 1989, Secondary circulation in the bottom boundary layer over sedimentary furrows, *J. Geophys. Res., 94(C7),* 9721-9730.

Volker, A., 1949, Bilan d'eau du lac Ijssel, Gen. Assoc. Int. Un. Geol, Oslo, Assoc. Int. Hydrol. Sci., 1, Trav. Comm. Potamol. Limnol.

Wiegand, R. C. and Carmack, 1981, A wintertime temperature inversion in Kootenay Lake, British Columbia, *J. Geophys. Res., 86*, 2024-2034.

Wiegand, R. C. and Carmack, 1986, The climatology of internal waves in a deep temperate lake, *J. Geophys. Res., 91(C3)*, 3951-3958.

Wiegand, R. C. and V. Chamberlain, 1987, Internal waves of the second vertical mode in a stratified lake, *Limnol. Oceanogr., 32(1),* 29-42.

Numerical Hydrodynamic Models of Lake Champlain

Kenneth Hunkins, Daniel Mendelsohn, and Tatsu Isaji

ABSTRACT

The surface of Lake Champlain undergoes basin-wide oscillations with amplitudes of a few centimeters and a period of four hours. The summer thermocline undergoes basin-wide vertical oscillations of 20 to 40 m with a period about four days. A hierarchy of numerical computer models has been developed to model these wind-driven motions. A one-dimensional, one-layer model reproduces the water movements of the wintertime lake when temperature and density are uniform throughout the lake. A one-dimensional, two-layer model reproduces the movements of the summer thermocline when it is forced by wind data derived from observations at Burlington Airport. Finally, a two-dimensional, two-layer model reproduces not only the large-scale thermocline movements along the length of this long, narrow lake but also the details of movements around headlands and within bays. This last model is designed for practical use by those involved in managing the resources of the lake.

INTRODUCTION

Currents in lakes are primarily driven by the atmosphere. Winds act directly on the surface to move the uppermost layers. Water then piles up against the downwind shore to create differences in surface elevation over the lake. These differences in surface elevation influence deeper layers and lead to currents at all depths. Wind-driven currents vary with changing weather systems and are the most important motions for time scales of hours to weeks. In addition there are seasonally-varying currents through lakes driven by river inflow along the shores and discharge through an outlet. In Lake Champlain inflow comes from the many rivers which drain the eastern Adirondacks and western

Vermont. Outflow is through the Richelieu River. For all currents the flow pattern is largely controlled by the shape of the lake basin.

There is a need for better knowledge of the flow patterns in Lake Champlain to aid in management of toxic contaminants and nutrient enrichment. Detailed measurement of circulation in a large lake would require an impractically large array of recording current and temperature sensors at closely-spaced locations and depths. So the recourse has been to deploy a limited array of instruments and to depend on interpolation and model studies to provide the details. Numerical models of hydrodynamic processes in the Great Lakes have made considerable progress in recent decades. However model studies of Lake Champlain are still in an early stage of development. Numerical models, which have advantages in flexibility over laboratory and analytical models, are the preferred choice for such studies. Here we review recent progress in numerical models designed to represent currents in Lake Champlain.

CHARACTERISTICS OF LAKE CHAMPLAIN

For the purposes of hydrodynamic modeling the most important features of a lake are density stratification, basin shape, driving and dissipative forces. Lake Champlain has a dimictic density structure, overturning twice each year as it alternates between winter well-mixed conditions and summer stratified conditions. This annual density cycle is driven by seasonal temperature changes in the atmosphere. Summer stratified conditions are typical for temperate lakes with a warm upper layer overlying a deep colder layer [Myer and Gruendling, 1979; Manley et al., this volume]. The principal basin of Lake Champlain is the Main Lake which extends a distance of 117 km northward from Crown Point to Rouses Point at the Canadian border (Figure 1). The Main Lake has an average breadth of 6.3 km, an average depth of 29 m and a maximum depth of 122 m. Although the shoreline of the Main Lake is irregular and marked by numerous bays and headlands, there is a more regular deep channel with a breadth of about 2 km which extends over nearly the entire length of the Main Lake. Other basins of Lake Champlain such as the Northeast Arm and the South Lake are connected to the Main Lake by narrow passages which restrict exchange between the basins.

Although air temperature is the principal influence on lake structure and currents at annual periods, wind stress is the principal driving force for lake motions over shorter periods of time. Wind systems over the lake are typically associated with synoptic weather systems which travel eastward. Synoptic wind systems are much larger in scale than lake dimensions and so winds tend to be fairly uniform over Lake Champlain. We may expect winds to be steered by shore topography so that wind directions tend to become aligned with the north-south lake axis. This assumed tendency still needs confirmation by comparison of winds over land with winds over the lake. Currents due to the throughflow of water from inlets to outlet are small in comparison with wind-driven currents. However, although throughflow is relatively small, it is nevertheless important since it is always in the same direction and flushes the lake, removing contaminants and nutrients.

Lake Champlain is partially covered each winter with ice which varies widely in extent and period of coverage. Ice-covered areas of the lake are decoupled from wind

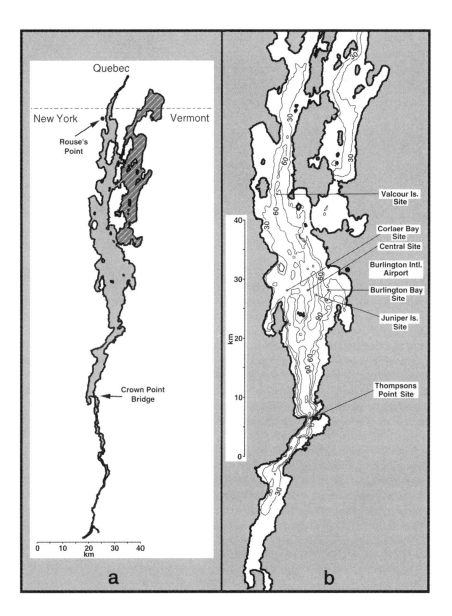

Figure 1. Lake Champlain shoreline and bathymetry. The Main Lake extends from Crown Point Bridge in the south to Rouses Point in the north. Shallow arms of the lake are connected to the Main Lake by restricted passages and not considered part of the Main Lake [Hunkins et al., 1998].

influence and winter current records from these regions show that currents beneath the ice are greatly reduced in velocity.

MODELING STRATEGIES

Knowledge of the physical structure of Lake Champlain was advanced by the five-year field study conducted by investigators from Middlebury College and NOAA [Manley et al., this volume]. The thermocline moves over a vertical range of about 30 m with periods of about four days (Figure 2). Movements at opposite ends of the lake are out of phase. In the figure, thermocline motions at moorings on the lake axis at a northern site (Valcour Island) and a southern site (Thompsons Point) are seen to move in opposite directions. Relations between wind and thermocline motion are not immediately obvious from the figure. An explanation of the movements leading to prediction must be based on physical theory.

Our understanding of currents and temperatures in lakes is based upon the physical laws governing mass, momentum and energy. These well-known laws are the cornerstones of macrophysics. They are expressed as a set of hydrodynamic equations which must be solved with appropriate boundary and initial conditions. In the case of a lake the boundary conditions at the surface are the driving forces of wind stress, air pressure and temperature. At the sides and bottom there is frictional resistance to flow. There can be no flow normal to a boundary surface. Since the forcing is continually changing, initial conditions are of less importance than boundary conditions.

Solution of the complete set of hydrodynamic equations is beyond present programming capabilities and simplifications must first be introduced before the equations can be solved. So far, models developed for Lake Champlain have only considered the conservation laws for mass and momentum. The thermodynamic equations for energy have not yet been applied to this lake although they govern the important seasonal heat cycle. Therefore present models are limited to time spans of a few months at most. Water movements in a lake cover a wide range of scales in time and space. Some selection must be made of the ranges to be modeled. The smallest scales in space and time constitute turbulent motion and are conventionally represented by an eddy viscosity term. The longest time scales are annual or longer and necessitate including thermodynamic processes. Presently available models of Lake Champlain are focused on the intermediate time scale of hours to several weeks.

In many research fields there are two levels on which problems can be approached. Simplified analytical and numerical models make drastic simplifications of the natural situation in order to explore and understand specific aspects of the problem. These approximate models are aimed at understanding the physical processes operating in the lake. They are useful in determining the most important processes in a given situation. These simple models provide guidance for developing more complex models which, in the case of lakes, incorporate complicated bottom topography and more complete equations. More complex models pay attention to societal needs by providing numerical simulations which will be of practical use to those involved with managing lake resources. Examples of both of these types of models have been developed for Lake Champlain and will be discussed here.

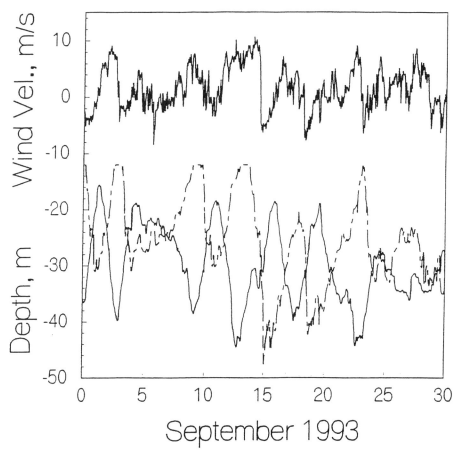

Figure 2. (Top) Wind velocity at Burlington Airport. North-south component. Positive velocity is directed northward. (Bottom) Depth of 10-degree isotherm at two sites in Lake Champlain during September 1993. The Valcour Island site (solid line) is near the northern end of the Main Lake and the Thompsons Point site (dashed line) is near the southern end [H.unkins et al, 1998].

The numerical models for Lake Champlain to be described here employ the forced shallow-water equations which are simplified versions of the hydrodynamic equations [Gill, 1982]. The horizontal scale of the motions is assumed to be large in comparison with water depth. Density structure is represented by either one layer for winter conditions or by two layers for summer conditions. Observed profiles of temperature in the lake show that this is a reasonable first approximation [Manley et al., this volume]. The models are driven by wind stress at the surface and damped by friction at the bottom. In other studies atmospheric pressure changes have been shown to contribute little to lake currents in comparison with wind stress. Therefore air pressure is neglected in these models.

These models have been designed to run on desk-top personal computers. Continuing increases in speed and memory for such computers now make them practical for numerical studies. The reasonable cost and wide usage of personal computers then allows the managers of lake resources to conveniently run the programs at their offices to solve practical problems.

SEICHES

Seiches are the most important type of lake motion in the period range of hours to days. Seiches are standing waves in which the surface and thermocline oscillate on a basin-wide scale and they often follow changes in wind speed or direction. Water is first piled up at the downwind end of the lake by wind drag. Later, after the wind drops or changes direction, lake waters oscillate freely with periods determined by the shape of the basin. Vertical displacements of the surface are in the millimeter to centimeter range with oscillation periods of minutes to hours. In stratified lakes, internal temperature surfaces may also oscillate as internal seiches. In this case amplitudes and periods are both much greater than those of surface seiches. Vertical displacements of meters to tens of meters and periods of hours to days have been observed for the thermocline in Lake Champlain [Manley et al, this volume]. An example is shown in Figure 2. Numerical models which attempt to model currents in Lake Champlain must be capable of reproducing seiche motions.

ONE-LAYER, ONE-DIMENSIONAL MODELS

The great length of Lake Champlain and its narrow width suggest the usefulness of models in which a long narrow channel represents the lake basin. In one-dimensional models it is assumed for simplicity that currents flow only in a north and south direction along the lake's long axis. The lake is divided into a string of 100 boxes, each 1.17 km in length but varying in cross-sectional area. Elevations of the surface and, for two-layered models, the thermocline interface are considered uniform within each box. Currents are uniform across each cross-section separating the boxes.

Periods of surface seiches derived from linear hydrodynamic theory for one-dimensional models are generally found to be in close agreement with measured periods of surface seiches in lakes. A numerical technique, devised by Defant [1961], solves the linear shallow-water equations by stepwise integration using the finite-difference method to yield the free period as well as the associated surface and interface movements for one-dimensional lakes of varying cross-section. The method has been applied successfully to many lakes. Before computers became available the numerical integration was carried out manually, a laborious procedure. Recently the technique has been programmed for personal computers and applied to the Main Lake to predict a period of 4.0 hours for the fundamental longitudinal mode of Lake Champlain [Prigo et al., 1995]. This predicted period agrees well with the range of 3.9 to 4.6 hours which was observed by Meyer and Gruendling [1979].

The observed oscillations are not truly free modes. The free oscillations predicted by Defant's method are equivalent to the ringing of a bell after being struck by the clapper. However the real wind stress acts continually on the lake surface and observed motions are forced oscillations. More complete models have been developed which include a wind stress which varies with time [Hunkins et al, 1998; Marinov, 1998]. In early experiments these models were forced with idealized wind pulses which arise instantaneously and then blow at a steady speed for some time. In this case damped oscillations are produced. This model can be used to predict currents for the wintertime lake when forced by more realistic winds based on observations. This has not yet been done since winter models require detailed knowledge of ice coverage which is not yet available. In the future when such knowledge does become available comparisons of this model with observations can be made.

Surface seiches involve only small displacements of water and, although interesting, are of minor importance for practical questions involving water movements. Internal seiches are of much greater importance to the understanding and management of lakes since a water parcel may be carried as far as 30 km up or down the lake and then back again to near its starting point during their four-day cycle.

TWO-LAYER, ONE-DIMENSIONAL MODELS

Summer stratification can be most simply represented by two homogeneous layers differing slightly in density. This approximation has proved effective in many theoretical studies and is used in Lake Champlain numerical models. Extension of the one-dimensional model to two layers raises questions about the relevance of two effects which are of negligible importance for one-layer models. These are the effects of nonlinearity and of the earth's rotation.

Although nonlinear effects are negligible for surface seiches, they may be of importance for internal seiches. The linearized equations apply to waves of small amplitude. When amplitudes are large relative to layer depth the complete equations incorporating nonlinear terms must be used. In early summer, when mean depth of the epilimnion is about 15 m, vertical excursions of the thermocline may reach 10 m or more in Lake Champlain. In autumn when the epilimnion is about 25 m deep vertical excursions of 15 m or more are frequently observed. These are evidently waves of large amplitude and nonlinear effects can be expected to be important for internal seiches.

Large-amplitude waves described by the nonlinear shallow-water equations have a tendency to steepen since propagation speeds for these waves are proportional to wave height. Higher, and therefore faster, waves overtake lower waves ahead of them, leading to steepening at the front of a wave packet. The resulting bores are analogous to shock waves in gases. This steepening tendency is limited by two factors. In the shallow-water formulation vertical motions, which are associated with dispersion, have been neglected and steepening is opposed by friction. When vertical motions are retained in the momentum equations the tendency of waves to steepen and overturn is counteracted by dispersion. When friction is negligible, the steepening effect is opposed only by dispersion. In this case there may be a balance between nonlinearity and dispersion with waves of permanent form, known as solitary waves, resulting. A special case of solitary waves is the train of solitons which follows a steepened wavefront [Whitham, 1974]. Solitons have been recorded in other long narrow lakes but not in Lake Champlain. In other lakes such as Seneca Lake in New York State and Babine Lake in Canada the

Korteweg-deVries equations have been suggested as a basis for modeling thermocline oscillations. The Korteweg-deVries equations balance nonlinear steepening effects with dispersion under conditions of weak nonlinearity. However, for Lake Champlain where wavefronts are not steep, the shallow-water equations appear to be more appropriate.

The reason for the different regime in Lake Champlain is not yet clear. Only a possible explanation can be offered here. It may be that topographic effects are responsible. At each end, the bottom of Lake Champlain slopes gently upward and there may not be a sufficient reservoir of accumulated water to produce a bore upon release after a wind change. Also the numerous shallow bays which provide storage for warm upper waters at the downwind end of the lake form a distributed source of water after a wind change, again inhibiting bore formation. The other mentioned lakes more closely resemble the rectangular basin of the model since they have straighter sides and deeper ends than Lake Champlain. In the central part of the Main Lake far from shore, current direction has been observed to rotate in a clockwise circle at the Coriolis period. Clearly the earth's rotation is important in the dynamics of this lake. Further evidence is found in cross-lake tilts of the thermocline which are observed to be in the sense expected for geostrophic motion. Rotational effects need to be included in two-layered models although they are not important for models with a single layer. The relative importance of Coriolis effects can be quantified by the radius of deformation, the ratio of long wave speed to the Coriolis frequency. For single-layer models the radius of deformation is about 170 km which is much larger than the lake width of about 6 km and Coriolis effects are not important. For two layers the internal radius of deformation is about 5 km, less than the lake width, and the earth's rotation must be considered.

Experiments with this model were first forced by idealized wind bursts in which the wind blows for either one hour or 12 hours at 10 m/s. The wind bursts blow upper warm water toward the downwind end of the lake and depress the interface at that end. At the upwind end the interface is elevated. The depression and elevation travel as waves toward the center of the lake where they cross each other. The two waves are mirror images for the shorter burst (Figure 3a). This is the small-amplitude, linear case. When the wind burst lasts for 12 hours the two waves are large in amplitude and differ in profile (Figure 3b). This is the large-amplitude, nonlinear case. The wave of depression for the interface travels upwind steepening as it goes becoming a bore at about the middle of the basin. The bore is analogous to a shock wave in compressible gases. It is a typical effect of nonlinearity on waves of large amplitude. The wave of elevation traveling downwind from the other end becomes less steep as it progresses. This corresponds to a wave of expansion.

Although internal seiches of large amplitude are observed in Lake Champlain they do not steepen to form the vertical wave fronts of bores. The reason for this was sought in the wind field which is not well modeled by wind bursts. In a second group of experiments the model was forced by the wind field recorded at Burlington Airport. This wind field was assumed to represent winds over the lake. Results from these runs showed surprisingly close agreement with observed changes in interface depth (Figures 4 and 5). Recently more representative wind data have been collected from a recording station on Colchester Reef but the modeling results using the reef wind data show little or no improvement in modeling results than those from the airport [Marinov, 1998]. The one-dimensional model was not originally intended to provide realistic results. It was intended to be a first step leading to more complex and realistic models. The reasonably

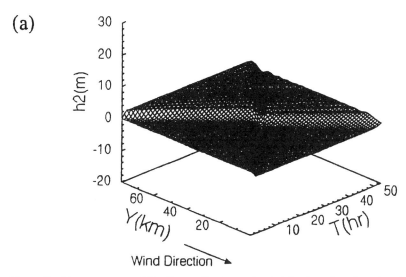

Figure 3a. Linear response of the thermocline to weak wind forcing. Interface elevation along the basin axis is shown as a function of time. A wind burst at 10 m/s blows for one hour in the negative y-direction shown by the arrow. After the wind drops, a wave of elevation propagates in the same direction as the wind and a wave of depression propagates in the opposite direction. These waves, propagating in opposite directions, are mirror images of each other. Rectangular model basin measures 80 km long, 5 km wide and 75 m deep with H1 = 25 m and H2 = 50 m. Reduced gravity, g' = 0.01 m s-2. Free period is 109 hours for the lowest internal mode [H.unkins et al, 1998].

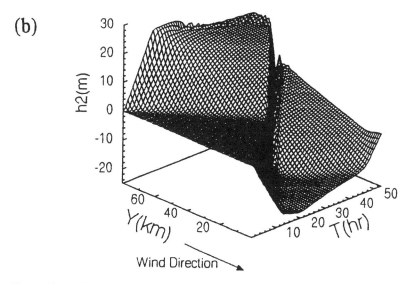

Figure 3b. Nonlinear response of the thermocline to strong wind forcing. Model parameters as in Figure 2a except that wind burst lasts for 12 hours. Note that the wavefront propagating in the upwind direction steepens with time while the wavefront propagating downwind becomes less steep with time [Hunkins et al., 1998].

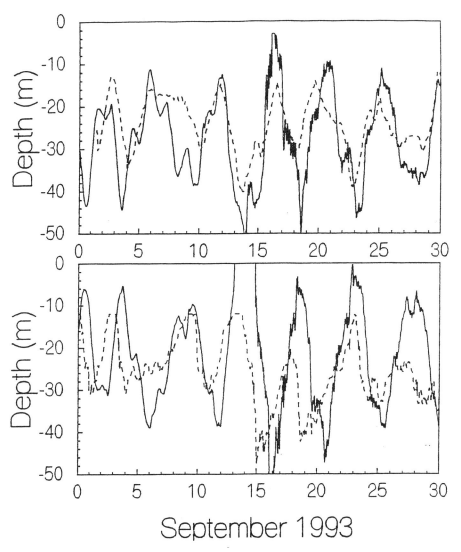

Figure 4. Linear model. Observed thermocline depths (dashed line) compared with model thermocline depths (solid line). Free period of model is 109 hours. Rectangular basin: 80 km long, 5 km wide and 75m deep. H1 = 25 m, H2 = 50 m. (Top) Valcour Island site. Thermocline represented by the 16-degree isotherm. (Bottom) Thompsons Pointsite. Thermocline represented by the 10-degree isotherm [Hunkins et al., 1998].

good agreement shown between observations and model results demonstrates that the simple one-dimensional model already contains the most important physical processes involved in wind-driven currents along the central axis of the lake. Evidently the smooth realistic changes in wind speed and direction do not generally lead to wave steepening and bores. Lake Champlain behaves as a continuously forced system. There are no

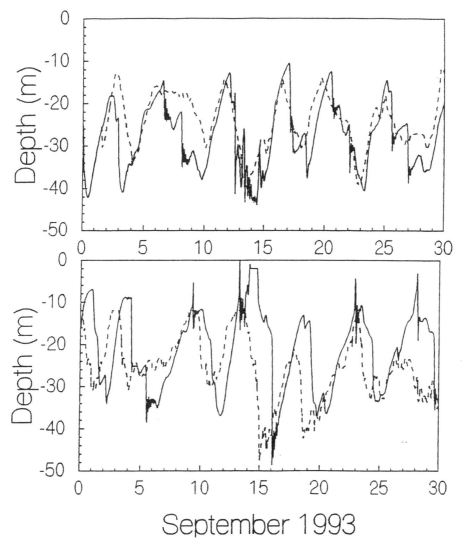

Figure 5. Nonlinear model. Observed thermocline depths (dashed line) compared with model thermocline depths (solid line). Parameters as in Figure 3 [Hunkins et al., 1998]

prolonged calm periods in which free waves might develop as they do in the models forced with idealized wind bursts. The fluctuating wind field has a broad spectral maximum associated with the passage of synoptic weather systems with energy concentrated in the three to five day period range. Since the first internal mode of the lake is about four days it is not surprising that large amplitude internal motions are generated by realistic winds. But the phase of the wind oscillation changes frequently and this is not an ideal forcing at the resonant frequency. These are not internal seiches insofar as that term indicates free oscillations. The lake is a continuously driven system with a response which is due to the irregular oscillation of the wind system as well as to the natural free period of the lake.

TWO-LAYER, TWO-DIMENSIONAL MODELS

Significant currents in Lake Champlain are generated by the movement of the internal interface, often larger than those generated near the surface even by strong winds. Currents are produced not only from the north-south, along-basin movement of the interface but also from the cross-basin variations associated with geostrophic rotation. With the extension of the two-layer model from one to two horizontal dimensions, the full motion of the thermocline can be modeled. For the two-layer model system including temperature and density differences between the layers, an internal radius of deformation is applicable, which for the Main Basin of the lake has been shown to be on the order of 5 km, less than the lateral dimension of the lake [Hunkins et al., 1998]. This implies that rotational terms in the model equations become important and therefor that rotation of the internal interface can be expected.

To produce realistic current circulation patterns more detailed information about the topography of the lake bottom and the shoreline geometry is needed in addition to the realistic local wind. The initial application of the 2-D model to Lake Champlain consisted of gridding (tiling) the body of the lake covering the area roughly between the 44th and 45th parallels, from Crown Point in the south to Rouses Point in the north, with 1 km rectangles as shown in Figure 6. The resulting grid is 32 cells across and 123 cells along the lake with about half of the total number of cells being 'water cells', where the model equations are solved. The remaining areas are 'land cells' and are not considered part of the computational domain. The rough outline of the various lake basins can easily been seen, including the South Lake, the Main Basin, Mallets Bay, the North East Arm and Missisquoi Bay. At the north and south ends of the main basin the cross-sectional area decreases, from both decreasing depth and decreasing width. It should be noted that grid resolution with respect to wavelength is a critical parameter. A particular grid resolution, (i.e. spatial dimensions of the cells) must be sufficient to resolve the waves in question or numerical degeneration of the wave may occur (Mendelsohn et al., 1995). For the longer wavelength, basin scale, first mode of the internal seiche this is not a problem, but may become important for shorter wavelengths that have been predicted as a result of high wind stresses of short duration. For these short duration events a sharp slope in the interface may develop, consisting of a number of harmonic waves, some of which may be short.

Each 'water cell' has a bottom depth associated with it and is divided into two layers. During calculations, water may flow from one water cell to another but only to the same layer, (e.g. top layer to top layer) and may not flow into or out of land cells. The thickness of each layer, as well as the height of the water surface is allowed to vary freely in time, based on the solution of the mass and momentum equations over the entire body of the lake. Although the thickness and depth of each layer is allowed to vary locally, the total water mass, temperature and density of each layer is held constant and development, growth and destruction of stratification is ignored for the period of simulation. This

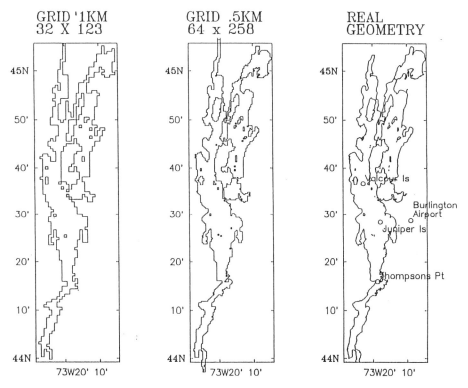

Figure 6. Two-dimensional, two-layer hydrodynamic model grid for Lake Champlain

simple two layer assumption, although overly restrictive for longer periods, allowed for some key simplifications in the model development and proved sufficient for the duration of the simulations.

For the experiments performed, only the Main Basin was considered, as the passages and flow between the main body of the lake and the smaller basins to the East are small and probably do not affect the motion of the thermocline appreciably. The data set for the summer of 1991 [Manley et al., this volume], focusing on August, was chosen as the period to be modeled for the quality of its duration and physical coverage in the main body of the lake. The data set contains long, continuous time series of both temperature and currents for at least two depths at stations at Valcour Island at the north end of the Main Lake, Thompsons Point at the south end south and Juniper Island in the central region. Figure 7 depicts a development of thermal structure as vertical location of maximum water temperature change and integrated temperature difference from these mooring locations. All plots indicated the existence of an oscillation peaked at a four day period. The north-south components of currents of the top and bottom meters were highly coherent at that period.

The thermal structure varied between two distinct features from the early to later summer. The thermal stratification had already developed at the start of field observation (June 1991) and intensified until the mid portion of August. After that the upper-lower

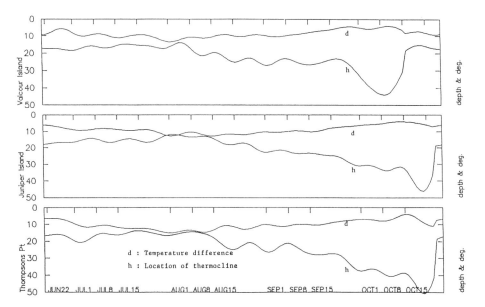

Figure 7. Observed thermal structure at three moorings in the main basin of Lake Champlain for summer 1991. Each mooring station had a chain of thermistors covering the depth of the water column. The Valcour Island station is in the north end of the Main Lake, Juniper Island near the middle and Thompsons Point at southern end. Filtered data.

temperature differences decreased and thermocline depth increased indicating increased mixing, possibly due to high wind events and consequent strong internal seiching. The current features appear to reflect the progress of thermal structure development; relatively weak currents in the early summer as thermal stratification was developing and strong currents (oscillation) in the late summer as the location of stratification deepened and the density difference reduced.

An initial stratification and temperature regime was assumed based on data for the summer period, thereby specifying the depths and density difference between the upper and lower layers. Turner (1973) has shown that the effective frequency for an internal seiche is directly proportional to the density difference between and layer thicknesses of the upper and lower layers, also critical parameters in determining the behavior of the internal seiche in Lake Champlain [Mendelsohn et al., 1995]. Hourly observations of wind stress taken at Burlington Airport were applied to the entire surface area of the model domain, as this was the best estimate of surface winds available at the time. There is most likely a great deal of variation in the surface winds over the length and breadth of the lake due to its dimensions but more importantly due to orographic steering (channeling of the wind by the mountainous local terrain). Very little is known about the influence of the orographic effects on the development and maintenance of the internal seiche however, speculation might lead one to conclude that alignment of the mountains along the length of the Main Lake serves to 'rectify' the winds to the north-south direction, enhancing stimulation to the internal seiche.

A number of simulations were made with different model parameters to assess the model response. The primary objective was to find the range of parameters that would produce a reasonable simulation of the internal seiche as well as the currents. Parameters varied included such basic numerical and physical functions as grid resolution, the minimum time step, interface friction, (at the water surface for wind as well as the internal interface), and horizontal diffusion to the critical initialization parameters including the layer thickness and the density difference between the surface and bottom layers. The results of the model simulation were compared to observations for a period covering the entire month of August. The model simulation was initialized and run for one month prior to the observation period to allow for the system to 'spin up', (i.e. for the internal interface to begin oscillating as a function of the wind forcing, from its initially flat, motionless state). Some general findings of interest were that the effect of bottom friction did not have a significant effect and that the fine grid (0.5 km) simulations did not show any improvement over the coarse grid (1.0 km) simulations.

Figure 8 shows a comparison of the model predicted and the observed currents for the upper and lower water column at the three current meter mooring sites and wind stress in stacked stick plot form. The August data moves between two distinct thermal periods. For the first part of the month the wind was relatively week and the thermal stratification was developing, whereas in the latter part of the month the winds became much stronger with a notable north-south component. This is clearly reflected in the observed currents as well as the model predictions and is most notable in the surface responses at both the Valcour and Juniper Island sites. Thompsons Point however, appears to maintain a more consistent, large amplitude fluctuation that is not particularly influenced by the changes observed elsewhere over the month and is more consistent than the model predicted currents at that location. The relative amplitude of observed and calculated currents agree with the exception of the surface observations at Thompsons Point. During strong wind periods, the model predicted and observed currents agree well in both amplitude and phase whereas during weak wind periods, current amplitudes agree with observations but the phase is often shifted.

It is clear that wind is an important forcing function as seen in all cases. The simulated currents responded strongly to strong wind events. It is questionable how well the lake surface wind can be represented by Burlington Airport wind, particularly for weak winds. It became apparent that a short-term simulation (of less than several internal seiche periods, 12 days) required detailed, appropriate initial conditions to reproduce the observed currents. The cases that were run for two months, one month longer than the short cases result in better agreement with the observations. The improvement was marginal. For periods of weak wind it may not be possible to synchronize the model predicted internal seiche with the naturally occurring oscillation observed in the lake without extensive and detailed initial conditions or a strong wind event that could force the oscillation.

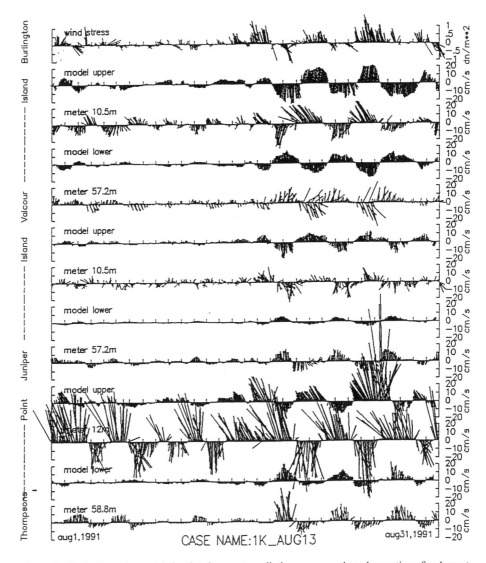

Figure 8. Hydrodynamic model simulated current predictions compared to observations for August 1991. The stick plot at the top is the wind stress recorded at Burlington Airport. The next four plots represent currents at the Valcour Island station for model predictions and observations near the surface and near the bottom respectively. The next set of four plots are for the Juniper Island station and the bottom four for the Thompsons Point station.

Acknowledgements. This work was supported by the U.S. Environmental Protection Agency, the National Ocean and Atmosphere Administration, the State of Vermont and Middlebury College.

REFERENCES

Defant, A., *Physical Oceanography, Vol. II,* Pergamon Press, Oxford, 1961.

Gill, A. E., *Atmosphere-Ocean Dynamics,* 662 pp., Academic Press, New York, 1982.

Hunkins, K., T. O. Manley, P. Manley and J. Saylor, Numerical studies of the four-day oscillation in Lake Champlain, *J. Geophys. Res., 103,* 18,425-18,436, 1998.

Manley, T.O., K. Hunkins, J. Saylor, G. Miller and P. Manley, Aspects of summertime and wintertime hydrodynamics of Lake Champlain, *this volume.*

Marinov, I., Wind-generated oscillations in Lake Champlain, Senior thesis, Middlebury College, Middlebury, Vt., 1998.

Mendelsohn, D. L., T. Isaji and H. Rines, Hydrodynamic and Water Quality Modeling of Lake Champlain. Final Report prepared for Lak Champlain Management Conference, c/o Water Management Division, U.S. Environ-mental Protection Agency, Region I, JFK Building, Boston, Mass. 02203.

Myer, G. E. and G. K. Gruendling, *Limnology of Lake Champlain, Lake Champlain Basin Study* No. 30, 407 pp., New England River Basins Commission, Burlington, Vt., 1979.

Prigo, R. B., T. O. Manley and B. S. H. Connell, Linear, one-dimensional models of the surface and internal standing waves for a long and narrow lake, *Am. J. Phys.,* 64(3), 288-300, 1996.

Turner, J. S., *Buoyancy Effects in Fluids,* 367 pp., Cambridge University Press, London, 1973.

Whitham, G. B., *Linear and Nonlinear Waves,* 636 pp., John Wiley, New York, 1974.

Gravity Currents and Internal Bores in Lake Champlain

James H. Saylor, Gerald Miller, Kenneth Hunkins, Thomas O. Manley, and Patricia Manley

ABSTRACT

The shape of lake basins, with a large variety of configurations, plays a strong role in determining the current and circulation patterns within them. Elongated and narrow basins such as Lake Champlain can exhibit extremely large thermocline displacements and oscillations in response to wind forcing during seasons of density stratification. Time series of currents and temperature variations in Lake Champlain were recorded during the season of thermal stratification for three consecutive years. The measurements were made using arrays of Acoustic Doppler Current Profilers and thermistor chains moored along the lake's thalweg. The long axis of the lake is oriented approximately north-south and is nearly 120 km long. The lake is about 6.3 km wide and has a mean depth of about 30 m. The depth decreases monotonically toward its north end from a maximum depth of 120 m at a location in the south. During seasons of weak thermal stratification in early summer and fall, strong winds from the south were observed to transport much of the less dense surface layer toward the north end of the basin and cause upwelling in the south. The resulting density distribution is similar to that observed in a lock exchange flow. A gravity current flows northward along the lake floor after the wind stress relaxes. The propagation of the current was monitored as it progressed through instrument moorings placed in the lake's shoaling north end. Evolution of the gravity currents into bore-like waves traveling on weak near bottom stratification occurred at the northernmost measurement location.

INTRODUCTION

Gravity currents, which are caused by the essential difference in density between two fluids, are common in the atmosphere, oceans, and lakes. The ubiquity of gravity currents observed in both nature and the laboratory is wonderfully illustrated by Simpson (1997). In this paper we present evidence of their frequent occurrence in Lake Champlain during seasons of weak density stratification of lake water during spring and fall. The gravity currents observed have been caused in all cases by similar wind forcing events that produce a water density gradient along the longitudinal axis of the lake.

The shape of lake basins plays a major role in determining current and circulation patterns within them. Elongated, deep, and narrow basins such as Lake Champlain are often found in formerly glaciated areas. These lakes have special interest because their currents are restricted to mainly one-dimensional flow, simplifying both their observational and theoretical investigation. Such lakes often exhibit extremely large thermocline displacements and oscillations in response to wind forcing along their longitudinal axis during seasons of thermal stratification.

It is now well known that the dominant cause of high velocity currents observed in Lake Champlain during the density-stratified season is the internal seiche. Studies reported by Myer and Gruendling (1979), Prigo et al. (1996), Hunkins et al. (1998), and T.O. Manley et al. (1999) have provided detailed information on lake basin configuration and bathymetry, observed and modeled internal seiche periods, and statistics on the measured and calculated current velocities. The internal seiche has a period of approximately four days (Figure 1), though varying somewhat with the seasonally-changing upper and lower layer thicknesses and temperatures. The internal seiche, together with lower frequency currents generated by meteorological forcing, cause a broad range of coherent motions between mooring locations. The measured amplitudes of the internal waves can be very large (5 to 10 m or more), and they can generate current speeds of more than 40 cm s^{-1} in either layer.

The main basin of Lake Champlain is nearly 120 km in length, averages 6.3 km in width, and has a mean depth of about 30 m (Figure 2). Its greatest depths are in the south end in the vicinity of Thompsons Point, and it shoals monotonically toward the north. The thalweg of the nearly north-south oriented basin forms the boundary between the states of New York and Vermont. Current meters and temperature recorders were installed along its major axis during the summer stratified season of several years, 1992 through 1994. Two moorings, one at Thompsons Point and the other at Valcour Island (Figure 2), were in place in 1992 and 1993, and each included an Acoustic Doppler Current Profiler (ADCP) to measure current profiles. In both years we observed what appeared as a gravity current propagating northward past the Valcour Island site (Figure 3). The gravity current always followed periods of strong wind stress directed toward the north and only occurred at times when the lake water stratification was weak. Two very similar-looking appearances of the current were observed in the early summer of each year, but the current passed only one station in the north end of the lake so there was not enough along-the-axis data to determine the nature and speed of this northward-propagating waveform.

In 1994, four moorings with ADCP meters were placed along Lake Champlain's major axis (Figure 2) to determine the speed of the advancing gravity current. The mooring at Valcour Island was redeployed, another was placed to the south of Valcour Island (Colchester), and two more were located to the north (Cumberland Head and Pt. AuRoche). Two events of cold water, more dense than the water above, traveling northward along the bottom were observed with enough resolution to estimate the wave speed. In one of these events the characteristics at the northernmost observation suggested an undular bore as the leading segment of flow. With the right set of conditions, bores are generated by gravity current flows (Simpson, 1997).

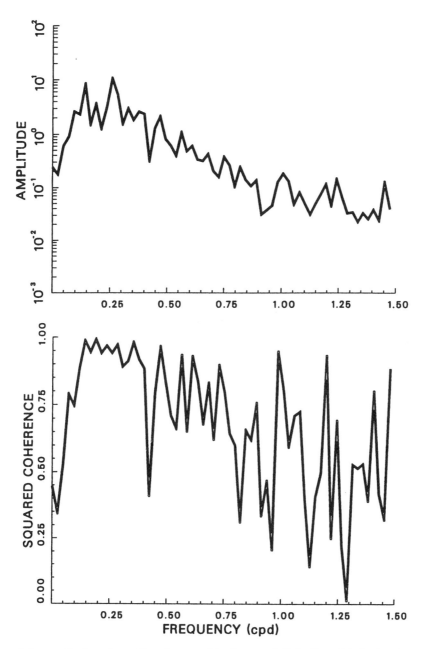

Figure 1. A normalized spectrum of low-frequency kinetic energy in Lake Champlain for the mid-June to mid-September recording period in 1994. The coherence is computed for velocity records from two of the closely spaced moorings. The internal seiche with period near 4 hrs and meteorological forcing provide a broad spectral peak of coherent motions.

138   Gravity Currents and Internal Bores

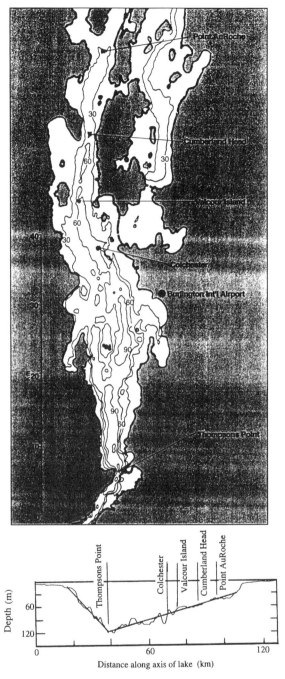

Figure 2. Location map showing lake basin bathymetry and mooring locations. At the bottom is a depth profile along the thalweg of Lake Champlain.

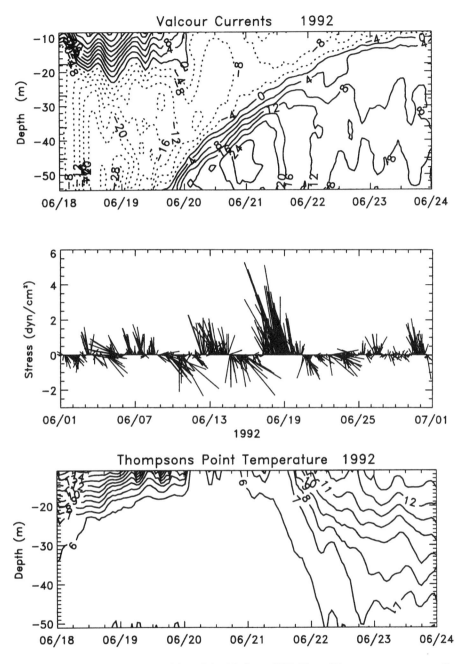

Figure 3. Gravity current passing Valcour Island in June, 1992. The solid contours are currents directed toward the north, the speeds are in cm s^{-1}. Also shown are meteorological observations at Burlington International Airport during June and upwelling at Thompsons Point associated with the gravity current excitation.

DATA COLLECTION

The data were collected using moorings essentially the same as those described in T.O. Manley et al. (1999) with one difference; in 1994, all of the ADCP meters were placed at the top of the taut-wire subsurface string, transmitting the acoustic signals toward the lake bottom. All of the moorings included thermistor strings, either in a single mooring line where depth permitted, or on a second upright leg of a U-shaped mooring. The temperature profilers had 11 thermistors spaced at 4-m separations. Locations of the moorings and lengths of the data recordings at each of them are given in Table 1.

The ADCP meters measure currents by analyzing time-gated reflected signals from acoustic pulses transmitted in a narrow beam into the water column. The pulses are backscattered from particles, animals, plants, or other inhomogeneities in the water column. The reflected acoustic signals are Doppler shifted proportional to the velocity of the backscattering materials. The ADCP's we used transmit acoustic signals from four transducers that are molded into a composite plastic transducer head. The transducers were oriented either 20° or 30° from the vertical in a convex configuration, varying with the particular instrument. The 90° azimuth increments between transducers allow determination of the north and east velocity components. The reflected acoustic signals also provide two estimates of vertical velocity. If the two estimates of vertical velocity are significantly different, the data is corrupt or the assumption of horizontal homogeneity is false (RD Instruments, 1989).

The operating frequency of the ADCP units in Lake Champlain in 1994 was 600 kHz, with a wavelength of 2.5 mm. The beam width of the signal was 2°. Maximum range of the measurements is the water depth between the transducer and the lake bottom multiplied by the cosine of the angle of the acoustic beam from vertical (RD Instruments, 1989). A 30° beam angle corresponds to about 85% of the data between the transducer head and the bottom being usable, the bottom 15% of the data for a downward facing instrument will be contaminated by the shortest path acoustic reflection from the bottom. With a beam angle of 20°, the corresponding data will be usable for 94% of the water column below the instrument. The depth cell size was 1 m, and the number of depth cells varied with the station depth. The pulse length of the signal was 1 m, and all ADCP's were downward facing in 1994 at a depth of 8 m below the surface. From the equations given by RD Instruments (1989), the random errors in ensemble averages were 3.2 cm s^{-1}, 1.9 cm s^{-1}, and 1.4 cm s^{-1} for Point AuRoche, Valcour Island, and Colchester, respectively. The data recorded onboard within the ADCP was downloaded for analysis by software provided by RD Instruments or other data-handling software packages.

A shortcoming of the data collections in all three years involves the lack of very close to the bottom information on the currents and temperature structure. During 1992 and 1993, the ADCP meters were moored as close to the lake floor as possible given the constraint of installing an acoustic release device between the meter and the mooring anchor for retrieval purposes. There were also limitations imposed by using the University of Vermont's Research Vessel *Melosira* for deployments, which required short lift strings to lift the equipment with a winch and A-frame over the stern rail of the vessel. To accomplish the lift, a short (1.3 m) piece of wire cable had to be placed between the release and the ADCP in order to tie off the load and reattach the winch wire to deploy the meter. The transducer head of the ADCP was therefore 4 m above the bottom (the length of the anchor, release, and added wire was about 3 m). With the first bin or two of sampling for the RDI meter not giving useful information, our closest to bottom current measurement was nearly 7 m above the lake floor. In most cases this height was above the front of the advancing gravity current and only in several observations were indications of an elevated, forward nose of the cur-

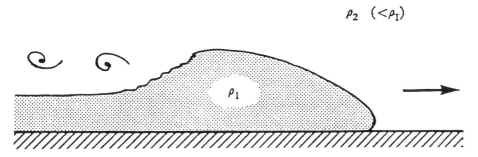

Figure 4. Typical gravity current as observed in laboratory studies (from Keulegan, 1958).

rent detected. Keulegan (1958) gave the shape of gravity currents as they are usually observed in natural or laboratory settings (Figure 4), and it is the close-to-bottom part of the current that is not covered in our observations. Using the ADCP meters in a downward looking mode was subject to the same limitation, the close to the bottom measurements were not available because the shortest path acoustic reflection from the bottom overwhelms the Doppler-shifted particle reflections.

RESULTS

The usual response of a density-stratified lake to small-amplitude, wind-driven upper layer displacements is the generation of internal seiches (Hunkins et al., 1998). The wave length and period of the seiches are governed by the length scales of the basin, layer thickness, density differences, and mode of the seiche initiated. Large, short-length-scale thermocline displacements, however, have been observed to drive nonlinear responses such as surges, bores, and solitary waves in Babine Lake (Farmer, 1978), Loch Ness (Thorpe, 1971), Lake Ontario (Mortimer, 1977), and Seneca Lake (Hunkins and Fliegel, 1973). The nonlinear effects of fully upwelled or downwelled thermoclines and their intersection of the surface and bottom have been reported more sparsely, although surges emanating from full south shore downwelling in Lake Ontario were reported in the aforementioned work of Mortimer (1977). Lake Champlain affords an opportunity to view some of these effects because full upwelling and downwelling occur frequently and predictably.

The depth profile along the thalweg of Lake Champlain is shown in Figure 2. The locations of the 1992 and 93 moorings (Thompsons Point and Valcour Island) and those deployed in 1994 are positioned along the profile. There is considerable vertical exaggeration in the figure, the constant northward slope approximating the actual profile is 0.08°. Response of the lake to strong northward directed wind stress is shown in Figure 3 for the gravity current observed in June 1992. South winds, recorded at the Burlington International Airport, were strong from 17-19 June. The wind direction was nearly constant and along the lake axis for the entire period, and this stress impulse was clearly the dominant wind-forcing event during the month. (The wind stress in this and following figures is simply estimated using a representative drag coefficient and air density, the same values for all months shown). Full upwelling in the weak prior stratification occurred at Thompsons Point, while at Valcour Island the thermistor within the close-to-the bottom ADCP measured a temperature greater than 9°C. The horizontal density differences produced the gravity current flow illustrated. Every occurrence of the gravity currents observed in the three year span originated from similar wind forcing events. The Champlain Valley, with the

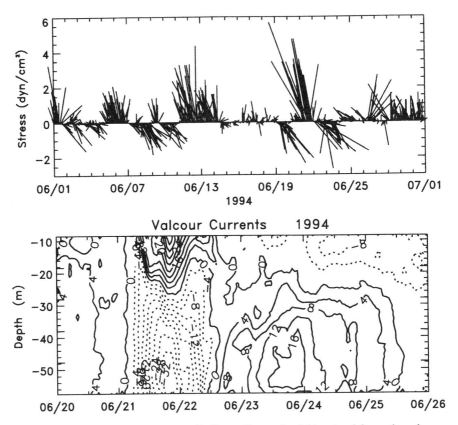

Figure 5a. Meteorological observations at Burlington International Airport and the northward propagation of a gravity current observed at Valcour Island following an episode of strong, northward-directed wind stress in June, 1994. Currents shown as in Fig. 3.

Adirondack Mountains to the west and the Green Mountains to the east, funnels wind along the lake axis. Upwelling at the deep end of the basin and downwelling in the north establishes the density gradient that drives the gravity current northward and up the gradually shoaling lake floor. The conditions that set up the density gradient are clearly illustrated in Figure 3, the south wind pushes the upper layer northward with return flow below. Two or three similar events occurred in the June through mid-July interval in each of the three years of observation.

Two episodes of gravity current excitation in 1994 afforded opportunity to measure the speed of northward propagation past all four moorings. The first of the episodes occurred in June, again in response to an interval of strong northward-directed wind stress (Figure 5). The current was first observed at the Colchester mooring at 0200 hrs on June 22, progressed past the Valcour Island and Cumberland moorings at 0945 and 1730 hrs, respectively, on the same day, and past the northernmost mooring at Point AuRoche, at 0945 on June 23. The speed of the northward moving gravity current over the 30.4 km distance from Colchester to Point AuRoche was 0.266 m s^{-1}. There was indication that the smaller northward stress on 12-13 June initiated a weaker current as well, but only two stations, Colchester and Point

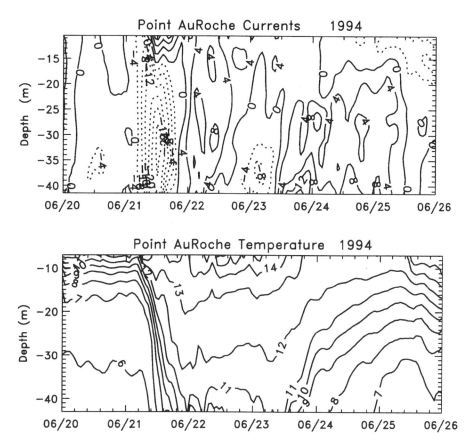

Figure 5b. The June, 1994 gravity current observed at Point AuRoche and water temperature variations associated with it. Currents shown as in Fig. 3.

AuRoche, were in place at the time, and the evidence for its full development was not as certain.

At each station, thermistor chains were also deployed (Figure 6). The shallowest thermistor on each chain was located just below the ADCP, some 8 m below the water surface The chains were 40-m long and therefore only the mooring at Point AuRoche had temperature measurements right at the bottom (Table 1). But, the approximate magnitude of the stratification, the thickness of the thermocline, and verification of the northward propagating gravity current is shown in the records. The stratification prior to the current was similar at all moorings, with the temperature across the thermocline ranging from 6 to 12°C. The thickness of the thermocline varied, but was 20 m or more in the weakly-stratified lake water. The closest-to-bottom thermistor on the chain shows an abrupt decrease in temperature as the gravity current arrives. Comparison of this thermistor also provides evidence of the temperature gradient along the axis of Lake Champlain that is established by south wind. The temperature profiles show the progression of the event as an elevating thermocline as the gravity current passes the mooring, and the following thermocline deepening in its wake after the head passes by.

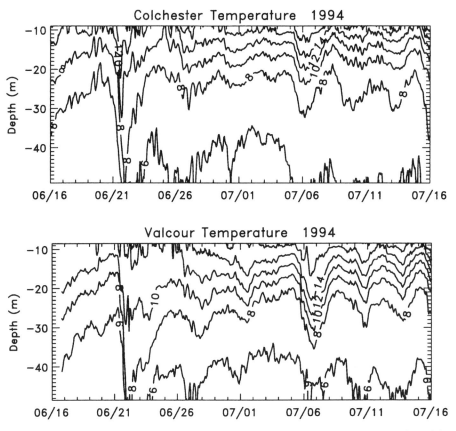

Figure 6a. Water temperatures observed at Colchester and Valcour Island during propagation of the June, 1994 gravity current.

In October 1994 another interval of strong north-directed wind stress initiated a fall season gravity current event. Figure 7 shows a 3-day-long interval of south wind with sufficient energy to force southward all of the hypolimnion water at Point AuRoche (Figure 8). The thermocline deepened at the Colchester and Valcour Island moorings as well, similar to the deepening noted in the June episode. The complete downwelling in the north of the lake was in part due to the weak stratification as fall cooling of the surface water had diminished the upper layer density. The ADCP current records and the temperature recordings again show the northward progression of the gravity current. It arrived at the Colchester mooring at 1500 hrs on October 9, passed Valcour Island at 2000 hrs, and arrived at Point AuRoche at 1930 hrs on October 10. The observed speed over the entire 30.4 km distance between stations 1 and 4 was 0.296 m s^{-1}.

Thermistor chain data indicate a stratification of 8 to 14°C at all of the stations prior to the wind storm. The thermocline was also noticeably thinner, with a thickness of 10 m or less, than it was in June. This fact, coupled with the larger density difference between layers, may account for the faster wave speed observed in October. The elevation of the thermocline as the gravity current passes by looks very much like the June observations. A gradual deepening of the thermocline following the current's passage was also observed.

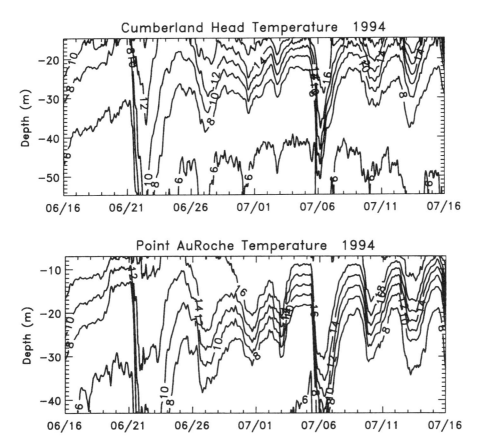

Figure 6b. Water temperatures observed at Cumberland Head and Point AuRoche during propagation of the June, 1994 gravity current.

COMPARISON WITH FIRST-ORDER SIMULATIONS

Gravity currents often occur in nature and can have substantial effects on air and water quality. Simpson (1977) provides examples of dust storms and smoke or haze fronts associated with gravity currents in the atmosphere, and P. Manley et al. (1998) link lake bottom sediment resuspension and transport with gravity current excitation. Experimental studies of salt wedges in rivers and estuaries have been the subject of numerous experimental studies in nature and in the laboratory. Theoretical analysis has also received much interest following the cutting edge work of Karman (1940) and Benjamin (1968). Benjamin referred to gravity currents as "a wedge of heavy fluid intruding into an expanse of lighter fluid", and his solutions giving their flow characteristics have been affirmed by numerous investigators (e.g., Wilkinson and Wood (1972) and Denton (1990) when necessary assumptions are made). The flow in these investigations was identified as a lock exchange flow or as Benjamin (1968) labeled it "a lock exchange density current". Much information has been published on gravity (or "density") currents, surges, and bores, with some of the

TABLE 1. Locations and water depths at mooring sites, length of recording intervals, and depth ranges of the current and water temperature measurements made in 1994.

Location Latitude Longitude	Water depth	Depth range of current measurement	Depth range of temperature measurement	Length of current measurement
Colchester 44 33.3450 N 73 20.7332 W	82 m	8-78 m	9-49 m	1200 6-14-94 to 1115 10-18-94
Cumberland Head 44 42.0455 N 73 21.6775 W	61.3 m	8-55 m	15-47 m	1200 5-20-94 to 0945 9-22-94
Valcour Island 44 36.1707 N 73 23.0226 W	64 m	10-60 m	9-49 m	1700 6-16-94 to 1330 10-17-94
Point AuRoche 44 47.7678 N 73 20.2208 W	44 m	8-41 m	7-39 m	1200 6-14-94 to 2300 10-15-94

differing terms being used to describe like phenomena. The difficulty arises in distinguishing and properly categorizing the wave forms in the natural environment.

Professor Yih (1980) defined an abrupt change in depth in a two-layered fluid as an internal hydraulic jump, and described the intrusion of a heavy fluid into a lighter one. The process was termed a density current in hydraulic texts, but Yih noted that the difference in specific weights caused these currents and that they should properly be termed "gravity currents". Kranenburg (1978) labeled the discontinuities as "density fronts", while Wood and Simpson (1984) defined moving hydraulic jumps as surges. Farmer (1978) described surges in Babine Lake and studied their northward propagation and other characteristics with instrument arrays. Farmer's observations are similar to our observations in Lake Champlain. How do the most fundamental theories compare with the Lake Champlain observations?

Benjamin (1968) analyzed the front of a frictionless gravity current in terms of a "cavity flow" displacing a fluid beneath it. If one end of a long, closed water-filled channel is removed, the water at the lower level will run out, being replaced by an air cavity flowing in above it. The air cavity will advance along the channel at a steady rate. Benjamin used two equations involving the velocity and depth of the outflowing layer (the continuity of mass equation and Bernouilli's equation applied along the interface) to describe the flow regime. Bernouilli's equation expresses the constancy of total energy (pressure, potential and kinetic) along a streamline. Because the fluid is assumed inviscid, the "flow force" or total pressure force plus the momentum flux per unit length is also conserved. Using of Bernouilli's equation, the only energy conserving solution possible occurs when the outflow fills exactly one-half of the channel. The case where the outflow fills more than one-half the channel was found to be impossible, while flows where the outflow filled less than one-half the depth resulted in a loss of more energy at the front than was available from wave radiation so that breaking at the front must occur. Benjamin also showed that the fractional depth of the outflow layer, i.e., h/H = the depth of the outflow / total depth = φ say, plays an impor-

TABLE 2. Comparison of observed and computed gravity current propagation speeds using lock-exchange and one layer energy conservation methods.

Event	June 1994	October 1994
Thermocline	6° - 12°C	8° - 14°C
	ρ_1=999.4996	ρ_1=999.2464
	ρ_2=999.9430	ρ_1=999.8509
Reduced gravity	g'=0.00435	g'=0.00593
Total Distance Traveled	30.4 km	30.4 km
Time Interval - Station 1 to 4	31.75 hr	28.5 hr
Avg. Height (h_4) / Avg. Depth (d)	$\frac{30 \text{ m}}{58 \text{ m}} = .52$	$\frac{35.66 \text{ m}}{57.33 \text{ m}} = .62$
Observed Speed	.266 m s^{-1}	.296 m s^{-1}
Yih	.251 m s^{-1}	.292 m s^{-1}
Denton Lower layer conservation	.247 m s^{-1}	.262 m s^{-1}
Denton Upper layer conservation	.240 m s^{-1}	.222 m s^{-1}

tant role in characterizing the flow. The Froude number, $U/(g'h)^{1/2}$, where U is the velocity of the air cavity along the channel and g' is the reduced gravity $g' = g \Delta \rho/\rho$, is $(2)^{-1/2}$ at $\varphi = 0.5$, the total energy conservation case. It is near 1.0 at $\varphi = 0.2$ and increases toward $(2)^{1/2}$ as φ approaches zero. Benjamin also calculated the approximate shape of the interface and found that the slope at the stagnation point on the ground was 60°.

Yih's (1980) lock exchange theory gives results shown in Table 2. His theory was a simple application of the principles of energy conservation. He assumed that two fluids of different density are separated by a barrier and that at time t = 0, the barrier is removed. The flow is analyzed assuming energy conservation at a time t = δt later. If the velocity of the exchange flow is U, the fronts propagate a distance Uδt. Using the changes in kinetic and potential energy and requiring that energy lost equals energy gained, Yih (1980) derived for U, the velocity of the exchange,

$$U = .707 \sqrt{gd \frac{\rho_2 - \rho_1}{\rho_2 + \rho_1}}. \quad (1)$$

This formulation uses several first-order approximations. First, and most obvious, there is no energy loss due to friction or turbulence, i.e., no dissipative effects. He also assumed that no mixing takes place between the two fluids and that each occupies half of the water column after the exchange. No surface deflection, which occurs experimentally and is present in most theories of hydraulic jumps, was considered. And finally, the Boussinesq approximation, allowing no variability in the horizontal except at the jump, is used, though we note that this approximation is universal to almost all gravity current theories. The model gives a front velocity the same as that obtained by Benjamin when the lower layer fills exactly one-half the channel. But even with these simplifying approximations, the Yih and Benjamin total energy conserving models fit our Lake Champlain observations very well. For the June gravity current, the difference between the observed and computed propagation speed is only 5.6%. For the October event, the difference is only 1.4%. The relative errors are remarkably small. We note that the thermocline thickness was less in October and the stratification more closely approximates a two layer model.

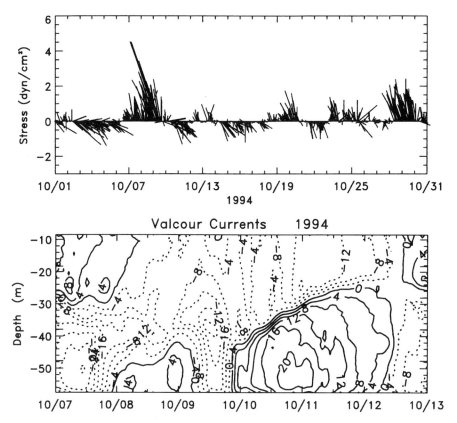

Figure 7a. Meteorological observations at Burlington International Airport and the northward propagation of a gravity current observed at Valcour Island following an episode of strong, northward-directed wind stress in October, 1994. Currents shown as in Fig. 3.

Yih (1980) also examined the situation where a gravity current propagates into an infinitely-deep layer. He used the same energy conservation principles in this study as well, following the method first applied by Karman (1940). Benjamin (1968) noted that Karman's approach of using two Bernoulli equations in the infinite depth intrusion should be replaced by a method using one energy equation and one momentum equation to conserve flow force. Denton (1990) used the approach of Benjamin, and of Wilkinson and Wood (1972), by taking a momentum balance over the fluid depth and assuming energy conservation in only one of the two layers. This method gave solutions for an intrusion with a height other than half the depth. Denton calculated the Froude number for energy conservation in the upper and lower layers separately. He also assigned conditions under which this approach was valid for a range of values of the bore height compared to the initial thickness of the bottom layer.

Denton(1990) examined three cases of "density front" propagation in two-layer fluids. His more robust analysis is also useful in understanding the Lake Champlain observations In all three examples he presented, there were two main dependent variables: the head velocity, U, and the free surface deflection, ε (Figure 9). An energy loss in one of the two

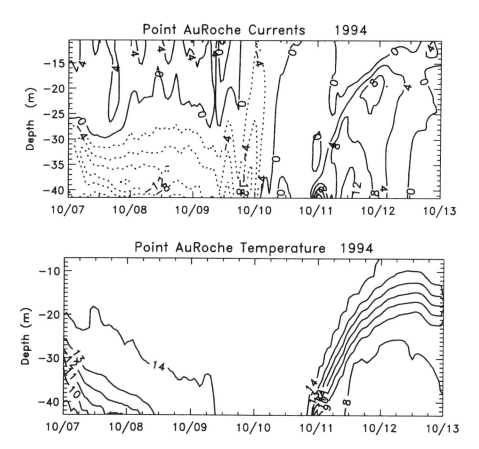

Figure 7b. The October, 1994 gravity current observed at Point AuRoche and water temperature variations associated with it. Currents shown as in Fig. 3.

layers was assumed. He first studied a bottom density current as analyzed by Simpson and Britter (1979), with a well-defined nose height, h_5, and mixing occurring behind the head. He also examined an internal bore propagating along the interface of an initially quiescent two-layer stratification, an example previously explored by Wood and Simpson (1984). And lastly, he examined the case of an internal hydraulic jump that is created behind a towed obstacle, again an analysis also considered by Wood and Simpson. Denton noted that an internal bore will behave like a density current if the initial thickness of the bottom layer, h_2, is small. We investigate his bore solutions for a thin bottom layer for application to Lake Champlain.

Denton (1990) did not consider either mixing or bottom friction individually; dissipative effects were lumped together in energy loss coefficients. He assumed hydrostatic pressure conditions upstream and downstream of the jump and applied the Boussinesq approximation when combining the momentum flow rate equations. Inertial terms of order ε/h_t and gravitational terms of order (ε/h_t), where h_t is the total depth, were neglected in his formulations. He then used a Bernoulli equation along a streamline in each of the upper or lower

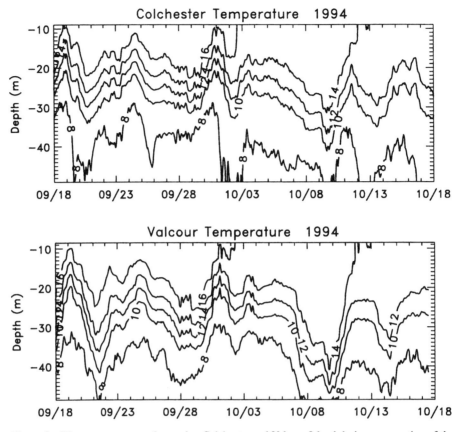

Figure 8a. Water temperatures observed at Colchester and Valcour Island during propagation of the October, 1994 gravity current.

layers, conserving energy in the appropriate layer. Solving the resulting equations for the frontal speed by eliminating the surface deflection, ε, gave the relations:

$$F_4 \sqrt{\frac{\left(h_4^* + h_2^*\right)\left(1 - h_4^*\right)^2}{\left(h_4^{*2} 2 h_2^* - 3 h_4^* h_2^*\right)}} \tag{2}$$

for upper layer energy conservation, and

$$F_4 \sqrt{\frac{\left(2 - h_4^* - h_2^*\right)\left(1 - h_4^*\right) h_4^*}{\left[h_4^*\left(1 + h_4^*\right) + h_2^* - 3 h_2^* h_4^*\right]}} \tag{3}$$

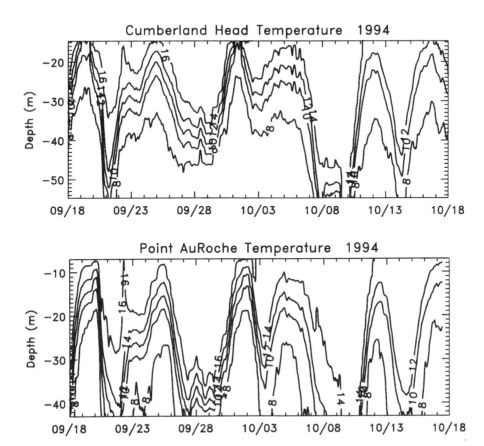

Figure 8b. Water temperatures observed at Cumberland Head and Point AuRoche during propagation of the October, 1994 gravity current.

for lower layer energy conservation, where $F_4 = U/(g'h_4)^{-1/2}$ is the Froude number, and the depths have been normalized as $h^* = h/h_t$. As noted earlier, the Bernoulli equations are based on the assumptions that the flow is steady, inviscid, and incompressible, and the equations are applied along streamlines only.

If the bottom layer is much smaller than the bore height, Denton showed that the method conserving energy in the lower layer gave results closest to those obtained experimentally. The density front propagation speeds predicted using his analysis are given in Table 2. This more complex analysis also gives results rather close to those observed. The computed speed of the June current front is within 7.1% of the observed speed while the computed speed of the October current front varies by 11.5% from the speed observed. Lower layer energy conservation was also used in the gravity current theories of Wilkinson and Wood (1972) and Kranenburg (1978). If we use Denton's model requiring energy conservation in the upper layer, the results are different. The differences are 9.8% and 25% for the June and October currents, respectively. Clearly, the selection of energy conservation in the lower layer gives a closer approximation of the propagation speed for both the spring and fall density fronts. The result is not surprising because Denton derived his lower layer energy

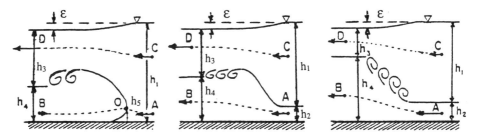

Figure 9. Three forms of a density front: (a) a bottom density current, (b) an internal bore, and (c) an internal hydraulic jump (from Denton, 1990).

conservation model with the assumption that the bore height was much larger than the initial lower layer height. Upper layer energy conservation, as originally proposed by Wood and Simpson (1984), gives speeds rather close to the observed speed of the June current, but more distant than the observed speed of the October current (Table 2). However, neither lower nor upper layer energy conservation assumptions, allowing for an energy loss in either the converging or diverging layer, approximate the Lake Champlain observations as well as Yih's (1980) lock-exchange (or Benjamin's h = .5d gravity current) theory based on total energy conservation.

Mass and momentum conservation across an internal bore with the assumption that energy is dissipated in both layers yields a range of possible bore speeds. Li and Cummins (1998) showed that the upper speed limit is given by the assumption of no energy dissipation in the contracting layer while the lower limit is given by the assumption of no energy dissipation in the expanding layer. The upper and lower limits agree within a few percent unless the expanding layer is shallow upstream and the internal bore propagates as a gravity current, as it does in Lake Champlain.

DISCUSSION

In the atmosphere, the presence of low-level stratification is particularly favorable for the evolution of gravity currents into many other phenomena: turbulent bores, undular bores, and ultimately solitary waves or solitons (Christie et al., 1978; Simpson, 1997). In laboratory and numerical experiments, bores are generated as gravity currents encounter strong stratification near the ground (e.g., Maxworthy, 1980; Smith, 1988; Rottman and Simpson, 1989). These can occur as a disturbance generated on the stable layer, as a gravity current is modified by interacting with a stable layer, and as a gravity current collapses. In Lake Champlain, the gravity current is initiated by the rush of less dense water toward the north end of the lake and the upwelling of dense water in the south end caused by strong south winds. Although lateral movement of the stratified water mass is substantial, we note that it is difficult to completely remove a stable layer from the lake floor. Bottom irregularities cause water to be trapped in holes and depressions. Sediment pore water and the sediment itself serves as a reservoir of low temperature water that contributes to stability. With a mechanism for formation of bores being the compaction, suppression, and thinning of the thermocline by a large magnitude gravity current, Denton (1990) provided an analysis to consider the effects of this stability. But the simulation of the density front speed is no better than that obtained using the simplest of lock-exchange models.

The initial evolution of a lock-exchange flow, where a fixed volume of dense fluid held behind a barrier is released into another fluid of lesser density, has also been studied experimentally by Huppert and Simpson (1980). In the initial slumping phase they found that the length of the current increased almost linearly with time, i.e., with constant velocity of the current's head. This phase precedes a self-similar regime in which a homogeneous gravity current that is dominated by inertial and buoyancy forces can be modeled by the shallow water equations, which admit a long-time asymptotic similarity solution. The approach to similarity was studied by Rottman and Simpson (1983) with experiments and a model based on the shallow-water equations. An interesting feature of their work was the formation of an internal bore if the initial height of the gravity current behind the lock was more than 70% of the depth of the ambient fluid. But until this bore overtook the front, the height of the current remained constant, resulting in a constant velocity of the nose, and therefore, a linear increase in the length of the current.

Many processes contribute to the evolution of a gravity current after initiation. Rottman and Simpson (1989) identified three types of bore that are found by towing an obstacle through a two-layer fluid: smooth undular, some mixing but still undular, and propagation as a gravity current bore. The classifications are bounded by values of the height of the bore to the height of the initial lower layer. Smooth undular had typical values between 1 and 2, some mixing but still undular had values between 2 and 4, and the appearance of a gravity current for values greater than 4. For the latter case they concluded that mixing dominates the motion, completely obliterating any undulations, and the motion appears as the typical gravity current. Rottman and Simpson also found that while generating a bore with the release of a gravity current, the bore speed was the same as the gravity current and that in some cases the two were indistinguishable. Smith (1988) and Fulton et al. (1990) emphasize the difficulty of distinguishing bores from gravity currents in atmospheric observations, despite the fact that the two phenomena are dynamically dissimilar, partly because of the wide spectrum of motions between bore and gravity current structures that can occur. Bores and gravity currents in the atmosphere are both accompanied by a wind shift toward the direction of movement of the disturbance. However, cooling associated with a gravity current is due to advection, whereas adiabatic ascent is the cause for cooling in case of a bore. Gravity currents and internal bores in lakes can also attain very similar forms; in nature it is not always possible to clearly separate the phenomena.

Each case of gravity current excitation in Lake Champlain is different because the winds forcing the initial thermocline set up vary as does the stratification characteristics. But in each of the three years of measurements examined, gravity currents occurred. The form of the current was always similar in passing the Valcour Island site, suggesting that the location is close to the origin of the current. As it propagates up the modest slope into the shallowing north end of the lake, we would expect some modifications of its shape as the deep channel's width varies and as the total water depth becomes limiting. Simpson (1997) observed in the laboratory that travel of a gravity current into shallow water can suppress the head waves of the current and slow or stop mixing at the interface between the two fluids. This occurred as the depth of the gravity current head became half as large as the local depth, and the current interface downstream becomes just about parallel with the water surface. An example of the changing form as the current progresses along the channel is shown in the ADCP recording at Point AuRoche on 6-24-94 (Figure 7). This more undular-bore-like appearance may have been caused by a reduction in the energy of the current as it approached the northern limit of the measurements.

Two intervals of high speed bottom currents occur when gravity currents in Lake Champlain are initiated. Wind toward the north forces an accumulation of less dense water in the lake's north end, depressing the thermocline and causing a surge of lower layer water southward. Southward directed lower layer currents can be quite large, sufficient in magnitude to initiate the suspension of bottom sediments (P. Manley et al., 1999). The surge of lower layer water has been modeled by Hunkins et al. (1998), who showed that as the surge moves southward after initiation in the north, a shock wave of lower layer expansion can grow rapidly. In the extreme cases we have examined in this paper, the lower layer ultimately fills the southern reaches of the lake. Farmer (1978) reported similar surges in Babine Lake and their subsequent evolution into solitary waves and solitons as had previously been observed by Hunkins and Fliegel (1973) in Seneca Lake. The surge in response to intense northward wind stress in Lake Champlain establishes massive redistribution of the two layers during seasons of weak stratification, generating a northward flowing gravity current that also drives strong bottom currents, again with velocities capable of mobilizing bottom sediments.

In summary, a strong pulse of south wind is responsible for establishing a water density gradient from south to north. During seasons of weak density stratification, nearly complete upwelling occurs in the south, and downwelling occurs in the north. The wedge-shaped bathymetry of Lake Champlain no doubt accentuates the ease and frequency with which this distribution can be accomplished. A gravity current flows northward when the wind force is relaxed. The speed with which the gravity current travels is well described by numerical predictions of the simplest lock-exchange experiments. This happens in spite of the heating and cooling of the two layers during the wind storms that ventilate the deep layer, and other energy dissipating processes. The absence of any observance of wave reflection from the north end attests to the dissipation. Utilizing a first order model that allows for energy conservation in only one of the two layers does not improve the prediction of frontal speed, although it does appear likely that an internal bore propagating on stable bottom stratification is a reasonable interpretation of the observations. A thin bottom layer would cause the bore to have the appearance of a gravity current.

Acknowledgements. The authors are indebted to Mr. Richard Furbush, captain of the University of Vermont's Research Vessel *Melosira*, for invaluable and expert assistance in the deployment and recovery of instruments used in the Lake Champlain studies. Brian Nowak provided much support in the collection and interpretations of data used in the paper. Drs. Charles Adams and C.R. Murthy provided very helpful comments in their reviews of the paper. The paper is Great Lakes Environmental Research Laboratory Contribution No. 1102.

REFERENCES

Benjamin, T. B., Gravity currents and related phenomena, *J. Fluid Mechanics*, 31(2), 209-248, 1968.

Christie, D. R., K. J. Muirhead, and A. L. Hales, On solitary waves in the atmosphere, *J. Atmos. Sci.*, 35, 805-825, 1978.

Denton, R. A., Accounting for density front energy losses, *J. Hydr. Engn.*, 116(2), 270-275, 1990.

Farmer, D. M., Observations of long non-linear internal waves in a lake, *J. Phys. Oceanography*, 8(1), 63-73, 1978.

Fulton, R., D. S. Zrnic, and R. J. Doviac, Initiation of a solitary wave family in the demise of a nocturnal thunderstorm density current, *J. Atmos. Sci.*, 47, 319-337, 1990.

Hunkins, K., and M. Fliegel, Internal undular surges in Seneca Lake: a natural occurrence of solitons, *J Geophys. Res.*, 78, 539-548, 1973.

Hunkins, K., T. O. Manley, P. Manley, and J. Saylor, Numerical studies of the four-day oscillation in Lake Champlain, *J. Geophys. Res.*, 103(C9), 18,425-18,436, 1998.

Huppert, H. E., and J. E. Simpson, The slumping of gravity currents, *J Fluid Mech.*, 99, 785-799, 1980.

Karman, Th. v., The engineer grapples with non-linear problems, *Bull. Am. Math. Soc.*, 46, 615-683, 1940.

Keulegan, G. H., The motion of saline fronts in still water, *Nat. Bur. Stand. Rept.* 5831, 1958.

Kranenburg, C., Internal fronts in two-layer flow. *J. Hydr. Div. ASCE*, 104(10), 1449-1453, 1978.

Li, M., and P. F Cummins, A note on the theory of internal bores. *Dynamics of Atmospheres and Oceans*, 28, 1-7, 1998.

Manley, P. L., T. O. Manley, K. Hunkins, and J. Saylor, Sediment deposition and resuspension in Lake Champlain, this volume, 1999.

Manley, T. O., K. L. Hunkins, J. H. Saylor, G. S. Miller and P. L. Manley, Aspects of summertime and wintertime hydrodynamics of Lake Champlain, this volume, 1999.

Maxworthy, T., On the formation of nonlinear internal waves from the gravitational collapse of mixed regions in two or three dimensions, *J. Fluid Mech.*, 96, 47-64, 1980.

Mortimer, C. H., Internal waves observed in Lake Ontario during the International Field Year for the Great Lakes (IFYGL) 1972: part 1, Descriptive survey and preliminary interpretation of near-inertial oscillations in terms of linear channel-wave models, Spec. Rept. No. 32, Center for Great Lakes Studies, Unv. Wisc. Milwaukee, 1977.

Myer, G., and G. K. Gruendling, *Limnology of Lake Champlain*, Lake Champlain Basin Study No. 30, New England River Basins Commission, Boston, Massachusetts 02109, 1979.

Prigo, R. B., T. O. Manley, and B. S. H. Connell, Linear, one-dimensional models of the surface and internal standing waves for a long and narrow lake, *Am. J Phys.*, 64(3), 288-300, 1996.

RD Instruments,. *ADCP principles of operation: A practical primer*, RD Instruments Inc., San Diego, California 92131, 38 pp. 1989.

Rottman, J. W., and J. E. Simpson, The initial development of gravity currents from fixed-volume releases of heavy fluids, *J. Fluid Mech.*, 135, 95-110, 1983.

Rottman, J. W., and J. E. Simpson, The formation of internal bores in the atmosphere: A laboratory model, *Quart. J. Roy. Meteor. Soc.*, 115(488), 941-963, 1989.

Simpson, J. E., *Gravity Currents in the Environment and in the Laboratory*, 2nd ed., 244 pp., Cambridge Press, 1997.

Simpson, J. E., and R. E. Britter, The dynamics of the head of a gravity current advancing over a horizontal surface, *J. Fluid Mech*, 140, 329-342, 1979.

Smith, R. K., Travelling waves and bores in the lower atmosphere: The "Morning Glory" and related phenomena. *Earth Science Reviews*, 25, Elsevier Science, 267-290, 1988.

Thorpe, S. A., Asymmetry of the internal seiche in Loch Ness, *Nature*, 231, 306-308, 1971.

Wilkinson, D. L. and I. R. Wood, Some observations on the motion of the head of a density current, *J. Hydraul. Res.*, 10(3), 305-324, 1972.

Wood, I. R., and J. E. Simpson,. Jumps in layered miscible fluids, *J. Fluid Mech.*, 140, 329-342, 1984.

Yih, C. S., *Stratified Flows*, Academic Press, New York, N. Y., 1980.

Sediment Deposition and Resuspension in Lake Champlain

Patricia L. Manley, Thomas O. Manley, James H. Saylor, and Kenneth L. Hunkins

ABSTRACT

High-resolution side-scan sonar surveys conducted in 1991, 1994 and 1996, permitted assessment of a large furrow field located east of Valcour Island, Lake Champlain. These furrows have a width-spacing ratio of 1:4 - 1:9 which classifies them Type 1A. Furrow lengths range from 16 to 828 m with over 50% of them less than 200 m. Morphological differences can be seen across the furrow field from west to east, with width-spacing ratios increasing to the west as a bathymetry becomes slightly deeper. Several mooring configurations and arrays were deployed within the furrow field including thermistor chains, sediment traps and current profilers. In addition, synoptic views from CTD/OBS surveys were obtained in 1991. The mooring in 1994 contained stereo cameras, sediment traps, thermistor chain, and an Acoustic Doppler Current Profiler (ADCP). The cameras took pictures for 23 days, at 4-hour intervals before instrument malfunction. All other apparatus operated for 4 months between June and October 1994. Correlation between the thermistor chain and the ADCP permit analysis of currents near the bottom boundary layer. Comparison of stereo images to current data gives quantifiable and visible information of erosion and deposition intervals. A high-speed current event was correlated to erosion within the bottom camera area. Other high-speed events, not documented by photographs, recorded by the ADCP suggest additional erosional events within the survey region. Documentation of these erosional events indicates that furrow development occurs via high-speed internal-seiche driven current activity within the bottom boundary layer separated by longer periods of deposition.

INTRODUCTION

Current-formed sedimentary bedforms have been studied in detailed within diverse settings deep-sea, continental rises, continental shelves, estuaries, large lakes and rivers. It is important to understand the development and evolution of these features as they form at the sediment-water interface and have been widely used to infer depositional environments in both modern and ancient sediments. The recognition of sediment bedforms in large lakes has been increasing in recent years as more swath-mapping and investigation of lake bottom morphology has been occurring. One particular bedform, sediment furrows, have been identified and studied in several of the Great Lakes in particular Lake Superior and Lake Ontario. Sediment furrows develop in areas of net deposition where directionally stable intermittent high-speed currents erode the bottom sediment (Flood, 1983; 1989). These strong bottom currents have the potential to resuspend sediments and associated contaminants.

Investigations of the thermal structure and hydrodynamics of Lake Champlain have documented the existence of a wind-forced internal seiche (Myer and Gruendling, 1979; Manley and Manley, 1993; Manley et al., 1993; Saylor et al., 1993; Connell, 1994; Prigo et al., 1996; Hunkins et al., 1998; Marinov, 1998; Marinov et al., 1988; Manley et al., this volume). The internal seiche oscillates along the long axis of the lake with a period of approximately 4.5 days (Myer and Gruendling, 1979; Manley et al., this volume). This long-period internal standing wave operates along the metalimnion that exists from the onset of stratification (early spring) through late fall.

The lake has a natural resonance period similar to that of the atmospheric disturbances (predominantly from the south) which drives lake circulation. During time periods when stratification is present, the internal seiche is the dominant driving mechanism of water movement. Periods of severe atmospheric storms causes depression of the metalimnion which can exceed depths of well over 60 m and can generate bottom current speeds of 20 - 50 cm/s. It is these high speed bottom current events that have the potential to erode and deposit bottom sediments of Lake Champlain.

Side-scan sonar investigations were conducted at hydrographic mooring locales to determine if any significant sedimentary bedforms were located on the lake bottom. The subject of this paper is the discovery and study of a large sedimentary furrow field east of Valcour Island, Lake Champlain.

VALCOUR ISLAND STUDY SITE

Lake Champlain is the sixth largest freshwater lake in the United States having a mean elevation of 28.2 m above sea level and a surficial area of 1133 km^2. Located along the New York - Vermont boarder, it has been historically and still is a major water source for the northeast as well as a transportation route and recreational source. A long linear lake oriented north-south, Lake Champlain is ~170 km long and has a maximum width of 19 km. Morphologically, the lake can be divided into three distinct sections that are all interconnected (Manley et al., this volume). The largest section is called the Main Lake which extends from Rouses Point near the Canadian-US border to the Crown Point Bridge. This section reaches a maximum depth of 122 m near Thompsons Point and is broadest near Burlington, VT. The study site, located east of Valcour Island, is located within the Main Lake (Figure 1). The Restricted Arm is located to the east of the Main Lake and is connected by three narrow passages. The third section resembles a river. Having an maximum depth of 7 m and barely 1.5 km wide, it extends from Crown Point Bridge to Whitehall (NY) where it is connected via the Champlain Barge Canal to the Hudson River (Figure 1).

Highly metamorphosed rocks of the Adirondack Mountains with sandstone and carbonate rocks are located on the western margin of the lake, whereas carbonates, sandstones and shales

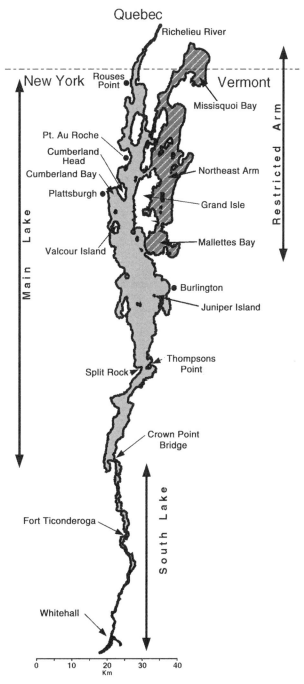

Figure 1. Location map of Lake Champlain. Lake is divided into three regions, the Main Lake, the Restricted Arm and the South Lake. This research was conducted in the Main Lake east of Valcour Island.

border the eastern side. The lake is underlain by several N-S oriented faults (Hunt, 1972; Hunt, 1977; Hunt, 1981) which are crossed by an east-striking sequence of high-angle faults. Little is known about the lake from the Paleozoic to the Lake Pleistocene, though it is assumed that the basin served as a river valley and was formed during the last glacial period (Chapman, 1942; Connally and Sirkin, 1973; Teller, 1987). During the last deglaciation, several stages occurred which preceded the present-day Lake Champlain. Approximately 13,000 ybp (Coveville stage), a large pro-glacial lake known as Lake Vermont (which filled the Champlain valley to an elevation of 183 m) drained southward through an outlet channel at Coveville, NY. As the glacier retreated northward, the lake level slowly dropped and a more northerly outlet near Fort Ann, NY formed (Fort Ann stage). After the glacier had moved significantly northward, the effect of isostasy aided the lowering of the lake level until it was continuous with the St. Lawrence lowlands and a marine excursion forming the Champlain Sea occurred at 10,200 ybp. The Champlain Sea existed for ~2,000 years until the northern portion of the valley rose through isostatic rebound shutting off the marine water and allowing for a freshening of the water and the formation of the present day Lake Champlain.

A natural narrowing of the lake occurs to the north of Valcour Island, as eastward protruding Cumberland Head reduces its width by 70% (Figure 1). Additionally, south of Cumberland Head and east of Valcour Island, steep fault-bound slopes (Hunt, 1981) cause the lake bottom to have a trough shape 2.5 km wide, which is narrower than the 6 km width of the lake to the south (Plate 1).

SEDIMENTARY FURROWS

Sedimentary furrows have been found in diverse sedimentary environments including the continental rise and deep sea (Hollister et al., 1974; Embley et al., 1980; Flood and Hollister, 1980), continental shelves (Belderson et al., 1972; McKinney et al., 1974), estuaries (Dyer, 1970; Flood, 1981; Allen, 1984; Flood and Bokuniewicz, 1986), large lakes (Flood and Johnson, 1984; Flood, 1989; Viekman et al., 1993; Manley and Manley, 1993; Viekman, 1994) and most recently in a non-tidal river (Singer and Manley, 1991; Manley et al., 1992; Manley and Singer, 1993). Furrows are longitudinal bedforms which are regularly spaced grooves oriented near parallel to the dominate bottom flow direction. Furrows can converge in the direction of flow to form a "tuning-fork" junction. Formed in fine-grained, cohesive sediments, they are several kilometers in length and have spacing and widths ranging from 10-350 m and 1-50 m respectively (Flood, 1983). Depth of these features is generally less than 5 m though depths of 27 m have been documented (Flood, 1983).

Previous studies on furrow formation (Flood, 1983; Viekman et al., 1989; Viekman et al., 1992; Viekman and Wimbush, 1993; Viekman, 1994) have suggested that furrows are initiated as large-scale secondary circulation in the bottom-boundary layer. This circulation pattern preferentially concentrates coarser materials in long linear trains (Figure 2). Speeds of at least 6 cm/s are needed to initiate this helical flow (Viekman et al., 1992). As the coarser-sized material is transported by traction or saltation, the underlying finer-grained sediment is abraded and resuspended, generating a trough. These troughs continue to grow during the intervals of strong abrasive flows or abate and fill in during weaker flows. Once established, however, furrow troughs can reinforce the position of the secondary circulation as well as continue to widen and deepen as the coarse-grained material is effectively trapped on the furrow floor (Flood, 1981; Flood, 1983). Flood and Johnson (1984) have noted that furrows are not observed in steep slope regions, but restricted to areas of low gradient or flat bottoms.

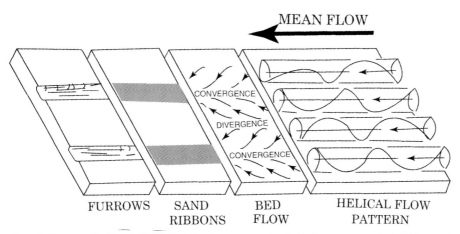

Figure 2. Cartoon after Flood (1983) depicting bottom secondary helical flow pattern which initiates and maintains sedimentary furrows.

Flood (1983) classified furrows into two types. Type 1 (classes 1A, 1B, and 1C) all have distinct troughs with steep walls and flat floors. These differ from Type 2 furrows which have indistinct troughs with gently sloped walls.

Recent numerical models have shown the flow within the vicinity of furrows to be more complex than the initially postulated by Flood (1983). Viekman (1994) showed that bottom helical flow in a furrow field will resemble that of Flood's model (1983) for Type 1 furrows with flow speeds of 8 cm/s. At these speeds an additional roll in and above the furrow was discerned extending ~ 5 meters above the lake bed. Vertical velocities within this helical roll were determined to be 0.01 cm/s, which may be sufficient to lift the fine fractions out of the furrow and allow for cross-stream Ekman flow to deposit the fines in the interfurrow region. At higher flow speeds, the circulation is confined to a small region above the furrow.

The region east of Valcour Island was studied from 1991 to 1994 and again in 1996 by side-scan sonar and moored instrumentation as outlined in Table 1. In this study we examine the role of bottom currents and associated sediment furrow fields to understand the relationship between current velocity and temperature observed in this region. In particular we can investigate the speed and direction of bottom currents and changes in suspended sediment of the bottom currents with regards to the internal seiche. Understanding these relationships will aid in understanding of furrow dynamics.

OBSERVATIONS AND RESULTS

Bathymetry and furrow Distribution

Side-scan studies were conducted east of Valcour Island using a Klein model 590 dual frequency side-scan sonar during three field seasons; 1991, 1994 and 1996. Analog records were obtained for all three surveys and digital sonar data and water depth was gathered during the 1996 survey. The 1991 survey used a 200 m slant range with a 170 m track spacing while the 1994 and 1996 surveys had similar spacing but at a 100 m slant range. The 1991 and 1994

TABLE 1: Location and data collected for the 4-year investigation of the furrow field near Valcour Island. Time periods are grouped according to deployments and recoveries. Dashes imply that data was not collected.

Date	S91 June - Oct	W91 Nov - May	S92 June - Oct	W92 Nov - May	S93 June - Oct	S94 June - Oct	1996
Latitude	44.61217	44.61217	44.61317	44.60983	44.60600	44.60284	------
Longitude	-73038017	-73.38017	-73.37850	-73.7983	-73.3780	-73.38371	------
	(Depth) (Duration)	(Depth) (Duration)	(Depth) (Duration)	(Depth) (Duration)	(Depth) (Duration)	(Depth) (Duration)	------
T-Chain	13-53 m 129 days	13-53 m 94 days	FAILED	14-54 m 203 days	11-51 m 137 days	9-50 m 122 days	------
RCM5-upper	11 m 129 days	12 m 201 days	------	------	------	------	------
RCM5-lower	57 m 129 days	58 m 129 days	------	------	------	------	------
ADCP			9-55 m 116 days	10-57 m 199 days	9-53 m 135 days	9-55 m 122 days	------
Sediment traps	(mbs) (Duration)	(mbs) (Duration)	(mbs) (Duration)	(mbs) (Duration)	(mbs) (Duration)	(mbs) (Duration)	------
upper	12 m 129 days	12 m 201 days	12 m 116 days	12 m 203 days	12 m 135 days	12 m 122 days	------
middle	32 m 129 days	32 m 201 days	32 m 116 days	32 m 203 days	32 m 135 days	32 m 122 days	------
lower	58 m 129 days	58 m 201 days	58 m 116 days	58 m 203 days	58 m 135 days	58 m 122 days	------
Bottom photos	------	------	------	------	------	23 days @ 4 hours	------
Ponar	3 sites	------	------	------	------	1 site	------
Side-scan	200 m range 170 m spacing	------	------	------	------	100 m range 170 m spacing	100 m range 170 m spacing

surveys used a short towing cable (100 m) which towed the sonar fish ~30 to 40 m above the lake bottom. The 1991 and 1994 surveys used Loran-C and GPS for navigating the track lines. In 1996, digital sonar data were collected using DGPS navigational system and a 300 m steel cable which allowed the sonar fish to be towed ~10 m above the bottom producing a better image. The digital data was acquired using ISIS software (® Triton Industries).

Water depths measured in this study were merged with existing NOAA bathymetry to produce a bathymetric map east of Valcour Island. The bathymetric map shows a broad north-south trough with axial depths decreasing from > 80 m in the south to 50 m to the north (Plate 1). The trough is bounded by steep slopes interpreted as faults (Chase and Hunt, 1972; Hunt, 1981) and continues to shallow northward to the east of Cumberland Head.

Analysis of the side-scan records (1991, 1994, and 1996) show that the sediment furrows are located at the central and western portion of the trough east of Valcour Island (Luecke, 1995; Figures 3 - 5). The furrows have a northwest-southeast orientation ranging from 343° to 355° (or 163° to 175°), which is nearly parallel with the sides of the bathymetric trough as defined by the 60 m contour. On average, furrows are ~300 m in length though several individual furrows are in excess of 700 m. Furrows are not continuous through the survey area. The field exists in five distinct groupings, Sectors 1 - 5 (Figure 3). As a furrow ends at the edge of a sector, it widens and terminates within a patchy region as imaged by side-scan sonar (Figures 4 and 5). Furrow statistics of averages of depth, width, and spacing for the

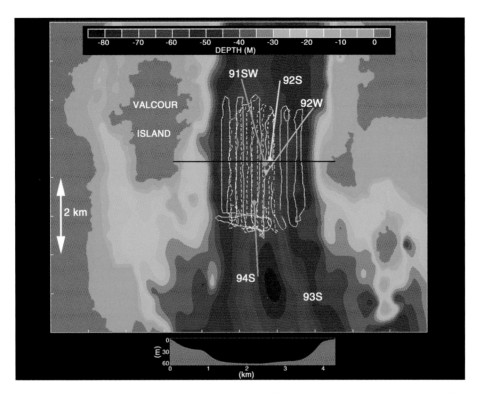

Plate 1. Site survey location east of Valcour Island, Lake Champlain. Ship track lines for the two side-scan sonar surveys are shown (1991 solid and 1994 dashed) and location of mooring positions for the four years of the study. S represents mooring deployment location and data gathered during the summer months (usually June to October) and W represents mooring deployment location and data gathered during the winter months (October to May). Depth contour is in 10 meter intervals with a maximum depth in the region of 80 meters. The cross section of bathymetry shows the trough-like nature of the bottom east of Valcour Island. Axis ticks are in 1 kilometer increments.

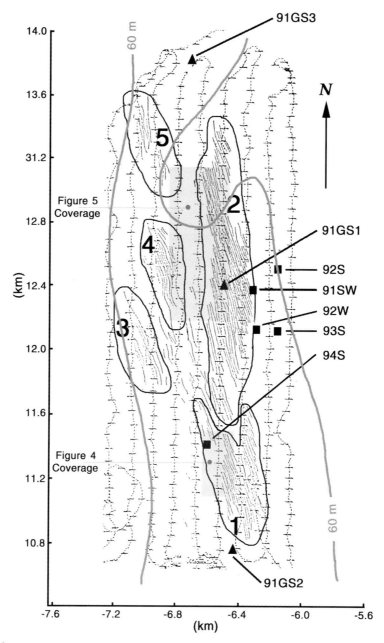

Figure 3. Furrow field east of Valcour Island as determined from side-scan sonar. Furrows have been grouped into Sectors 1-5. Statistics for furrow width, depth and spacing are found in Table 2. Furrows show tuning forks joining to the NNW and SSE suggesting bi-directional flow. Axis scale is in km. Locations of Figures 4 and 5 are indicated by shaded boxes. Mooring locations are indicate by squares and identification labels. 91SW and 94 S mooring locations are within Sectors 2 and 1, respectively. Grab sample sites are identified by triangles and labels 91GS1, 91GS2, 91GS3 and at mooring location 94S.

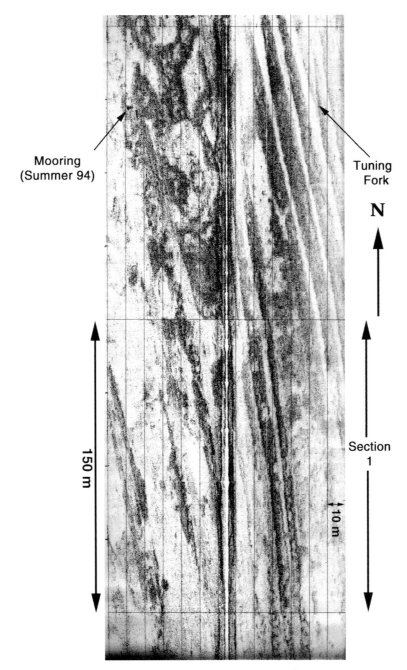

Figure 4. Side-scan image of furrows located in Sector 1. Also imaged is the position of the S94 mooring. Mooring is located in a furrow trough near its end. Furrows are aligned N15W (345° or 165°). High reflectivity is associated with dark returns and low reflectivity and shadow zones are bright returns.

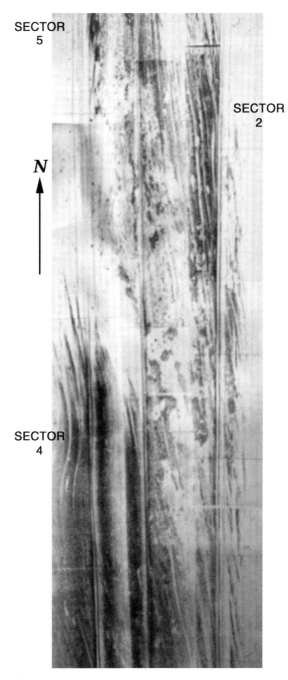

Figure 5. Mosaic of side-scan sonar profiles of the furrow field east of Valcour Island. This mosaic encompasses part of Sector 2, 4 and 5. Note the patchy nature of the bottom at the terminus of the furrow field.

TABLE 2: Furrow statistics. Calculated average width, depth and spacing measurements from side-scan sonar records for each Sector. Width:spacing ratio of 1:4 - 1:9 classifies these furrows as Type 1A using Flood's (1993) classification scheme.

Average values	Sector 1	Sector 2	Sector 3	Sector 4	Sector 5
width (m)	2.65	3.38	3.41	3.42	4.17
depth (m)	1.29	1.83	2.11	1.63	2.67
spacing (m)	9.04	8.67	23.34	13.49	10.91
width: spacing ratio*	4.12	3.66	9.02	6.28	3.18

* width:spacing ratio is an average of all individual width:spacing ratios within a sector and not using the average values of width and spacing listed above.

furrows within a sector provide an indication of variability encountered (Table 2). The furrow width to spacing ratio is ~1:4 in the eastern portion of the survey (Sectors 1 and 2) and changes to 1:6 to 1:9 in the western side of the field area (Sectors 3 and 4). This identifies the furrows as Type 1A using Flood's characterization (Flood, 1983). Type 1A in general have steep walls, flat trough floor and have spacings more than 5 to 15 times their widths. Several joining patterns or "tuning forks" (e.g. Figure 4) are present in the furrow fields and the sense of joining is bi-directional.

Grain Size Results

Grain size analysis of ponar bottom samples (from 1991, and 1994) were performed by a Malvern MasterSizerE at Hamilton College (Clinton, NY). Approximately 1g of sediment sample was dispersed in a solution of sodium hexametaphosphate and filtered water and vigorously stirred within an ultrasound bath. The sample was then introduced into the Malvern MasterSizer E which uses a laser-diffraction technique to determine the grain size distribution. For this study, a He-Ne laser, a poly-disperse model, and a 300-mm lens were used to enable detection with the greatest range (1.2-600 µm).

Bottom samples were collected at four locations within the survey region (3 samples in 1991 and 1 sample in 1994). Malvern results show there is an increase in the grain size from a silt-size outside the furrow field (mean = 5.25 φ) to medium-sized sand inside the furrow field (mean = 3.4 φ) (Figure 6).

Sediment Trap Analysis

Sediment trap containers used on the long-term moorings were cylindrical polypropylene core liners having a 5:1 aspect ratio (e.g. Gardner, 1980). No preservatives for preserving organic matter were used. Sediment traps were placed at ~5 m above the lake bottom, 12 m below the lake surface and at 50% water depth. The trap samples were filtered using numerous 0.01 micron filters. Wet and dry weights were measured and accumulation rates in terms of grams/m^2/day were calculated.

Sediment traps were collected for summer and winter months of both 1991 and 1992 (Table 1; 91SW and 92SW on Plate 1) and only during the summer months for 1993 and 1994 (93S and 94S on Plate 1). For all collection times, the bottom trap had the highest calculated flux.

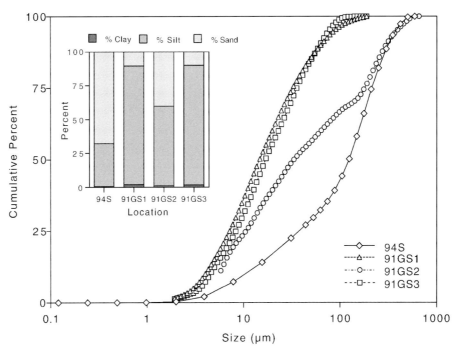

Figure 6. Cumulative percent for grab samples taken during 1991 (91GS1, 91GS2, 91GS3) and at the mooring site in 1994 (94S). Only the 94S site is the coarsest sized material. The inset histogram indicated the sand-silt-clay percents for the four sites.

The winter month bottom traps contained the largest amount of flux, with the exception of S91 (10.3 grams/m^2/day) (Figure 7 and Table 3). Excluding the winter trap values, the second highest flux was the bottom trap collected during S94 (2.7 grams/m^2/day).

Sedimentation rates were calculated using the bottom trap accumulations and an average sediment density of 2.65 g/cm^3. These varied between summer and winter months (Table 4). The winter rate was determined from bottom trap data obtained during W91 and W92 and yields a sedimentation rate of ~0.09 cm/yr. Likewise, bottom trap S92 and S93 data gives an average sedimentation rate of 0.005 cm/yr for a summer rate. These specific traps were chosen because they were outside the furrow field. Extrapolating these rates for an entire year, the sediment influx to the bottom in this region is approximately 0.05 cm/yr. This compares well with Chase and Hunt (1972) isopach values near Valcour. They determined a total thickness of Lake Champlain sediment to be 6 meters. Taking the interface between Champlain Sea and Lake Champlain to be 10,200 ybp (Elson, 1969; Fillion and Hunt 1974; McDonald, 1968), yields a sedimentation rate of 0.06 cm/yr.

Optical Backscatter and CTD Surveys

During June and September 1991, CTD and optical backscatter surveys were conducted near the current meter mooring site at Valcour Island using a Sea Bird 'Sea Cat' CTD with a Sea-Bird SBE 24 optical backscatter and turbidity monitor (OBS). The sensitivity of the OBS

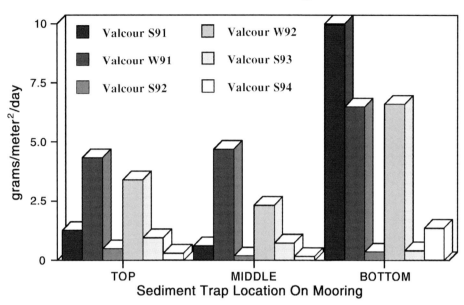

Figure 7. Sediment trap histogram. Trap locations were top (~12 meters below lake surface), middle (32 meters below lake surface), and bottom (58 meters below lake surface). Winter trap data shows a higher accumulation from top to bottom and overall from the summer data. Exception is the high accumulation rate of the bottom trap during the summer of 1991. S91, S92, S93, and S94 are mooring deployment locations and data gathered during the summer months (usually June to October) and W91 and W92 are mooring deployment locations and data gathered during the winter months (October to May).

unit is reported in mg/l and is based on a factory calibration. Measured data were contoured using a three-dimensional minimum tension gridder and displayed using EarthVision software (® Dynamic Graphics). Outside the furrow area, values range from 10 to 30 mg/l, however, near the mooring site, higher values of 60 to 80 mg/l were observed with a high of 90 mg/l (Plate 2). Comparison between the surveys conducted in June and September, show a decrease in the range of optical backscatter from 20 - 80 mg/l to 2 - 7 mg/l (Plate 2).

Hydrodynamics

Long-term moorings emplaced east of Valcour Island have had various configurations and instrumentation (Table 1). In 1991 the basic configuration of the mooring included two Aanderaa RCM5 current meters located at approximately 12 m and 58 m below the surface, an Aanderaa temperature chain (T-chain) located between the two current meter locations and three sediment traps at levels as described above. Location of the moorings is shown on Plate 1. The mooring was pulled, data retrieved and then redeployed each June and October. In 1992, 1993 and 1994, an upward looking RDI Acoustic Doppler Current Profiler (ADCP) current meter replaced the Aanderaa RCM5s. The ADCP uses doppler technology to measure current speeds along three vectors, north-south, east-west and vertical. Measurements were taken at one meter depth intervals between 9 meters and 55 meters every fifteen minutes. The

TABLE 3: Sediment trap data. Particle flux rates are in grams/m²/day.

Trap Level	S91	W91	S92	W92	S93	S94
Top	1.3	4.3	0.5	3.4	0.9	0.6
Middle	0.6	4.7	0.2	2.3	0.7	0.3
Bottom	10.3	6.5	0.4	6.6	0.4	2.7
Bottom-Top	9.0	---	-0.1	---	-0.5	2.1

TABLE 4: Bottom trap data used for determining sedimentation rates

Trap ID	Days duration	Particle flux (g/m²/da)	Sedimentation Rate (cm/yr)
S91	130	10.3	0.142
W91	202	6.5	0.089
S92	122	0.4	0.005
W92	204	6.6	0.091
S93	135	0.4	0.006
S94	124	2.7	0.037

thermistor chain recorded temperatures every four meters over a 40 m length every hour (see Table 1 for exact depth range). During the 1994 field season a pair of Benthos Photosea 5000 stereo cameras were added to the mooring package. This mooring was placed on the northwestern edge of the furrow field in Sector 1 and near the terminus of a furrow. Its location with respect to the furrow field was verified by the side-scan sonar survey (Figure 4).

Though current data were obtained from 1991 through 1994, only the 1991 and 1994 moorings were within the furrow fields, Sectors 2 and 1 respectively. Analysis of all current meter data shows a consistent bi-directional dominance of flow regimes above and below the metalimnion as observed in the polar histograms of Figure 8 for stratified periods. Surface currents showed oscillating flow nearly N-S (having average speed of ~13 cm/s when wind forcing was from the SW and ~8 cm/s, when wind direction was from the NW) (see Manley et al., this volume). Bottom currents were also bi-directional showing similar trends and average speeds of ~7 cm/s. This N - S bi-directionality in the surface current was significantly altered during the winter months (unstratified period) as there was no preferential direction, but the bottom flow still maintained a strong N-S alignment with slower speeds of 2-3 cm/s (Figures 8 and 9, see Manley et al., this volume).

Current data (Manley et al., this volume) has shown that atmospheric forcing during stratified conditions causes a depression of the metalimnion and forces the bottom water to flow opposite to the dominant wind direction. In extreme wind forcing, the bottom currents can exceed 20 cm/s. We term these time periods as high speed events (HSE). The occurrence of HSE was investigated for the data collected over the four year study (Figures 9 and 10). HSEs with speeds exceeding 20 cm/s occur only during the period when the lake is stratified. Bottom current speeds rarely exceed 10 cm/s during unstratified periods (Table 5). During times of stratification and external forcing by strong atmospheric storms, bottom speeds in excess of 20 cm/s can occur and can even obtain speeds > 53 cm/s (occurred during fall 1993). Though some high speed currents flow toward the north, over 60% of the HSEs (those >20

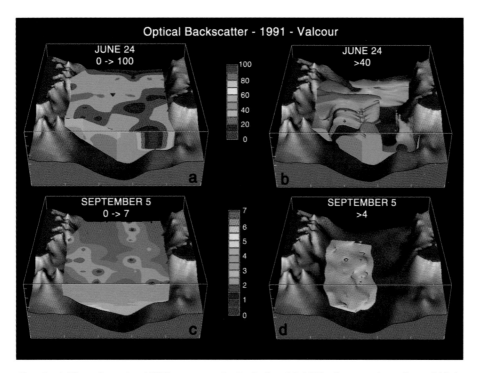

Plate 2. a) Three dimensional OBS survey conduction in June 24, 1991. Concentrations of material is in mg/l. Largest concentration was 90 mg/l. b) Three dimensional OBS survey of June 24, 1991 showing only those concentrations higher than 40 mg/l. c) Three dimensional OBS survey conduction in September 5, 1991. Concentrations of material is in mg/l. Note the low concentration in comparison to the June survey. d) Three dimensional OBS survey of September 5, 1991 showing only those concentrations higher than 4 mg/l.

Polar Histograms of Direction at the Valcour Island Moorings
1991 - 1994

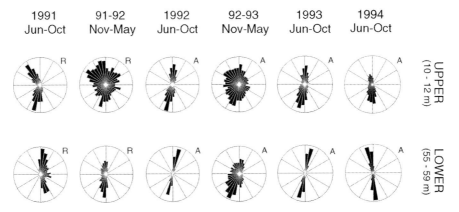

Figure 8. Polar histograms of current directions for the 1991-94 surveys. Top row of rose diagrams shows composite directions for the epilimnion and the bottom row of rose diagrams are the composite directions for the hypolimnion. Label A indicates data obtained from an ADCP unit and R indicates data obtained from Aanderaa RCM5s. Polar histograms are in 10° increments for cumulative flows. Bottom currents show a bi-directional flow with dominant flow occurring within 20° of either N or S. Surface flows are bi-directional only during stratified periods.

cm/s) flow toward the south (Table 6). At least three to four HSEs occurred each year having speeds in excess of 30 cm/s. All HSEs with speeds >30 cm/s flow to the south with the exception of one event in fall of 1991.

ADCP data obtained during 1994 was located within Sector 1. Starting on June 21, 1994 and extending into June 22, 1994, one such high speed event occurred (Plate 3). Southward moving currents in the bottom boundary layer reached speeds of 36.6 cm/s. Temperature data for the same time period showed a rapid depression of the metalimnion. Other than this single event in June, the remainder of the summer months, July and August, showed no speeds reaching 30 cm/s, although, on five separate occasions, southward flowing currents reached speeds between 20 and 25 cm/s. From mid-September to mid-October, stratification decreased (Manley et al., this volume) due to continual atmospheric cooling prior to fall turnover. During this time period, five HSEs occurred. Magnitudes during these events exceeded 30 cm/s. The greatest event occurred during September, when bottom speeds reached 38.9 cm/s. The vertical component of speed for all current data was not found to be significant. Over the four-month study, vertical speeds were less than 2 cm/s even during the high speed events.

Bottom Photography Images

The Photosea cameras (field season 1994) were oriented to image a 70 cm by 70 cm square quadrant of the bottom surface within a marked frame. The cameras took pictures for 23 days at 4 hour intervals prior to failure of the internal clock. This timing circuit malfunction caused pictures to be taken at a variable rate of four to twelve images per minute and exposed the remainder of the film before the retrieval date. All other apparatus operated for 4 months between June and October 1994.

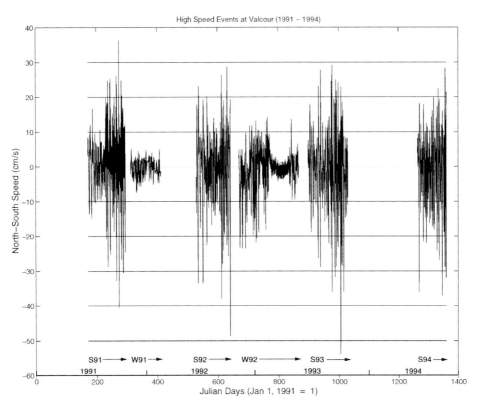

Figure 9. North-South speed of four years of bottom current data. Horizontal lines are in increments of 10 cm/s. Negative values show flows towards the south while positive values are flows towards the north. All flows greater than 30 cm/s flow southward with the exception of one event in fall of 1991.

Five photographs which were taken during a 1994 HSE are described here (Plate 4). Plate 4a was taken on June 17 at 4:40 am, two days after deployment of the mooring. The bottom surface appears relatively flat with the exception of a diamond-shaped feature in the left-center of the image. The white blotches on the image are in the water column and not objects on the lake bottom. Plate 4b, taken June 21 at 12:40 shows the present of aquatic vegetation called *Potamogeton robbinsii* and the diamond-shaped feature persists. This aquatic vegetation is photosynthetic and usually is found in water depths less than 17 meters. Its occurrence at 60 meters water depth indicates that this vegetation has been transported by currents into the region. Plate 4c, taken 8 hours later shows the photogrammetric study area with extremely low visibility. The bottom is not discernible, only the uppermost section on the left vertical post is identifiable. Plate 4d was taken June 23 at 4:40 a.m. The diamond-shaped feature is no longer visible. The bottom is covered with organic debris and woody material. A large layered fragment, approximately 16.2 cm by 6.5 cm (enclosed by small box in Plate 4d), is identifiable as well as smaller indurated fragments adjacent to it. Four hours later the large layered fragment showed evidence of movement as it was rotated counterclockwise 15°. The images taken from June 23 onward show increasing and decreasing amounts of *Potamogeton robbinsii* migrating in and out of the camera frame but no other periods of high sediment suspension in the bottom waters.

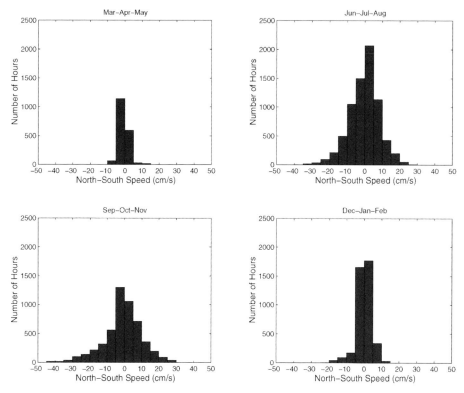

Figure 10. Histograms of four years of bottom currents separated by seasons (Spring, Summer, Fall, Winter). Negative values are currents flowing toward the south while positive values are currents flowing toward the north. Divisions are in increments of 10 cm/s. During winter and spring months (no stratification) speeds are mostly under 10 cm/s and fluctuate between north and south. During stratification periods (summer and fall) higher bottom currents are observed have a preferred south-directed flow.

Stereophotographic pairs of bottom images were analyzed using a Seagle Model 90 stereophotocomparator. The comparator simultaneously measures horizontal parallax within the overlapping sections of the camera pairs. This orientation enables measurements along the x, y, and z axes (with respect to the frame) from which depth measurements within the frame can be determined. A time series of photos were analyzed to observe any change in deposition/erosion occurring within the furrow trough. Two photo pairs were chosen for preliminary comparison (Plate 4a and Plate 4d). Initial results suggest that the area imaged by the photographs has undergone a minimum of 3-6 mm of erosion during the HSE.

DISCUSSION

Active furrow fields have been identified in large rivers such as the Hudson River, Buffalo River, and in large lakes including Lake Superior and Lake Ontario (e.g. Flood and Bokuniewicz, 1986; Flood and Johnson, 1984; Viekman et al., 1992). A furrow field thought to be active if there are episodes of erosion between periods of deposition. The task is to characterize the depositional environment.

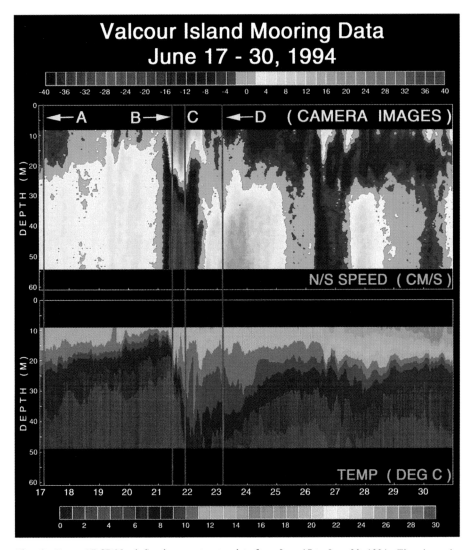

Plate 3. Top - ADCP North-South current meter data from June 17 to June 30, 1994. Time is on the horizontal axis and depth in the water column on the vertical. Yellow and red colors show currents traveling in the north direction (positive values) whereas blue and purple colors show currents traveling to the south (negative values). A high speed event occurs on June 21 flowing to the SE. Letters A, B, C, and D correspond to the camera images in Plate 4. Bottom - Temperature data from a thermistor chain for the same days. Time is on the horizontal axis and temperature with depth is on the vertical. Red colors represent warmer temperatures and blue colors show lower temperatures. On June 21, when the high speed event occurs, the thickness of the epilimnion is rapidly increasing.

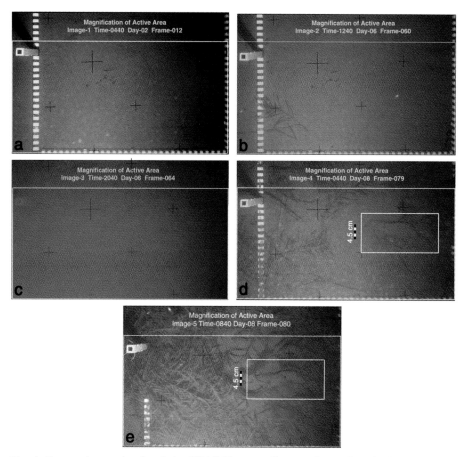

Plate 4. Bottom photographs taken during 1994 field season. Camera reference frame is 70 cm by 70 cm; a portion of which is shown in the photograph. Each white section of the camera reference frame is 0.9 cm in length. The vertical post (left side of picture) is 30.5 cm in high. North direction is approximately along the bottom camera frame and to the right. Large and small crosses are fiducial marks used to correlate the stereo pairs. The large white box is a reference guide for comparison purposes for the four frames. Time sequence of frames are as follows A) June 17 at 4:40 a.m. B) June 21 at 12:40 p.m. C) June 21 at 20:40 p.m. D) June 23 at 4:40 p.m. E) June 23 at 8:40. In frame D a small white box outlines the large layered fragment exposed during the high speed event. A small scale of 4.5 cm is superimposed on the image for reference. E) Note the rotation of the layered fragment by ~15° counterclockwise.

TABLE 5: Percentage of bottom currents having various speeds for the 1991 - 1994 data. Less than 4% of all recorded bottom speeds are in excess of 20 cm/s.

	0 - 10 cm/s	10 - 20 cm/s	> 20 cm/s	Number of Obs (hourly)
Fall	69.9	21.3	8.7	5164
Winter	95.8	4.1	0.1	4105
Spring	99.0	1.0	0	1849
Summer	78.6	18.6	2.9	7303
All Measurements	82.0	14.4	3.6	18421

TABLE 6: Dominate direction for high speed bottom current events. For speeds > 30 cm/s the dominate flow direction is towards the south.

	20-30 cm/s N	20-30 cm/s S	>30 cm/s N	>30 cm/s S
1991	4	8	1	3
1992	3	12	0	4
1993	8	13	0	4
1994	2	9	0	3
Average	4	10	0	3.5

Few long term field investigations of flow characteristics near sedimentary furrows have occurred in large lakes. The most comprehensive work has been done in Lake Superior (Flood and Johnson, 1984; Flood, 1989; Viekman et al., 1989; Viekman et al., 1992; Viekman and Wimbush, 1993; Viekman, 1994). In Lake Superior, the prominent furrows are located in cohesive sediments having a low accumulation rate (0.061 - 0.072 cm/yr) and the bottom current flows in a direction which is within 25 degrees of the furrow alignment. At this orientation Viekman et al. (1989) documented that the secondary helical circulations were the most vigorous and supported Flood's (1983) model of furrow dynamics. These flows were episodic and occurred for a small percentage of the current record (Viekman et al., 1992).

Furrows east of Valcour Island are aligned NW-SE (343 - 355°). Current meter data from 1991 - 1994 has documented that major movement of the water is predominantly N-S and created by the internal seiche (Manley et al., the volume). Bottom currents are oriented within 20° of N or S. Depending on the time of year, the speeds of these currents can fluctuate greatly but for stratified periods they tend to exceed 6 cm/s (Figure 9). Viekman et al., (1992) has documented in Lake Superior, that bottom current speeds of > 6 cm/s are needed for secondary helical flow to be initiated within a furrow field. Because of furrow alignment within 25 degrees of the dominant flow direction and near-bottom speeds are commonly > 6 cm/s, the sediment furrows near Valcour Island appear active and not relic features.

Active furrows are a source of sediment resuspension when near-bottom currents reach high enough speeds to erode sediment. For cohesive sediment, currents need to be > 16 cm/s for this to occur. Though only a synoptic view, the CTD site surveys done during June and September of 1991 suggest the presence of resuspended sediment. During September the optical backscatter data shows less than 6.5 mg/l of particulate material in the water column, with the highest values located on the western section of the deep trough next to Valcour Island (Plate 2). In contrast, the June survey showed particle concentrations were an order of magnitude more than in September. The highest concentrations were located within the bottom 16 meters of the water column, and predominantly in the central to western portion of the trough (Plate 2).

Material collected in sediment traps can include a large amount of resuspended matter depending on the depth of the trap and the time of year (Eadie et al, 1984). Eadie et al. (1984) has shown that near surface traps deployed during stratified periods represent the flux of new matter entering Lake Michigan and bottom traps collect material which has been resuspended. Thus an approximation of the amount of resuspension can be assumed to be the flux measured in the deepest trap minus the upper trap rate. During unstratified periods, the traps can give an estimate of the amount of resuspension which is reequilibrating in the lake and fluxes tend to be 5 to 10 times those during stratified periods.

Lake Champlain flux rates range from 0.6 to 1.25 grams/m^2/day in the surface traps and 0.4 to 10.3 grams/m^2/day in traps located 6 meters above the lake bottom during stratified times. This compares favorably with the rates obtained during the stratified months in Lake Michigan (0.24 - 1.2 grams/m^2/day) (Eadie et al., 1984). The exception to this is observed for the bottom traps of S91 and S94 having rates of 10.3 and 2.7 grams/m^2/day respectively (Table 4). The location of the S91 mooring (Plate 2 and Figure 3) is at the edge of the Sector 2 furrow field and the S94 mooring location is adjacent to Sector 1 furrow field (Plate 2 and Figures 3 and 4). Note that the summer moorings for 1992 and 1993 were located outside the furrow fields towards the east and these traps show small amount of particulates collected for the summer months.

Gloor et al. (1994) studied seiche activity in Lake Alpnach in Switzerland and concluded that resuspension within the bottom boundary layer (2 to 7 m thick) could be directly attributed to burst-like events associated with seiche activity. Gloor et al. (1994) showed that in the bottom boundary layer, particle concentration was 3 to 4 times that of the overlying layers.

The bottom cameras located on the S94 mooring actually documented a period of high sediment concentration in the bottom boundary layer on June 21 when the camera frame became obscured (Plate 4c). After the particulate material dissipated, the bottom showed evidence of erosion and a great deal of organic and physical debris brought into the region (Plate 4d). We can correlate this period of large sediment concentrations with a high speed event recorded by the ADCP at the mooring location. The ADCP data shows that during June 21 - June 22 a strong SE current occurred ranging in speeds from 16 cm/s and peaking at 36 cm/s occurred (Plate 3). Hunkins et al. (1998) defined this high speed event as being accompanied with a southward propagating surge which then was followed by a northward flowing gravity current along the sediment water interface (see Saylor et al., this volume).

Analysis of the photographic images documents only one period of high suspended particle concentrations and this correlates with a HSE (>30 cm/s, southward flowing) on June 21 (Plate 4c). After this HSE, flow reversed and an image taken on June 23 (Plate 4d) documents the erosional effect that the HSE had on the lakebed within the furrow field. This northward flowing bottom current reached a speed maximum of 16 cm/s. This speed was strong enough to rotate the layered fragment (Plate 4e) but caused no further erosion of the sediment. Remaining photographic images show changes in the location of the aquatic vegetation but no other period of high sediment concentration. There was no effective change in bottom morphology even when bottom speeds reached 21 cm/s on July 6.

Other furrow studies have documented that speeds of at least 6 cm/s are needed to initiate helical flow within the bottom boundary layer (Viekman et al., 1992). This is apparently the minimum speed necessary for furrow development (Flood and Hollister, 1980; Flood, 1989; Viekman et al., 1992). This study suggests that speeds of 30 cm/s or greater are needed to generate peak erosive activity also necessary for furrow development. Speeds of 10 to 20 cm/s are more likely to affect depositional patterns rather than eroding furrow trough sediments as evidenced by the rotation of the indurated fragment. The movement of the aquatic vegetation may be an abrasive which could cause some erosion but photographic images could not detect any significant change in water clarity associated with this movement.

In general, bottom current speeds at Valcour Island fall within the range of 6 - 10 cm/s during months of stratification and tend to move light aquatic vegetation along the bottom. It is only with strong atmospheric forcing that bottom speeds exceed 20 cm/s. During those periods the predominate bottom current direction is to the S or SSE. When sufficiently strong southerly winds occur, it is this initial set up phase which generates the largest bottom southward currents capable of eroding and resuspending sediment within the furrow field.

CONCLUSIONS

Located at the northern end of the Main Lake, the trough east of Valcour Island is a natural constriction within the lake which funnels the water into a narrow region. The lake floor at this location is nearly flat with steep sides. Sediment furrows exist within the western side of this trough and are grouped into several distinct fields each having variable width:spacing ratios. Active sediment resuspension has been observed with synoptic backscatter surveys as well as year-long sediment trap information. The most dramatic documentation for resuspension is in the bottom camera photographs which captured an extreme loss of view due to sediment within the water column. The correlation of the high sediment resuspension with high-speed bottom currents allows us to calculate the duration of sedimentation periods versus times of active erosion from long-term current meter data.

Type 1A furrows have been suggested to indicate long periods of sedimentary deposition interspersed with periods of intense erosion (Flood, 1983). The furrow field east of Valcour Island experiences only 5 to 10 high speed events during a year (an average of 1 to 3% of a year) the remainder of the time the field experiences sedimentation. Thus the data presented supports that furrows are typical of an environment where there are short-lived, intense erosional events separated by long intervals of deposition. This study also suggests that speeds in excess of 30 cm/s are needed to effectively erode the furrows and resuspend large amounts of sediment. Periods of unstratified water are times when the majority of sedimentation occurs within a furrow field.

Our study shows that lakebeds can be impacted by vigorous currents capable of eroding and resuspending sediment. The fate of this material and any associated contaminants is our ultimate concern (see McIntosh et al., this volume). One of the highest concentrations of polychlorinated biphenyl (PCB) substances in the surface sediments occurs near the Wilcox Dock area of Cumberland Bay just northwest of the survey area. The presence of *Potamogeton robbinsii*, a shallow water aquatic plant, at the furrow site indicates that there is a connection between the shallower bay waters to the deep lake. It is important for us to understand the hydrodynamics of the lake and the interaction bottom currents have at the sediment-water interface. Knowing more about the lakebed morphology and present-day sediment dynamics will aid us in our management of Lake Champlain.

Acknowledgments. This Publication is a result of work sponsored by the Cooperative Institute of Limnology & Ecosystems Research (CILER) under cooperative agreements from the Environmental Research Laboratory (ERL), National Oceanic & Atmospheric Administration (NOAA), U.S. Department of Commerce under Cooperative Agreement No.

NA90RAH00079. The U.S. Government is authorized to produce and distribute reprints for governmental purposed notwithstanding any copyright notation that may appear hereon. This study was support in part by the funds of U.S. Geological Survey, Grant No. 14-08-0001-G2050. The Bottom Camera equipment and comparator were purchased through a grant from the Lintilhac Foundation. We wish to acknowledge Captain Fred Fayette of the research vessel *Neptune* for his support over several years of this project. We also thank Captain Dick Furbish of the UVM Research Vessel *Melosira* for assistance with the mooring deployments. We thank Dr. Eugene Domack for the use of Hamilton College's Malvern MasterSizer E for the grain-size analysis. 2-D and 3-D modeling was accomplished using EarthVision, of Dynamic Graphic, Alameda, CA.

REFERENCES

Allen, J. R. L., *Sedimentary Structures: their character and physical basis*, in New York, Elsevier, 592, 663p., 1984.

Belderson, R. H., N. H. Kenyon, A. H. Stride, and A. R. Stubbs, *Sonographs of the sea floor*, in New York, Elsevier, 1972.

Chase, J. S. and A. S. Hunt, Sub-bottom profiling in central Lake Champlain- a reconnaissance study, Conference Great Lakes Research, Proc. 15., 1972.

Chapman, C. H., Late-glacial and post-glacial history of the Champlain valley, Report of the State Geologist of Vermont, 1942.

Connally, G. G. and L. A. Sirkin, Wisconsinan history of the Hudson-Champlain Lobe: The Wisconsinan Stage, *Geological Society of America Memoir 136*, 47-69, 1973.

Connell, B., Numerical model of the surface and internal seiche on Lake Champlain, unpublished thesis, Middlebury College, Physics Department, 1994.

Dyer, K. R., Linear erosional furrows in Southampton water, *Nature*, 255, 56-58, 1970.

Eadie, B. J., R. L. Chambers, W. S. Gardner, and G. L. Bell, Sediment trap studies in Lake Michigan: resuspension and chemical fluxes in the southern basin, *J. Great Lakes Res.*, 10, 307-321, 1984.

Elson, J. A., Radiocarbon dates, *Mya arenaria* phase of the Champlain Sea. *Canadian Journal of Earth Sciences*, 6, 367-372, 1969.

Embley, R. W., P. J. Hoose, P. Lonsdale, L. Mayer, and B. E. Tucholke, Furrowed mud waves on the western Bermuda Rise, *Geological Society of America Bulletin*, 91, 731-740, 1980.

Fillion, R. H. and A. Hunt, Late Pleistocene benthic foraminifera of the southern Champlain Sea as paleotemperature and paleosalinity indicators, *Maritime Sediments*, 10, 14-18, 1974.

Flood, R. D., Distribution, morphology, and origin of sedimentary furrows in cohesive sediments, Southampton Water, *Sedimentology*, 28, 511-539, 1981.

Flood, R. D., Classification of sedimentary furrows and a model for furrow initiation and evolution, *Geol. Soc. Amer. Bull.*, 94, 630-639, 1983.

Flood, R. D., Submersible studies of current-modified bottom topography in Lake Superior, *J. Great Lakes Res.*, 15, 3-14, 1989.

Flood, R. D. and H. J. Bokuniewicz, Bottom morphology in the Hudson River Estuary and New York Harbor, *Northeastern Geology*, 8, 130-140, 1986.

Flood, R. F. and C. D. Hollister, Submersible studies of deep-sea furrows and transverse ripples in cohesive sediments, *Marine Geology*, 36, M1-M9, 1980.

Flood, R. D. and T. C. Johnson, Side-scan targets in Lake Superior-evidence for bedforms and sediment transport, *Sedimentology*, 31, 311-333, 1984.

Gardner, W. D., Field assessment of sediment traps, *J. Marine Research*, 38, 41-52, 1980.

Gloor, M., A. Wuest, and M. Munnich, Benthic boundary mixing and resuspension induced by internal seiches, *Hydrobiologia*, 284, 59-68, 1994.

Hollister, C. D., R. D. Flood, D. A. Johnson, P. Lonsdale, J. B. Southard, Abyssal furrows and hyperbolic echo traces on the Bahama Outer Ridge, *Geology*, 2, 395-400, 1974.

Hunkins, K. L., T. O. Manley, P. Manley, and J. Saylor, Numerical studies of the four-day oscillation in Lake Champlain, *J. Geophys. Res.*, 103(C9), 18425-18436, 1998.

Hunt, A. S., Bottom sediments of Lake Champlain 1965- 1971, *GSA Bulletin*, 9, 630-639, 1972.

Hunt, A. S., Sediment thickness, Eastern Lake Champlain, Vermont Water Resources Research Center and U.S. Department of Interior, 1977.

Hunt, A. S., Detailed investigation of suspected Holocene fault movement in Central Lake Champlain, Vermont, 1-47, 1981.

Luecke, M., Sediment dynamics in a furrow field east of Valcour Island, Lake Champlain, unpublished thesis, Middlebury College, Geology Department, 83 p., 1995.

Manley, P. L., L. Fuller, and J. K. Singer, Bottom morphology and environmental implications for the Buffalo River: Evidence from side-scan sonar, *International Association for Great Lakes Research, Program and Abstracts*, 1992.

Manley, P. L. and T. O. Manley, Sediment-current interactions at Valcour Island, Lake Champlain- A case of helical flow in the bottom boundary layer, Northeastern Section GSA, *Abstracts with Programs, 25(2)*, 36, 1993.

Manley, P. L. and J. K. Singer, Potential for sediment resuspension of contaminated sediment, *GSA Abstracts with Programs*, 25, A-290, 1993.

Manley, T. O., P. L. Manley, J. Saylor, and K. L. Hunkins, Lake Champlain hydrodynamic monitoring program - An overview, Northeastern Section GSA, *Abstracts with Programs, 25(2)*, 61, 1993.

Manley, T. O., K. L. Hunkins, J. Saylor, and P. L. Manley, Aspects of summertime and wintertime hydrodynamics of Lake Champlain, this volume.

Marinov, I., Wind-generated oscillations in Lake Champlain, unpublished thesis, Middlebury College, Physics Department, 53 p., 1998.

Marinov, I., R. Prigo, and T. Manley, 1998, Wind-generated oscillations in Lake Champlain, *Lake Champlain Research Consortium Abstracts with Programs*, 1, 31, 1998.

McDonald, B. C., Deglaciation and differential postglacial rebound in the Appalachian region of southeastern Quebec, *J. of Geology*, 76, 664-677, 1974.

McIntosh, A., M. Watzin, and J. King, Toxic substances in Lake Champlain: an overview and a case study, this volume.

McKinney, T. F., W. L. Stubblefield, and D. J. P. Swift, Large-scale current lineations on the central New Jersey shelf: investigations by side-scan sonar, *Mar. Geology*, 17, 79-102, 1974.

Myer, G. and G. K. Gruendling, *Limnology of Lake Champlain*, SUNY Plattsburgh, 1979.

Prigo, R., T. O. Manley, and B. S. H. Conell, Linear, one-dimensional models of the surface and internal standing waves for a long and narrow lake, *Am. J. Phys.*, 64, 288-300, 1996.

Saylor, J., G. Miller, K. Hunkins, T. Manley, and P. Manley, Gravity currents and internal bores in Lake Champlain, this volume.

Saylor, J., T. O. Manley, P. L. Manley, and G. Miller, Physical processes driving high-speed currents in Lake Champlain bottom water, Northeastern Section *GSA, Abstracts with Programs, 25(2)*, 76, 1993.

Saylor, J., G. Miller, T. Manley, and P. L. Manley, Observations of high-speed bottom currents in Lake Champlain, *EOS, Trans. Amer. Geophys. Union*, 75, 230, 1994.

Singer, J. K. and P. L. Manley, A side-scan survey of the Buffalo River, International Association for Great Lakes Research, *Programs and Abstracts*, 83, 1991.

Teller, J. T., Proglacial lakes and the southern margin of the Laurentide Ice Sheet, in Ruddiman, W. F. and H. E. Wright, J., eds., *North America and adjacent oceans during the last deglaciation*. Boulder, Geological Society of America, 39-69, 1987.

Viekman, B.E., Streamwise vortices in the oceanic bottom boundary layer, Ph.D. dissertation, University of Rhode Island, Kingston, RI. 232 pp., 1994.

Viekman, B. E., R. D. Flood, M. Wimbush, M. Faghri, Y. Asako, and J. C. Van Leer, Sedimentary furrows and organized flow structure: A study in Lake Superior, *Limnol. Oceanogr.*, 37, 797-812, 1992.

Viekman, B. E., M. Wimbush, and J. C. Van Leer, Secondary circulations in the bottom boundary layer over sedimentary furrows, *J. Geophys. Res.*, 94, 9721-9730, 1989.

Viekman, B. E., and M. Wimbush, Observations of the Vertical structure of the Keweenaw Current, Lake Superior, *J. Great Lakes Res.*, 19(2), 470-479, 1993.

Response of St. Albans Bay, Lake Champlain to a Reduction in Point Source Phosphorus Loading

Scott C. Martin, Richard J. Ciotola, Prashant Malla, and Subramanyaraje N.G. Urs

ABSTRACT

Point source phosphorus (P) loading to St. Albans Bay, Lake Champlain (Vermont) was reduced by 94% during the 1980's. However, recovery of the water quality was delayed by release of accumulated phosphorus from the bottom sediments. Sediment cores were collected from 43 locations in summer of 1992 and subjected to a phosphorus extraction sequence of NH_4Cl–NaOH–HCl. Total P concentrations averaged 1239 µg/g. Biologically available inorganic P (BAIP) accounted for 37.0% of this, organic P for 20.3%, and HCl-P for 42.7%. Results were compared to those of a similar survey conducted in 1982. Between 1982 and 1992, total sediment P decreased by an average of 366 µg/g. Of this, 52.5% came from the organic P fraction, 28.7% from BAIP, and 18.8% from HCl-P. A mass balance model for total phosphorus in the water column and bottom sediments was calibrated using historical data. The model was used to analyze phosphorus cycling in the bay, and to predict future trends in water quality. The results indicate that the effective phosphorus loading from the bottom sediments has decreased substantially, and that the bay-wide total phosphorus goal of 17 µg/L will be met with no further loading reductions.

INTRODUCTION

St. Albans Bay is located on the Northeast Arm of Lake Champlain, about 40 km north of Burlington, Vermont. For many years, the Bay exhibited highly eutrophic conditions, including intense summer algal blooms, resulting from excessive nutrient loading. The bay receives nonpoint source loading from agricultural runoff and point source loading from the City of St. Albans wastewater treatment facility (WWTF) discharge. In 1983, a Hood Dairy facility in St. Albans was closed, and in 1987 the WWTF was upgraded to improve phosphorus removal. These events, along with the Vermont Phosphorus Detergent Ban (1978), led to a 94% reduction in the point source phosphorus loading, and a 46% reduction in the total loading to the bay [*Hyde, et al., 1993*].

Water quality in St. Albans Bay has been monitored extensively by several programs, including the Vermont Lay Monitoring Program (LMP; 1979-present), the St. Albans Bay Rural Clean Water Program (RCWP; 1981-90), and the Lake Champlain Diagnostic Feasibility Study (LCDFS; 1990-92). The data indicate a sharp spatial gradient in phosphorus concentrations, from over 100 µg/L in the inner bay to under 50 µg/L in the outer bay. Monitoring results from the late 1980's and early 1990's suggested that total phosphorus levels dropped significantly throughout the bay [*RCWP, 1991*]. However, phosphorus concentrations were still high enough to maintain eutrophic conditions and cause periodic summer algal blooms.

A steady-state mass balance modeling analysis [*Smeltzer, 1983*] indicated that the point source phosphorus loading reductions should result in substantial decreases in phosphorus concentrations and trophic level in St. Albans Bay. When severe algal blooms persisted through the early 1990's, it was concluded that internal loading of phosphorus from the bottom sediments was delaying recovery of the bay [*Smeltzer, 1991; Smeltzer, et al., 1993*]. This study was designed to provide further insight into this phenomenon. The goals of the study were:

1) to conduct phosphorus fractionation analyses on an extensive set of sediment core samples for comparison to similar results from a 1982 survey; and
2) to predict the future response of St. Albans Bay by applying a coupled, dynamic mass balance model for phosphorus in the water column and bottom sediments.

SEDIMENT PHOSPHORUS SURVEY

Background

An extensive study of phosphorus levels and chemical forms was performed in 1982 on sediment cores collected at 26 locations in St. Albans Bay and the adjacent Stevens Brook wetland [*Ackerly, 1983*]. Sediments were retained from depths of 0-1 cm, 1-2 cm, 3-4 cm, and 7-8 cm. Total sediment phosphorus was measured on 23 samples using a perchloric acid digestion, and a phosphorus fractionation sequence of NH_4Cl, "Vermont buffer", NaOH, and HCl was applied to 20 cores. These reagents were assumed to extract soluble reactive P, readily available P, potentially available mineral P, and residual inorganic P, respectively. In St. Albans Bay, total sediment phosphorus ranged from 276 to 3234 µg/g (mean of 1198 µg/g), and was dominated by the HCl-extractable fraction, which averaged 51.3% of the total. In Stevens Brook wetland, total sediment P ranged from 1378 to 3255 µg/g (mean of 2340 µg/g); the potentially bioavailable fractions (NH_4Cl, "Vermont buffer", and NaOH) accounted for 48.5%, while HCl-P averaged only 24.5% of total P. Total sediment P varied spatially within the system, following the trend: Stevens Brook Wetland > Outer St. Albans Bay > Inner Bay > Middle Bay. Surface sediments (0-2 cm) were enriched in all forms of phosphorus (particularly the bioavail-

TABLE 1. Analytical methods.		
Analysis	Method	Original Reference
Sediment P Fractionation:		
Soluble Reactive P	1N NH$_4$Cl	*Williams, et al. [1967]*
Mineralizable P	1N NaOH	*Jackson [1970]*
Residual Inorganic P	0.5N HCl	*Williams, et al. [1976]*
Total Sediment P	Perchloric acid	*APHA, et al. [1975]*
P in Extraction Solutions	Ascorbic acid	*Harwood, et al. [1969]*
Percent Organic Matter	Ignition at 380 C	*Ackerly [1983]*

able forms) compared to deeper sediments (3-8 cm). The mean total sediment P concentrations for all cores were 1646 µg/g at a depth of 0-1 cm, 1449 µg/g at a depth of 1-2 cm, 1321 µg/g at a depth of 3-4 cm, and 1180 µg/g at a depth of 7-8 cm.

Methods

In August of 1992, sediment core samples were collected at 43 locations using a Wildlife Supply Co. KB core sampler (5 cm diam. X 50 cm length). Of these, 13 were located in the Stevens Brook wetland, 15 in inner St. Albans Bay, 12 in the middle bay, and three in the outer bay (Figure 1). A Loran system was used to locate sampling sites, which included 25 locations visited by Ackerly in 1982. Sediments from depths of 0-1 cm, 1-2 cm, 3-4 cm, 7-8 cm, and 11-12 cm were removed using a teflon-coated steel spatula and placed in plastic screw-cap containers. Sediments were dried at 60 C for 48 hr, ground using a glass mortar and pestle, and passed through a #140 sieve. Moisture content of each resulting sediment sample was determined from the weight loss upon drying at 105 C for 8 hr. Total sediment P and percent organic matter were measured on all samples. A phosphorus fractionation sequence of NH$_4$Cl, NaOH, and HCl was applied to samples from the 25 sites also visited in 1982. In order to obtain a valid comparison between 1982 and 1992 conditions, the analytical procedures used by Ackerly [*1983*] were adopted, with one exception. The Vermont buffer step was eliminated from the phosphorus fractionation scheme based on the observation that the sum of phosphorus extracted by NH$_4$Cl and NaOH was not significantly different than that removed by NaOH alone [*Ackerly, 1983*]. Analytical methods are summarized in Table 1.

Results and Discussion

For convenience in the discussion of results, the sum of NH$_4$Cl-P and NaOH-P is designated as biologically available inorganic phosphorus (BAIP). It should be noted that, while NaOH does extract organic phosphorus from sediments, this is not detected by the subsequent analysis. The difference between total sediment P and the sum of all extractable fractions is considered to be organic P. Mean concentrations are summarized by sediment depth and by region in Table 2. The results showed little variation with depth except in the outer bay, but did show significant horizontal differences within the bay/wetland system.

Mean total sediment P levels in wetland samples were roughly double, and BAIP levels roughly four times, those in the inner and middle bay samples. Concentrations of these

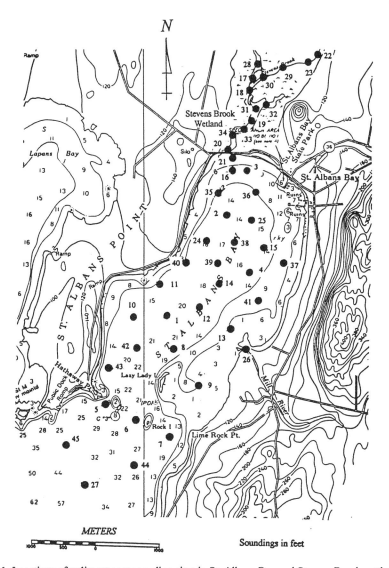

Figure 1. Locations of sediment core sampling sites in St. Albans Bay and Stevens Brook wetland.

two parameters in the outer bay were intermediate between those found in the wetland and in the inner bay. The mean BAIP was 49.9% of total P in the wetland, 25.2% in the inner bay, 23.6% in the middle bay, and 36.2% in the outer bay. Organic P accounted for a relatively constant fraction (about 20%) of total sediment P regardless of location. The concentration of HCl-P showed little variation with location, averaging 530 µg/g for all samples. In the inner and middle bay, HCl-P is the dominant form of sediment phosphorus, accounting for over 50% of total sediment P. It is believed that phosphorus associated with fine particulate matter and algal biomass is flushed out of the shallow inner and

TABLE 2. Summary of mean analytical results for St. Albans Bay sediments – 1992.

Depth/Region	BAIP (μg/g)	HCl-P (μg/g)	Organic P (μg/g)	Total P (μg/g)	% Organic Matter
Sorted by Depth:					
0-1 cm	481	529	271	1281	5.21
1-2 cm	448	512	254	1214	5.63
3-4 cm	431	532	227	1190	5.22
7-8 cm	433	532	243	1208	4.87
11-12 cm	401	549	238	1188	6.25
Sorted by Location:					
Stevens Brook Wetland	908	535	376	1819	7.69
Inner St. Albans Bay	252	544	205	1001	4.53
Middle St. Albans Bay	209	515	161	885	4.27
Outer St. Albans Bay	485	499	356	1340	6.51
All Data	456	530	253	1239	5.51

middle bay by wind-driven currents and deposited in the deeper outer bay. Based on these results, the 1992 pools of total P stored in the top 8 cm of bottom sediments were estimated as 16,170 kg in Stevens Brook wetland, 110,600 kg in the inner bay, 231,300 kg in the middle bay, and 250,000 kg in the outer bay.

Direct comparisons between 1992 and 1982 conditions were possible for phosphorus fractionations and total sediment P on 71 samples from 18 core locations. Statistical results of this comparison are presented in Table 3 for both the entire data set combined and for the data sorted by sediment depth. The changes in BAIP, HCl-P, organic P, and total sediment P could all be described by normal distributions. When data from all locations and depths were combined, a t-test for paired observations showed the

TABLE 3. Statistical comparison of phosphorus fractions – 1982 vs. 1992

Parameter	Depth (cm)	1982 Mean (μg/g)	1992 Mean (μg/g)	Mean Δ (μg/g)	St. Dev. (μg/g)	p
Total P	0-1	1809	1315	-494	413	<.001
BAIP	0-1	690	498	-192	295	.0113
HCl-P	0-1	634	525	-109	201	.0305
Organic P	0-1	484	291	-193	187	<.001
Total P	1-2	1579	1178	-401	348	<.001
BAIP	1-2	562	451	-111	258	.0816
HCl-P	1-2	584	504	-80	113	.0070
Organic P	1-2	434	223	-211	269	<.001
Total P	3-4	1446	1119	-327	277	<.001
BAIP	3-4	469	383	-86	199	.0811
HCl-P	3-4	570	529	-41	103	.0992
Organic P	3-4	406	296	-200	163	<.001
Total P	7-8	1310	1074	-236	307	.0087
BAIP	7-8	389	360	-29	265	.653
HCl-P	7-8	585	542	-43	136	.200
Organic P	7-8	336	172	-164	174	.0031
Total P	All	1539	1173	-366	346	<.001
BAIP	All	530	424	-106	258	<.001
HCl-P	All	594	525	-69	144	<.001
Organic P	All	416	224	-192	199	<.001

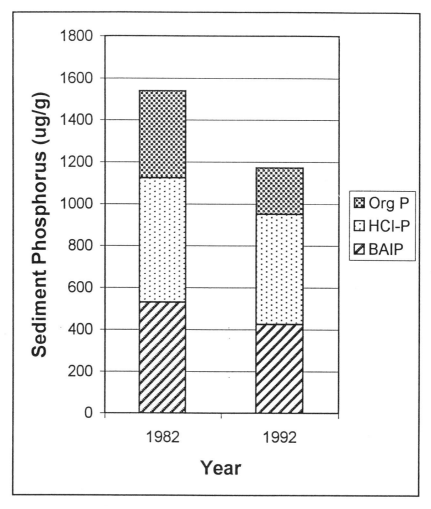

Figure 2. Changes in sediment phosphorus fractionation in St. Albans Bay between 1982 and 1992.

probability that 1992 and 1982 were identical to be less than 0.001 for all parameters. Total sediment P decreased by an average of 366 µg/g, or 23.8 % of the mean 1982 level. Of this, 106 µg/g (28.7%) came from the BAIP fraction, 69 µg/g (18.8%) from the HCl-P fraction, and 192 µg/g (52.5%) from the organic P fraction. The differences between 1982 and 1992 conditions are presented graphically in Figure 2. The large contribution of the organic P fraction is consistent with the observed decrease in mean organic matter content of the bottom sediments from 7.18% in 1982 to 4.89% in 1992. This suggests that, since the WWTF upgrade, organic matter is decomposing in the bottom sediments of the bay/wetland system at a faster rate than it is being deposited from the water column. This, in turn, suggests that primary productivity has declined on an annual average basis.

The depth-sorted data in Table 3 indicate that the loss of total sediment P was greatest for surface sediments and decreased rapidly with depth. Organic P accounted for the largest fraction of this loss at all depths, but did not show any clear trend with depth. The loss of both total sediment P and organic P were statistically significant at the p<.001 level at depths of 0-1 cm, 1-2 cm, and 3-4 cm, and at the p<.01 level at 7-8 cm. Much of the variability in the loss of sediment P with depth was due to the BAIP fraction. BAIP accounted for 39.0% of the loss in total sediment P in the 0-1 cm layer, and differences between 1982 and 1992 concentrations were highly significant (p=.0113). At 7-8 cm, BAIP accounted for only 12.2% of the loss in total sediment P, and the differences between 1982 and 1992 concentrations were not statistically significant (p=.653). The loss of HCl-P exhibited a modest decrease with depth. Based on this analysis, it was concluded that sediments at depths greater than 10 cm probably play a limited role in the exchange of phosphorus with the water column.

MASS BALANCE MODELING STUDY

Model Development

Chapra and Canale [*1991*] developed a mass balance model for phosphorus in lakes that accounts for the effects of sediment phosphorus recycle on water column concentrations by explicitly modeling phosphorus in the bottom sediments. The processes included in the mass balance equations are external loading, outflow, settling, recycle, and burial. This model was developed for lakes where spatial gradients in phosphorus levels can be neglected and the water body can be treated as one completely mixed spatial segment. Since St. Albans Bay and Stevens Brook wetland exhibit considerable spatial variability, modification of the model was necessary in order to divide the system into several spatial segments. Application of the model to a multi-segment system required that terms describing inflow from upstream segments and dispersive exchanges with adjacent segments be added.

The modified model equations applied for total phosphorus in the water column and bottom sediments, respectively, are:

$$V_{1i}\frac{dP_{1i}}{dt} = W_i + \sum_{j=1}^{m}(Q_{ji}P_{1j}) - (\sum_{j=1}^{m}Q_{ji})P_{1i} + \sum_{k=1}^{n}E_{ik}(P_{1k}-P_{1i}) - v_{si}A_{1i}P_{1i} \quad (1)$$
$$- v_{ri}A_{2i}P_{2i}$$

$$V_{2i}\frac{dP_{2i}}{dt} = v_{si}A_{1i}P_{1i} - v_{ri}A_{2i}P_{2i} - v_{bi}A_{2i}P_{2i} \quad (2)$$

where V_{1i} and V_{2i} are the volumes of water column segment i and underlying bottom sediment segment, respectively (m^3); P_{1i} and P_{2i} are total phosphorus (TP) concentrations in water column segment i and the underlying bottom sediment segment, respectively (mg/m^3); W_i is the external loading rate of TP to water column segment i (mg/d); m is the

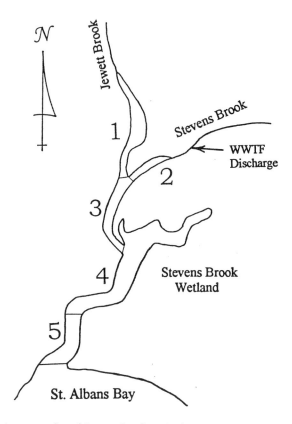

Figure 3. Spatial segmentation of Stevens Brook wetland.

number of upstream segments flowing into segment i; Q_{ij} is the flow from segment j to segment i (m³/d); P_{1j} is the TP concentration in the water column of upstream segment j (mg/m³); n is the number of segments adjacent to (i.e. sharing a boundary with) segment i; E_{ik} is the dispersive exchange flow between segments i and k (m³/d); P_{1k} is the TP concentration in adjacent segment k; v_{si} is the settling velocity of TP in the water column of segment i; A_{1i} and A_{2i} are the surface areas of water column and bottom sediments, respectively, in segment i (m²); v_{ri} is the resuspension velocity of bottom sediments in segment i (m/d); and v_{bi} is the burial velocity of bottom sediments in segment i (m/d).

These coupled differential equations were solved using a fourth-order Runge-Kutta algorithm. The model calculations were performed by a program written using Microsoft QuickBASIC (Version 4.5). A spreadsheet preprocessor was developed using Quattro Pro (Version 3.0, Borland International) to construct input data files for the model.

To apply the model to St. Albans Bay, the bay and adjacent wetland were divided into a total of nine spatial segments (Figures 3 and 4) based on observed trends in water quality. Each segment was considered to be completely mixed for purposes of model application. Key physical characteristics of the segments are presented in Table 4.

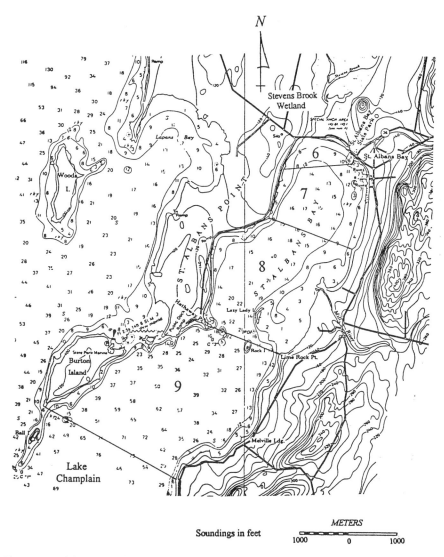

Figure 4. Spatial segmentation of St. Albans Bay.

Hydrodynamic Calibration Using Chloride

The hydrodynamic coefficients in finite-segment models for bays are frequently calibrated using chloride as a conservative tracer [*Richardson, 1974; Martin, et al., 1995*]. This approach was taken to estimate the dispersive exchange rate between adjacent segments in the bay/wetland system. Although loading and calibration data were only available for 1992, chloride was modeled for the years 1983 to 1992 in order to ensure stable model predictions for the calibration year (1992). Chloride loading estimates for all

TABLE 4. Morphometry of model segments

Segment	Volume (m^3)	Surface Area (m^2)	Mean Depth (m)
1	5.98 X 10^4	7.48 X 10^4	0.80
2	9.98 X 10^3	1.25 X 10^4	0.80
3	4.00 X 10^4	5.34 X 10^4	0.75
4	7.93 X 10^4	9.32 X 10^4	0.85
5	4.44 X 10^4	3.06 X 10^4	1.45
6	1.21 X 10^6	6.42 X 10^5	1.88
7	8.08 X 10^6	2.46 X 10^6	3.29
8	1.78 X 10^7	5.26 X 10^6	3.38
9	8.28 X 10^7	8.20 X 10^6	10.10

years were obtained for Stevens Brook, Jewett Brook, the St. Albans WWTF, and Mill River by taking the product of gaged flow rates and average annual chloride concentrations reported by Smeltzer, *et al.* [1993]. The chloride data base was not extensive enough to characterize the inverse relationship between chloride concentration and flow rate typically observed in streams. However, the goal in this case was to develop a reasonable representation of annual average hydrodynamic conditions in the bay/wetland system. It is believed that the approach utilized was suitable to meet this goal. The resulting chloride loading rates are summarized in Table 5. The chloride concentrations used for model calibration were the averages of all measurements taken within each spatial segment during the 1992 survey by Smeltzer, *et al.* [1993].

Because dispersive exchange is an important mass transport process in St. Albans Bay, model predictions are sensitive to boundary concentrations in Lake Champlain, just outside the mouth of the bay. Boundary concentrations of chloride were measured in several studies by the Vermont Department of Environmental Conservation (DEC) [*Smeltzer, 1983, 1991 and 1993*], and ranged from 7.7 mg/L in 1983 to 9.65 mg/L in 1992.

Dispersive exchange rates were set initially to the average of monthly values reported by Smeltzer, *et al.* [1993], and then adjusted to provide the best possible fit to the 1992 data. The chloride profile predicted by the calibrated model is compared to measured concentrations in Figure 5. The calibrated dispersive exchange rates are listed in Table 6. Agreement is excellent in the lower reaches of the wetland and in the bay (segments 4-9) where dispersion is very active. Here the predicted concentrations are heavily influenced

TABLE 5. Estimated chloride loadings (kg/d) to St. Albans Bay, 1983-1992

Year	Jewett Brook	Stevens Brook	Mill River	St. Albans WWTF
1983	1009.6	3335.5	4436.1	534.7
1984	813.7	2757.1	5924.6	509.4
1985	300.4	1038.1	2210.4	519.8
1986	335.3	2785.1	5651.7	503.4
1987	420.4	1337.3	2164.4	672.2
1988	157.2	1199.6	1582.5	681.4
1989	297.3	1820.2	2797.8	834.2
1990	521.7	3041.4	7954.5	1072.1
1991	499.3	2296.6	4061.5	674.0
1992	525.0	1485.0	1704.5	520.0

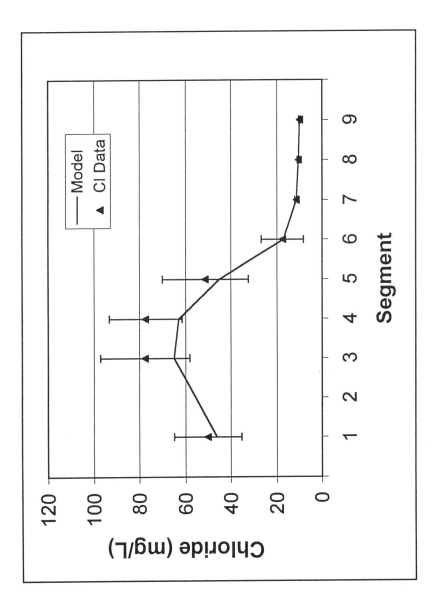

Figure 5. Comparison of model predictions of chloride with data from 1992 St. Albans Bay Diagnostic Feasibility Study.

TABLE 6. Calibrated dispersive exchange rates between model segments

Segment Boundary	Dispersive Exchange Rate (m^3/d)
1-3	1.000×10^4
2-3	0
3-4	1.000×10^5
4-5	1.000×10^4
5-6	3.000×10^4
6-7	4.000×10^5
7-8	2.880×10^6
8-9	6.480×10^6
9-Lake	1.741×10^7

by the boundary concentration in Lake Champlain. In the upper reaches of the wetland (sections 1-3), predicted chloride levels were slightly low, although still within one standard deviation of the mean of measured chloride concentrations. Since mass transport in these sections is dominated by advection, the only possible explanation for this discrepancy is uncertainty in the estimates of chloride loading.

Total Phosphorus Modeling

The total phosphorus model was calibrated by applying it for the 93 year period from 1900 to 1992. It was necessary to use a long time period for calibration in order to properly simulate the historic buildup of phosphorus in the bottom sediments. This long simulation period also provided a more rigorous test of the kinetic parameters in the model. The parameters required as model input were: initial conditions; boundary concentrations in Lake Champlain; inflows and total phosphorus loadings from tributaries and point sources; and kinetic coefficients (settling, resuspension, and burial rates).

Input data. No phosphorus measurements were available from the early 1900's. However, Hyde, *et al.* [1993] estimated that both point and nonpoint phosphorus loadings to the bay/wetland system had already increased significantly over undeveloped background conditions. For this study, it was assumed that water column total phosphorus concentrations in the year 1900 were 50-60% lower than measured 1992 levels in Stevens Brook wetland (except segment 2) and 30-40% lower in St. Albans Bay. The initial total phosphorus concentration in segment 2 was assumed to be 35-40% greater than 1992 levels since the point source loading from the St. Albans WWTF was greater in 1900. Initial total P levels in the bottom sediments were assumed to be about 50% less than measured concentrations in 1992.

Boundary conditions for total phosphorus in Lake Champlain were based on data collected from 1979 to 1981 at Vermont Lay Monitoring Program station #16, which lies about 0.5 km outside St. Albans Bay (i.e. from the outer boundary of segment 9). Concentrations were assumed to increase linearly from 10.0 µg/L in 1900 to 18.5 µg/L in 1985, then decrease linearly to 13.5 µg/L in 1990 and remain constant thereafter.

Flow rates for Jewett Brook, Stevens Brook, Mill River, and the City of St. Albans

TABLE 7. Mean inflows (m^3/d) to St. Albans Bay

Year	Jewett Brook	Stevens Brook	Mill River	St. Albans WWTF
1982	21,266	37,284	109,464	9,234
1983	33,652	37,061	126,747	6,684
1984	27,123	30,634	169,274	6,368
1985	10,104	11,534	63,155	6,498
1986	11,117	30,946	161,477	6,293
1987	14,012	14,859	61,839	8,403
1988	5,239	13,329	45,214	8,518
1989	9,910	20,224	79,937	10,428
1990	17,391	33,793	227,270	13,401
RCWP Mean	16,643	25,518	116,042	8,425
1992	17,500	16,500	48,700	6,500

WWTF were gaged from 1982 to 1990 as part of the St. Albans Bay Rural Clean Water Program [*1991*]. Intermittent gaging continued during 1992 [*Smeltzer, et al., 1993*]. The mean flows from these studies are listed in Table 7. For the tributaries, measured flows were used as model input when available, and the nine-year RCWP means were used for all other years. WWTF flows were assumed to increase linearly from zero in 1900 to the RCWP mean of 8425 m^3/d in 1930, and then remain constant until 1982. After 1982, measured flows were used, except for 1991, when the RCWP mean was used.

Total phosphorus loading rates for the period 1900-1981 were taken from Hyde, *et al.* [*1993*]. Loading rates were kept constant for each decade during this period. For 1982-1989, loading estimates from the RCWP were used. Combining the data on flow and total phosphorus loading from the RCWP, eight-year mean total phosphorus concentrations were calculated for each source, yielding the following results: 1153 µg/L for Jewett Brook; 674 µg/L for Stevens Brook; 340 µg/L for Mill River; and 3478 µg/L for the St. Albans WWTF. To estimate tributary loadings for 1990, the gaged flow rates were multiplied by the eight-year mean total P concentrations. The WWTF loading rate was estimated from the product of gaged flow and the two-year post-upgrade mean total P concentration. The same approach was used for 1991, except that mean RCWP flow rates were used in place of gaged flows. Loading rates for 1992 were taken from Smeltzer, *et al.* [*1993*]. All total phosphorus loading rates used as model input are summarized in Table 8.

The results of the sediment phosphorus survey indicated that sediments deeper than 8 cm were much less active in the exchange of phosphorus with the water column than sediments in the top 4 cm. Based on this observation, 10 cm was selected as the depth of the active sediment layer for the model application.

Model calibration. Water column total phosphorus data used in model calibration were taken from three studies - the St. Albans Bay Rural Clean Water Program [*1991*], the Lake Champlain Diagnostic Feasibility Study (LCDFS) [*Vermont DEC and New York State DEC, 1992*], and the Diagnostic-Feasibility Study for Control of Internal Phosphorus Loading from Sediments in St. Albans Bay (SABDFS) [*Smeltzer, et al., 1993*]. Each model segment contained at least one sampling station from these studies. Calibration was evaluated by comparing mean annual measured total P concentrations from each

TABLE 8. Estimated phosphorus loadings (kg/d) to St. Albans Bay

Year(s)	Jewett Brook	Stevens Brook	Mill River	St. Albans WWTF
1900-1909	13.38	13.38	29.85	24.28
1910-1919	12.52	12.52	27.94	26.50
1920-1929	15.44	15.44	34.44	30.85
1930-1939	19.38	19.38	43.24	54.24
1940-1949	16.46	16.46	36.74	46.39
1950-1959	17.15	17.15	38.27	62.30
1960-1969	19.21	19.21	42.86	62.95
1970-1979	16.98	16.98	37.88	68.33
1980-1981	17.15	17.15	38.27	60.00
1982	29.32	20.01	25.26	58.97
1983	24.03	21.90	18.36	33.50
1984	29.17	21.64	81.57	28.29
1985	11.01	8.14	15.04	32.91
1986	1.03	28.68	103.61	26.51
1987	17.58	9.52	22.76	15.77
1988	6.11	9.49	12.43	3.72
1989	15.94	11.53	27.12	4.10
1990	20.05	22.77	77.28	5.56
1991	19.19	17.20	39.44	3.50
1992	6.70	2.00	8.40	3.30

study with model predictions. Calibration data for total P in the bottom sediments were obtained from Ackerly [*1983*] and this study. Measured concentrations in µg/g were converted to mg/m^3 by equation (3). Data from all sediment cores taken in each spatial segment were averaged.

$$C_S = TP\ (1 - \phi) \cdot 2600 \qquad (3)$$

where C_S is the volumetric total P concentration in bottom sediments (mg/m^3); TP is the measured total P concentration in bottom sediments (µg/g, or mg/kg); ϕ is the porosity of bottom sediments; and 2600 is the assumed density of solid matter in the bottom sediments (kg/m^3).

During the review of input and calibration data, an apparent discrepancy was discovered between total phosphorus data collected by the RCWP and that collected by the LCDFS and SABDFS. For example, total phosphorus levels reported at RCWP station #11 for 1990 are more than twice those reported at LCDFS station #40, even when samples were collected on the same date. Since RCWP-11 is further from the mouth of Stevens Brook, total phosphorus concentrations should actually be lower than at SCDFS-40. In addition, estimates of total phosphorus loading via tributaries to St. Albans Bay for 1991 from the LCDFS [*Smeltzer, 1993*] and for 1992 from the SABDFS [*Smeltzer, et al., 1993*] were less than 25% of the eight-year (1982-89) mean RCWP estimates. These discrepancies cannot be explained by natural factors, such as climatic differences or changes in water quality. Rather, the most likely cause is differences in sampling and analytical methods. It is believed that the RCWP tributary loading estimates are more reliable due to more complete temporal coverage. However, the water column data from the LCDFS and SABDFS appear more reliable. For model calibration, all available data

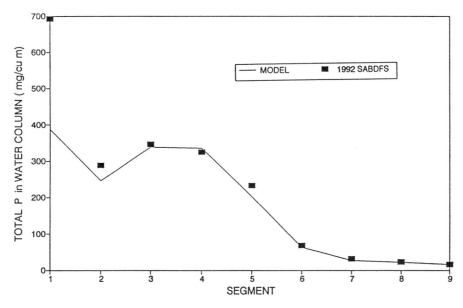

Figure 6. Comparison of calibrated model predictions for total phosphorus in the water column with average measured concentrations from the 1992 St. Albans Bay Diagnostic Feasibility Study.

sets were used for tributary loading estimates, while model predictions of total P in the water column were compared primarily to the LCDFS and SABDFS data.

The model calibration process was begun by setting values of the settling, resuspension, and burial velocities to those used by Chapra and Canale [1991] for Shagawa Lake. These were then adjusted within the ranges suggested by Bowie, et al. [1985] to obtain reasonable agreement between model predictions and measured total P concentrations in the water column and bottom sediments of the bay/wetland system. The model's numerical algorithm was run with a time step of 0.5 days.

Total phosphorus profile plots comparing model predictions to the data for all nine spatial segments were found to be an effective tool for evaluating the quality of model calibration. Model predictions of total P in the water column for 1992 are compared with SABDFS data in Figure 6. Predicted total sediment P for 1982 and 1992 are compared with data from Ackerly [1982] and this study, respectively, in Figure 7.

For the water column, agreement is excellent except for segment 1 (Jewett Brook). One possible explanation is that several samples taken from SABDFS station #16 (in segment 1) when the flow in Jewett Brook was very low showed unusually high total P levels. The calibration data reflect the influence of these samples. Since annual average flows and phosphorus loading rates were used as model inputs, the model was not able to reproduce this seasonal phenomenon. However, the ability to accurately model downstream segments was not adversely affected.

Model predictions of total P in the bottom sediments agreed closely with the 1992 data in all segments, and with the 1982 data in St. Albans Bay (segments 6-9). Agreement was not as good with the 1982 data from Stevens Brook wetland (segments 1-5). However, Ackerly [1983] analyzed only four sediment cores from the wetland. The data points

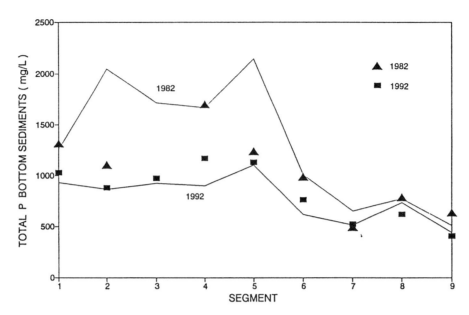

Figure 7. Comparison of calibrated model predictions for total phosphorus in the bottom sediments with average measured concentrations from Ackerly [*1983*] and this study for 1982 and 1992, respectively.

shown for segments 1, 2, 4 and 5 each represent results from only one sediment core, while no data were available for segment 3. Thus, it is possible that the data do not adequately represent the overall 1982 conditions in these segments.

Model predictions for the entire 93-year calibration period are shown for the water column and bottom sediments of St. Albans Bay (segments 6-9) in Figures 8 and 9, respectively. Dramatic decreases in total phosphorus levels were predicted for the period 1980-92 in segment 6. Since the extreme inner bay is strongly influenced by advective transport from the Stevens Brook wetland, it is expected that the response of segment 6 would closely parallel the decrease in total phosphorus loading via Stevens Brook. The model predicted much smaller (but nevertheless important) decreases in total P in the water column and sediments of segments 7, 8 and 9 following the point source loading reductions. This is also reasonable, considering the strong influence of wind-driven dispersion on water quality in the bay.

Calibrated values of settling velocity ranged from 0.05 m/d in the wetland to 2.00 m/d in the outer bay. Resuspension velocities ranged from 2.0×10^{-5} m/d in segment 8 to 7.0×10^{-5} m/d in segment 9, and burial velocities ranged from 1.0×10^{-6} m/d in segment 8 to 2.5×10^{-5} m/d in segments 2 and 5. The set of kinetic coefficients obtained through calibration is not unique - other combinations of values may also yield comparable agreement between the model predictions and field data. The uncertainty associated with these coefficients could be reduced by performing system-specific experimental measurements of the process kinetics.

Mass balance analysis. Using the calibrated model, the magnitudes of individual mass

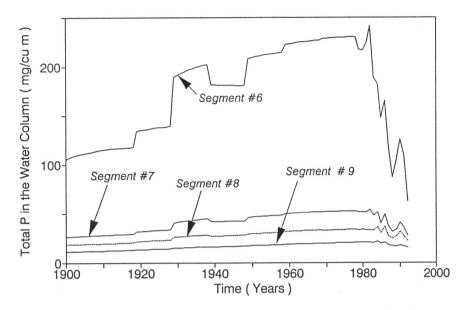

Figure 8. Calibrated model predictions of total phosphorus in the water column of St. Albans Bay for the period 1900-1992.

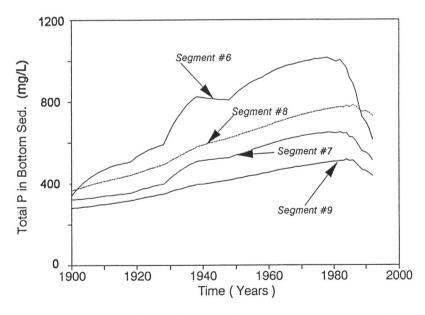

Figure 9. Calibrated model predictions of total phosphorus in the bottom sediments of St. Albans Bay for the period 1900-1992.

TABLE 9. Mass balances terms (kg/d) predicted for St. Albans Bay and
Stevens Brook wetland by calibrated total phosphorus model

Mass Balance Term	Bay 1982	Wetland 1982	Bay 1988	Wetland 1988
External loading	25.26	108.29	12.43	19.32
Upstream inflow	93.93	0.00	17.23	0.00
Outflow	-3.65	-96.88	-1.24	-17.23
Dispersion in	3.19	0.00	2.19	0.00
Dispersion out	-39.72	-3.19	-45.29	-2.19
Settling	-520.14	-19.29	-395.96	-9.45
Resuspension	433.26	13.00	406.74	9.83
(Net Settling)	-86.88	-6.29	10.78	0.38
Total	-4.86	1.93	-3.90	0.28

balance terms were compared for 1982 conditions (before point source loading reductions) and 1988 (after loading reductions). To avoid overstating the importance of dispersion between segments, the mass balance analysis was performed for the whole wetland by grouping segments 1-5 and for the whole bay by grouping segments 6-9. The results are presented in Table 9.

The wetland was dominated by advection, with the outflow of total phosphorus to St. Albans Bay nearly equaling the external load for both years. Net settling and dispersion were small relative to advection. In the bay, prior to the loading reduction, 72.9% of the incoming total phosphorus was deposited in the bottom sediments (i.e. net settling) and 33.3% was lost to Lake Champlain by dispersion. Advective loss of total P to the lake was negligible by comparison. The model indicated that, in both the bay and wetland, the bottom sediments changed from being a net sink (loss) or total P before the point source reduction to a net source after the WWTF upgrade. In 1988, 26.7% of the total phosphorus entering the water column of St. Albans Bay originated from the bottom sediments.

Prediction of future water quality trends. The calibrated model was applied to predict future trends in total P concentrations in the bay and wetland to the year 2050 under the assumption that loading rates would remain constant at 1992 levels. Unfortunately, there was considerable uncertainty regarding the actual 1992 loading rates. To account for this uncertainty, the model was run for two different loading scenarios. In the first future scenario (designated F1), loadings were considered to remain constant from 1992 to 2050 at the rates estimated by the 1992 SABDFS [*Smeltzer, et al., 1993*]. In the second scenario (F2), nonpoint source loadings from the tributaries were set at the eight-year (1982-1989) mean from the RCWP, and point source loading from the WWTF was set at the average of rates obtained by the RCWP for 1988 and 1989. Flows from all sources were set at the nine-year (1982-90) RCWP averages. The total P loading rates and flows used for the future simulations are summarized in Table 10. The boundary concentration of total P in Lake Champlain was assumed to remain at 13.5 µg/L indefinitely.

Sample model predictions of future conditions are presented for segment 6 in Figures 10 and 11. With the approach taken, future predictions take the form of total phosphorus concentration ranges that bracket the expected response of the system. Because the loading estimates for the two scenarios differ dramatically, the predicted concentration ranges are often very wide. Under scenario F1, water column total P concentrations would continue to decrease, reaching levels well below those observed in 1992. The predicted

TABLE 10. Total phosphorus loading rates (kg/d) and flows (m³/d) used in modeling future conditions in St. Albans Bay and Stevens Brook wetland

Quantity/Scenario	Jewett Brook	Stevens Brook	Mill River	St. Albans WWTF
Total P load - F1	6.70	2.00	8.40	3.30
Total P load - F2	17.93	16.36	38.27	3.91
Flow - F1 and F2	16,634	25,518	116,042	8,425

response of the bottom sediments was somewhat slower than the water column. Under scenario F2, the model predicted that, while bottom sediment total phosphorus levels would decline slightly for 5-10 years, water column concentrations would show very little change.

In May of 1991, the Vermont Water Resources Board adopted a set of water quality standards for total phosphorus in various areas of Lake Champlain. The goal set for St. Albans Bay was 17 µg/L. To evaluate likely future progress toward this goal, the model predictions were used to calculate volume-weighted average total P concentrations for the water column of the entire bay and wetland at several points in time. The results are presented in Table 11. If the average of scenarios F1 and F2 is assumed to provide the best available estimate of future conditions, the model predicted that the water quality standard will be met by the year 2005 with no further loading reductions. The model also predicted that, under scenario F1, the water quality goal would be met by 1996, while under scenario F2, it would not be met. In order to reduce the uncertainty associated with this prediction, it would be necessary to initiate a new study of tributary loading rates combining the temporal coverage of the RCWP study with a thorough evaluation of sampling and analytical methods.

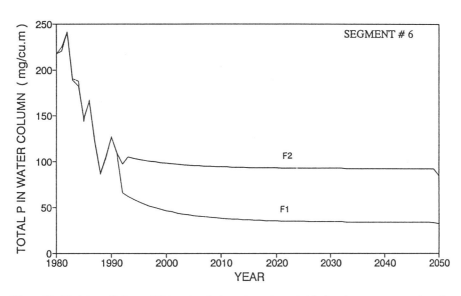

Figure 10. Model predictions of future trends in water column total phosphorus concentrations for inner St. Albans Bay, segment 6.

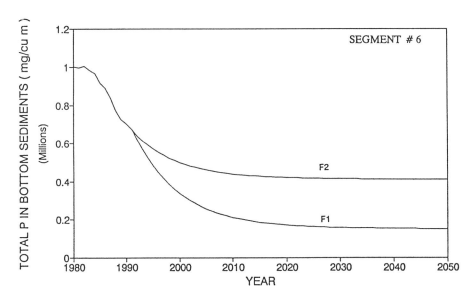

Figure 11. Model predictions of future trends in bottom sediment total phosphorus concentrations for inner St. Albans Bay, segment 6.

The results presented above should be interpreted with caution. Whole bay average total P concentrations may be somewhat misleading. Even if the water quality standard is met, total P in the wetland and inner bay may be high enough at times to cause significant algal growth. The combined inflows from the wetland typically contain 100-200 µg/L of dissolved P. During periods when tributary flows and dispersive exchange rates are low (e.g. mid- to late-summer), this inflow would mix slowly with the more dilute waters of the outer bay, maintaining elevated phosphorus levels capable of supporting algal blooms in the inner bay. As the rate of phosphorus recycle from the bottom sediments decreases in the future, the severity, frequency, and spatial extent of algal blooms should decrease. However, they are not likely to disappear totally unless steps are taken to reduce nonpoint source phosphorus loading.

SUMMARY

This study involved a detailed review of data from previous studies on St. Albans Bay, extensive sampling and analysis of phosphorus (P) levels and fractionation in St. Albans bay bottom sediments, and development and application of a mass balance model for total phosphorus in St. Albans Bay and Stevens Brook wetland.

The results of this study confirmed the suspicion that recycling of phosphorus from the bottom sediments of St. Albans Bay and Stevens Brook wetland was acting to delay the recovery of the bay following reductions in point source phosphorus loading. Under 1988 conditions (after the WWTF upgrade), the bottom sediments accounted for an estimated 26.7% of the phosphorus loading to St. Albans Bay. Between 1982 and 1992, total P

TABLE 11. Model predictions of future volume-weighted average total phosphorus concentrations (µg/L) in water columns of St. Albans Bay and Stevens Brook wetland

Year	Scenario F1 Bay	Scenario F1 Wetland	Scenario F2 Bay	Scenario F2 Wetland	Average Bay	Average Wetland
1982	27.4	1312	27.4	1312	27.4	1312
1988	20.2	645	20.2	645	20.2	645
1993	17.8	278	20.3	693	19.0	485
1998	16.3	245	19.5	682	17.9	463
2005	15.1	228	18.9	675	17.0	452
2015	14.2	221	18.5	673	16.3	447
2025	13.7	220	18.2	673	16.0	446
2049	13.3	219	18.0	672	15.7	446

levels in the bottom sediments decreased by an average of 366 µg/g, or 23.8%. The organic P fraction accounted for 52.5% of this loss, while biologically available inorganic P accounted for 28.7%. The decline in total sediment P was much more pronounced in the surface sediments (0-2 cm) than in deeper sediments (7-8 cm). The organic matter content of the bottom sediments decreased from 7.2% in 1982 to 4.9% in 1992, suggesting that the reduction in point source loading had resulted in a decline in annual average primary productivity in St. Albans Bay.

A mass balance model for total phosphorus in the bay/wetland system was developed and calibrated using historical data. The model accounted for the accumulation of phosphorus in the bottom sediments during several decades of high external loading, and the subsequent recycling of phosphorus following point source loading reductions. Advection was the dominant term in the phosphorus mass balance for spatial segments in Stevens Brook wetland; net sediment-water exchange was relatively unimportant. In St. Albans Bay, however, the net sediment-water exchange was a major loss mechanism for total P prior to point source reduction, and a significant source of phosphorus after the WWTF upgrade. In the bay, dispersion was found to be a much more important mass transport mechanism than advection.

The calibrated model was applied for the period 1992-2050 to predict future trends in water quality and the role of sediment phosphorus recycle in St. Albans Bay. Nonpoint total phosphorus loading estimates obtained by the Lake Champlain and St. Albans Bay Diagnostic-Feasibility studies were about 76% lower than those from the St. Albans Bay Rural Clean Water Program. This discrepancy introduced considerable uncertainty into the model simulations of future conditions. Using the average of these two loading estimates, the model predicted that the bay-wide average total phosphorus goal of 17 µg/L established for St. Albans Bay would be met by the year 2005 with no further loading reductions. The results indicate that the bottom sediments are likely to continue retarding the recovery of St. Albans Bay until at least the year 2005, and possibly longer.

Despite the recent and potential future improvements in the water quality of St. Albans Bay, it is expected that periodic summer algal blooms will continue due to the high nonpoint source phosphorus loading. Phosphorus concentrations in Stevens Brook wetland and inner St. Albans Bay can become extremely elevated during periods when tributary flows and dispersive exchange rates are low. Although the frequency and severity of these blooms are expected to decrease in the future, it is likely that they will only be eliminated through aggressive efforts to reduce nonpoint source phosphorus loading.

Acknowledgments. This work was supported by the Vermont Department of Environmental Conservation, Waterbury, VT. We would like to thank E. Smeltzer for providing assistance with sample collection and background data on St. Albans Bay.

REFERENCES

Ackerly, S., *Sediment-Phosphorus Relationships, St. Albans Bay, Lake Champlain*, M.S. thesis, University of Vermont, Burlington, VT, 1983.

APHA, AWWA, and WPCF, *Standard Methods for the Examination of Water and Wastewater*, 14th edition, American Public Health Association, Washington, DC, 1975.

Bowie, G. L., W. B. Mills, D. B. Porcella, C. L. Campbell, J. R. Pagenkopf, G. L. Rupp, K. M. Johnson, P. W. H. Chan, and S. A. Gehrini, *Rates, Constants, and Kinetics Formulations in Surface Water Quality Modeling*, 2nd ed., EPA/600/3-85/040. USEPA Environmental Research Laboratory, Athens, GA, 1985.

Chapra, S. C., and R. P. Canale, Long-term phenomenological model of phosphorus and oxygen for stratified lakes, *Water Res.*, 25, 707-715, 1991.

Harwood, J. E., R. A. Van Steenderen, and A. L. Kuhn, A rapid method for orthophosphate analysis at high concentrations in water, *Water Res.*, 3, 417-423, 1969.

Hyde, K., N. Kamman, and E. Smeltzer, History of phosphorus loadings to St. Albans Bay, 1850-1990, Vermont Department of Environmental Conservation, Waterbury, VT, 1993.

Jackson, M. L., *Soil Chemical Analysis*, Prentice-Hall, 1970.

Martin, S. C., S. C. Hinz, P. W. Rodgers, V. J. Bierman, J. V. DePinto, and T. C. Young, Calibration of a hydraulic transport model for Green Bay, Lake Michigan, *J. Great Lakes Res.*, 21, 599-609, 1995.

Richardson, W. L., Modeling chloride distribution in Saginaw Bay, *Proc. 17th Conf. Great Lakes Res.*, 462-470, Internat. Assoc. Great Lakes Res., 1974.

St. Albans Bay Rural Clean Water Program (RCWP), Final Report submitted by Vermont RCWP Coordinating Committee, Burlington, VT, 1991.

Smeltzer, E., Progress report on St. Albans Bay water quality modeling studies, Lake and Reservoir Modeling Series Report No. 6, Vermont Dept. of Water Res. and Environ. Eng., Montpelier, VT, 1983.

Smeltzer, E., The response of St. Albans Bay, Lake Champlain to phosphorus loading reductions, Vermont Dept. of Environmental Conservation, Waterbury, VT, 1991.

Smeltzer, E., Personal communication, Vermont Dept. of Environmental Conservation, Waterbury, VT, 1993.

Smeltzer, E., N. Kamman, K. Hyde, and J. C. Drake, Dynamic mass balance model of internal phosphorus loading in St. Albans Bay, Lake Champlain, Vermont Dept. of Environmental Conservation, Waterbury, VT, 1993.

Vermont Department of Environmental Conservation and New York State Department of Environmental Conservation, *Lake Champlain Diagnostic Feasibility Study Interim Progress Report*, submitted to USEPA Region I, Boston, MA, and Region II, New York, NY, 1992.

Williams, J. D. H., J. K. Syers, and T. W. Walker, Fractionation of soil inorganic phosphate by a modification of Chang & Jackson's procedure, *Soil Sci. Soc. Am. Proc.*, 31, 936-739, 1967.

Williams, J. D. H., J. M. Jaquet, and R. L. Thomas, Forms of phosphorus in the surficial sediments of Lake Erie, *J. Fish. Res. Bd. Can.*, 33, 413-429, 1976.

Importance of Instream Nutrient Storage to P Export from a Rural, Eutrophic River in Vermont, USA

Deane Wang, Suzanne N. Levine, Donald W. Meals Jr.
James P. Hoffmann, John C. Drake, and E. Alan Cassell

ABSTRACT

A three year study was conducted evaluating the potential of the LaPlatte River (VT) to store and release phosphorus. A mass balance approach was employed in which P stocks in water, surface sediments (upper 5 cm), epilithon, macrophytes + epiphytes, and detritus were measured at each season in two reaches representing slow and rapid sections of the river. In addition, P exchange between compartments was measured with radiotracers. Total phosphorus storage averaged 33 g m^{-2} at the soft-bottom, slow-flowing pool site, but was an order of magnitude lower (3.1 g m^{-2}) at the cobble-dominated, fast-flowing reach. The majority of P was stored in sediments at both reaches (97% and 71%, respectively), but P uptake rates by sediments were one-tenth those of biota. Phosphate and Rhodamine WT dye were added directly to the LaPlatte River to determine the degree of P retention over a 48 hr period. In winter, retention was negligible, but during the growing season, 30% of added P remained in the stream for at least 33 hr. These results support the concept that the biological component of the river has an important seasonal role in modifying P concentrations in stream water. Extrapolating to the LaPlatte River network, we estimate that about the same magnitude of P is stored in the river (range of 1.4-14 Mg, point estimate 8 Mg P) as is discharged from the river annually, 7.6 Mg P. In this system, potential for long-term P storage appears limited.

INTRODUCTION

Nutrients in streams draining human-dominated landscapes are often high in phosphorus (P), thus threatening surface waters with eutrophication. The dynamics associated with the emission, transport, transformation, storage, deposition and uptake of phosphorus in surface waters of concern are complex and variable given both the heterogeneity in the landscape (geology, soil, land use) and the intricate hierarchy of the stream network in which a nutrient such as P may move [Hunsaker and Levine 1995]. Despite our lack of knowledge about the relative importance of specific processes that govern the export of phosphorus into specific rivers and lakes, managers have implemented a wide variety of strategies to reduce phosphorus loading to surface water [Sharpley et al. 1994]. To maximize returns on resources and effort expended to meet the goal of P reduction, scientists and managers must use limited data interpreted in the context of conceptual and/or mathematical models of watershed P dynamics to prioritize the selection of strategies.

Management of P export to Lake Champlain has been identified as a priority activity by a U.S.-Canada, multi-stakeholder Management Conference convened to consider environmental issues in the Lake Champlain Basin [Lake Champlain Management Conference 1996]. Understanding river P dynamics in the region was subsequently identified as a need to support this priority. The objective of the research reported in this paper was to evaluate the role of in-stream P dynamics of rural rivers in modifying P export to Lake Champlain. Earlier research in two intensively-monitored watersheds where best management practices were implemented found that reductions in P emissions to small drainages and streams did not result in immediate reduction in P fluxes further downstream [Meals 1990, 1996, VT RCWP Coord. Comm. 1991]. This suggested that release of P over months to years from storage in the stream could mask reductions in contributions of P from the terrestrial component of the watershed system. The reverse may also be true, with the river retaining P and thus acting as an important buffer between land inputs of P and lake P loading. Other studies [Hill 1982] have documented the potential for stream retention of P, especially in waters carrying high P loads.

We evaluated the validity of the notion that streams and rivers are important sources or sinks of anthropogenic P by studying two common reach types in the LaPlatte River and using a mass balance approach to evaluate stocks of P in water, sediment, epilithon, macrophytes, epiphytes, and detritus in these reaches and their change in magnitude over seasons. The LaPlatte River was selected for study because phosphorus concentrations in this stream have been measured for decades [Meals 1990, 1996]. Some results of the study have been published previously [Hoffmann et al. 1996, Pelton et al. 1998].

METHODS

Overall Approach

We hypothesized that P concentration and flow regime are among the most important factors determining how P is processed within a reach. Thus a useful organizational framework for stream analysis and management might be classification of stream reaches by P concentration and physical environment (flow regime plus bottom characteristics). A P concentration vs. physical environment matrix might be formed, and each cell subsequently analyzed for storage and flux characteristics. Rivers could then be analyzed as composites of stream reach types. For the current study, we were faced with a high diversity of stream segment types. We chose to focus on an area with a relatively eutrophic river, typically exporting significant amounts of P to surface waters. Within this watershed we chose two stream reach types that are well-represented in the LaPlatte River: 1) a high P, low velocity and soft-bottomed condition and 2) a contrasting environment where P is high but the bottom cobble-dominated and water velocity rapid.

A mass balance approach was used to estimate P dynamics in the river reaches. The reach was conceptualized to be comprised of P stocks contained in five compartments (water, sediment, epilithon, macrophytes + epiphytes, and detritus). Seasonal stock assessments were conducted to estimate uptake or release from each compartment.

To provide additional perspectives on P cycling and retention in streams, other experimental studies were conducted, including: 1) measurements of ^{33}P- or ^{32}P- PO_4 uptake from the water by various compartments, 2) tracking of a plume of added phosphate past a downstream site, and 3) sorption-desorption studies using sediments brought into the laboratory. To integrate the estimates of stock and storage potential of phosphorus over space and time, two approaches were taken: 1) a simulation model of important P dynamics was built and parameterized with the values obtained through intensive study of the selected reaches, and 2) stream morphology (length and width of the entire river network) was determined from aerial photographs and a geographic information system (GIS), and then these data were used to extrapolate P stock sizes to the river as a whole.

To establish a whole-river context for the studied reaches, a morphometric study of the river network including stream orders, reach lengths, reach widths, and stream grade was conducted. Rough order-of-magnitude estimates of potential storage and uptake in the river were made from these data.

Study Sites

The LaPlatte River flows 24 km from headwaters in the forested foothills of Vermont's Green Mountains through a rural, agricultural watershed dominated by dairy farms and increasing suburban development. The river drains 13,815 ha and has a mean annual discharge of 1.2 m^3 s^{-1} [U.S. G. S. 1995]. For the water years 1993 and 1994 annual discharge was 1.1 m^3 s^{-1} and 1.2 m^3 s^{-1} thus reflecting fairly typical hydrologic conditions during the period of our study. It is moderately eutrophic with mean annual total phosphorus (TP) and total nitrogen concentrations of 0.14 and 0.75 mg/l [E. Smeltzer, Vermont Department of Environmental Conservation].

The two stream reaches selected for study were 150 m in length, and were named after the nearest access road. Bacon Dr., about 6 km from the mouth of the river, had a bottom substrate that was a diverse mixture of sand, silt, and clay, with interspersed areas of pebbles and cobbles. Suitable substrate for epilithon represented about 35% of the total area of this reach. A macrophyte community, dominated by Elodea canadensis Michx (Waterweed), and two species of Pondweed, Potamogeton pectinatus L. and Potamogeton natans L., covered approximately 75% of the reach by late summer. The Spear Street site was located about 10 km from the mouth of the river. Its substrate was mostly cobble with interspersed boulders and small patches of gravel mixed with sand, silt, and clay. Suitable substrate for epilithon represented about 75% of the total area of this reach. A sparse macrophyte community, dominated by the same genera as at Bacon Dr., covered less than 1% of the reach.

Stock Assessment

Stocks of P in various compartments in the reach were measured during each season beginning in August 1993 and ending April 1995. The major biotic and abiotic compartments of the stream were assumed to be water, sediment, epilithon, macrophytes + epiphytes, and detritus. Detritus was defined as detached organic material >1 cm in length and lying on the sediment surface or suspended in the water. Smaller detritus in suspension was included in water column TP while small detritus on the sediment surface or any-sized detritus within sediments was considered part of the sediment compartment. The periphyton compartment (epilithon and epiphytes) included sediment trapped in the matrix. Temporal variation in stock size was used as one estimate of the storage potential of the stream reach.

Grab samples of water were collected at the upper and lower ends of each reach. Sediment samples were obtained with a 4.7 cm diameter core and cut into three depth intervals, 0-2 cm, 2-5 cm, and 5-10 cm prior to analysis. A map was made (using a transit and meter tape) of the sediments in the reach, and a stratified random sampling regime employed during the August and November 1993 surveys. Subsequent sample locations were located randomly in the reach. For the Spear St. site, 24 sample locations for sediments were randomly selected at each survey. Because of the armored nature of this reach, coring was not feasible, and bulk grab samples were taken from the interstices between pebble, cobble and boulder material by scooping this material into pre-cleaned, pre-capped, core barrels. To quantify the area of interstitial fine grained sediments versus cobble/boulder armored stream bottom, 50 random locations were point counted. The point count consisted of identifying stream bottom substrate as "fine grained interstitial" or "boulder/cobble" at 10 cm intervals (100 points) within a 1 m^2 frame.

For epilithon, 20 random sample sites were selected within the reach as a whole. The proportion of the 20 locations that yielded samples was used to estimate the percentage of the reach supporting epilithon. Sampling involved scraping and brushing the epilithon from a measured area on a chosen rock into polypropylene cups. A random sampling design with 0.1m^2 quadrats was used for macrophyte collection whenever plants were scarce; and a stratified random sampling scheme when plants were abundant and patchy (the latter applied to Bacon Dr. in summer and in fall 1994). A surveyed map of plant coverage in the reach allowed for extrapolation of stratified sampling results to the stream. Quadrat number was normally 40, but in winter only 8 quadrats were sampled. Detritus sampling also employed 8-40 random 0.1 m^2 quadrats. The leaves, twigs, fragments of dead macrophytes and occasional invertebrate carcasses hand-picked from within the quadrats were > 1 cm in size, and thus could be classified as "coarse particulate organic matter" (CPOM) [Galarneau 1998]. All samples collected were kept in a cooler on ice and returned to the laboratory for processing within 2 hours.

Chemical Analyses

Water samples were analyzed for soluble reactive P (SRP), total dissolved P (TDP) and TP using standard methods (persulfate digestion, EPA method 365.4), and within recommended holding times, i.e. 48 hours for SRP and 28 days for TDP and TP [US EPA, 1983]. TDP samples were filtered (0.45 μm filter) and acidified with concentrated H_2SO_4 to pH<2 on the day of collection and stored at 4°C. Samples for TP analysis were acidified and stored without filtration.

Sediment samples were analyzed for total phosphorus using a combination of ignition and a hot HNO_3-H_2SO_4 digestion, followed by spectrophotometric analysis according to EPA 365.2. This procedure is a modification of the digestion for TP in water as presented in Standard Methods [APHA, 1985]. Sediment grain size was obtained by sieving for gravels (> 2 mm), sands (2 mm - 0.063 mm) and silts plus clays (< 0.063 mm). The last two fractions subsequently were analyzed for total phosphorus.

Epilithon was dried at 85° C to constant weight. For each site and date, the ash (450° C) and P content of the epilithon were estimated using 3-5 samples randomly chosen from field samples whose dry weight exceeded 150 mg. Epilithon P content was determined either by the ascorbic acid method, or by inductively-coupled plasma atomic emission spectroscopy (Leeman Labs, Inc., Lowell, MA, model PlasmaSpec 2.5) after wet digestion with concentrated sulfuric and nitric acids [APHA 1992].

Macrophytes (including epiphytes) and detritus were dried to constant weight at 105 °C, weighed, and then ground in a Wiley mill. Weighed subsamples were digested in hot H_2SO_4-HNO_3 for subsequent TP analysis using EPA Method 365.2. In some cases, epiphytes and macrophytes were separated prior to analysis. This was done by placing stems and leaves in bottles and shaking for 1.5 minutes [Jones, 1980]. The plant material then was removed from the bottle, dried and weighed, and the detached epiphytes collected by filtering them onto pre-weighed 0.45 μm membrane filters. These were subsequently dried and re-weighed. The filters and epiphytes were digested in hot acid and analyzed for TP employing the same methods as used for macrophytes.

Bioavailable P (BAP) was estimated on three occasions (summer 1994, fall 1994 and spring 1995) with the Selenastrum capricornutum Printz bioassay by measuring increase in cell density according to standard procedures [Miller et al. 1978]. Samples were autoclaved prior to bioassay. Cell density increased linearly in the graded series of external phosphate standards (typical $R^2 = 0.98$); however, recovery of the internal standards indicated inhibition on some occasions and stimulation on others which necessitated corrections of + 10 to 40%. In June 1995, BAP was estimated with ^{33}P-PO_4 by using the Rigler assay [Rigler 1966].

Uptake Studies

The basic approach for examining the removal of P from solution was to add radiolabeled phosphate to water in laboratory models of the stream and monitor its appearance in other compartments. The calculation of a rate constant for P transfer used a general equation for a two-compartment system:

$$y_t = y_{asymp.t} + (y_o - y_{asymp.0})e^{-kt} \qquad (1)$$

where y_t is the radioactivity in the water at time t, y_o is the radioactivity in the water at the initiation of the experiment, $y_{asymp.0}$ and $y_{asymp.t}$ are the asymptotic values for radioactivity in the water at times zero and t, k is the rate constant for P transfer, and t is time. Because the amount of P in particulate form (in organisms) greatly exceeded the amount of PO_4-P in the water, the asymptotic value for P in the water ($Y_{asymp.0}$ or $_t$) was assumed to be zero. Rate constants, therefore, could be obtained through linear regressions of either ln % (^{33}P or ^{32}P in solution) on time, or ln (^{33}P or ^{32}P activity in solution) on time. This approach was not appropriate for the studies we conducted with macrophytes and epiphytes because water and sediments were included in the stream microcosms that we used. Consequently, a more complex simulation model was developed to describe and quantify uptake dynamics. This model is fully described in Pelton et al. [1998].

Short-term uptake of phosphate by epilithic periphyton was determined using $^{33}PO_4$ and the method of Steinman et al. [1991]. Eight to sixteen rocks were selected from the study reach (average surface area: 37.1 (± 4.3, 1 SE) cm^2) and placed into eight 2 L polycarbonate containers with 1.0 L of 0.45 μm filtered river water. Adsorption controls (no rocks) and killed controls (autoclaved rocks) were also included. Each container was stirred with a magnetic stir bar set to 500 rpm creating a flow rate over the rocks of about 12 cm/s. Approximately 185 kBq of carrier-free $^{33}PO_4$ was added to each container. Light intensity was set at 300 μE m^{-2} s^{-1} (PAR) and average temperature was within 2 to 3 C of ambient river water. One ml samples were collected at timed intervals for 6 hr, filtered and added to CytoScint™ liquid scintillation solution (ICN Radiochemicals, Irvine, CA) for counting.

To examine P uptake by detritus and its microbial community, box elder leaves (*Acer negundo*), macrophyte debris (leaves from *Potamogeton natans*), and small pieces of wood were collected from the stream and brought into the laboratory. Each specimen was placed in a beaker containing 150 ml of stream water and a tracer amount of $^{32}PO_4$ and maintained at ambient temperatures in a water bath. Ten-milliliter aliquots of water were removed at intervals over a 24-h time course, filtered through GFF filters (to remove any detrital fragments), and counted on a liquid scintillation counter (Cerenkov counting).

Macrophytes and epiphytes were studied in plexiglass microcosms (79 x 40 x 40 cm) with a recirculating throughflow of river water. Each microcosm contained a removable tray which was filled in the field with cores of sediments and plants. All three of the plants common to the LaPlatte River (*Elodea canadensis*, *Potomogeton pectinatus*, and *Potomogeton natans*) were included in each tray. A recessed chamber in the microcosm held the tray of sediments so that the sediment surface lay flush with the floor of the larger chamber, and water flowed through plant stems and over sediments. Plants and sediments were collected exclusively from the Bacon Dr. reach, as collections from Spear St. would have depleted the scant stock present there. The microcosms were maintained on a wet table in the laboratory under a bank of VHO fluorescent lights (~300 $\mu E/m^2/s$; with light-dark cycles). Water flow rates (~30 cm/s, created by centrifugal pumps) were in the range observed at the Bacon Dr. site during summer low flow.

^{32}P was added to the water of 6 microcosms at average ambient river conditions (TP and TN:TP). ^{32}P loss from water and accumulation into macrophytes and epiphytes was monitored for 3 days (0.5, 1, 2, 6, 24, 48, and 72 h). Water samples (15 ml; 3 per microcosm per sampling) were collected directly into scintillation vials using a 5 ml automatic pipette. Plant samples (one or two from each species of plant present) were obtained by cutting 4-6 cm of stem and leaf from the top of plants in the chambers, placing the fragments in a bottle for shaking to remove epiphytes, and then transferring the plants to tins. The epiphytes released into the bottle were concentrated by filtering them onto 0.45 μm preweighed filters. On the third day, the artificial streams were broken down. Water was drained from the microcosms, with care taken to avoid disturbing the sediment-water interface, and samples of sediment and tank wall algae were collected with a cut-off 30 cc syringe and filter paper, respectively. Exchange rates were obtained from parameter estimation fitting observed data to a first-order, linear, differential compartment model. Methods are described in more detail in Pelton et al. [1998].

Sediment Sorption/desorption

Sediment samples from the Bacon Dr. reach were collected for sorption/desorption studies by hand-inserting 9 cm diameter plastic core barrels into bottom sediments and capping them *in situ*. To determine the adsorption isotherm, known weights of wet sediments were placed in centrifuge tubes, and 30 ml of phosphate solution added to each. The solutions consisted of filtered river water (FRW) spiked with KH_2PO_4 to several different concentrations, including a FRW control. The sediment slurries were shaken for 24 hours, centrifuged at 10,000 rpm for 12 minutes (Jarrel Ash J-21, relative centrifugal force and the supernatant decanted for analysis. To assess P release from sediments, the tubes, wet sediment and residual supernatant were re-weighed and 30 ml of unspiked FRW were added to each. This slurry was shaken for 8 hour, re-centrifuged, and the supernatant decanted for P analysis.

To ascertain rates of adsorption, wet sediment was placed in centrifuge tubes, and 30 ml of FRW (spiked to yield a standard addition of 1000 ug P/liter) were added. Samples were agitated for time intervals of 0.05, 0.33, 0.66, 1.25, 3, 6, 12, and 24 hours, after which they were filtered through 0.45 μm membrane filters or centrifuged at 10,000 rpm for 12 minutes. The supernatant was then analyzed for SRP. To those samples that had been shaken for 24 hours, 30 ml of FRW without a P spike were added, samples were again shaken for 24 hours, centrifuged and the supernatant analyzed to determine the amount of P released.

Spike Addition

To examine in-stream P retention during periodic export of P characteristic of runoff events, soluble P was added to the river and its movement downstream monitored over a 3 km reach (from Carpenter Road to Spear St.). Rhodamine dye (1 liter of 5%, FWT Red Liquid 50, Formulabs, Piqua, OH) detectable with a fluorometer to 1 ppb was added as a water tracer along with the P. It allowed for the estimation of spike dilution and indicated the arrival of the pulse at the downstream site. A winter low-flow experiment took place on December 20, 1994. Its P spike consisted of 0.89 kg of P as 0-50-0 P_2O_5 fertilizer (yielding an initial, peak concentration of about 14 mg L^{-1} P at the top of the reach, and a maximum concentration at the end of the reach of 0.21 mg L^{-1}). A summer low flow experiment was conducted from June 11-13, 1995, with the addition of 1.14 kg P as KH_2PO_4 (yielding an initial, peak concentration of about 28 mg L^{-1} P at the top of the reach, and a maximum concentration at the end of the reach of 0.19 mg L^{-1}). For both experiments, the P and the dye were mixed together in approximately 50 - 80 L of river water and dumped simultaneously into the stream at a point of concentrated flow in the stream cross-section. The P mass was chosen to provide an elevated P concentration at the end of the reach of sufficient duration and magnitude to allow a mass balance to be calculated with adequate statistical resolution.

Grab samples were collected at the downstream site (Spear St.) every ten minutes following the spike and immediately tested for fluorescence in a Turner-Sequoia Model 111 Fluorometer. Once fluorescence was detected, all subsequent samples were retained for P analysis. Sampling continued after the dye plume passed for 10 and 20 h during the first and second experiments, respectively. All samples were stored on ice while awaiting transport to the laboratory.

Phosphorus uptake dynamics were evaluated by comparing TP concentration with the temporal patterns of dye concentration (a conservative tracer). Overall TP recovery was estimated by integrating the area under the time vs. concentration*flow curve and comparing it to the known spike amount.

Whole-river extrapolations

The linear extent of the LaPlatte River surface water network was estimated using USGS single-line GIS data digitized from 1988 1:20,000 orthophotographs for Chittenden County, VT. Average width for streams greater than second order was estimated from the orthophotographs using randomly selected stream cross-sections. For first and second order streams, widths were measured in the field at road/stream intersections for eight randomly selected streams.

Stream surface area was estimated by multiplying length X width for each stream section in the database. This yielded an estimate of planar surface of the stream network at the time of the width evaluation. This planar surface area is thus directly applicable to the area unit used to establish quadrats used for measurement of stocks of periphyton, sediment, macrophytes, etc. (mg m^{-2} of stream surface).

RESULTS

The two studied reaches, representing a soft-bottom, slow-flowing pool site and a cobble-dominated, fast-flowing reach, were general similar in the chemistry of the water, but differed in their physical characteristics. A comparison of site characteristics for the two reaches is given in Table 1.

Phosphorus Stocks

The results of the seasonal stock assessments are summarized in Table 2 and in Figures 1a - 1b. P storage patterns in the two reaches differed markedly. The Bacon Dr. reach stored more than ten times as much P as the Spear St., 33 g m^2 vs. 3.1 g m^2, on average. In both reaches the majority of TP was stored in sediments, 97% at the more quiescent Bacon Drive site and 71% at the quick flowing Spear Street site. Our estimates of sediment P stock were highly sensitive to the depth of sediment assumed to be "active" in sediment-water exchange. We assumed that the active depth was 5 cm.; inclusion of greater or lesser sediment depth in our stock assessment would markedly change the apparent distribution of P among stream compartments. It is likely that the sediment does not behave as a single layer P stock, but rather as a gradient of interactive zones, with the surface most active, followed by a rapid drop off in activity. Seasonal

TABLE 1. General characteristics of the two studied reaches. Values are from this study.

Reach characteristics	Spear St.	Bacon Dr.
mean width at lowest flow (m)	8.6	12.3
mean width at highest flow (m)	10.1	14.1
mean depth at lowest flow (m)	0.15	0.25
mean depth at highest flow (m)	0.24	0.36
range flow (cfs)	4-23	5.5-26
range %gravel	52 - 96	8 - 24*
range % sand	4 - 48	44 - 86*
channel slope (rise/run)	0.0085	0.0025
range TP concentration (mg/l)	0.03 - 0.19	0.02 - 0.12
range SRP concen. (mg/l)	0.01 - 0.10	0.01 - 0.09

* 0-2 cm depth

variation in P concentration in the 0-2 cm sediment layer was high with a range of 0.25 mg P g-soil^{-1} (difference of the maximum - minimum = 44% of mean value), reflecting an active connection to seasonal fluctuations in P in the water column. The 2-5 cm layer only varied 0.07 mg P g-soil^{-1} over the two-year study period. Thus the 5 cm depth may be a conservative estimate of the actively exchanging P stocks, overestimating total system P.

TABLE 2. Summary of total phosphorus (TP) stocks (mg m^{-2}, per unit area of stream reach) by major compartments in the two studied reaches. Means are for TP stock for each compartment averaged over all stock assessment dates (with percentage of total P stock). Standard errors of the means appended in brackets.

Season	Water	Sediment	Epilithon	Macrophytes & epiphytes	Detritus	Total
	Spear St.	- - - - -	- - - - -	- - - - -	- - - - -	- - - - -
Summer 93	-	-	680 [280]	-	-	
Fall 93	20 [<1]	2540 [360]	1120 [140]	2 [2]	65 [78]	3740
Winter 93	16 [<1]	1600 [250]	260 [60]	0 [0]	160 [70]	2040
Spring 93	14 [<1]	2510 [570]	600 [290]	3 [2]	160 [100]	3280
Summer 94	34 [<1]	1610 [210]	560 [260]	4 [2]	1800 [2500]	4010
Fall 94	12 [<1]	2900 [440]	430 [110]	19 [9]	67 [27]	3420
Winter 94	11 [<1]	1960 [390]	430 [40]	0 [0]	33 [33]	2440
Spring 94	8 [0]	2420 [300]	120 [10]	0 [0]	52 [28]	2600
Means	16	2200	530	4	330	3080
% of total	0.6	71.4	17.2	0.1	10.7	100
	Bacon Dr.	- - - - -	- - - - -	- - - - -	- - - - -	- - - - -
Summer 93	-	23000 [2400]	370 [120]	1400 [240]	110 [80]	24900
Fall 93	23 [<1]	26500 [1800]	190 [190]	510 [170]	470 [180]	27700
Winter 93	22 [5]	36200 [7100]	0 [0]	0 [0]	340 [30]	36500
Spring 93	22 [<1]	37200 [2400]	240 [100]	11 [17]	630 [460]	38100
Summer 94	42 [1]	30300 [2500]	220 [80]	940 [230]	500 [630]	32000
Fall 94	15 [1]	38300 [5800]	240 [80]	320 [100]	170 [70]	39000
Winter 94	12 [<1]	36400 [3800]	70 [10]	0 [0]	170 [90]	36600
Spring 94	7 [1]	31400 [2300]	100 [10]	0 [0]	150 [90]	31700
Means	20	32400	180	400	320	33300
% of total	0.1	97.2	0.5	1.2	1.0	100

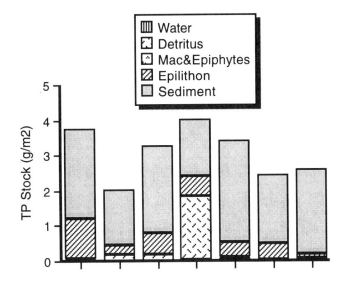

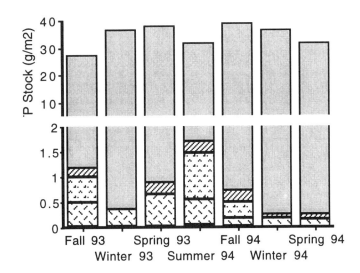

Figure 1. Distribution of total phosphorus (TP) stocks (mg m^{-2}) among all compartments for each assessment period: a) Spear Street reach, b) Bacon Drive reach. Note break in scale on the vertical axis for the Bacon Drive reach.

Epilithon, on average, contained 17% of the P stored at the Spear St. site, but accounted for no more than 1% of stored P in the Bacon Dr. reach. Macrophytes and epiphytes stored a negligible amount of P at Spear St. where plants were scarce, but contained as much as 6% of the P stored at Bacon Dr. when plants were abundant

(Summer 1993). A very small proportion of P was stored in detritus, except during one stock assessment at Spear St., when a large log happened to be sampled. The water compartment also contained little of the instantaneous standing stock of P (<1%).

The two reaches also differed with regard to seasonal variation in P stocks. Total P stocks at Bacon Dr. were relatively consistent over the study period, with little apparent seasonal variation. Total P stocks at Spear St., however, peaked in summer and fall and were smallest during winter. There was some seasonal variation in the distribution of P stocks among compartments at Spear St., while the distribution of P among compartments at Bacon Dr. was consistent except for an increase in macrophyte P in summer and fall (Figure 1b). Detritus P was highest in the fall, consistent with peak leaf fall in this region. Epilithon P stocks were low in winter, but otherwise relatively constant at Bacon Dr.

Total phosphorus standing stocks ranged from 2.0-4.0 g m^2 at the Spear St. reach, and from 25-39 g m^2 at the Bacon Dr. reach over the period of the study (Table 2). Sediment accounted for the large difference in total P stored between the two sites (average of 2.2 g m^2 vs. 32 g m^2), as well as for the large range in P stored within each sites (2.9-1.6 = 1.3 g m^2 at Spear and 38-23 = 15 g m^2 at Bacon).

The range in total P in all compartments in a reach is not a good estimate of the P uptake rate because the maximum difference in stock size does not occur in adjacent time periods (see Table 2). A rough approximation of this uptake rate can be estimated, however, by finding the maximum increase between adjacent seasonal samples. For the Spear St. reach, this rate is roughly 11 mg m^2 day^{-1} (winter to spring 1994, 1240 mg m^2 divided by 118 days). For Bacon Dr., it is roughly 96 mg m^2 day^{-1} (fall 1993 to winter 1994).

Bioavailable P (BAP) concentrations in the water phase of the reaches were normally much lower than total P concentrations, and frequently lower than SRP as well. The highest concentrations (120 - 140 μg L^{-1}) were measured during the summer of 1994 when we sampled after a modest storm, and the lowest during fall 1994 and spring 1995 (4 - 22 μg L^{-1}). On all dates, BAP was greater at Bacon Drive than at Spear Street (on average by about 2-fold). The percentage of TP that was BAP varied seasonally with lowest values in the fall (4 and 12% at Spear St. and Bacon Drive, 4 and 8 μg L^{-1} BAP, respectively) and highest values in the summer (19 and 35% at Spear St. and Bacon Drive, 25 and 55 μg L^{-1} BAP, respectively). The percentage of TP that was BAP varied seasonally with lowest values in the fall (4% at Spear St. and 12% at Bacon Drive, 4 and 8 μg L^{-1} BAP, respectively), and highest values in the summer (19 and 35%, 25 and 55μg L^{-1} BAP, at Spear St. and Bacon Drive respectively).

Short-term uptake

Four experimental approaches were employed to estimate short-term uptake rates by compartments in the stream ecosystem: 1) measurement of ^{32}PO$_4$ uptake by macrophytes, epiphytes, and sediments in microcosms in which water was labeled, 2) measurement of ^{33}PO$_4$ uptake by epilithon in stirred containers with labeled ambient river water, 3) sorption-desorption studies on sediments in the laboratory, and 4) our whole-river study of the downstream movement of a P amendment.

Uptake by epilithon and detritus. Phosphate flux to epilithon was measured in summer and fall 1994, and in spring 1995 (Table 3). For the Spear St. reach, average summer phosphate-P uptake was 66 (+10) mg m^2 day^{-1}, falling to a mean of 8 (+1) mg m^2 day^{-1} in autumn, and 5.6 (+0.3) mg m^2 day^{-1} in the following spring (per unit rock surface). In the summer about 43% of the uptake was abiotic, probably from adsorption onto the mucilaginous epilithon matrix and associated silts and clays. This dropped to 20-30% for spring and fall. Uptake was measured in the fall only at Bacon Dr., with an

TABLE 3. Epilithon uptake rates (mg m^{-2} day^{-1} of rock surface) for various conditions during three seasons. Treatments: a) description of flow (< 5 cm s^{-1} and >20cm s^{-1}) and light conditions from which rocks were removed for summer 1994; b) the visible amount of epilithon on rocks for fall 1994 (>35 g m^2 AFDM and <20 g m^2 AFDM); and c) amount of added orthophosphate for spring 1995. Values for the reach average with total dry weight and ash-free dry weight (AFDM) are also provided. BAP is bioavailable P (μg L^{-1}) determined by Selenastrum bioassay for summer and fall, and by Rigler assay for spring. See text for details.

Site	Season	Treatment	BAP	SRP	Uptake
Spear St.	Sum94	Slow flow, unshaded	25	100	77
Spear St.	Sum94	Slow flow, shaded	25	100	34
Spear St.	Sum94	Fast flow, unshaded	25	100	104
Spear St.	Sum94	Fast flow, shaded	25	100	51
Spear St.	Sum94	Killed control	25	100	30
Spear St.	Sum94	Adsorption control	25	100	0.2
Spear St.	Sum94	Reach average (biomass= 16 g/m^2 of rock surface AFDM)	25	100	66
Spear St.	Fall 94	High biomass	3	60	7.1
Spear St.	Fall 94	High biomass	3	60	8.3
Spear St.	Fall 94	Low biomass	3	60	2.3
Spear St.	Fall 94	Killed control, high biomass	3	60	2.8
Spear St.	Fall 94	Killed control, low biomass	3	60	1.8
Spear St.	Fall 94	Reach average (biomass= 47 g/m^2 of rock surface AFDM)	3	60	8.0
Bacon Dr.	Fall 94	High biomass	7	50	17
Bacon Dr.	Fall 94	High biomass	7	50	18
Bacon Dr.	Fall 94	Low biomass	7	50	11
Bacon Dr.	Fall 94	Reach average (biomass= 13 g/m^2 of rock surface AFDM)	7	50	17
Spear St.	Spr 95	Ambient P	3	15	5.6
Spear St.	Spr 95	Ambient P	3	15	6.2
Spear St.	Spr 95	Ambient + 4μg PO$_4$-P/L	7	19	12
Spear St.	Spr 95	Ambient + 10μg PO$_4$-P/L	13	25	30
Spear St.	Spr 95	Ambient + 20μg PO$_4$-P/L	23	35	80
Spear St.	Spr 95	Ambient + 40μg PO$_4$-P/L	43	55	80
Spear St.	Spr 95	Killed control, ambient P	3	15	1.5

estimated rate of 17 (+3) mg m^2 day^{-1} (per unit rock surface), of which about 15% was abiotic. The highest uptake rates were associated with high levels of bioavailable phosphorus (BAP) as indicated in Table 3.

Uptake per unit area of stream reach (as opposed to exposed rock substrate) was somewhat lower than the rates indicated above because exposed rock substrate made up about 80% and 40% of the stream surface at Spear St. and Bacon Dr., respectively. Thus, on a stream reach basis, spring and fall rates ranged from 4 to 7 mg m^2 day^{-1}. Summer rates at Spear ranged from 25 to 80 mg m^2 day^{-1} per unit area of stream reach. P uptake by tree leaf and dead macrophytes detritus was relatively low, and uptake by dead wood negligible (Table 4).

Uptake by macrophytes and epiphytes in microcosms. Two experiments in August and September 1994 demonstrated that both macrophytes and associated epiphytes removed P from the water column (Table 4). However, epiphyte uptake was more rapid, ranging up to 18 mg m^2 day^{-1} under ambient phosphorus levels. In this experimental design, with the water column as the P source, sediment uptake rates were much lower, <1 mg m^2 day^{-1}. Total P uptake (macrophytes, epiphytes, sediment) ranged from 13 to 30

TABLE 4. Macrophyte, epiphyte, sediment, and detritus uptake rates (specific flux in mg g^{-1} day^{-1} and uptake in mg m^{-2} day^{-1} of stream surface area) as determined by ^{32}P radioactive tracer experiments on Bacon Drive microcosms. Associated bioavailable phosphorus (BAP, μg L^{-1}) is also included. See text for experimental details.

Compartment	Month 1994	BAP	Specific flux	Uptake
Macrophyte	August	21	0.051	9.8
Epiphyte	August	21	0.14	18
Sediment	August	21		0.68
Macrophyte	August	500	1.0	170
Epiphyte	August	500	1.4	180
Sediment	August	500		14
Macrophyte	September	12	0.029	5.0
Epiphyte	September	12	0.062	8.0
Sediment	September	12		0.33
Tree leaves	October	4	0.011	0.35
Dead macrophytes	October	4	0.035	1.3
Dead wood	November	12	0.000002	0.004

mg m^2 day^{-1}. Addition of available P dramatically altered equilibrium conditions, increasing uptake rates by more than an order of magnitude, to >300 mg m^2 day^{-1} including all compartments (Table 4).

Sediment Adsorption/Desorption. Sediment adsorption/desorption studies suggested that the LaPlatte River sediments are close to P equilibrium at typical ambient TP levels. Langmuir and Freundlich isotherm equations fitted to the observed adsorption/desorption data were as follows:

$$\text{Freundlich: } S = K_F * C^n, \tag{2}$$

where S is the μg P adsorbed /g sediment, and C is the equilibrium solution concentration (μg/l)
parameters: $K_F = -0.0003$, $n = 0.634$
fit: $r^2 = 0.951$

$$\text{Langmuir: } S = S_T * C/(K_L + C), \tag{3}$$

where S_T = maximum adsorption capacity; S and C are as above.
parameters: $K_L = 1088$, $S_T = 246$ μg P g^{-1} sediment

In adsorption rate experiments, using filtered river water containing an initial P spike equivalent to 1000 μg L^{-1}, agitated sediment reached an equilibrium in approximately 12 hours, with about half the total adsorption taking place in less than 1 hour.

These results demonstrate that the studied sediment is far from saturated under ambient conditions, having a high and rapid capacity to take up phosphate at high phosphate concentrations. Because laboratory conditions were so different from the river environment, rates were not extrapolated to a unit area basis. We assume that during periods of high flow with sediment resuspension, there is adequate time to reach adsorption equilibrium, leading us to believe that the potential sediment P uptake capacity is high in the LaPlatte.

Spike addition. Spike addition of orthophosphate to the LaPlatte River in winter revealed little or no P retention within the 3 km study reach over 20 hr. By contrast,

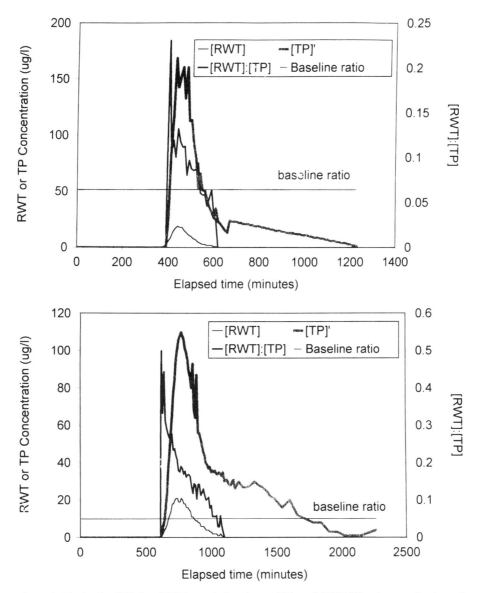

Figure 2. Rhodamine WT dye (RWT), total phosphorus (TP), and RWT:TP ratio over the time of the spike addition experiments in the LaPlatte River, VT. Concentration units are $\mu g\ L^{-1}$. Theoretical ratio is the ratio of dye to TP in the original addition. Elapsed time is from time of spike addition 3 km upstream, a) December 1994, b) June 1995.

when the same reach was spiked in summer, 31% of the added P was retained for at least 33 hr. The dynamics of P uptake in the two experiments were different because of the greater development of a biological community in summer (Fig. 2a and 2b). In winter, as well as summer, the P plume in the river lagged the Rhodamine WT dye plume, attesting to the uptake of P by sediments, as well as by biota. This lag is also captured by plotting the dye to P ratio (Figs. 2a and 2b).

Sediment sorption due to the high phosphate concentrations present in the stream immediately after spike addition appears to have been followed by desorption of P into the low P water that followed passage of the P plume. This pattern, as predicted by the P isotherm and rate measurements observed in the laboratory sediment studies, seemed to be augmented by biotic uptake during the summer experiment. High biotic uptake rates following P enrichment were also documented in radioactive tracer studies for epilithon, macrophytes, and epiphytes, as well as for sediment.

In the winter experiment, 582 g P, or 65% of the added P, passed the downstream end of the study reach during the passage of the dye plume. This indicates that 35% or 308 g of the added P was retained at least temporarily in the reach. In summer, 759 g or 66% of P was retained. Thus, initial uptake over the 30,000 m^2 reach was 10 mg m^2 for winter and 25 mg m^2 for summer. Using the time of passage of the dye plume to provide a time frame for this uptake, crude estimates of winter and summer uptake rates are 60 and 78 mg m^2 day^{-1}, respectively.

Whole-river extrapolations

The LaPlatte River is a fifth order stream (stream order identification following Strahler [1957]) with reasonably typical morphometric relationships. Its bifurcation ratio of 3.77 falls within a range for watersheds with homogeneous lithology of 3 and 5 [Summerfield 1991], and its drainage density is 1.52 km k m^2, compared with typical densities for US watersheds of between 0.41 and 2.90 [Smith and Stopp 1978]. Stream surface area increases with stream order going from 8.14 ha for a first order stream to 13.52 ha for a fifth order stream (Fig. 3). Total stream surface area was 50.95 ha. This value, if associated with a storage or uptake rate per unit area, could provide a rough estimate of the potential P storage in the whole river. The studied reaches represent only a small component of the whole river diversity represented in a 4th order, fairly shallow grade stream (Fig. 3). However, if the selected reaches can be considered to be on the low and high end of potential P storage or uptake, then the results from these two reaches may bracket the value for the river as a whole. Taking the point estimates of seasonal increase in river storage for the two reaches (1.2 and 8.9 g m^2 for winter to spring 1994 for Spear, and fall 1993 to winter 1994 for Bacon, respectively, see Table 2) and multiplying this by total river surface area, we obtain for the whole river a range of 0.6 to 4.5 Mg of stored P from one season to the next.

DISCUSSION

Current role of in-stream P storage on P loading

A fair amount is known about the dynamics of P in streams, including uptake by epilithon and macrophytes; sorption-desorption in sediments; input, storage, and flushing of detritus; and the input of dissolved and particulate diffuse pollutants. This is especially true for forested upland or headwater streams [e.g., Elwood et al. 1981, Mulholland et al. 1985, Graham 1988, 1990, Munn and Meyer 1990, Rosemond 1994]. In this study, the goal was to quantify P dynamics, especially uptake and storage, in rural/agricultural rivers in the Lake Champlain Basin. The task was difficult because of the high spatial and temporal complexity of streams and rivers in the 21,326 km^2 basin. Factors operating to diversify river networks over river continua include: river complexity (topographic, physiographic, human influence), seasons (temperature influenced chemical equilibria, biotic activity, human activities), stochastic events (storms, duration, frequency, intensity), and spatial heterogeneity. The best hope for a study of this kind is to provide a

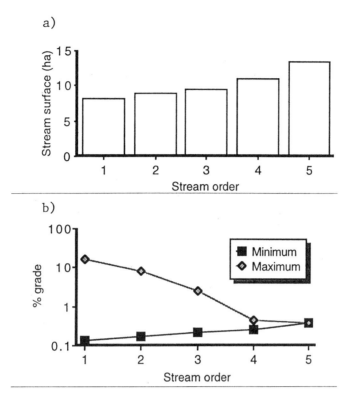

Figure 3. a) Area of stream surface (ha) and b) minimum and maximum percent grade by stream order (note log scale for % grade), for the LaPlatte River, Vermont.

conceptual framework for organizing observations and establish a first approximation of the range of possible values.

A variety of approaches was used in this study to estimate potential P storage and uptake rates for two reaches in the LaPlatte River. First, a mass balance approach with periodic stock assessments was used to estimate change in P storage in various stream reach compartments. An increase in the P content of a reach results from both uptake of P (biotic uptake, sediment absorption, adsorption onto detritus) or influx of P-containing material (sediment bedload or deposition, detritus deposition). For the cobble-dominated, fast-flowing reach at Spear St., the highest observed increase in P content translates to 11 mg m^2 day^{-1} (Table 5). For the soft-bottom, slow-flowing pool environment at Bacon Dr., the estimate was 96 mg m^2 day^{-1}.

Uptake estimates from laboratory process studies suggested that when P concentrations (especially BAP) are low, only moderate amounts of phosphorus, 2 to 8 mg m^2 day^{-1}, are retained at the Spear St. reach, with the uptake being accomplished primarily by epilithon (Table 5). Epilithon control over P dynamics is not unusual for a stream where sufficient light and/or minimal grazing occurs [Mulholland et al. 1995, Steinman et al. 1995]. The epilithon biomass (20-40 g AFDM m^2) achieving this uptake rate was markedly higher than the 5-15 g AFDM m^2 reported for most other studies [Nelson et al. 1969, Corning et al 1989, Paul and Duthie 1989, Rosemond 1994], with the exception of Graham's study [1988] in New Zealand, which reported 40 g AFDM m^2.

TABLE 5. Comparison of uptake rates (mg m^{-2} day^{-1}, per unit area of stream reach) from a) stock measurements, b) radiotracer uptake experiments, and c) in-situ spike addition in the LaPlatte River.

Method	Spear St.	Bacon Dr.	Compartment	Notes
Stock assessment	11	96	whole ecosystem	from Table 2 max. estimate
Spike addition	60-78 (3 km upstream reach)		whole ecosystem	BAP ~ 3 μg L^{-1}
Radiotracer labelling		14	microcosm ecosystem	from Table 4 BAP ~ 12 μg L^{-1}
Radiotracer labelling		30	microcosm ecosystem	from Table 4 BAP ~ 20-30 μg L^{-1}
Radiotracer labelling	24-80		epilithon	converted from Table 3 BAP ~ 20-30 μg L^{-1}
Radiotracer labelling		4-8	epilithon	converted from Table 3 BAP ~ 7 μg L^{-1}
Radiotracer labelling	2-6.5		epilithon	converted from Table 3 BAP ~ 3 μg L^{-1}

However, the maximum uptake rates observed in these studies (15-30 mg m^2 day^{-1}) were similar to ours when bioavailable P was higher.

Uptake studies indicate a high sensitivity of the uptake rate to BAP for all biological compartments (Table 5). These values increase dramatically if high bioavailable P (>20 ug L^{-1}) is present as was the case in the August 1994 survey, with epilithon uptake from Spear St. reaching as high as 100 mg m^2 day^{-1} and whole ecosystem uptake from Bacon Dr. as high as 30 mg m^2 day^{-1}. These values are similar to the estimates of average short-term uptake of P in the spike addition experiments (60 to 78 mg m^2 day^{-1}). In those in-situ tests, average SRP river concentrations were high due to the addition of phosphate in a single high dose.

All of the estimates provide some general consistency (Table 5). The stock assessment approach, which integrates changes in storage over a season of high and low flows and high and low available P levels, yields a range from perhaps 10-100 mg m^2 day^{-1}. The uptake studies suggest a range from 2-8 mg m^2 day^{-1} during the greater part of the year when bioavailable P is a lower percentage of total P, but reaching to as high as 24-80 mg m^2 day^{-1} during summer periods when bioavailable P is high. Combining these values over a season with variable bioavailable P, results in an range not so different from the 10-100 mg m^2 day^{-1} obtained via the stock assessments.

The Spear St. reach may represent the low end of the P storage and uptake range due to its low sediment, high cobble composition. In contrast, the Bacon Dr. reach may reflect the high end of the spectrum with its high, fine-sediment content and slow pools allowing vigorous macrophyte growth. Thus these two reaches may bracket the potential P storage and uptake for the river as a whole. Extrapolating the 10-100 mg m^2 day^{-1} rate to the whole 50.95 ha stream surface yields a range for P storage between seasons of 0.6 to 4.5 Mg P. This estimate for the current range of P levels in the river is somewhat lower than the estimated annual diffuse or nonpoint P load to Lake Champlain from the LaPlatte River, 7.6 Mg for 1991 [VT DEC, NYS DEC, 1997]. Thus, our various studies support the notion that under current river conditions, in-stream processes do not comprise a dominant sink for phosphorus coming from the watershed (predominantly agricultural diffuse P). Hill *et al.* [1998] had a similar conclusion in their evaluation of in-stream and hyporheic environments.

Extrapolation of the Spear St. and the Bacon Dr. stocks to the entire river indicates that the total P stock in all compartments ranges from 1.4 to 14 Mg (middle point-estimate of 8 Mg). By observation, we know that the true mix of reach characteristics lies somewhere between those for these two sites, and thus the total P stock in the river should also lie between these two values. From this we conclude that roughly one annual load of P (about 8 Mg) may be currently stored in the various in-stream compartments in the LaPlatte River. This P load to stock ratio of about 1 then also frames the potential role of the river in influencing the current estimated loading of P to Lake Champlain.

Potential In-stream P Uptake - Implications for Management

Under the current concentrations of total P (0.03-0.14 mg/L^{-1} during each of the stock assessment periods), in-stream storage of P is apparently not a major factor influencing the current influx of total P from diffuse sources to Lake Champlain. However, this does not imply that the in-stream environment is not a potential "buffer" for particular forms of P or different P loads. Bioavailable P was generally a small percentage of total P. Uptake rates as measured in experimental work demonstrated high uptake when BAP was high. Uptake of P as BAP could thus be an important percentage of this pool of P moving through the river under conditions of high P availability, especially during our summer field period (Tables 3 and 4). High summer concentrations were observed to follow different P dynamics in other river studies [e.g., Klotz 1991, Mulholland 1992, Rosemond 1994].

Short-term P dynamics under elevated levels of bioavailable P can reflect much higher rates of uptake in most of the compartments studied (periphyton and macrophytes, see Tables 3, 4; sediment, see Langmuir and Freudlich equations). Periodic runoff inputs of available P during summer storms or lapses in performance of point sources can result in these abnormally high P levels in the stream. Our laboratory data suggest that for such short-term events, the stream has a moderate capacity for taking up phosphorus, especially the bioavailable forms. While short-term abiotic uptake/sorption by fine sediments is also apparently high as demonstrated by the spike addition experiments, this removal is followed by a quick release of P back to the water when P concentrations decline. Sustained emission of effluent with a high BAP content (e.g., from a conventional sewage treatment plant with no P controls) would be ameliorated over longer periods of times as the in-stream system, and especially the sediment, shifted to a higher equilibrium condition. Other studies [e.g., Hill 1982, Mulholland et al. 1990, Hart et al. 1992] have documented a similar P dynamic with higher levels of available P in stream water.

Thus the function of the river as source/sink is dependent on the pollution history of the river and the levels and forms of P in the water. For the relatively eutrophic LaPlatte River, reduction of P inputs to Lake Champlain must occur through modification of land practices responsible for P export to the stream. The concept that management would be most effective if distance of sources from the stream mouth were taken into consideration does not appear to make sense, except for cases of substantial P contributions from point sources, which tend to have highly bioavailable forms of P. Most such sources in the LaPlatte River have already been the subject of P reduction strategies (tertiary treatment). Diffuse sources of P, wherever they are with respect to the river network, need equal attention for P reduction. Management priorities must be established with other criteria.

Can our results be extended beyond the LaPlatte River? Just as information about processes is extended beyond the specific research sites, our conclusions about P retention may provide a useful perspective on other rivers in the Lake Champlain Basin that contribute substantial amounts of P to the lake. In rural basins, moderate levels of P dominated by non-bioavailable forms, may result in an equilibrium situation where the in-stream stocks are not a major factor in movement of P from the land to the lake. Urban basins or those with high levels of bioavailable P from sewage treatment facilities may have a fundamentally different relationship, with P storage in stream compartments serving as an important temporary buffer for elevated P levels. Because of the complexity and variability of the river ecosystems in the Basin, these tentative notions about P dynamics need to be evaluated under actual conditions using an adaptive management framework.

Acknowledgments. This project was funded by a grant from the Lake Champlain Basin Management Conference, and supported in part by the Vermont Water Resources and Lake Studies Center. The authors would like to thank two anonymous reviewers for their careful and insightful comments to improve this manuscript.

REFERENCES

American Public Health Association, Standard methods for the examination of water and wastewater, 472-495, 1985.

American Public Health Association, Standard methods for the examination of water and wastewater, 18th ed., APHA, AWWA, and WEF, Washington, D.C., 1992.

Corning, K.E., H.C. Duthie, and B.J. Paul, Phosphorus and glucose uptake by seston and epilithon in boreal forest streams, *J. of the North Amer. Benthol. Soc.* 8: 123-133, 1989.

Elwood, J.W., J.D. Newbold, A.F. Trimble, and R.W. Stark, The limiting role of phosphorus in a woodland stream ecosystem: effects of P enrichment on leaf decomposition and primary producers, *Ecology* 62: 146-158, 1981.

Galarneau, H.L., Storage, transport, and decomposition of detritus in the LaPlatte River, VT, MSc. thesis, Univ. VT, 1998.

Graham, A.A., The impact of fine silt on epilithic periphyton, and possible interactions between periphyton and invertebrate consumers, *Verhandlungen Internationale Vereinigung fuer Theoretische und Angewandte Limnologie* 23: 1437-1440, 1988.

Graham, A.A., Siltation of stone-surface periphyton in rivers by clay-sized particles from low concentrations in suspension, *Hydrobiologia* 199: 107-115, 1990.

Hart, B.T., P. Freeman, and I.D. McKelvie, Whole-stream phosphorus release studies: variation in uptake length with initial phosphorus concentration, *Hydrobiologia* 235/236: 573-584, 1992.

Hill, A.R., Phosphorus and major cation mass balances for two rivers during summer flows, *Freshwater Biology* 12: 293-304, 1982.

Hill, A.R., The potential role of in-stream and hyporheic environments as buffer zones, in *Buffer Zones: Their Processes and Potential in Water Protection,* edited by N.E. Haycock, T.P. Burt, K.W.T. Goulding, and G. Pinay, pp. 115-127, *Quest Environmental, Harpenden*, Hertfordshire, UK., 1997.

Hoffmann, J.P., E.A. Cassell, J.C. Drake, S.N. Levine, D.W. Meals, and D. Wang, Understanding phosphorus cycling, transport and storage in stream ecosystems as a basis for phosphorus management, Technical Report No. 20, Lake Champlain Basin Program, 1996.

Hunsaker, C.T. and D.A. Levine, Hierarchical approaches to the study of water quality in rivers, *BioScience* 45: 193-203, 1995.

Jones, R.C., Primary production, biomass, nutrient limitation, and taxonomic composition of algal communities epiphytic on the submerged macrophyte *Myriophyllum spicatum* L. in a hardwater, eutrophic lake, Ph.D. Thesis, Univ. Wisconsin, Madison, WI, 1980.

Klotz, R.L., Temporal relation between soluble reactive phosphorus and factors in stream water and sediments in Hoxie Gorge Creek, New York, *Can. J. Fish. Aquat. Sci.* 48: 84-90, 1991.

Lake Champlain Management Conference, Opportunities for action - An evolving plan for the future of the Lake Champlain Basin, Lake Champlain Basin Program, Grand Isle, VT, 1996.

Meals, D.W., LaPlatte River Watershed - water quality monitoring and analysis program,

Comprehensive final report 1979-1989, Program report No 12. Vermont Water Resources Research Center, University of Vermont, Burlington, VT, 1990.

Meals, D.W., Watershed-scale response to agricultural diffuse pollution control programs in Vermont, USA, *Water Science and Technology* 33(4-5): 197-204, 1996.

Miller, W.E., J.C. Greene, and T. Shiroyama, The *Selenastrum capricornutum* Printz algal assay bottle test - experimental design, application and data interpretation protocol, U.S. EPAEPA-600/9-78-018, 1978.

Mulholland, P.J., Regulation of nutrient concentrations in a temperate forest stream: roles of upland, riparian, and instream processes. *Limnol. Oceanogr.* 37: 1512-1526, 1992.

Mulholland, P.J., E.R. Marzolf, S.P. Hendricks, R.V. Wilkerson, and A.K. Baybayan, Longitudinal patterns of nutrient cycling and periphyton characteristics in streams: a test of upstream-downstream linkage, *J. of the North American Benthol. Soc.* 14: 357-370, 1995.

Mulholland, P.J., J.D. Newbold, J.W. Elwood, and J.R. Webster, Phosphorus spiraling in a woodland stream: seasonal variations, *Ecology* 66: 1012-1023, 1985.

Mulholland, P.J., A.D. Steinman, and J.W. Elwood, Measurement of phosphorus uptake length in streams: comparison of radiotracer and stable PO_4 releases, *Can. J. Fish. Aquat. Sci.* 47: 2351-2357, 1990.

Munn, N.L., and J.L. Meyer, Habitat-specific solute retention in two small streams: an intersite comparison, *Ecology* 71: 2069-2082, 1990.

Nelson, D.J., N.R. Kevern, J.L. Wilhm, and N.A. Griffith, Estimates of periphyton mass and stream bottom area using phosphorouos-32, *Water Res.* 3: 367-373, 1969.

Paul, B.J., and H.C. Duthie, Nutrient cycling in the epilithon of running waters, *Can. J. Bot.* 67: 2302-2309, 1989.

Pelton, D.K., S.N. Levine, and M. Braner, Measurements of phosphorus uptake by macrophytes and epiphytes from the LaPlatte River (VT) using ^{32}P in stream microcosms, *Freshwater Biology* 39: 285-299, 1998.

Rigler, F.H., Radiobiological analysis of inorganic phosphorus in lake water, *Verhandlungen Internationale Vereinigung fuer Theoretische und Angewandte Limnologie* 16: 465-470, 1966.

Rosemond, A.D., Multiple factors limit seasonal variation in periphyton in a forest stream, *J. of the North Amer. Benhol. Soc.* 13: 333-344, 1994.

Sharpley, A.N., S.C. Chapra, R. Wedepohl, J.T. Sims, T.C. Daniel, and K.R. Reddy, Managing agricultural phosphorus for protection of surface waters: Issues and options, *J. Environ. Qual.* 23: 437-451, 1994.

Smith, D.I. and P. Stopp, *The river basin*. Cambridge University Press, New York, NY, 1978.

Steinman, A.D., P.J. Mulholland, and D.B. Kirschtel, Interactive effects of nutrient reduction and herbivory on biomass taxonomic structure, and P uptake in lotic periphyton communities, *Can J. Fish. Aquat. Sci.* 48: 1951-1959, 1991.

Steinman, A.D., P.J. Mulholland, and J.J Beauchamp, Effects of biomass, light, and grazing on phosphorus cycling in stream periphyton communities, *J. of the North Amer. Benthol. Soc.* 14: 371-381, 1995.

Strahler, A.N., Quantitative analysis of watershed geomorphology, *Trans. Am. Geophys. Union*, 38(6): 913-920, 1957.

Summerfield, M.A., *Global geomorphology*. Longman Scientific & Technical, New York, NY, 1991.

United States Environmental Protection Agency. 1983.*Methods for chemical analysis of water and wastes*, EPA-600/4-79-020, 1991.

U.S. Geological Survey, Water flow data for the LaPlatte River at Shelburne Falls, VT, 06/01/95, Station number 04282795, 1995.

Vermont Department of Environmental Conservation and New York State Department of Environmental Conservation, A phosphorus budget, model, and load reduction strategy for Lake Champlain, Final Report of the Lake Champlain Diagnostic-Feasibility Study, VT DEC, Waterbury, VT, 1997.

Vermont RCWP Coordinating Committee, St. Albans Bay Rural Clean Water Program, Final Report. 1991, Vermont Water Resources Research Center, University of Vermont, Burlington, 1991.

Slurry Sidedressing and Topdressing Can Improve Soil and Water Quality in the Lake Champlain Basin

Denis Côté, Aubert Michaud, Thi Sen Tran, and Claude Bernard

ABSTRACT

Since recent years in Québec, animal slurry is spread more frequently on growing crops like corn, cereal grains and meadow, in late spring and summer, as the main source of nutrients. Research on manure nutrient efficiency for crop production, on long term nutrient accumulation in the soil and on its impacts on surface water, have demonstrated the advantages of corn sidedressing over fall and spring surface broadcasting. Increased nitrogen efficiency reduces the rate of application and prevents excessive accumulation of P, K, Cu and Zn in the plow layer. Risk of nutrient loss in the runoff is lowered since slurry is incorporated simultaneously to the soil at 10 cm depth. Cereal grains are also successfully topdressed or banded with slurry at early growth stages. Topdressed or banded slurry on grass shows more consistent N efficiency than high pressure surface broadcasting. From a watershed rehabilitation perspective, topdressing, sidedressing and banding techniques increase the acceptance of surplus manure by potential receiver farms due to minimal risk of soil compaction, increased nitrogen fertilizer efficiency and corresponding economy on mineral fertilizer imports. In the Pike River watershed, manure and mineral phosphorus inputs generally exceed crop uptake. Higher manure nitrogen fertilizer value, together with a sound nutrient management plan, enables the control of phosphorus build-up in soils, which is a critical issue of an eutrophication-focussed rehabilitation project such as the Pike River.

INTRODUCTION

Over 20 million tons of solid manure and 15 million tons of liquid manure, mostly pig manure, is spread annually on agricultural land in Québec (Couture et al., 1992). Liquid

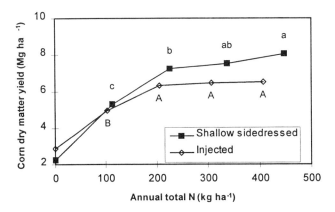

Figure 1. Corn yield response to pig slurry N rate following injected or shallow sidedressed application.

manure is rapidly taking over solid manure on larger beef and dairy farms. While liquid manure is stored safely in storage tanks, solid manure is often piled on the ground on smaller dairy farms, beef and poultry operations, causing point source pollution. Non-point source pollution by manure nutrients, related to its use as a fertilizer, is probably more important. A large proportion of manure spreading, up to 80% in some watersheds is done in September and October, once corn, cereal grain and last hay cut have been harvested. Most of this manure spread is eventually incorporated into the soil by chisel or moldboard plowing, except for manure spread on grassland. In addition, 500 000 tons of mineral fertilizer are also applied annually. Corn is grown on 350 000 ha, spring cereals and soybeans, are cropped on 300 000 ha, while the remainder 1.3 millions hectares is almost entirely used for hay and pasture. On many farms and in some cases, entire sub-watershed, nutrients are applied in excess of crop needs, causing nutrient accumulation in the soil and increasing risks of water contamination. Excessive nutrient application is sometimes the result of inadequate method and timing of manure spreading, in which cases the very low effective nitrogen availability for the crop is compensated by higher rate of manure application. On other farms, mineral fertilizers are added without accounting for manure nutrients. Even at optimum rate of application, nutrients and pathogens in the manure can move to surface water if method and timing of spreading are inadequate.

New regulations (Gouvernement du Québec, 1997) concerning manure application to cropland have been adopted in order to further protect water quality from pollution related to agriculture activities. Under this regulation, total annual rate of manure spreading in the period between March 31st and October 1st is now limited by a risk index of environmental phosphorus losses and crop nitrogen requirements. The phosphorus risk index is related to the level of soil available P and the saturation of P fixation capacity. Total available nitrogen from manure application must not exceed annual crop needs. Use of sprinkler type irrigation system for liquid manure spreading will also be forbidden, starting fall 1998. Manure must not be spread at less than 30 m from a well, an identified river or lake, and at less than 5 m from ditches and non-identified lakes.

By October 1998, spreading in October and November, will be restricted partially or totally under this new regulation. Restrictions are based on the size of the herd, types of crop grown on the farm, the existing manure surplus on the farm if the farm is situated in one of the three designated watersheds (Yamaska, L'Assomption and Chaudière), and on the farmers obligation to have an agro-environmental fertilization plan supervised by a professional agronomist.

These recent regulations has put more pressure on farmers to consider new spreading techniques such as sidedressing, sub-canopy banding and topdresing. This is especially needed where surplus manure and rich P and K soils are present. Research on better manure spreading time and techniques adaptable to corn and cereals grain and grassland have been pursued in Québec over the last twenty years in order to increase nutrient efficiency and to improve soil and water quality. Results from some of these studies are discussed in this paper.

CORN SIDEDRESSING WITH PIG SLURRY – NUTRIENT EFFICIENCY AND ACCUMULATION

For optimum corn yield, mineral nitrogen recommendation in Québec range from 150 to 180 kg N ha^{-1}. Pig slurry usually contains 60 to 70% of its total N as NH_4-N and the remainder as organic N. A portion of the organic N will mineralize during the year of application and the rest of it will accumulate in the soil. On year after year application, the organic N from manure builds up so that its contribution as mineralized N, also called back-effect, increases gradually. It is expected that all the NH_4-N and a portion of the organic N will be available to meet corn needs, if properly incorporated in the soil. When supplying nitrogen to corn using only pig slurry, P and K needs are often exceeded due to a higher P:N ratio in the slurry than the plant needs. In a long term trial, Côté et al. (1996) and Tran et al. (1996) studied the capacity of pig slurry to supply corn N needs, and the resulting nutrient accumulation in the soil. Five rates of pig slurry (0, 30, 60, 90, and 120 Mg ha^{-1}), were sidedressed to growing corn at the 6 to 8-leaf growth stage, during 16 successive years, beginning in 1980, on a poorly drained Le Bras silty loam, 25 km south of Québec City. The 120 Mg ha^{-1} application rate, which is frequently applied in the fall by farmers, is expected to meet total N requirements for next year corn, while contributing to enrichment of poor P and K soil. This excessive rate of application was included in the sidedressing field trial for this last reason. Additional treatments with complete mineral fertilization at rates 150 and 180 kg N ha^{-1} and uniform 150 kg P_2O_5 and K_2O ha^{-1} were set up during the last 6 years of the field trial. All manure treatments were distributed in a completely randomized experimental design with four replications. To ease the statistical interpretation of following figures (1, 2 and 3), different letters within each curves indicate significant differences between N level; capital and small letters are not to be compared between each other.

The yield response to slurry nitrogen was better with shallow sidedressing on 80% of the inter-row space than with injecting in a narrow and 20 cm deep slot in the center of the inter-row space (fig. 1). On this figure, injected slurry curve is a 5 years average (1980-1984), while shallow sidedressing curve is also a 5 years average (1985-1989). The 60 Mg ha^{-1} rate provided an annual average of 208 kg total-N, 83 kg P_2O_5 and 94 kg K_2O, for injected slurry, and 225 kg total-N, 125 kg P_2O_5 and 106 kg K_2O for shallow sidedressing, which is close to the corn needs for a preplant application. The shallow sidedressing equipment we used, as described by Jokela and Côté (1994), is the most

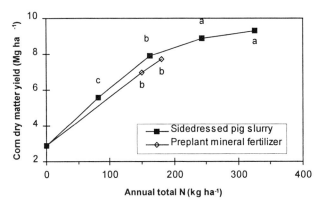

Figure 2. Corn yield response to mineral fertilizer and pig slurry N rate.

common multiple cultivator tine soil incorporating tool built by Québec spreader manufacturers. The very small yield increase with the 120 Mg ha^{-1} is not justified on an agronomic and neither an environmental basis. For the period extending from the 11th to the 16th year, the corn yields were similar for 150 and 180 kg ha^{-1} mineral N and with the 162 kg total-N provided by slurry (fig. 2). Yields were also higher than those of the 1985-1989 period for the 60 Mg ha^{-1} rate, probably because of slurry back-effect. Direct measurement of the back-effect was done the 14th year of the trial, showing a 2 Mg ha^{-1} increase in corn yield, over control plots, for the 60 Mg ha^{-1} slurry subplots not receiving slurry that particular year. N efficiency coefficient for pig slurry, including 10 to 15 years back-effect, measured yearly from 11th to 16th year, varied between 0.8 to 1.0 in shallow sidedressing. The N efficiency coefficient is based on the N required per unit yield increase from slurry relative to that of mineral N fertilizer. Futhermore, a one year trial on starter mineral N and P fertilizer demonstrated that their use, for a 60 Mg ha^{-1} sidedressed

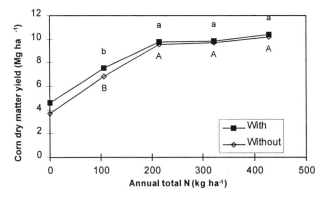

Figure 3. Corn yield response to N rate in shallow sidedressed pig slurry with and without starter application.

TABLE 1. Yearly P and K apparent recovery by silage corn sidedressed with pig slurry, over a 16 year period, on a Le Bras silty loam

Slurry application rate (Mg ha^{-1})	P recovery (%)	K recovery (%)
0	--	--
60	19	60
120	13	44

slurry, was not required in spite of the 30 to 40 day period between planting and slurry sidedressing, and medium-rich P soil test (fig. 3).

The P and K recovery in the corn harvest (table 1) is respectively 46 and 36% higher with the 60 Mg ha^{-1} application rate than with the 120 Mg ha^{-1} one. The unrecovered nutrients accumulated in the soil over the 16 years period to levels shown in table 2. For the 60 Mg ha^{-1} application rate the P and K levels are both classified as rich, while P is classified as extremely rich for the 120 Mg ha^{-1} application rate. The resulting mean annual rate of accumulation expressed as kg ha^{-1} of Mehlich-3 available forms was respectively 7.9 and 4.8 for P an K, at the 60 Mg ha^{-1} rate. These low rates of accumulation can be easily controlled by using crop rotations incorporating species with low N/P requirements. It is not as easy with 120 Mg ha^{-1} slurry which shows very high accumulation rates, in the order of 20 kg ha^{-1} yr^{-1} for P and K. The accumulation of micronutrients like Cu and Zn at the 60 Mg ha^{-1} slurry application is well under the 100 mg kg^{-1} level considered as limiting in Canada (Standish, 1981).

As an extended conclusion, sidedressing a growing row crop with animal slurry is more economically and agronomically sound than fall spreading. Moreover, it prevents air pollution from ammonia, pathogens and odor associated with high pressure broadcasting (aeroaspersion) commonly used in fall application.

TABLE 2. Soil P, K, Cu and Zn enrichment after 16 years of sidedressed pig slurry application on corn, on a Le Bras silty loam

Slurry rate (Mg ha^{-1})	P			Mehlich3-K (0-20 cm) (kg ha^{-1})	Mehlich3-Cu (0-10 cm) (mg kg^{-1})	Mehlich3-Zn (0-10 cm) (mg kg^{-1})
	Mehlich3 (0-20 cm) (kg ha^{-1})	Water soluble (0-10 cm) (mg kg^{-1})	Saturation§ (%)			
0	76	0.92	3.0	144	1.5	0.7
60	202	2.87	9.1	221	3.9	3.0
120	397	6.26	21.5	427	6.7	7.3
Rate effect (F)¶	**	**	**	**	**	**

§ P saturation = (Mehlich3 P / P fixation capacity) * 100
¶ ** significant at P≤ 0,01level

CEREAL GRAIN AND GRASS SUB-CANOPY BANDING AND TOPDRESSING WITH PIG SLURRY

Early spring cereal grain seeding date is imperative for maximum grain yield. Unfortunately at early date of seeding, soil water content and water table are usually too high for good traffic conditions and risks of nutrient runoff and leaching are high also. With fall spreading, because of low N efficiency expected for next year crop, farmers apply higher rates, with resulting over application of P and K. For those reasons, research on N efficiency of pig slurry applied on growing cereal grain, at differents growth stages, have been done on field trials with a regular size tank spreader, equipped with special spreading devices. The tractor pulled tank spreader was equipped with a rear mounted spreading boom, 50 cm above the ground, operating at very low pressure. Openings on the boom deliver slurry, through drop tubes spaced 35 cm apart, to the ground in narrow bands beneath the crop canopy. The intent of sub-canopy banding is to provide less surface exposure, to create wind protection and to prevent contamination of foliage with slurry, resulting in reduced ammonia volatilization (Jokela and Côté, 1994). Sub-canopy banding is done between 2-leaf and 6 to 8-leaf growth stage for optimum fertilizing efficiency. At the 6-leaf growth stage, Seydoux (1993), has demonstrated that the barley yield reduction in tank spreader wheel path for an 8 m wide spreading boom is negligible. The slurry N efficiency coefficients obtained from comparison of subcanopy banding of slurry with mineral N, in one year trials on cereals, ranged between 0.7 to 1.15 (Seydoux, 1993; Marmen, 1995).

Pig slurry topdressing, using multiple low pressure deflectors mounted on a boom, was studied by Côté and Couture (1995), for applications made either at post seeding (before barley emergence), or at 2 or 6-leaf growth stages, on a sandy soil with low N fertilility. A similar N efficiency coefficient (0.7) for post seeding and 2-leaf growth stage was observed. At the 6-leaf growth application date, barley was showing N-deficiency symptoms that took a few weeks to disappear, resulting in a lower grain yield and N-efficiency coefficient of 0.6.

Usually, 25 to 35 Mg ha^{-1} of pig slurry (110 kg total N), either topdressed or sub-canopy banded at the 2 to 6-leaf growth stage, can replace the usual 80 kg mineral N ha^{-1} application. At these low rates of slurry application, there is an over application of 20 to 30 kg P ha^{-1}, which is controled either by a companion crop like clover (Marmen, 1995) or by crop rotation with potato (Côté and Couture, 1995) or any other plant species with a low N/P requirement.

Topdressing and band spreading slurry on grass stands with low pressure spreading booms, is done frequently on early stands in the spring or between hay cuts in the summer. These spreading techniques are gradually replacing high pressure broadcasting (aeroaspersion). In 1997, about 20% of pig raising farms were using a low pressure boom spreader. High pressure broadcasting is usually done in the fall, after last hay cut, because farmers want to prevent transmission of bad odor to hay. They use a higher rate of application to compensate resulting low N efficiency for next year crop. Many research trials are under way to study slurry N efficiency comparisons between high pressure broadcasting and low pressure booms on grass and legume-based hay. The slurry N efficiency coefficients for low pressure boom spreading vary from 0.6 to 0.7 for spring and summer applications. However this efficiency coefficient is very inconsistent (0.1 to 0.7 range) for slurry broadcasting at high pressure, because of large variations in ammonia losses. Ammonia losses depend on the occurence of rainfall during the first week following spreading (Jokela and Côté, 1994).

RUNOFF QUALITY

In hog production, 97% of the manure is managed as liquid. Besides, the concentration of this industry in some watersheds has resulted in acute livestock concentrations and manure management problems. Consequently, much of the research effort on manure impacts on runoff waters has been directed towards pig slurry. Applying slurry to agricultural soils modifies their hydrological properties, and may result in increased pollutant loadings to nearby water bodies if some good management practices are not implemented.

Impacts on Hydrological Processes

Pig slurry contains 95% water on the average. Applying large volumes on sloping or unprotected fields increases the runoff potential. Gangbazo et al. (1992) applied 0, 27.3 and 54.6 Mg ha^{-1} of slurry on a sandy-clay loam, with a 3% slope. The slurry was either left on the soil surface or incorporated in the top 20 cm. Twenty-six millimeters of artificial rainfall were applied at two rates (11 and 22 mm h^{-1}), 1, 24 or 48 hours after manure spreading. The results indicate that runoff was increased and infiltration reduced after manure applications, proportionally to the the amount applied. These effects were reduced by the incorporation of the manure and by larger time intervals between manure spreading and the occurrence of the rainfall.

Grando (1996) surface applied 37 m^3 ha^{-1} of pig manure to a barley crop on a silty loam. He showed that runoff was increased by 43 and 24% for the first two rains occuring 1 and 24 hours after spreading respectively. Subsequent rains did not result in significant runoff increases.

Sediment and Nutrient Losses - Short Term Experiments

The short-term hydrological impacts influence the dynamics of pollutant release. In Grando's experiment, the plots receiving manure released 72% more suspended solids than the control, 1 hour after slurry application. This difference decreased in time, with successive rains. The 4th simulated rainfall, 120 hours after manure application, produced runoff waters with 14% more suspended solids than the control (fig. 4).

The losses of nitrogen and phosphorus followed the same trend. The first rainfall (1 hr after manure spreading) generated a loss of 4.9-5.9 kg ha^{-1} of total N, 65% of which was soluble and 53% ammonium nitrogen. Total N losses dropped to 1 kg ha^{-1} or less for the three following rains, with the particulate forms accounting for 60% and more.

For phosphorus, the loss generated by the first rain ranged from 0.6 to 0.75 kg ha^{-1}, a 600% increase over the control. On the average, 35% of the total P (Pt) was under soluble forms. The losses for the second rainfall (24 hrs after spreading) were greatly reduced to a little more than 0.1 kg Pt ha^{-1} (fig. 5).

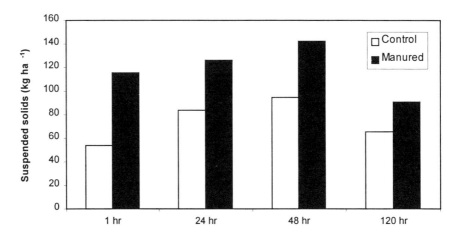

Figure 4. Impact of surface manure spreading on the loss of suspended solids (Grando, 1996).

Gangbazo et al. (1993) also concluded that a rainfall occuring within 1 hour after slurry spreading greatly increases the nutrient losses in surface runoff. They observed that the NH_4-N, Pt and orthophosphate losses increased quadratically with the application rate for a 1 hr delay, while the increase was linear for longer delays. Incorporating the manure in the first 20 cm of the soil reduced the losses to levels similar to those from the control plots.

These results indicate that manure spreading may seriously impair surface water quality in some circumstances, particularly if rain occurs shortly after spreading and if manure if left exposed on the soil surface. However, Gangbazo's results suggest that it is possible to minimize these impacts by turning the manure under right after spreading.

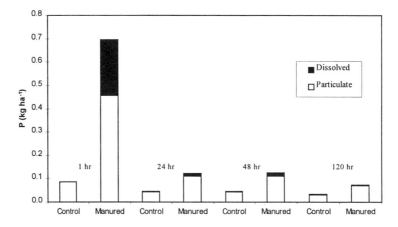

Figure 5. Impact of surface manure spreading on the loss of phosphorus (Grando, 1996).

TABLE 3. Cumulated impact of surface application of liquid manure on bacteria losses (Grando, 1996)

Treatment	Simulated rains			
	1 hr (n ml^{-1})	24 hr (n ml^{-1})	48 hr (n ml^{-1})	120 hr (n ml^{-1})
Total coliforms				
Control	424a[§]	256a	232a	245a
Manured	3045b	465b	436b	362b
Fecal streptococci				
Control	381a	362a	295a	192a
Manured	1607b	737b	478b	384b

[§] for the same rain, different letters indicate a significant difference at the 0,05 level

Bacterial Pollution - Short Term Experiments

Grando's work indicate that a rain occuring shortly after manure spreading (1 hr) produces a sharp increase of the bacteria counts in surface runoff. The increase was 6-8 fold for total coliforms and 3.5-5 fold for fecal streptococci (Table 3). The impact decreased steadily with time, with successive rainfalls, due to dieback of bacteria and exhaustion of the bacterial stock available for export.

Nutrient Losses - Long Term Experiments

Gangbazo et al. (1996a) studied the impacts of overfertilizing hay and corn with pig manure. Besides the recommended N and P fertilization (control treatment), the plots received slurry volumes such as to bring two times the N fertilizer recommendations, i.e. 540 and 165 kg N ha^{-1} for corn and hay respectively. The total (surface and drainage runoff) N losses reached more than 100 kg ha^{-1} for the overfertilized plots, a 5 fold increase over the control. The P loadings were not influenced. Spreading the pig manure in the spring or the fall slightly modified the distribution of the losses between different forms (NH_4-N vs NO_3-N; dissolved vs particulate P).

Bernard and Côté (1997) compared fall to spring spreading of pig slurry on silage corn. Manure volumes providing 150 and 250 kg N ha^{-1} were applied in the spring and fall respectively. Despite greater nutrient inputs, the fall applications did not produce higher corn yields. However, they did produce N and P losses by surface runoff that were increased 50 and 150% respectively during the growing season. These results suggest that the environmental losses were also increased before and after the growing season, even though this was not actually measured.

In an other experiment, Gangbazo et al., (1998) compared three fertilization strategies, spring preplant slurry, summer sidedressed slurry to a standard mineral fertilizer preplant application. Slurry rate was calculated from N needs for grain corn and slurry N content, using a 0.7 N efficiency coefficient for the slurry. They reported an increase of 0.13 kg ha^{-1} yr^{-1} of total P lost in runoff from sidedressed plots compared to both other treatments

TABLE 4. Nutrient losses under two springtime manure spreading strategies at agronomic rates (Gangbazo et al., 1998)

Parameter	Annual loading (kg ha^{-1} yr^{-1})		
	Surface runoff	Drainage water	Total
Nitrogen			
Mineral-preplant	4.96	37.7	42.7
Slurry-preplant	3.48	42.8	46.2
Slurry-sidedressed	4.88	55.1	60.0
F test	**¶	ns	ns
Phosphorus			
Mineral-preplant	0.29 (38)§	0.09 (11)	0.39 (33)
Slurry-preplant	0.30 (50)	0.07 (14)	0.37 (43)
Slurry-sidedressed	0.43 (33)	0.08 (13)	0.51 (31)
F test	**	ns	*

¶ **, * : significant at p≤0.01, 0.05; ns : non significant
§ Fraction of Pt as orthophosphates

(Table 4). This may be caused by the experimental set-up, caracterized by 15 m long plots with 7% slope ending up immediately in runoff collector. During spreading, part of the slurry just sidedressed could immediately flow close to the colector. Such conditions not usually found in large fields. Total-N losses in surface runoff were lower with preplant slurry than with preplant mineral fertiliser. Corn yield were similar for all treatments.

These experiments indicate that fertilizing soils with liquid manure may alter runoff waters under poor management conditions. However, the results from Gangbazo et al. (1996b) suggest that it is possible to obtain good yields when using liquid manure as a fertilizer, without impairing the quality of receiving water bodies.

WATERSHED PERSPECTIVE

The downstream portion of Pike River watershed, on the Québec side of the border, covers an area of 661 km^2 draining into the Missisquoi Bay at the Northeast end of Lake Champlain (figure 6). Monitoring from 1979 to 1992 has shown that the annual total nitrogen and phosphorus loads average 383 and 22 metric tons respectively, which represents global surface area contributions of 5.8 kg N ha^{-1} and 0.33 kg P ha^{-1} (Simoneau, 1993). Prevalent agricultural land use and systematic investment in urban and industrial point sources control as well as manure storage facilities over the past two decades suggest that current nutrient loads are dominated by agricultural non-point sources. Watershed rehabilitation efforts now focus on nutrient management planning in order to reduce phosphorus loads and associated eutrophication of Missisquoi Bay.

Watershed Agricultural Nutrient Balance and Soil Phosphorus Build-up

Farm-scale nutrient balance have been computed for the 352 farms located in the Québec downstream portion of the Pike River watershed using 1995 census data

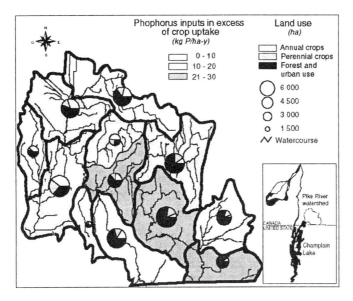

Figure 6. Land use and phosphorus balance in the Pike River subwatersheds.

(MAPAQ, 1995). Crop uptake and nutrient requirements have been derived from crop production data and fertilizer recommendations (CPVQ, 1996) assuming medium soil P and K levels. Manure inputs have been derived from animal production data and standardized manure nutrient content and efficiency indices (CPVQ, 1995). Nitrogen and phosphorus inputs have been derived from annual fertilizer purchases included in the census data. These monetary values have been transposed in N, P and K units based on farm-scale computed N-P-K ratios of mineral fertilizer needs. The latter ratios accounts for manure nutrient fertilizer values produced on the farm (MAPAQ, 1995; CPVQ, 1996; AFEQ, 1996). Phosphorus balances aggregated at the subwatershed scale are presented in table 5.

Globally, watershed phosphorus balance shows inputs nearly two times greater than crop uptake, while nitrogen inputs remains below crop nutrient requirements. Surplus annual phosphorus balance tend to be more important in the subwatersheds crossed over by Pike River main stream, where they reach the 20-30 kg range based on total cropped area (figure 6). This constant positive phosphorus balance on agricultural land implies a continuous build-up of soil phosphorus levels, which in turn has been linked to increased levels of bioavailable P in runoff from cropland (Pote et al., 1996). Under Québec standardized methods of soil testing, an annual input of 3.5 kg P ha^{-1} in excess of crop uptake has been shown to increase soil Mehlich-3 P test by 1.0 kg P ha^{-1} yr^{-1} (Tran et al., 1990; Giroux et al., 1996). This suggests that most intensive agricultural subwatersheds of Pike River have an average annual increase in Mehlich-3 soil P test of 7 kg P ha^{-1} approximately. Using Mehlich-3 available P/Al ratios as environmental index of P, soil test values from the downstream portion of Pike River show that 30% of land base has already reached a 10% P saturation level, considered as a critical level at which P balance should be stabilized (Giroux and Tran, 1994) (figure 7).

TABLE 5. Phosphorus balance in Pike river subwatersheds

Subwatershed	Total area (ha)	Cropped area (ha)	Annual crop uptake (Mg)	Annual mineral inputs (Mg)	Annual manure inputs (Mg)	Annual P balance (Mg)	Annual P balance (kg ha^{-1})
A	6 425	4 163	77	74	81	78	19
B	2 952	1 403	28	26	15	12	9
C	5 521	2 442	47	43	47	43	18
D	5 626	2 068	38	29	40	30	15
E	3 094	2 250	43	30	54	41	18
F	3 990	1 468	27	29	32	34	23
G	1 676	753	14	11	10	7	9
H	5 861	1 765	30	27	16	12	7
I	6 013	1 633	27	14	40	28	17
J	6 344	1 457	19	19	35	35	24
K	3 528	423	3	5	8	10	24
L	4 295	1 012	14	7	9	2	2
Total	55 325	20 837	366	314	385	333	16

Manure Management Strategy and Nutrient Balance

Topdressing and sidedressing management strategies present many advantages over fall application of animal slurry, namely minimal risk of soil compaction and improved manure nitrogen efficiency with the corresponding economy on mineral fertilizer purchase. Consequently, the acceptance of surplus manure by potential receiver farms is increased. Slurry nitrogen efficiency nearly doubles when a dominantly (70%) fall broadcast scenario is replaced by a topdressing and sidedressing application. Making the assumption that most of the manure is managed as slurry, the hypothetical gain in nitrogen value would represent 260 metric tons a year over the whole Pike River

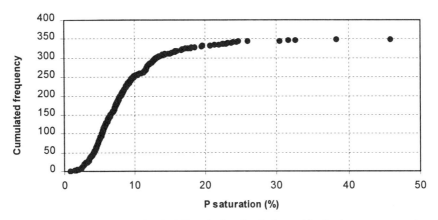

Figure 7. Soil phosphorus saturation in 349 agricultural soils located in the downstream portion of the Pike River watershed.

TABLE 6. Implications of manure management strategy for nitrogen balance and phosphorus rates in Pike River watershed

Nitrogen Balance	
Crops nitrogen requirements (Mg N ha^{-1}yr^{-1})	2 210
Current mineral nitrogen import (Mg N ha^{-1}yr^{-1})	1 158
Manure nitrogen fertilizer value, conventionnal management (Mg N ha^{-1}yr^{-1})	292
Manure nitrogen fertilizer value, topdress/sidedress management (Mg N ha^{-1}yr^{-1})	553
Theoritical gain in nitrogen efficiency (Mg N ha^{-1}yr^{-1})	260
Phosphorus rates	
Recommended nitrogen rate (kg N ha^{-1}yr^{-1})	141
Resulting P rate, conventionnal management (kg P ha^{-1}yr^{-1})	92
Resulting P rate, topdress/sidedress management (kg P ha^{-1}yr^{-1})	49
Theoritical reduction in phosphorus rate (kg P ha^{-1}yr^{-1})	43
Recommended phosphorus rate (kg P ha^{-1}yr^{-1})	19

watershed (Table 6). This is equivalent to 26% of the current annual mineral N import (1160 Mg N yr^{-1}).

Phosphorus load is the major issue of the watershed rehabilitation project. Consequently, controlling phosphorus saturation levels in soils is critical. Higher manure nitrogen fertilizer value associated with a topdressing and sidedressing management reduce the phosphorus application rate. The annual area-weighted recommended manure nitrogen rate is 141 kg N ha^{-1} (Table 6) in the Pike River watershed. Under a topdressing and sidedressing management strategy, this agronomic target nitrogen application rate results in a phosphorus annual input of 49 kg P ha^{-1}, which is nearly half of a corresponding phosphorus annual input from a conventional broadcasted fall-based scenario (92 kg P ha^{-1} yr^{-1}). Even under optimal slurry nitrogen efficiency, phosphorus input still exceeds crop needs, since phosphorus recommended rate for the crops grown in the watershed averages 19 kg P ha^{-1} yr^{-1}.

The perspective of continued phosphorus accumulation in agricultural soils calls for furthermore reduction in phosphorus content of applied liquid manure. Up to 95% of slurry's phosphorus is concentrated in the solid phase (Tengman et al., 1995). Readily applicable solid/liquid separation processes should be considered as manure's best management practice for eutrophication-focussed projects such as the Pike River watershed.

REFERENCES

Association des fabricants d'engrais du Québec (A.F.E.Q.), Guide de fertilisation, 3ième édition, A.F.E.Q. Québec, 1996.

Bernard, C., and D. Côté, Rapport final de projet, Centre de recherche et d'expérimentation en sols, 1997.

Conseil des productions végétales du Québec Inc (C.P.V.Q. Inc), Grilles de référence en fertilisation, Agdex 540, 2e édition, 1996.

Conseil des productions végétales du Québec Inc (C.P.V.Q. Inc), Sols : coefficients d'efficacité des engrais de ferme, Bulletin technique no.22, Agdex 538, 1995.

Côté, D., T.S. Tran and A. N'Dayegamiye, Efficacité fertilisante du lisier de porc épandu en postlevée du maïs, *Agrosol*, 9(1), 14-20, 1996.

Côté, D. and B. Couture, Du lisier de porc dans une rotation céréale - pomme de terre, *Grandes Cultures*, 5 (4), 10-12, 1995.

Couture, J.-N., M. Trudelle, and M. Perron, Situation de la valorisation des engrais de ferme, in *Colloque sur la gestion des fumiers*. CPVQ inc., pp. 183-201, 1992.

Gangbazo, G., A.R. Pesant, D. Cluis, and D. Couillard, Étude en laboratoire du ruissellement et de l'infiltration de l'eau suite à l'épandage du lisier de porc, *Can. Agric. Eng.*, 34, 1-9, 1992.

Gangbazo, G., A.R. Pesant, D. Cluis and D. Couillard, Effets du lisier de porc sur la charge d'azote et de phosphore dans l'eau de ruissellement sous des pluies simulées, *Can. Agric. Eng.*, 35, 97-103, 1993.

Gangbazo, G., A.R. Pesant and G.M. Barnett, Effets de l'épandage des engrais minéraux et de grandes quantités de lisier de porc sur l'eau, le sol et les cultures, Ministère de l'Environnement et de la Faune, Dir. Écosystèmes aquatiques, 1996.

Gangbazo, G., D. Côté, A.R. Pesant and G.M. Barnett, Effets de l'épandage du lisier de porc en présemis et en postlevée sur l'eau, le sol et le maïs-grain, Ministère de l'Environnement et de la Faune, Dir. Écosystèmes aquatiques, 1998.

Gouvernement du Québec, Le règlement sur la réduction de la pollution d'origine agricole en bref, Ministère de l'Environnement et de la Faune, 1997.

Giroux, M. and T.S. Tran, Critères agronomiques et environnementaux liés à la disponibilité, la solubilité et la saturation en phosphore des sols agricoles du Québec, *Agrosol*, 9(2), 51-57, 1996.

Giroux M. and T.S. Tran, Étude des facteurs affectant l'évolution des teneurs en P et K des sols agricoles, *Agrosol*, 7(2), 23-30, 1994.

Giroux M., D. Carrier and P. Beaudet, Problématique et méthode de gestion des charges de phosphore appliquées aux sols agricoles en provenance des engrais de ferme, *Agrosol*, 9(1), 36-45, 1996.

Grando, S., Effets de deux modes d'épandage de lisier de porc sur la qualité de l'eau de ruissellement, Mémoire de fin d'études, ÉNITA de Bordeaux, 64 p + annexes, 1996.

Jokela, W. and D. Côté, Options for direct incorporation of liquid manure, in *Liquid manure application system, design, management, and environmental assessment*, Proceedings from the liquid manure application systems conference, Rochester, New York, December, 1-2., pp. 201-215, 1994.

Marmen, S., 1995. Évaluation de la fertilisation de l'orge de printemps par épandage de lisier de porc en postlevée, Mémoire de maîtrise, Faculté des sciences de l'agriculture et de l'alimentation, Université Laval, 1995.

Ministère de l'Agriculture, des Pêcheries et de l'Alimentation du Québec (MAPAQ), Fiches d'enregistrements des exploitations agricoles, 1995.

Pote, D.H., T.C. Daniel, A.N. Sharpley, P.A. Moore, Jr., D.R. Edwards, and D.J. Nichols, Relating extractable soil phosphorus to phosphorus losses in runoff, *Soil Sci. Soc. Am. J.*, 60, 855-859, 1996.

Standish, J.F., Concentrations de métaux lourds dans les matières d'égoût traitées et les sous-produits, Agriculture Canada, Direction générale, Production et inspections des aliments, Trade memorandum T-4-93, 1981.

Seydoux, S., Effet d'un épandage par rampe de lisier de porc en postlevée sur une culture d'orge de printemps. Mémoire de maîtrise, Faculté des sciences de l'agriculture et de l'alimentation, Université Laval, 1993.

Tengman, C., H. Person, and D. Rozeboom, On site separation of liquids and solids: technology to concentrate swine manure phosphorus. *J. Anim. Sci.*, 73 (Suplement 1), 58, 1995.

Tran, T.S., M. Giroux, J. Guilbeault, and P. Audesse, Evaluation of Mehlich 2 extractant to estimate the available P in Quebec soils, *Commun. Soil Sci. Plant Anal.*, 21, 1-28, 1990.

Tran, T. S., D. Côté, and A. N'Dayegamiye, Effets des apports prolongés de fumier et de lisier sur l'évolution des teneurs du sol en éléments nutritifs majeurs et mineurs, *Agrosol*, 9 (1), 21-30, 1996.

Simoneau, M., Qualité des eaux du bassin de la rivière Richelieu, Direction des écosystèmes aquatiques, Ministère de l'Environnement et de la Faune du Québec, 1993.

Toxic Substances in Lake Champlain: an Overview

Alan McIntosh, Mary Watzin, and John King

ABSTRACT

Assessment of toxic substances in Lake Champlain has historically been limited. Monitoring efforts have focused such contaminants as mercury and polychlorinated biphenyls in fish. Additional monitoring of toxic substances in lake water, sediments, and biota by state and federal agencies has identified several sites where contaminants are elevated. Prior to 1990, research on toxic substances in Lake Champlain was largely limited to measurements of trace elements in surface sediments. Recent efforts have measured the deposition of atmospheric mercury in the basin and assessed the effects of toxic substances on fish. The most extensive evaluation, initiated in 1990, focused on sediment-associated contaminants lakewide. Substantial PCB contamination was found near Wilcox Dock in Cumberland Bay; efforts to remediate this site are underway. Sediments in Inner Burlington Harbor, which has received inputs from a sewage treatment facility and from stormwater runoff, contain both inorganic and organic contaminants. Biological evaluations at this site have been inconsistent. Outer Malletts Bay produced the most dramatic results in sediment toxicity tests. Levels of arsenic, iron, manganese and nickel in the bay's surface sediments were the highest in Lake Champlain, often exceeding sediment quality guidelines. A Toxicity Identification Evaluation indicated that manganese was the agent responsible for the toxicity. Trace elements entering Outer Malletts Bay via the Lamoille River are concentrated in deep sediments due to the bay's shape and the restricted exchange between the bay and the lake. Hypolimnetic oxygen depletion in the bay's lower waters also influences the behavior of trace elements.

INTRODUCTION

The issue of toxic substances in Lake Champlain has historically received less attention from the research community than other concerns. Prior to 1990, only limited monitoring and isolated research were conducted, and there was little, if any, attempt to determine whether or not toxic substances posed any threat to the lake ecosystem. In fact, the New England River Basin Commission [1979] concluded in its Level B study that the lake's water was safe to drink and its fish were safe to eat. The report stated that the levels of toxic substances present posed no immediate threat but acknowledged that improved monitoring efforts were needed, given the unknown effects of toxic substances in the lake.

In the 1980s, state monitoring resulted in fish consumption advisories because of mercury (Hg) contamination of walleye and polychlorinated biphenyl (PCB) contamination of lake trout. Public concerns about toxic substances in Lake Champlain began to be addressed with the passage of the Lake Champlain Special Designation Act in 1990, which provided funding for the first lake-wide assessment of toxic substances.

The following paper provides a brief historical perspective on toxic substances assessments in the Lake Champlain basin and presents a case study of one unique area, Outer Malletts Bay in northwestern Lake Champlain.

HISTORICAL OVERVIEW

Monitoring

Most of the early monitoring efforts concerning toxic substances in Lake Champlain focused on concerns about contamination of fish flesh. Measurements of Hg levels in fish collected from the lake between 1988 and 1994 indicated that, while concentrations in most species were below Food and Drug Administration (FDA) tolerance levels, large walleye collected from some tributaries to the lake consistently exceeded the FDA guideline [Lake Champlain Basin Program, 1996]. Comprehensive testing of organic contaminants in fish, which began in 1987, revealed that large lake trout frequently contained PCB levels in excess of FDA tolerance levels. Several pesticides were detected in fish tissues as well.

Other components of the lake's ecosystem have been monitored more sporadically. New York State has included several Lake Champlain stations in its Rotating Intensive Basin Survey (RIBS), which measures toxic substances in water, sediments and biota collected from river mouths. Data from the 1987-1988 survey showed water quality to be good at all sites tested except the Richelieu River [New York State Department of Environmental Conservation, 1990].

In 1992, the U.S. Geological Survey measured the concentrations of toxic substances in stream bed sediments collected from 34 major and 39 minor tributaries in the Lake Champlain basin. Their findings included the discovery of elevated levels of PCBs in Scomotion Creek, a tributary to Cumberland Bay [Lake Champlain Basin Program, 1996]. Earlier monitoring of toxic substances in various biota collected from the Richelieu River by Quebec's Ministere de l'Environnement detected elevated levels of pesticides and PCBs in mollusks and fish [Watzin, 1992].

In the early 1990s, the States of Vermont and New York also monitored for toxic substances in 12 relatively developed urban watersheds. Levels of several trace elements, including zinc (Zn), nickel (Ni), lead (Pb), copper (Cu) and chromium (Cr), were elevated in stream water and sediments, with more highly developed watersheds typically characterized by higher contaminant levels. Caged mussels used as bioindicators in this assessment routinely showed measurable levels of polycyclic aromatic hydrocarbons (PAHs) but no pesticides or PCBs [Quackenbush, 1995].

Levels of trace elements are now routinely measured in air samples collected at two sites in Vermont and two in New York. Findings include elevated levels of Zn in Burlington (VT) samples and periodically elevated arsenic (As) levels, probably emanating from a metals smelter in Quebec [Lake Champlain Basin Program, 1993]. Volatile organic compounds have been measured at a site at Willsboro Bay by New York State since 1990. Only benzene has consistently exceeded state guidelines at this location [New York State Department of Environmental Conservation, 1993].

Research

The earliest known research on toxic substances in Lake Champlain was conducted in the 1970s by Dr. Allen Hunt of the University of Vermont. Hunt [1976] measured levels of selected trace elements in sediments at 200 sites lake-wide and found that Pb and Zn were above background levels throughout the lake.

A major lake-wide assessment of sediment-associated toxic substances was initiated by a team of scientists from the University of Vermont, SUNY-Plattsburgh and the University of Rhode Island in 1991 in response to the Lake Champlain Special Designation Act [McIntosh, 1994]. The early years of the assessment focused on chemical measurements and toxicity tests on surface sediments from all segments of the lake, while later investigations centered on areas with documented contamination, including Cumberland Bay, Inner Burlington Harbor and Outer Malletts Bay.

Historic activities near the Wilcox Dock area of Cumberland Bay have resulted in significant PCB contamination in the upper sediments of a small portion of the bay. Interestingly, standard toxicity tests and an examination of the benthic community in the area failed to produce strikingly positive results, despite the substantial degree of contamination present at the site. Further assessment of the site by the New York State Department of Environmental Conservation has confirmed the existence of high levels of PCBs at the site, and remedial actions to restore the site are underway.

The discovery of high concentrations of PCBs in sediments in Cumberland Bay in 1992 stimulated a variety of research and assessment activities focused on these contaminants. Lester and McIntosh [1994] assessed the role of the freshwater shrimp, *Mysis relicta*, in accumulating PCBs from lake sediments. Their findings suggested a key role for mysids in transferring PCBs to higher levels of the food web. Robert Fuller and colleagues at SUNY-Plattsburgh and Cliff Callinan and his colleagues from the New York State Department of Environmental Conservation have assessed the distribution and fate of PCB congeners contained in Cumberland Bay sediments. They have documented a clear gradient in water column PCB levels, with concentrations decreasing at increasing distances from the area containing contaminated sediments [Callinan *et al.,* 1998].

The contamination in Inner Burlington Harbor presents a relatively typical case for an urban harbor. Stormwater runoff from the city of Burlington and the discharge from a sewage treatment plant have doubtlessly contributed a variety of toxic substances to the harbor; the presence of a breakwater running along most of the length of the harbor has served to reduce lakeward flow and promote the accumulation of contaminants in harbor sediments. Evaluation of harbor sediments by a team of scientists at the University of Vermont and the University of Rhode Island indicated a highly heterogeneous environment, with wide ranging values for particle size and percent organic matter content. Contaminant levels were also highly variable, with highest levels generally associated with the most highly organic sediments. High-level sediment guidelines were exceeded at some locations for Hg and silver (Ag) among the inorganic substances and for PCBs, PAHs, and p,p'-DDE [McIntosh et al., 1997].

Biological assessments of Inner Harbor sediments were varied. A suite of measures, including such standard approaches as the 48-hr acute toxicity test with *Ceriodaphnia dubia,* the chronic 10-day *Chironomus tentans* growth test and the 7-day larval *Pimephales promelas* test, were used to assess sediment quality. Test results failed to reveal a consistent pattern at sites where all three tests were employed. In fact, the *C. dubia* test failed to produce comparable results when repeated during different years. Both spatial and temporal variability in sediment composition likely contributed to the results noted [Watzin et al., 1997].

Analysis of the benthic macroinvertebrate community revealed a fairly healthy community harbor-wide, with only one site adjacent to a sewage treatment plant discharge demonstrating the classic pattern of tolerant-species dominance commonly seen in areas with severe organic enrichment. Multivariate statistical analysis revealed that sediment grain size was a better predictor of benthic community structure than contaminant level in the sediments [Watzin et al., 1997].

Overall, the assessment of Inner Burlington Harbor sediments underscored the challenges for those confronting highly heterogeneous sediment regimes. Because both temporal and spatial changes in contaminant behavior may be the rule in such areas, scientists must exercise great caution in establishing and evaluating assessment programs.

Substantial research has begun on the fate of Hg in the Lake Champlain basin. T. Scherbatskoy (University of Vermont) and several colleagues, including G. Keeler at the University of Michigan and J. Shanley of the U.S. Geological Survey, have been measuring dry and wet deposition of atmospheric Hg at a site near Underhill, VT. Their findings have suggested that levels of Hg deposition in the Lake Champlain basin are similar to those in non-urban regions in the Great Lakes and that snow melt events may carry a substantial fraction of the Hg deposited within the watershed into surface waters [T. Scherbatskoy, personal communication].

Relatively little research has been done on the effects of toxic substances on the Lake Champlain ecosystem. D. Facey of St. Michael's College and V. Blazer of the U.S. Biological Survey [Blazer et al., 1994] have studied the condition of bottom-associated fish collected at several contaminated and non-contaminated sites in the lake. They found an increased incidence of external lesions and enlarged livers in fish collected from such areas as Cumberland Bay. Additional work by Facey has focused on fish collected from Inner Burlington Harbor. Several fish species have shown symptoms of physiological stresses which may be related to exposure to toxic substances [Facey et al., 1998].

M. Watzin of the University of Vermont and A. Friedmann of the Dartmouth Medical School investigated the sublethal effects of mercury on juvenile walleye. They found that juvenile fish fed a diet contaminated with methylmercury at levels bracketing those found in the field in other lake systems grew less than fish fed an uncontaminated

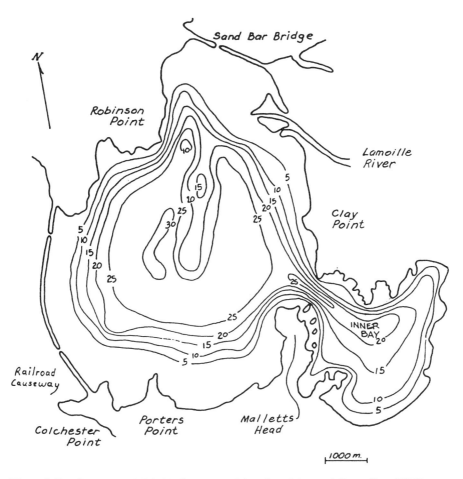

Figure 1. Depth contours (m) in Malletts Bay. Taken from Myer and Gruendling (1979).

diet. Juvenile males showed a variety of reproductive impairments, including testicular atrophy [Friedmann et al., 1996].

CASE STUDY: OUTER MALLETTS BAY

Introduction

The most surprising results of the lake-wide assessment came from Malletts Bay [McIntosh et al., 1997]. This isolated portion of the lake has no obvious major point sources of toxic contaminants, although several minor inputs do exist along the Lamoille River, which drains into the northeastern corner of the Bay (Figure 1).

Early assessments [McIntosh, 1994] demonstrated that pore waters extracted from sediments throughout the Bay produced mortality in standard 48-hour acute toxicity tests done with *Ceriodaphnia dubia*. Chemical analyses of sediments indicated elevated levels of several trace elements which exceeded available sediment guidelines, but no substantial concentrations of organic contaminants were found. Based on chemical and biological findings, an intensive investigation was undertaken to understand the source of the toxicity. Key components of the investigation included extensive chemical analyses of outer bay sediments and water, additional toxicity testing, and a Toxicity Identification Evaluation (TIE).

Site Description

Malletts Bay is the only region of Lake Champlain to develop both pronounced thermal stratification and progressive oxygen depletion during the late summer and early fall. The Lamoille River, with the fourth largest watershed in the Lake Champlain basin, flows into Outer Malletts Bay, discharging its load of particulate and dissolved materials directly to the bay. A major talc mining and processing operation and seven sewage treatment plants are known point sources within the Lamoille River watershed [Schuck, 1995]. In addition, agricultural runoff is a non-point source of nutrient loading to Malletts Bay [Hyde, 1991]. Several marinas on the south shore of Inner Malletts Bay and residences may also contribute pollutant loadings to the bay.

A unique characteristic of Outer Malletts Bay is that circulation between the bay and the Main Lake is extremely restricted by the presence of engineered structures on the western and northern shores. The Sand Bar Bridge Corporation built a causeway bridge across the northern shore of Outer Malletts Bay in 1847. The only opening in this causeway is 1 to 2 m deep and 10 to 20 m wide [Hyde, 1991]. The Rutland and Canada Railroad completed a railroad causeway across the western shore of Outer Malletts Bay in 1901. This causeway has two openings, one on the north that is 54 m wide and about 10 m deep and one on the south that is 24.7 m wide and about 1 m deep [Hyde, 1991]. The only water exchange with the rest of the lake occurs through these openings and is dependent upon the wind direction. For example, a north wind causes southward flow from the Inland Sea through the Sand Bar causeway into Outer Malletts Bay and westward flow from Outer Malletts Bay into the Main Lake through the railroad causeway openings. The flow is reversed with a southerly wind [Hyde, 1991; Myer and Gruendling, 1979]. The extremely restricted circulation of Outer Malletts Bay causes most of the material entering the bay to be trapped. For example, Potash and Henson [1976] estimated that 72% of the phosphorus annually discharged into Malletts Bay is trapped within the bay.

Chemical Analyses

Methods. Surface sediments were collected with a Smith-MacIntyre grab sampler from the UVM Research Vessel *Melosira* at 26 sites in the Outer Bay, the Inner Bay and at a "reference" site in the Main Lake adjacent to the western outlet of the outer bay (Figure 2). At each location, the upper 2 cm of sediment were analyzed for trace element concentrations. In addition, the impacts of seasonal oxygen depletion on the remobilization of iron (Fe), manganese (Mn) and phosphorus (P) from the deep

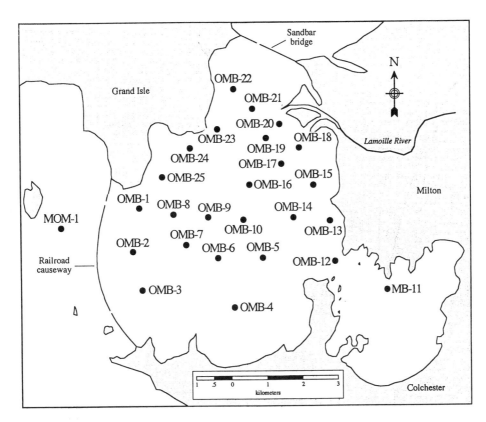

Figure 2. Malletts Bay surface sediment sampling locations.

sediments of Outer Malletts Bay were assessed by collecting water samples along a vertical profile at four sites of varying depths (Figure 3). Each site was sampled at four times from July 7, 1994 to June 5, 1995; these dates were chosen to represent different stages of thermal stratification. At each site, a CTD cast was made to measure conductivity, temperature and oxygen, and, based on dissolved oxygen levels and temperature, water samples were collected simultaneously using a series of 2.5 L Niskin bottles attached to a sampling line at selected intervals. If a sudden decrease in dissolved oxygen was noted, water samples were collected to bracket the area of decrease. Typically, a total of five to six samples were collected at each site. For full details of sampling and analytical procedures, the reader is referred to McIntosh [1994] or McIntosh *et al.* [1997].

Results and Discussion. Analyses of trace metals in surface sediments from the 26 sites indicated that several occurred at levels above background. Cadmium, Cr, Cu, Pb and Zn concentrations exceeded available low-level sediment quality guidelines from NOAA [Long *et al.,* 1995] and the Province of Ontario [Persaud *et al.,* 1993] at a number of the sites in Outer Malletts Bay but failed to exceed higher level guidelines at any location. Much higher relative values were noted for Fe, Mn, Ni, and As (Figures 4-7), with concentrations for these elements frequently exceeding severe effects levels.

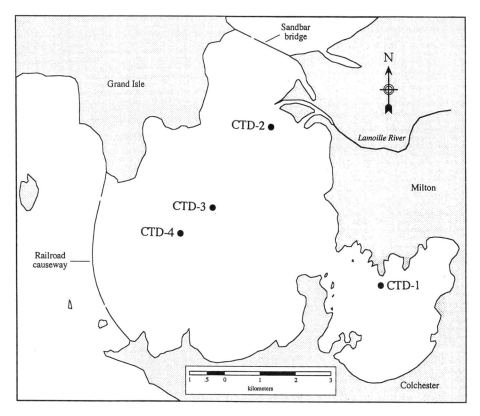

Figure 3. Malletts Bay water column sampling locations.

Maximum concentrations of elements were generally found at sites in the southwestern section of Outer Malletts Bay. Given the positioning of the mouth of the Lamoille River and the bowl-shaped morphology of the bay, such results were not unexpected. Arsenic and Mn also occurred at comparatively high levels at locations adjacent to the mouth of the Lamoille River. Lesser amounts were typically found at more northerly sites and at the "reference" location.

The source of the high levels of the four trace elements is unknown. Possible origins include spoils from abandoned talc mines lying in the upper reaches of the watershed near Johnson, VT [Shuck, 1995] and transport of atmospherically deposited metals by the Lamoille River. Historic data [NASQUAM, 1991] have demonstrated the existence of a plume of atmospheric As carried from a smelter in Canada over portions of Vermont. Deposition of such materials over the extensive drainage basin of the Lamoille River (1,024 square miles) might lead to subsequent elevated levels in Outer Malletts Bay sediments. The contribution of various As-bearing geological formations within the basin must be considered as well.

Analysis of water column samples indicated that thermal stratification had been established in late spring or early summer and that the depth of the epilimnion had increased to 20 m by late summer. At all sites except CTD-2 (see Figure 3), progressive

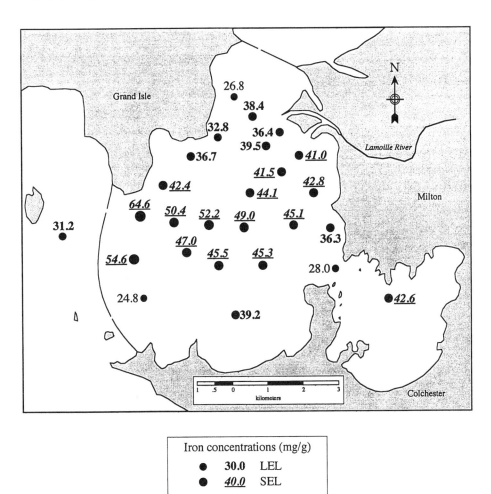

Figure 4. Comparison of iron concentrations (mg/g) in Malletts Bay surface sediment samples to low effects level and severe effects level guidelines, Persaud et al. (1993).

oxygen depletion occurred until late summer, when minimal oxygen levels of 1 to 2.5 mg/L were detected. Fall turnover and mixing occurred by late October.

A closer examination of temperature and oxygen measurements at Site CTD-4 revealed a typical pattern of oxygen depletion in the deepest waters, peaking in the samples collected on September 22, 1994 (Figure 8). At the same time, clearly elevated levels of Mn were detected in water samples. Interestingly, release of Fe was not nearly as severe as that of Mn. Prior studies by Davison et al. [1982] and Engstrom et al. [1985] demonstrated that seasonal oxygen depletion in hypolimnetic waters of lakes can lead to preferential release of Mn from sediments, while complete anoxia may be necessary to facilitate release of iron. The model of Engstrom et al. predicts that Mn precipitates as fine oxide particles and moves downslope to deeper water by the process

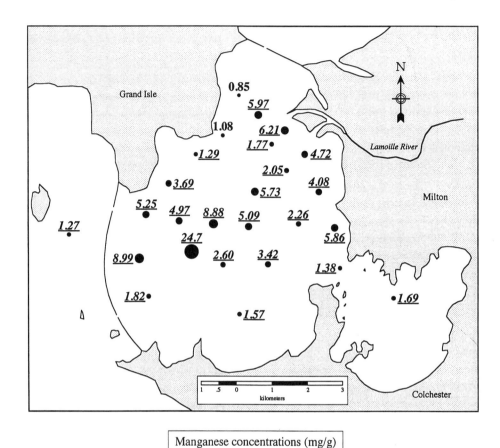

Figure 5. Comparison of manganese concentrations (mg/g) in Malletts Bay surface sediment samples to low effects level and severe effects level guidelines, Persaud *et al.* (1993).

of sediment focusing. The repetition of this process annually likely accounts for the elevated levels of Mn found in the deep sediments of Outer Malletts Bay.

Biological Assessments

Methods. Toxicity tests were conducted on sediments from a total of nine sites in the inner and outer bay and the reference area located in the main lake to the west of Outer Malletts Bay (Figure 9). Two types of tests were performed. Forty-eight hour sediment pore water toxicity tests using *Ceriodaphnia dubia* were conducted on sediments from six locations during May 1993 (pre-stratification), July 1993 (mid-stratification/pre-anoxic conditions), and September 1993 (post-stratification/anoxic

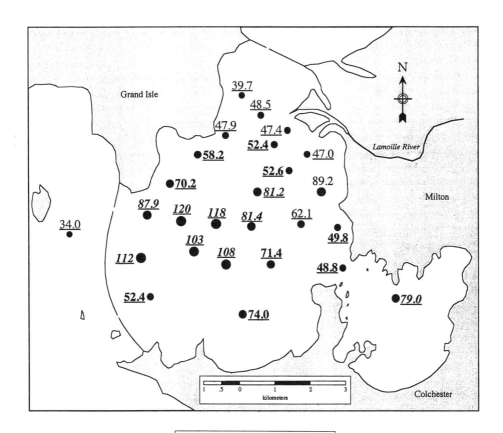

Figure 6. Comparison of nickel concentrations (μg/g) in Mallets Bay surface sediment samples to low (ER-L) and moderate (ER-M) contamination guidelines low (LEL) and severe (SEL) effects levels guidelines, Long *et al*. (1995) and Persaud *et al*. (1993).

conditions) and from four locations during pre-anoxic conditions in June and July 1994. Details on test techniques are contained in McIntosh *et al.* [1997].

In an effort to determine the potential for effects of sediment-associated trace elements on fish, a series of 7-day sediment pore water toxicity tests using eggs and larvae of the fathead minnow *Pimephales promelas* were conducted during 1995. Sediments from three sites (Sites OMB 1,7, and 8 in Figure 9) were evaluated on four dates. In choosing these three Outer Bay sites, we were able to examine surface sediments containing elevated levels of trace elements for three different depths. Any effects noted at the shallower locations, especially Site 1, might suggest the potential

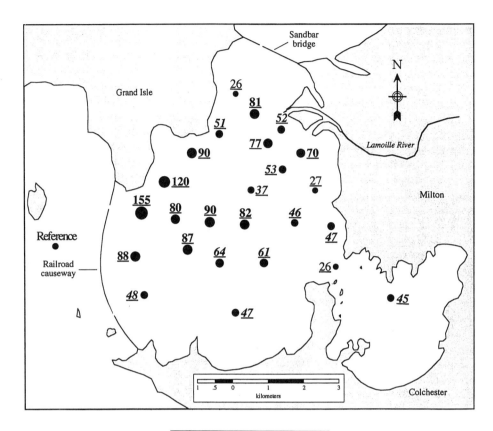

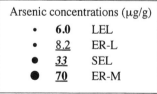

Figure 7. Comparison of arsenic concentrations (µg/g) in Mallets Bay surface sediment samples to low (ER-L) and moderate (ER-M) contamination guidelines low (LEL) and severe (SEL) effects levels guidelines, Long et al. (1995) and Persaud et al. (1993).

for effects on fish reproduction, as opposed to the deepest site, where exposure of the larval stages of most fish species would be limited.

Results and Discussion. Pore water tests conducted during 1993 yielded a fairly consistent pattern of toxicity in *Ceriodaphnia dubia* (Figure 10 top). In the deep hole in Outer Malletts Bay (Site OMB-7), mortality was complete or nearly so on all three dates. Partial, but statistically significant, mortality was noted on all three dates at Sites OMB-3 (about 13 m in depth) and OMB-5 (about 30 m in depth; see Figure 1 for depth contours).

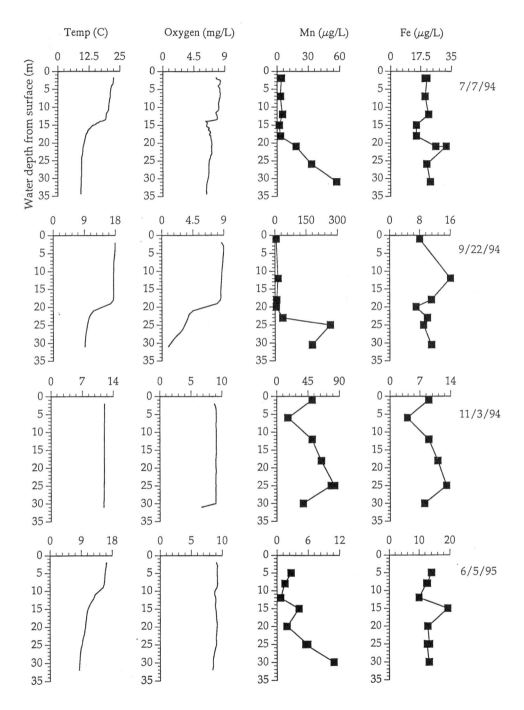

Figure 8. CTD-4: Temperature, Oxygen, Manganese and Iron vs. Depth in Outer Malletts Bay.

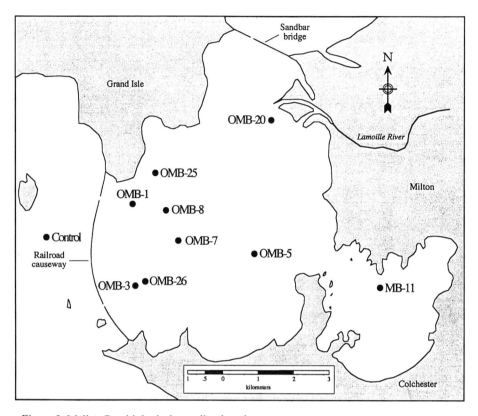

Figure 9. Mallets Bay biological sampling locations.

The results from Site OMB-5 are not surprising, since this site is located in the deeper section of Outer Malletts Bay close to Site OMB-7 (Figure 9), where the greatest accumulation of contaminants has occurred. However, the findings of substantial mortality in pore water from sediments collected at Site OMB-3, which lies near the railroad bridge in the southwestern part of the bay, are more difficult to explain because chemical analyses of surface sediments collected on another date from this site did not indicate particularly high concentrations of trace elements. It is possible that currents moving from the northeast toward the openings in the railroad bridge have carried contaminants to the vicinity of Site OMB-3, where they have been deposited in patches in the shallow sediments. Because surface sediments for the toxicity tests and the chemical analyses were collected on different days, they may have been taken from different patches with varying chemical composition.

The occurrence of partial, but statistically significant, mortality on two of the three dates at Site MB-11, located in the deep hole of Inner Malletts Bay, confirmed findings from earlier studies. The significant mortality at Site OMB-26 during the summer series and at Site OMB-2 during the fall series may have resulted from changes in bioavailability or from the variability inherent in resampling at a particular site. In general, the occurrence of occasional (Site OMB-20) or consistent (Site OMB-3) partial mortality at several sites in shallower areas of Outer Malletts Bay suggests that the

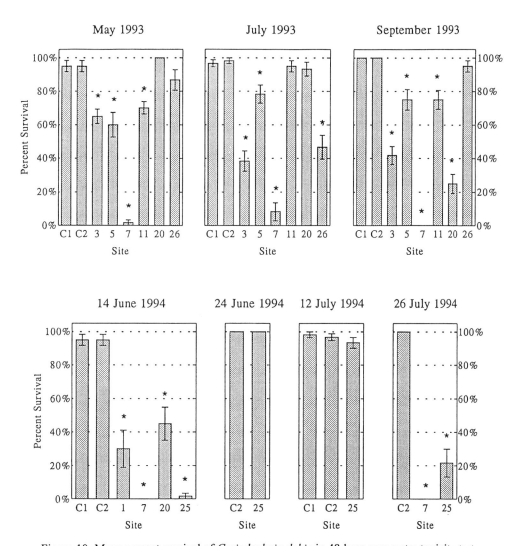

Figure 10. Mean percent survival of *Ceriodaphnia dubia* in 48-hour pore water toxicity tests. Bars represent standard error. C1 indicates sediment pore water control; C2 indicates reconstituted water control. Asterisk indicates mean survival significantly different from control(s).

effects of trace elements may be fairly widespread in Outer Malletts Bay sediments, extending beyond the zone of highest contamination in the deep sediments of the bay.

In general, no temporal pattern was obvious in the 1993 results; that is, toxicity did not dramatically increase at most sites during the fall sampling, when release of trace elements from anoxic sediments might be expected to be the greatest. The nearly complete mortality occurring at Site OMB-7 in any season obscured any seasonal trends in the deepest portions of the bay.

In June and July of 1994, Site OMB-7 again showed complete mortality, while Site OMB-20, near the mouth of the Lamoille River, showed partial, but statistically significant, mortality (Figure 10 bottom). Site OMB-25 showed highly inconsistent results. Again, this might represent patchiness in the distribution of contaminants in the sediment because this site is also located on the shallow, western rim of the bay. Because all 1994 sample collections were made under similar conditions, it is unlikely that changes in bioavailability were driving this pattern.

The effects of pore water collected from Outer Malletts Bay sediments on larval fathead minnows were dramatic (Figure 11). Pore water from Site OMB-7 sediments was consistently toxic, with no larvae surviving the 7-day test on any of the four sampling dates. Site OMB-8 sediments, tested in March and August, produced complete or nearly complete mortality on both dates. Finally, sediment pore water from Site OMB-1, the shallowest site (Figures 9 and 10), produced inconsistent results. Survival was 45% when pore water was tested in April; 0% in August; and 80% in October.

The patterns of fathead minnow response to sediment pore water over the 7-day test are illustrated in Figure 12. In August, the larvae in pore water from Site OMB-1 suffered very rapid mortality. This pattern was noticeably different from the other time periods, and may have resulted from low dissolved oxygen levels in the test water.

Toxicity Identification Evaluation (TIE)

Because the sediments of Outer Malletts Bay have been contaminated with As, Mn and Ni at levels that exceeded sediment quality guidelines, it was unclear which toxic substance or substances were the cause of the toxicity observed. Although a number of sites from throughout the bay showed toxicity, only one site in the deepest part of the bay (Site OMB-7; latitude 44° 34.85'; longitude 73° 18.08'; Fig. 1) was consistently acutely toxic to *Ceriodaphni dubia* . To determine the agent or agents of toxicity at Site OMB-7 in Outer Malletts Bay, we used Toxicity Identification Evaluation (TIE), a procedure developed by the U.S. Environmental Protection Agency (US EPA) in the late 1980s to determine the toxic agents in complex effluents, but now widely applied to ambient waters and sediments [Ankley *et al.,* 1990; Hoke *et al.,* 1992; Norberg-King *et al.,* 1991; and Schubauer-Berigan 1993].

Methods. Guidelines describing the three parts of a TIE have been published [US EPA 1991a; Durhan *et al.,* 1993; Mount *et al.,* 1993]. Phase I involves characterizing the class or classes of contaminants contributing to toxicity. In Phase II, specific toxicants that could be responsible for the observed toxicity are identified. Finally, in Phase III, various approaches are used to confirm that the toxicants identified in Phase II are the cause of the toxicity observed in Phase I. Because TIE procedures were originally designed for effluent samples, aqueous samples are needed to conduct TIEs on contaminated sediments. Evidence has shown that pore water extracted by centrifugation provides a conservative estimate of bulk sediment toxicity when metals are the contaminants of concern [Giesy and Hoke 1989; Schubauer-Berigan and Ankley 1991; Ankley *et al.,* 1991]; therefore, sediment pore water was used for this TIE. *Ceriodaphni dubia* was chosen as the toxicity test species. Full details of the analytical procedures are presented in Boucher and Watzin [1998].

Results and Discussion. The evidence collected in this TIE overwhelmingly implicated Mn as the primary agent of toxicity in Site OMB-7 pore water [Boucher and

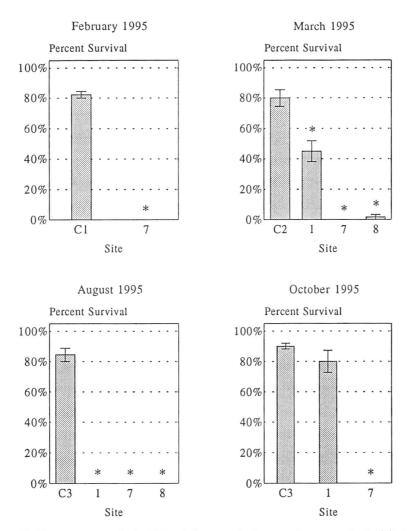

Figure 11. Mean percent survival of *Pimephales promelas* in seven-day pore water toxicity tests. Bars represent standard error. C1 indicates sediment control; C2 indicates reconstituted water control; C3 indicates commercial culture water control. Asterisk indicates mean survival significantly different from the contro (p<0.05).

Watzin, 1998]. Phase I results implicated divalent metals as the class of compounds responsible for toxicity. Phase II revealed concentrations of Mn that could be harmful to aquatic life. Phase III experiments verified that ambient concentrations of Mn in Site OMB-7 pore water were acutely toxic to *Ceriodaphni dubia* under laboratory conditions. Toxic unit calculations indicated that Mn concentrations could account for the majority of the toxicity observed in Site OMB-7 pore water. The concentrations of the other contaminants were not sufficiently high to cause toxicity in spiking experiments performed in the laboratory, whereas Mn spiking proportionally increased

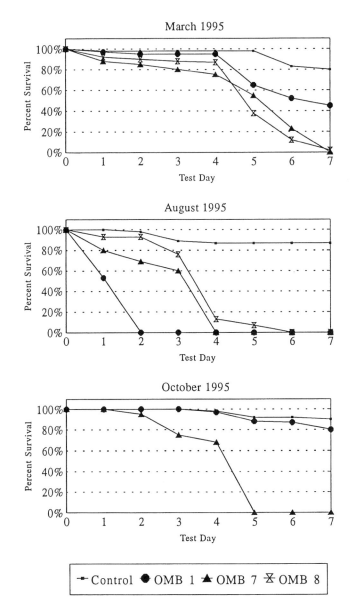

Figure 12. Mean percent survival of *Pimephales promelas* over time during seven-day pore water toxicity tests.

the toxicity of the sample. The response of clean pore water spiked with Mn to Phase I manipulations was similar to that observed when these manipulations were performed on Site OMB-7 pore water during Phase I.

Rarely has Mn been determined to be a toxic agent in lake sediments, even though

Mn is often a major component of freshwater sediments. The situation in Malletts Bay may be unique. The bedrock in the watershed draining into Malletts Bay is enriched with Mn, and sediments containing Mn are carried by the Lamoille River into the bay [Schuck 1995]. Causeways across the openings of the bay (Figure 1) prevent the water of the bay from mixing with water in the main lake [Myer and Gruendling 1979]; therefore, the sediments settle to the bottom and are focused into the deepest part of the bay. Thermal stratification regularly occurs in the bay in late summer or fall; as a result, oxygen becomes depleted in the hypolimnion. This most likely results in reductive dissolution of manganese oxyhydroxides, which comprise a large proportion of the Mn in the sediments [Balistrieri *et al.*, 1992; Davison 1993]. Manganese has been detected in the bottom water and sediment pore water at high levels as stratification and anoxia develop. Manganese ions most likely remain in the sediment pore water after turnover because of a lag in the time it takes oxygen to diffuse from the water column into the sediments and the relatively slow oxidation kinetics of Mn(II). Therefore, Mn levels remain sufficiently high in surface sediments to cause acute toxicity to *Ceriodaphni dubia* at all times of the year.

The results of this TIE revealed that Mn is responsible for the acute toxicity in Site OMB-7 sediments. The extent to which the Mn contamination at this site contributes to toxicity in the water column or has an effect on the ecology of the bay is not clear. At other sites, investigators have found that compounds responsible for toxicity in pore water do not always cause toxicity in the overlying water column [Ankley *et al.*, 1991; Burgess *et al.*, 1993]; but the situation in Outer Malletts Bay is undoubtedly complex. The implication of Mn as a toxic agent at this site underscores the need to look broadly for potential contaminants in any hazard assessment.

REFERENCES

Ankley, G.T., A. Katko, and J. W. Arthur, Identification of ammonia as an important sediment-associated toxicant in the Lower Fox River and Green Bay, Wisconsin, *Environ. Toxicol. Chem.* 9:313-322, 1990.

Ankley, G.T., M.K. Schubauer-Berigan, and J.R. Dierkes, Predicting the toxicity of bulk sediments to aquatic organisms with aqueous test fractions: pore water vs. Elutriate, *Environ. Toxicol. Chem.* 10: 1359-1366, 1991.

Balistrieri, L.S., J.W. Murray, and B. Paul, Biogeochemical cycling of trace metals in the water column of Lake Sammamish, Washington: response to seasonally anoxic conditions, *Limnol. Oceanog.* 37:529-548, 1992.

Blazer, V.S., D.E. Facey, J.W. Fournier, L.A. Courtney, and J.K. Summers, Macrophage aggregates as indicators of environmental stress, in *Modulators of Fish Immune Responses*, vol. 1, SOS Publications, Fair Haven, NJ, 1994.

Boucher, A.M. and M.C. Watzin, A toxicity identification evaluation of metal-contaminated sediments using an artificial pore water containing DOC as the dilution media, *Environ. Toxicol. and Chem.* In press, 1998.

Burgess, R.M., K.A. Schweitzer, R.A. McKinney, and D.K. Phelps, Contaminated marine sediments: water column and interstitial toxic effects, *Environ. Toxicol. Chem.* 12:127-138, 1993.

Callinan, C.W., R. Bonham, and J. Racette, Cumberland Bay PCB Study, New York State Department of Environmental Conservation, Draft Report prepared for the Lake Champlain Management Conference, 1998.

Davison, W, Iron and manganese in lakes, *Earth-Sci. Rev.* 34:119-163, 1993.

Dawson, W., C. Woof, and E. Rigg, The dynamics of iron and manganese in a seasonally anoxic lake; direct measurement of fluxes using sediment traps, *Limnol. Oceanog.* 27:137-150, 1982.

Durhan, E.J., T.J. Norberg-King, L.P. Burkhard, G.T. Ankley, M.T. Lukasewycz, M.K. Schubauer-Berigan, and J.A. Thompson, Methods for aquatic toxicity identification evaluations: phase II toxicity identification procedures for samples exhibiting acute and chronic toxicity, EPA/600/R-92/080, U.S. Environmental Protection Agency, Washington, DC, USA, 1993.

Engstrom, D.R., E.B. Swain, and J.C. Kingston, A paleolimnological record of human disturbance from Harvey's Lake, Vermont: Geochemistry, pigments and diatoms, *Freshwater Biol.* 15:261-288, 1985.

Facey, D., C. Leclerc, D. Dunbar, D. Arruls, and J. Shaw, Physiological indicators of stress among fishes exposed to contaminated sediments from Lake Champlain, abstract from Lake Champlain Research Consortium 1998 Spring Conference. Burlington, VT, 1998.

Friedmann, A.S., M.C. Watzin, T. Brinck-Johnsen, and J.C. Leiter, Low levels of dietary methylmercury inhibit growth and gonadal development in juvenile walleye (*Stizostedion vitreum*), *Aquat. Toxicol.* 35:265-278, 1996.

Giesy, J.P. and R.A. Hoke, Freshwater sediment toxicity bioassessment: rationale for species selection and test design, *J. Great Lakes Res.* 15:539-569, 1989.

Hoke, R.A., J.P. Giesy, and R.G. Kreis Jr., Sediment pore water toxicity identification in the Lower Fox River and Green Bay, Wisconsin, using the Microtox assay, *Ecotox. Environ. Safety.* 23:343-354, 1992.

Hunt, A.S., Trace metal concentrations in Lake Champlain sediments, Proceedings of Lake Champlain Basin Environmental Conference, July 15, 1976, I.M.E., SUNY Plattsburgh-Miner Center, Chazy, NY, 1976.

Hyde, K.M., Internal cycling of phosphorus in Malletts Bay, Lake Champlain, Masters Thesis, U. of Vermont, 1991.

Lake Champlain Basin Program, Quarterly report for December, 1993.

Lake Champlain Basin Program, Background Technical Information for: Opportunities for Action—An Evolving Plan for the Future of the Lake Champlain Basin, 213 pp, 1996.

Lester, D.C. and A. McIntosh, Accumulation of polychlorinated biphenyl congeners from Lake Champlain sediments by *Mysis relicta*, *Environ. Toxicol. Chem.* 13(11):1825-1841, 1994.

Long, E.R., D.D. MacDonald, S.L. Smith, and H.D. Calder, Incidence of adverse biological effects within ranges of chemical concentrations in marine and estuarine sediments, *Environ. Manage.* 19:81-97, 1995.

McIntosh, A., ed, Lake Champlain Sediment Toxics Assessment Program: An Assessment of Sediment-associated Contaminants in Lake Champlain, Phase I, Lake Champlain Basin Program Technical Report No. 5, US EPA, Boston, MA, 325 pp, 1994.

McIntosh, A., M. Watzin. *et al.*, Lake Champlain Sediment Toxics Assessment Program: an assessment of sediment-associated contaminants in Lake Champlain, Phase II. Lake Champlain Basin Program Technical Report 23B, 1997.

Myer, G.E. and G.K. Gruendling, Limnology of Lake Champlain, Lake Champlain Basin Study Technical Report No. 30, New England River Basins Commission, Burlington, VT, USA, 1979.

Mount, D.I., T.J. Norberg-King, G.T. Ankley, L.P. Burkhard, E.J. Durhan, M.K. Schubauer-Berigan, and J.A. Thompson, Methods for aquatic toxicity identification evaluations: phase III toxicity confirmation procedures for samples exhibiting acute and chronic toxicity, EPA/600/R-92/081, U.S. EPA, Washington, DC, USA, 1993.

New England River Basin Commission, Shaping the Future of Lake Champlain: Final Report of the Lake Champlain Basin Study, Final Report of the Lake Champlain Basin Study, New England River Basin Commission, 1979.

Norberg-King, T.J., E.J. Durhan, G.T. Ankley, and E. Robert, Application of toxicity identification evaluation procedures to the ambient waters of the Colusa Basin Drainage, California, *Environ. Toxicol. Chem.* 10:891-900, 1991.

New York State Department of Environmental Conservation, Biennial Report: Rotating

Intensive Basin Studies water quality assessment program, 1987-88, Department of Environmental Conservation, 1990.

New York State Department of Environmental Conservation, Ambient Air Monitoring Network Summary Report for Volatile Organic Compounds, 1990/1991, Department of Environmental Conservation, Division of Air Resources, Bureau of Air Quality Surveillance, 1993.

Persaud, D., R. Jaagumagi, and A. Hayton, Guidelines for the protection and management of aquatic sediment quality in Ontario, Ontario Ministry of Environment and Energy, Toronto, ON, Canada, 1993.

Potash, M. and E.B. Henson, Oxygen depletion patterns in Malletts Bay, Lake Champlain, Great Lakes Res. Div. Publ. No. 15:411-415, 1966.

Quackenbush, A., Identifying toxic constituents of urban runoff from developed areas within the Champlain Basin, Lake Champlain Toxic Source Characterization Interim Report, Results of Screening Activities, Vermont Department of Environmental Conservation, Water Quality Division, 1995.

Schubauer-Berigan, M.K. and G.T. Ankley, The contribution of ammonia, metals and nonpolar organic compounds to the toxicity of sediment interstitial water from an Illinois River tributary, *Environ. Toxicol. Chem.* 10:925-939, 1991.

Schubauer-Berigan, M.K., J.R. Amato, G.T. Ankley, S.E. Baker, L.P. Burkhard, J.R. Dierkes, J.J. Jenson, M.T. Lukasewycz, and T.J. Norberg-King, The behavior and identification of toxic metals in complex mixtures: examples from effluent and sediment pore water toxicity identification evaluations, *Arch. Environ. Contam. Toxicol.* 24:298-306, 1993.

Schuck, R., An historical record of arsenic contamination in the sediments of Arrowhead Mountain Lake, Milton, Vermont, Masters Thesis, University of Vermont, Burlington, VT, 1995.

U.S. Environmental Protection Agency, Methods for aquatic toxicity identification evaluations: phase I toxicity characterization procedures, 2nd ed, EPA/600/6-91/003, Washington, DC, USA, 1991a.

Watzin, M.C., A Research and Monitoring Agenda for Lake Champlain: Proceedings of a Workshop, December 17-19, 1991, Burlington, VT, Lake Champlain Basin Program Technical Report No. 1 U.S. EPA, Boston, MA, 1992.

Watzin, M.C., A.W. McIntosh, E.A. Brown, R. Lacey, D.C. Lester, K.L. Newbrough, and A.R. Williams, Assessing sediment quality in highly heterogeneous environments: a case study of a small urban harbor in Lake Champlain, *Environ. Toxicol. Chem.* 16:2125-2135, 1997.

Ecological Effects of Sediment-Associated Contaminants in Inner Burlington Harbor, Lake Champlain

J. M. Diamond, A.L. Richardson, and C. Daley

ABSTRACT

We analyzed and compared current sediment and benthic ecological conditions in the harbor with data collected previously. Twenty-two samples collected from 12 sites (5 samples from each of 2 relatively clean sites and one sample from each of 10 possibly impaired sites in the harbor with 2 sites replicated) were sampled in summer 1997 for whole sediment toxicity, polynuclear aromatic hydrocarbons (PAHs), select metals, several physicochemical parameters, vertical profile characteristics, organism tissue PAHs, and benthic macroinvertebrate community integrity. Fathead minnow growth in sediment toxicity tests corresponded with benthic macroinvertebrate data and both the fish and *Hyalella* 10 day tests correctly predicted areas of highest contaminant concentrations. *Hyalella* was more sensitive to sediment characteristics than the fish and generally predicted greater impacts than that measured using benthic macroinvertebrate analyses. Chronic larval fish growth in laboratory tests was inversely related to sediment PAH concentration while *Hyalella* survival was related to sediment zinc and lead concentrations. Benthic assemblage integrity was related to sediment PAHs and to a lesser extent, metals such as copper and nickel. Spatial pattern of contaminants was consistent with earlier results, however, there were: (a) significantly lower concentrations of most contaminants in surficial sediments presently as compared to 3-4 years ago and (b) a substantial increase in the number of zebra mussels in the harbor. The decrease in sediment contaminants is coincident with the relocation of a sewage outfall. Interpretations of benthic biological data were limited because of highly heterogeneous sediment conditions in the harbor and the lack of clear reference sites.

INTRODUCTION

Inner Burlington Harbor of Lake Champlain has received numerous toxicants from point and nonpoint sources in its watershed. Previous sediment sampling and analyses [McIntosh et al. 1996; Watzin et al. 1997] demonstrated relatively high concentrations of silver, lead, and PAHs in the harbor, especially in the southern end, compared to sites outside the breakwater. Much of this area corresponds to an old sewage outfall and oil dolphins but could also represent migration of inputs from the old rail yard and nonpoint sources in and around Burlington. Because the surficial sediment (top 2-3 cm) at most sites had lower pollutant concentrations than sediments at greater depths, inputs of pollutants in recent history (past 30 years) may be declining. However, these studies also indicated substantial temporal and spatial heterogeneity with respect to sediment contaminant concentrations and toxicity [Watzin et al. 1997]. Because the case for toxic effects within Burlington Harbor is still weak, strong ecological and biological stress components, in addition to bioavailability and toxicity tests, were incorporated into the present study to establish a weight of evidence about the effects of the toxic contaminants in Inner Harbor.

Biological assessments, using benthic macroinvertebrates, were used in conjunction with other field and laboratory analyses to help determine the effects of sediment contamination and other stressors on the biota of Burlington Harbor. Bioassessments have been used successfully to assess sediment quality in the Great Lakes [Reynoldson et al. 1995; Gerritsen et al. 1997], to assess overall biological quality of large reservoirs [TVA 1995], and to assess effects of non-point source pollution on lakes in Florida [FLDEP 1993]. Benthic macroinvertebrates live on or in the sediment for most or all of their life cycle, they have limited mobility, and they are longer-lived than microscopic organisms such as algae or plankton. This means that they cannot escape exposure, and they integrate stresses occurring throughout their life cycle. Benthic organisms are therefore, widely considered the best indicator organisms for sediment pollution [e.g., Holland 1990].

The overall objective of this project was to assess the hazard resulting from toxic contaminants in the sediments of Inner Burlington Harbor using a sediment quality triad approach [Chapman et al. 1992]. Because certain potentially toxic contaminants are known to occur in Burlington Harbor, the objective of this project was divided into three major component questions.

- Have toxic sediments altered benthic communities of Burlington Harbor?
- Could such changes affect other ecological components of Lake Champlain?
- Do the toxic contaminants in Burlington Harbor sediments accumulate up the food chain and cause risks to higher terrestrial and aquatic trophic levels and human health?

METHODS

Sampling Design

Earlier work [McIntosh et al. 1996] indicated that most of the sites with high silver, lead, and PAH concentrations (the chief pollutants recognized) were located in the southern end of the harbor, from site BH20, south (Figure 1). Sampling locations in the

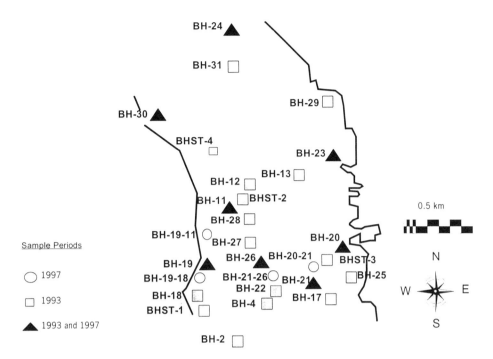

Figure 1. Map showing sites sampled in inner Burlington Harbor in 1993 only (McIntosh et al. 1996), in the present study only, and those sampled in both studies.

present study were identified by reanalyzing the 1993-94 data from the harbor with a spatial statistical model known as kriging [Myers 1988] to estimate contaminant concentrations and uncertainties throughout the harbor. Kriging is a geostatistical estimation method which incorporates a model of the spatial variability of data directly. A weighted moving average technique is used to interpolate values of a particular parameter from a data set onto a grid of points (latitude and longitude) for contouring. The technique used here was ordinary kriging which produces minimum-variance estimates by taking into account the variogram generated. A variogram is a plot of the average squared differences between data values as a function of separation distance and shows the general pattern of variability in a graphical information system (GIS). There are four standard variogram models that could be generated: spherical, exponential, Gaussian and power models. For each chemical, the variogram was calculated using USEPA's software Geo-EAS [version 1.2.1, USEPA 1990] and fitted with each of the four models by a non-linear least-squared procedure.. Due to the small number of data points provided in the Phase 2 study for the inner harbor area (McIntosh et al. 1996), only 8 chemicals were successfully modeled: copper, aluminum, iron, manganese, nickel, silver, nitrogen, and phosphorous. Because silver was highly correlated with PAH in the Phase 2 study (Pearson Correlation coefficient $r = .761$, $p = .0001$, $N = 20$; McIntosh et al. 1996), we could infer (with some uncertainty) PAH spatial variability from that derived for silver.

The sampling sites selected for the present study were those with the greatest uncertainty (using existing data), and the highest likelihood of contamination (Figure 1). Tetra Tech sampled 10 sites in the harbor and collected 10 replicate samples from two different sites (reference sites) with relatively low contaminant concentrations and/or toxicity to help assess sediment quality in the harbor, particularly with respect to biological and toxicological measures. Five replicate samples were collected from site BH30 inside the harbor to take into account the background disturbance regime (due to ferry and other boat traffic as well as natural disturbances). The remaining 5 reference samples were collected from site BH24 outside the harbor, but nearby geographically, to yield relatively undisturbed background sediment conditions (Figure 1). The five replicate samples collected at sites BH24 and BH30 were tested separately for all toxicity and biological analyses, yielding five individual measures for toxicity and macroinvertebrate community structure at these two sites. Subsamples from each of the five samples collected at both BH24 and BH30 were composited into one sample from each site for physicochemical analyses. Two sites, BH11 and BH23 were replicated once as well to obtain a measure of the variability or uncertainty surrounding chemical measures obtained in this study. A total of eight sites were sampled both in this study and in previous work. At each site, the depth profile (at 1 ft depth intervals) for temperature, dissolved oxygen, pH, and conductivity was measured using a Hydrolab model H_2O multiprobe following the standard procedures given by the manufacturer. The equipment was calibrated at the beginning and mid-way through sampling. The equipment was rinsed and acclimated in site water at each new site prior to taking depth measurements. Measurements were recorded on pre-prepared log sheets. Trained volunteers assisted in field sampling, sample processing, in-situ depth profile analyses, and sample preservation.

Sediment Sampling and Analyses

Benthic fauna and sediments were sampled in August 1997 similar to the timing of previous studies. Sites were identified using differential global positioning and checked frequently during sampling to ensure proper sampling location. Each site was sampled using five-seven petite Ponar grabs, depending on the amount of sediment collected in each grab sample. This sampling method was similar to that previously done [McIntosh et al. 1996]. Acceptability of sediment samples was judged using several criteria outlined in Tetra Tech's Quality Assurance Project Plan [Tetra Tech 1997]. If these criteria were not met, the sample was discarded and the site was re-sampled. Contents of the Ponar samples from the site were composited and homogenized in the field using Teflon or high density plastic equipment to obtain a representative sample from each site for chemical, toxicological, and biological analyses. Biological samples used sediment from the first three grab samples collected at each site. The remaining sediment collected was composited and used for chemical and toxicological analyses. At each site, depth and sediment characteristics were recorded on field log sheets. Samples were handled and stored following standard operating procedures [Tetra Tech 1997] and samples were recorded on sample chain-of-custody forms for shipping off-site for analysis.

Table 1 summarizes the analyses performed in this study, the methods used, and detection limits. Sediment chemical analyses included PAHs, simultaneously extracted metals (SEM), total organic carbon (% TOC), acid volatile sulfides (AVS), total organic nitrogen (TON), ammonia, particle size, and pH. Five metals (those previously showing the highest levels: silver, nickel, copper, lead, and zinc) were measured. Zebra mussels (*Dreissena polymorpha*) were collected from several sites (BH20, BH21, BH20-21,

TABLE 1. Summary of parameters analyzed, methods used, and detection limits.

Parameter	Method	Detection Limit
Sediment		
particle size	ASTM D4882 [1992]	1%
total organic carbon	EPA - 9060	0.1%
total organic nitrogen	EPA - 350.2 and 351.3	100 Fg/Kg
ammonium	EPA - 350.1	100 Fg/Kg
pH	APHA 423	0.2 units
AVS	FGS - 0036	1.33 mg/Kg
SEM	FGS - 0036/EPA 1638 or 1639	1 mg/Kg
PAHs	EPA 8270, 8310	5 Fg/Kg
Water Chemistry		
temperature	BRF 050, APHA 2550	0.1 C
dissolved oxygen	BRF 050, APHA 4500-0(G)	0.2 mg/L
conductivity	BRF 050, APHA 2510	20 Fhos/cm
pH	BRF 050, APHA 4500-HA+	0.2 units
Biology/Toxicity		
Fathead minnow survival and growth	EPA [1991]	N/A
Hyalella 10d survival	EPA [1994]	N/A
Benthic macroinvertebrate assessment	Gerritsen [1997]	N/A
Organism Tissue Analytical Chemistry		
PAHs	EPA 8270/8310	20 Fg/Kg
lead	EPA 6010/200.7	250 Fg/Kg

BH23, BH24, and BH30) and analyzed for tissue PAHs and percent lipid content on a composite sample of organisms collected at each site.

A portion of the sample from three inner harbor sites (BH20-21, BH23, and BH21) were sieved to isolate the fine fraction less than 63 and also analyzed for PAHs, TOC, and TON because recent research has suggested that PAH bioavailability is much better correlated with the concentration present in the fine fraction than in the sediment as a whole, particularly if cellulose fibers (such as from plant material) or gravel, are present [Landrum pers. comm.].

Pearson's Product Moment correlation analysis was used to infer relationships between pairs of physicochemical characteristics using a significance level of 0.05. Principal components and Factor Analysis (Statistica, version 4.0) were used to determine relationships among sediment physicochemical variables and to relate physicochemical variables to biological and toxicological measures. Chemical data were log-transformed as necessary to meet assumptions of variance homogeneity and data normality prior to analysis.

Toxicity Tests

Samples from each site (a total of five samples from each of the two "reference" sites BH24 and BH30 and one sample from each of the 10 test sites, two of which, BH23 and BH11, were duplicated) were used in 10-day *Hyalella* acute [USEPA 1994] and in 7d fathead minnow survival and growth [USEPA 1991] whole sediment toxicity tests. Tests were conducted using a manual renewal system in which overlying water in each chamber was replaced twice each day. Differences in endpoint values among sites were determined using ANOVA and post-hoc HSD multiple means tests ($p < 0.05$). Survival data were analyzed using an arcsine square root transformation and fish weight data were log-transformed to satisfy variance homogeneity assumptions of the analysis. Toxicity endpoints were related to physicochemical properties of the sediment using forward stepwise multiple regression analysis (Statistica, version 4.0). In an effort to control Type 1 errors due to the relatively small sample size (N = 14 for most regression analyses in this study), independent variables (physicochemical measures) were included in the model only if their F statistic exceeded 1.0, and they improved the overall R^2 of the model.

Biological Analyses

Samples were mixed and placed in white enamel pans equipped with a grid frame (20 individual cells). All organisms were picked with the aid of a dissecting microscope (15X). Due to relatively few individuals collected at some sites, we sorted each sample in its entirety. Total number of individuals ranged from 10 to 246 in a given sample (3 ponars). Organisms were enumerated and identified by macroinvertebrate specialists at Tetra Tech's Biological Research Facility. For this project, specimens were generally identified to genus (or lower) using the most current literature available. Several indicator variables, or metrics, were calculated based on the benthic species data collected at each site including: taxa richness, number of Chironomid (midge) genera, percent mollusc and crustacean taxa, percent Oligochaetes (worms), percent dominant taxon, and several other taxonomic and feeding guild metrics. Analyses were conducted for sites at which at least 80 organisms were collected (19 of the 22 samples collected) to minimize the effect of sample size on taxa richness measures. Using sediment chemistry results to define "good" or "poor" sites, we compared the distribution of values for various metrics for the two classes of sites. Frequency distributions were plotted for those metrics that showed significantly different ranges of values for the two types of sites. Each metric distribution was then trisected and the lower third was given a score of 1, the mid-third a score of 3, and the upper third, a score of 5 [Barbour et al. 1996; Gerritsen et al. 1997]. As an example, the metric number of Chironomid genera, exhibited reasonable discrimination between supposedly toxic and non-toxic sites. Trisecting the distribution of values observed for this metric yielded the following scoring system: sites with 3 or fewer Chironomid genera were given a score of 1 (poorest), sites with 4-8 genera were given a score of 3 (intermediate) and sites with greater than 8 genera were given a score of 5 (best). A total of six metrics were identified as having the greatest discriminatory potential: taxa richness, percent dominance within the Chironomid genera, the percent dominance of Chironomid genera as compared to all taxa collected, crustacean-mollusk taxa richness, and percent Tanytarsini Chironomide. These six metrics were scored and summed for each site resulting in a range in the total biotic score of 6 and 30. Although some of these metrics were correlated with each other and not independent, there is ample evidence from studies in both lakes and streams that demonstrate the utility of a multi-metric scoring approach such as that used here [FDEP 1994; Barbour et al 1996].

Forward stepwise multiple regression analysis was used to determine the strength of relationships between either specific metrics or overall biotic scores and toxicity test results or sediment physicochemical characteristics. Independent variables (physicochemical measures) were included in the model if their F statistic exceeded 1.0 and they improved the overall R^2 of the model.

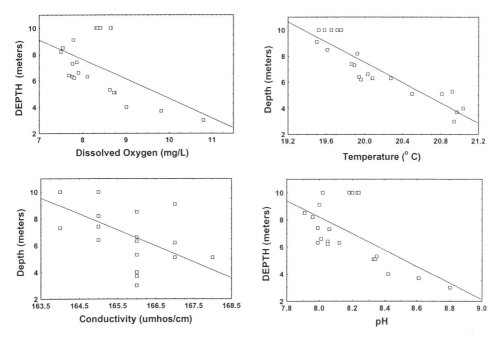

Figure 2. Bottom water temperature, dissolved oxygen, pH, and conductivity as a function of depth for each site sampled in inner Burlington Harbor, August 1997.

RESULTS AND DISCUSSION

Water and Sediment Physicochemical Characteristics

Minor differences in pH, temperature, dissolved oxygen, and conductivity were observed between the surface and bottom at most locations. As expected, deeper sites had lower pH, temperature, and dissolved oxygen concentrations in the bottom waters (Figure 2). Minimum dissolved oxygen concentrations were above 7.0 mg/L at all locations and pH was generally between 7.5 and 8.5 indicating well-oxygenated and well-buffered conditions in the harbor.

Percent solids in sediments was inversely related to depth and percent fines was inversely related to percent solids ($r = -.68$ and $r = 0-.73$, respectively, $p < .05$). Sediment ammonia concentration increased with increasing depth ($r = 0.69$, $p < .05$) and total organic nitrogen, and total organic carbon concentrations were correlated with each other and with ammonia. Sediment ammonia concentrations in this study were probably not high enough to be toxic [< 700 mg/kg; Ankley et al. 1990; USEPA 1994]. Higher ammonia concentrations and lower pH were especially prevalent near the breakwall, where less hydraulic flushing and fewer sediment disturbances are likely to occur.

Higher concentrations of PAH's were observed primarily in the southern part of the harbor (sites BH20, BH20-21, and BH21-26), similar to previous findings, although site BH23 (near mid-shoreline) also had a relatively high PAH concentration (Figure 3A).

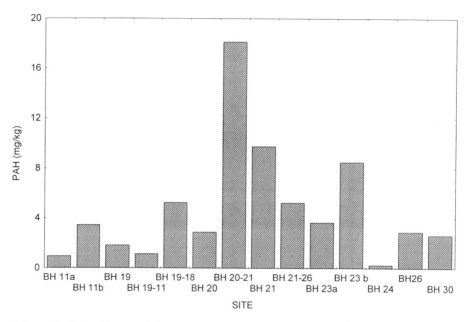

Figure 3A. Polycyclic aromatic hydrocarbon (PAH) concentration in sediments by site in August 1997.

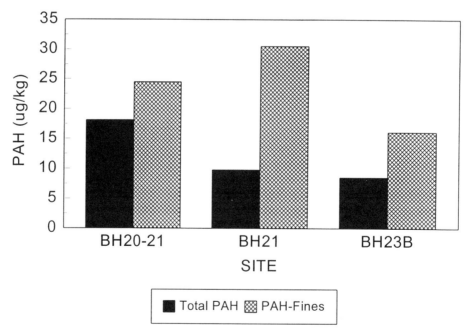

Figure 3B. Total PAH measured in whole sediment samples and in only the fine sediment fraction of those same samples from Burlington Harbor, August 1997.

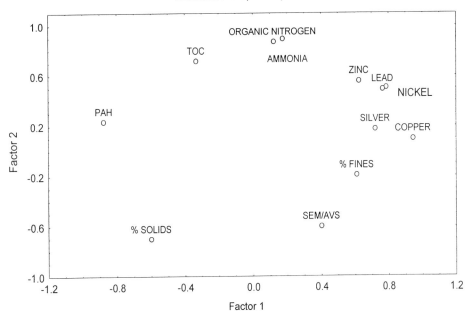

Figure 4. Summary of Principal Components and Factor Analysis of physicochemical data collected for sediments in Burlington Harbor, August 1997.

PAH concentrations were higher if based on the fine sediment fraction only (Figure 3B, t-test, $p < .01$). The most prevalent sediment PAH compounds in this sampling (in decreasing order) were Fluoranthene, Pyrene, Phenanthrene, Benzo(a)pyrene, and Benzo(K)fluoranthene. Sediment effect concentrations (SECs), derived from sediment toxicological and chemical data in the Great Lakes [USEPA 1996], indicated that all of the above compounds, and/or total PAHs, had a high probability of causing toxicity to aquatic life at sites BH20-21, BH21, BH19-18, BH21-26, and BH23. These same sites also exceeded Effect Ranges Low [ERL; Long and Morgan 1991; Long et al. 1995] for these same PAH compounds and for total PAHs. Previous work [McIntosh et al. 1996] also identified these sites, or those in the same vicinity of the harbor, as having relatively high sediment PAH concentrations.

PAH analysis of zebra mussel tissue indicated a significant relationship between sediment and tissue PAH (Wilcoxon Rank Sum, $p > 0.10$) at the same sites. These results suggest that sediment PAHs are mobilized and bio-concentrated by aquatic invertebrates in the harbor.

Copper, zinc, nickel, and lead concentrations were highly correlated with each other suggesting there may be a common source of these metals. Silver concentration was uncorrelated with other metals (as previously observed) suggesting either a different transport/fate mechanism or different source of this metal in the harbor. Factor Analysis of the chemical data yielded 2 Factors: Factor 1 was correlated with metals and percent fines and Factor 2 correlated with nutrients, TOC, PAH, and percent solids (Figure 4).

Thus PAH concentration was often inversely related to metal concentrations in the harbor suggesting either different sources for these two classes of pollutants or different transport and depositional mechanisms. SEM/AVS ratios were < 1.0 at most sites suggesting a low probability of metal toxicity in the harbor, similar to results reported in previous work [McIntosh et al. 1996].

Based on measurements made at the 8 sites sampled in both the current study and in previous work, concentrations of several metals and PAHs were lower in this survey, particularly in the area near the old wastewater outfall (t-tests, N = 8, p < 0.05; Figure 5). We attribute this decrease in surficial sediment pollutants to the relocation of the wastewater treatment plant discharge outside of the harbor and the implementation of certain storm water runoff controls in the intervening years. However, we cannot rule out temporal variability in chemical concentrations resulting from natural and man-made disturbances in the harbor, as noted in previous research [McIntosh et al. 1996; Watzin et al. 1997]. Although absolute pollutant concentrations were not the same at the 8 sites sampled in both 1993 and 1997, the variation in relative magnitude of pollutant concentrations among these sites was generally similar between the two studies (Wilcoxon Rank Sum, p < .05). Kriging spatial analysis, using the present results along with those collected in 1993, suggests that the area between sites BH25 and BH19, and between sites BH18 and BH11, have a high probability of elevated sediment pollutant concentrations and consequently, toxicity to aquatic life. The area around site BH23 also appears to have a high probability of elevated sediment pollutants.

Toxicological Results

Figure 6 summarizes results for the 7-day chronic fathead minnow and 10-day acute *Hyalella* whole sediment toxicity tests. Fish survival was at least 70% at all sites and fish growth exceeded the minimum individual weight criterion required by USEPA for control exposures [Figure 6A; USEPA 1991]. However, there was a substantial range in fish growth among sites with sites BH20, BH20-21, and BH23 exhibiting the lowest individual fish weight. *Hyalella* survival appeared to be a more sensitive indicator than fish growth (Figure 6B). Except for the two "reference" sites sampled in this study, (sites BH24 and BH30), most sites exhibited substantial reductions in *Hyalella* survival. Forward stepwise multiple regression analyses indicated that fish weight (growth) in sediment toxicity tests was significantly related to PAH concentration while *Hyalella* survival was related to zinc and lead concentration ($R^2 = 0.61$; F = 15.67, and $R^2 = 0.74$, F = 13.06, respectively; p < .01). Given that SEM/AVS ratios were generally < 1.0 in this study, *Hyalella* may have been responding to some other factor correlated with metal concentration. Similar results were observed by Canfield et al. (1996) and in the Great Lakes [Assessment and Remediation of Sediment Contaminants; ARCS, USEPA 1996]. Both total PAH and Fluoranthene concentrations at sites BH19-18, BH20-21, BH23, and BH21 were above the lowest observed effect concentrations (similar to an Apparent Effects Threshold Value) calculated for Great Lakes sediment and *Hyalella* [USEPA 1996] and also above the Effect Range Median (ERM) values calculated for this species based on Long and Morgan=s (1991) entire dataset. Alternatively, SEM/AVS may not be a completely reliable indicator of metal toxicity in the harbor. Indeed, Ankley et al. [1996] reported that AVS may not play a significant role in metal binding in aerobic systems, or those having low productivity, either of which will limit sulfide reduction in

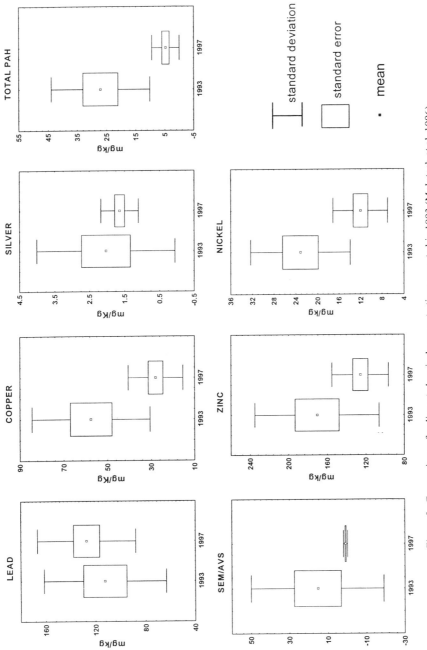

Figure 5. Comparison of sediment chemical concentrations reported in 1993 (McIntosh et al. 1996) and in the present study for the same locations in Burlington Harbor.

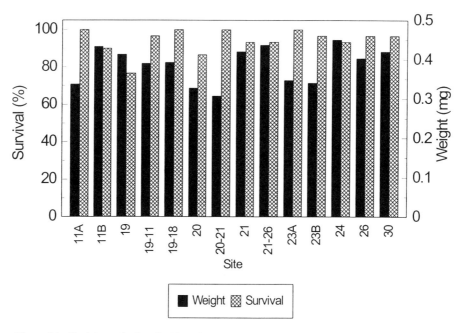

Figure 6A. Toxicity endpoints for Pimephales promelas (fathead minnow) chronic sediment test on samples collected in August 1997.

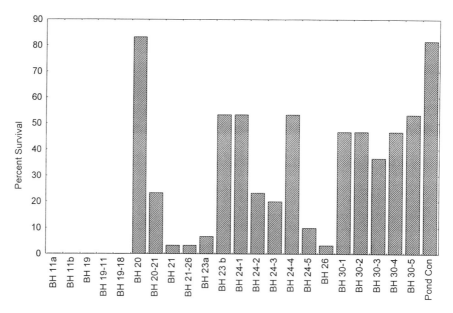

Figure 6B. Survival results for the Hyalella azteca 10-day whole sediment test on samples collected in August, 1997 from Burlington Harbor.

TABLE 2. Summary of forward Stepwise Multiple Regression Analysis on total biotic score of benthic macroinvertebrates collected in August 1997 and physicochemical parameters measured at the same sites and time. Log (PAH)= natural logarithm of polycyclic aromatic hydrocarbon concentration.

	BETA	Std Error of BETA	B	Std Error of B	t(5)	p - level
Intercept			21.231	4.625	4.590	0.006
Log (PAH)	-0.747	0.210	-3.228	0.909	-3.550	0.016
Silver	0.177	0.219	1.445	1.788	0.808	0.456
Kjeldahl	0.408	0.222	0.001	0.001	1.839	0.125
% Fines	-0.291	0.188	-0.056	0.036	-1.546	0.183
SEM/AVS	0.251	0.225	1.003	0.899	1.115	0.316

sediments. Burlington Harbor, and Lake Champlain as a whole, are generally aerobic, low productivity waters [McIntosh et al. 1996]. Thus, toxicity responses exhibited by *Hyalella* may have been due to cumulative effects of metals, PAHs and perhaps other pollutants present in these sediments.

Biological Results

Site BH24, outside the harbor, had the fewest invertebrates and this result appeared to be directly related to the especially coarse sandy substrate observed there. Forward stepwise multiple regression analysis indicated that PAH and nickel concentrations were negatively related, and ammonia, zinc, percent solids, and Kjeldahl nitrogen were directly related to the total biotic score ($R^2 = 0.866$, F = 6.4508, p for the model = 0.03, Table 2). However, the only significant (p< 0.05) explanatory physicochemical factor was sediment PAH concentration (Table 2). Benthic macroinvertebrate species distribution may have also been responding to available nutrients (i.e., nitrogen) and the percent fine substrate available as observed previously [Watzin et al. 1997].

One major difference between biological results gathered in the present study and those obtained previously is the recent invasion, and growing abundance, of *Dreissena polymorpha* (zebra mussels) in the harbor. Zebra mussels were collected at most nearshore sites (e.g., sites BH23, BH21, BH20-21) and appeared to be limited to < 7m depth. Forward stepwise multiple regression analysis indicated that depth was the most significant explanatory variable for mollusc and crustacea abundance (primarily zebra mussels and *Asselus* or isopods, p = .013) and secondarily, ammonia and TOC (overall model $R^2 = .93$, p < .015). Zebra mussels and isopods were often collected in association with submerged macrophytes. There was a significant inverse relationship between mussel abundance and the number of Chironomid genera observed (r= -95, p < .05), however, it is not clear whether this was a result of biological interactions or the result of other factors correlated with zebra mussel abundance.

Fathead minnow weight in chronic sediment toxicity tests was significantly related to the total biotic score (p < .001) and appeared to be an efficient predictor of biological condition (Figure 7). *Hyalella* survival in sediment tests was also significantly related to biotic score (p < .001) but appeared to overestimate effects on benthos at some locations.

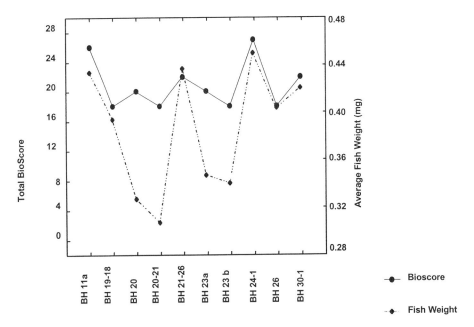

Figure 7. Relationship between fathead minnow growth in sediment toxicity tests and corresponding benthic macroinvertebrate assessment score.

CONCLUSIONS

Results of the present research support earlier work conducted between 1990 and 1993 in the harbor. Specifically, the southern end of the harbor and the breakwall area appear to have generally higher metal and PAH sediment concentrations, more sediment toxicity, and poorer benthic assemblage than other sites. The present sampling at selected new locations generally confirmed and lend more certainty to the pollutant spatial patterns observed earlier. Toxicity and biological results of multiple samples taken at supposed reference sites inside and outside of the harbor in this study also confirmed the high degree of spatial heterogeneity reported in earlier studies. The fact that sediment samples from site BH24, outside the harbor, showed poorer biological scores than site BH30 inside the harbor, suggests that sediment disturbance and pollutant accumulation are important factors immediately outside the harbor as well as inside the harbor.

Biological and toxicological results appeared to be related primarily to sediment PAH concentration. Similar relationships were observed in previous studies, however, it is not clear to what extent temporal variability affects this relationship. Temporal variability may also be a factor behind the lower sediment pollutant concentrations observed in this study as compared to 1993. However, the relocation of the sewage outfall outside the harbor and implementation of some stormwater runoff controls are consistent with the observed decrease in pollutants. Follow-up toxicity monitoring in the spring 1998 will help address chronic sediment toxicity potential and the extent of temporal variability.

One of the biggest differences observed between the current study and studies conducted earlier is the presence of zebra mussels in the harbor. In 1997, this species appeared to be confined to sites < 7m depth and was especially associated with aquatic macrophytes. Preliminary tissue PAH data were consistent with sediment PAH data indicating that this species is useful as a sentinel indicator of bioaccumulation potential. Follow-up protein expression analyses on field collected zebra mussels, and *Lumbriculus* bioaccumulation testing on select sediment samples taken in the spring of 1998, will be used to further evaluate long-term toxicity potential and associated ecological risks to aquatic life, and wildlife that depend on aquatic resources in the harbor.

Acknowledgements. The authors wish to thank the Lake Champlain Basin Program for generously supplying laboratory space and logistical support for field sampling and organism sorting. Doug Burnham (VTDEC) was especially helpful providing technical oversight and assisting in field sampling. This work is being conducted under grant number LC-X 0018-0-01 through the New England Interstate Water Pollution Control Commission.

REFERENCES

Ankley, G., D. DiToro, D. Hansen, and W. Berry. 1996. Technical basis and proposal for deriving sediment quality criteria for metals. Environ. Toxicol. Chem. 15:2056-2066.

Ankley, G.T., A. Katko, and J. Arthur. 1990. Identification of ammonia as an important sediment-associated toxicant in the lower Fox River and Green Bay, Wisconsin. Environ. Toxicol. Chem. 9:312-322.

Barbour, M., J. Gerritsen, G. Griffith, R. Frydrenborg, E. McCarron, J. White, and M. Bastian. 1996. A framework for biological criteria for Florida streams using benthic macroinvertebnrates. J. N. Amer. Benthol. Soc. 15:179-184.

Bradley, B.P. 1990. Stress Proteins: Their detection and uses in biomonitoring ASTM 13th Symposium on Aquatic Toxicology. ASTM STP 1096 (W.G. Landis and W.H. van der Schalie, Eds.). American Society for Testing and Materials, Philadelphia.

Bradley, B.P. and J.B. Ward. 1989. Detection of a major stress protein using a peptide antibody. Marine Env. Res. 28:471-475.

Bradley, B.P., Brown, D.C., Lamonte, T.N., Boyd, S.M. and O'Neill, M.C. 1996. Protein Patterns and Toxicity Identification in "Biomarkers and Risk Assessment" ASTM STP 1306, David A. Bengtson and Diane S. Henshel, Eds., American Society for Testing and Materials, Philadelphia.

Canfield, I., F. Dwyer, J. Fairchild, C. Ingersoll, N. Kemble, D. Mount, T. LaPoint, G. Burton, and M. Swift. 1996. Assessing contamination of Great Lake sediment using benthic invertebrates and the sediment quality triad approach. J. Great Lakes Research.

Chapman, P., E. Power, and G.A. Burtora, Jr. 1992. Integrative Assessments in Aquatic Ecosystems. In: Sediment Toxicity, Assessment. G.A. Burton, Jr. (Ed.), pp. 313-340, Lewis Publishers, Chelsea, MI.

Florida Department of Environmental Protection. 1994. Lake bioassessments for the determination of nonpoint source impairment in Florida. Florida Department of Environmental Protection, Biology Section, Tallahassee, FL.

Gerritsen, J., R. Carlson, D.L. Charles, D. Dycus, C. Faulkner, G.R. Gibson, R.H. Kennedy, and S.A. Markowitz. 1997. Lake and reservoir bioassessment and biocriteria. Technical guidance document. Darft. Office of Water, USEPA, Washington, DC.

Holland, A.F., A.T. Shaughnessy, L.C. Scott, V.A. Dickens, J. Gerritsen, and J.A. Ranasinghe. 1989. Long-term benthic monitoring and assessment program for the Maryland portion of Chesapeake Bay: Interpretive report. Maryland Department of Natural Resources, Power Plant and Environmental Review, Annapolis, MD. Report #CBRM-LTB-EST-89-2.

Kung, S.Y. 1993. Digital Neural Networks. PTR Prentice Hall, Englewood Cliffs, NJ.

Long, E., D. MacDonald, S. Smith, and F. Calder. 1995. Incidence of adverse biological effects within ranges of chemical concentrations in marine and estuarine sediments. Envir. Mgmt. 19:81-97.

Long, E.R. and L.G. Morgan. 1991. The potential for biological effects of sediment-sorbed contaminants tested in the National Status and Trends Program. NOAA Technical Memorandum, NOS OMA 52, Seattle, WA.

McIntosh, A., M-Watzin, et al. 1996. Lake Champlain Sediment Toxics Assessment Program. Phase II. Section B. Burlington Harbor. Lake Champlain Basin Program, Burlington, VT.

Myers, D.E. 1988. Some aspects of multivariate geostastical analysis. In: C. F. Chung (ed) Quantitative Analysis of Mineral and Energy Resources, D. Reidel Publishing Co., Dordrecht, Germany, pp 669-687.

Reynoldson, T.B., R.C. Bailey, K.E. Day, and R.H. Norris. 1995. Biological guidelines for freshwater sediment based on Benthic Assessment of Sediment (the BEAST) using a multivariate approach for predicting biological state. Austral. J. Ecol. 20:198-219.

Tetra Tech. 1997. Quality Assurance Project Plan: Chronic Sediment Contamination in Inner Burlington Harbor, Lake Champlain. Tetra Tech, Inc., Owings Mills, MD 21117.

TVA (Tennessee Valley Authority). 1995. Aquatic ecological health determinations for TVA reservoirs - 1994: An informal summary of 1994 vital signs monitoring results and ecological health determination methods. Water Management Division, Tennessee Valley Authority, Chattanooga, TN.

USEPA 1990. Geo-EAS (Geostatistical Environmental Assessment Software) ; version 1.2.1. Robert S. Kerr Environmental Research Center, US Environmental Protection Agency, Ada, OK

USEPA. 1991. Short-term Methods for Estimating Chronic Toxicity of Effluents and Surface Waters to Freshwater Organisms. EPA-600/4-91/002. Third Edition, U.S. Environmental Protection Agency, Washington, D.C.

USEPA. 1994. Methods for measuring the toxicity and bioaccumulation of sediment-associated contaminants with freshwater invertebrates. EPA/600/R-94/024. U.S. Environmental Protection Agency, Washington, D.C.

USEPA. 1996. Assessment and remediation of contaminated sediments (ARCS) program. Calculation and evaluation of sediment effect concentrations. EPA/905/R96/008. U.S. Environmental Protection Agency, Washington, D.C.

Watzin, M., A. McIntosh, E. Brown, R. Lacey, D. Lester, K. Newbrough, and A. Williams. 1997. Assessing sediment quality in heterogeneous environments: a case study of a small urban harbor in Lake Champlain, Vermont USA. Environ. Toxicol. Chem. 16:2125-2135.

Mercury Cycling and Transport in the Lake Champlain Basin

James B. Shanley, Andrea F. Donlon, Timothy Scherbatskoy, and Gerald J. Keeler

ABSTRACT

Mercury contamination and its potential effects on human health is an issue of growing concern in the Lake Champlain basin. This paper is an effort to review and synthesize the current state of understanding of sources, sinks, and movement of Hg within the basin. We compile and review existing information and present some new data on Hg concentrations and fluxes in atmospheric deposition, throughfall, foliage, surface water, soil water, soil, and biota, then use these data to formulate a conceptual model of Hg cycling in the basin. Mercury is primarily of anthropogenic origin and enters the basin in atmospheric deposition. At a site near the center of the basin, Hg deposition during a two-year period averaged 444 mg ha^{-1} yr^{-1}, of which about 70% was dry deposition. Most of this Hg reaches the land surface in throughfall or litterfall and is retained in the soil. A significant percentage of the Hg may be re-volatilized to the atmosphere. Stream export of Hg averaged only 27 mg ha^{-1} yr^{-1} for the same two-year period. Mercury appears to move to stream channels primarily in association with DOC, whereas Hg in streamwater tends to be associated with organic suspended sediment. Our review identified a need for more research on the speciation and bioaccumulation of mercury, and more research on Hg cycling in non-forested areas of the basin.

INTRODUCTION

Mercury (Hg) contamination of surface waters and biota has emerged as a global problem [Nriagu and Pacyna, 1988; Fitzgerald, 1993; U.S. EPA, 1997]. Concern for mercury contamination in the environment stems from its potent neurotoxicity to humans. Mercury concentrations in fish such as walleye in Lake Champlain are elevated to the point where consumption of greater than 250 g of fish per month exceeds the maximum U.S. EPA recommended Hg dosage. The State of Vermont has issued an advisory on fish consumption that warns of this danger, and cautions children and pregnant women not to consume any walleye, nor any more than 2-3 servings per month of other fish species [State of Vermont, 1990]. A recent study [River Watch Network, 1998] suggests a possible link between neurological disorders among the native Abenaki and consumption of fish from the Missisquoi River. Public health concerns prompted passage of a 1998 law in Vermont to phase in limitations on Hg emissions. Discussion continues among other U.S. state and Canadian provincial leaders to enact similar measures.

On a global basis, Hg is increasing in the environment [Swain et al., 1992; NESCAUM et al., 1998]. Upward trends in Hg have been indicated by sediment cores in Alaska [Engstrom and Swain, 1997], an increase in Hg in air over the North Atlantic Ocean between 1977 and 1990 [Slemr and Langer, 1992], and as described throughout the global biogeochemical Hg cycle by Fitzgerald et al. [1998]. Localized areas, however, have undergone recent declines in Hg. Engstrom and Swain [1997] used sediment cores from eastern Minnesota lakes to demonstrate that Hg deposition increased between 1920 and 1960, but decreased after 1960. They suggested that the Hg decrease in their midwestern sites may represent a decrease in Hg emissions in the industrial areas of eastern North America.

In the Lake Champlain basin, as in many areas of North America, the primary Hg source is emissions from anthropogenic activity, primarily from coal combustion and waste incineration, and subsequent wet or dry deposition to the landscape [Watzin, 1992; Vasu and McCullough, 1994; Scherbatskoy et al., 1997]. Industrial and geologic inputs of Hg to Lake Champlain basin waters are minor. Although most of the Hg deposited in the Lake Champlain basin probably originates from outside the basin, there is growing evidence to suggest that local industry and waste incineration as well as emission of Hg from vegetation, soils, landfills, and land-applied sludge contributes up to half of the total deposition [NESCAUM et al., 1998].

The Lake Champlain basin has a large forested land area that has the potential to capture atmospheric pollutants and transfer them to soils, surface waters, and groundwater by several mechanisms. Terrestrial ecosystems accumulate atmospheric pollutants through wet and dry deposition, throughfall (precipitation passing through the forest canopy), litterfall (deposition of senescent foliage), and interception of cloud water in higher elevations. Forest canopies in this region have foliar surface areas up to 4.6 times greater than the land area they cover [Rea et al., 1996], providing Hg concentration in throughfall that averages twice that of precipitation and an annual Hg flux in litterfall that is greater than annual wet deposition [Rea et al., 1996]. Swain et al. [1992] showed the importance of watershed:lake area ratio in Hg loading to lake sediments. Examining the behavior of Hg in forested catchments, therefore, is important to understanding the present and long-term role of atmospheric Hg in the Lake Champlain basin.

From a human health perspective, it is the concentration of methylmercury (Me-Hg) that is critical because this form bioaccumulates in fish and is toxic to humans. In aquatic systems, Me-Hg is generally present at concentrations only 1 to 10% of total Hg concentration [Lee and Iverfeldt, 1991; Driscoll et al., 1994; U.S. EPA, 1997]. However, Me-Hg can comprise a greater percentage in organisms. Mason and Sullivan [1997] found in Lake Michigan that by the third trophic level, all Hg in aquatic organisms was Me-Hg. Water column Me-Hg is a result of methylation of ionic Hg which occurs in sediment and the water column by microbial action and abiotic processes. Methylation rates increase under anaerobic conditions [Branfireun et al., 1996], high DOC [Miskimmin et al., 1992], warm temperatures, and low pH [U. S. EPA, 1997]. Wetlands tend to be sources of Me-Hg [Kelly et al., 1995; Krabbenhoft et al., 1995; Branfireun et al., 1996].

The objectives of this paper are to summarize the somewhat disparate information on Hg pools, concentrations, and fluxes in waters, sediments, and biota in the Lake Champlain basin. We will draw on monitoring data and research results from a deposition station and catchment monitoring site near the center of the basin, as well as data from other relevant studies and monitoring efforts. We will synthesize the available information into an overview of Hg cycling in the Lake Champlain basin. Based on this assessment we will identify gaps in information and understanding and suggest directions for further research.

METHODS

The Lake Champlain basin comprises 21,150 km^2, including much of western Vermont, northeastern New York, and part of southern Québec (Figure 1). The ratio of land to lake surface area in the basin is 19:1. Land cover is 64% forest, 16% agricultural, 14% water/wetlands, and 6% urban/developed [personal communication, Eric Pyle, Vermont Center for Geographic Information]. The largest population centers in the basin are Burlington and Rutland, Vermont and Plattsburgh, New York. The topography is generally mountainous, with peaks exceeding 1500 m in the Adirondack Mountains of New York and 1300 m in the Green Mountains of Vermont. Agricultural activity is concentrated in the gently sloping lands around the shores of Lake Champlain. Despite its rural character, the area receives pollutants transported from the industrial midwest via prevailing westerly winds [Scherbatskoy et al., this volume; Olmez et al., 1998]. With respect to Hg, however, anthropogenic sources from within the region appear to be more important than sources from outside the northeast (61% and 39%, respectively) [NECSCAUM et al., 1998].

This paper draws extensively from research at the Proctor Maple Research Center (PMRC) and the Nettle Brook catchment, both in Underhill Center, Vermont, 34 km east of Lake Champlain (Figure 1). We summarize and update the deposition and streamflow record discussed in two earlier papers [Scherbatskoy et al., 1997; Scherbatskoy et al., 1998]. We also report on results from other headwater streams and larger rivers in the Lake Champlain basin, and present new data on Hg in soils, soil water, and plankton. Information for this paper has also been assembled from reports of the Vermont Agency of Natural Resources, U.S. Geological Survey, and other publications. The remainder of this section provides background information and methods for previously unpublished data newly presented in this paper.

PMRC and the Nettle Brook catchment are on the western slope of Mt. Mansfield. The sites are operated by the Vermont Forest Ecosystem Monitoring program (VForEM) as part of their long-term integrated studies on responses of forested ecosystems to environmental change [Wilmot and Scherbatskoy, 1994]. PMRC, at an elevation of 400

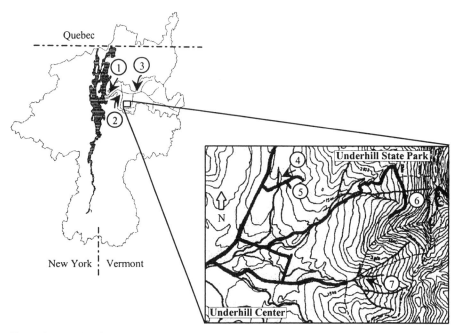

Figure 1. Map showing sampling locations. Left panel shows the Lake Champlain drainage basin with sites along the Lamoille River. Right panel shows a close-up of sites in Underhill Center, Vermont. Sampling sites are as follows: (1) Lamoille River at West Milton, (2) Browns River, (3) Lamoille River at Jeffersonville, (4) Harvey Brook, (5) Proctor Maple Research Center, (6) Stream 10, and (7) Nettle Brook.

m, has a 30-m tower and a full meteorological station. Since 1991, wet Hg deposition has been determined daily and dry deposition weekly, as described in Landis and Keeler [1997]. PMRC is the only site in the Lake Champlain basin where Hg deposition is routinely monitored.

Nettle Brook is an 11-ha mixed hardwoods catchment 3 km southeast of PMRC. Elevation ranges from 445 to 664 m, aspect is southwest, and slope averages 35% (range 5-75%). Soils are moderately well-drained Peru extremely stony loam underlain by mica-albite-quartz schist bedrock. Average annual rainfall is 1100 mm. January daily minimum temperature is -16C and July daily maximum temperature is 24C. Annual sulfate and nitrate deposition averages 21 and 16 kg ha^{-1}, respectively [Scherbatskoy et al., 1998]. Nettle Brook is the only site in the basin where streamwater Hg is regularly monitored and Hg biogeochemical process research is conducted [Scherbatskoy et al., 1998]. Since October 1993, stream water samples have been collected at Nettle Brook at a continuously gauged V-notch weir for analysis of Hg, DOC, and suspended sediment. Sampling frequency has varied, with higher priority given to high-flow periods, particularly the snowmelt periods of 1994, 1995, and 1996. In total, 106 samples were collected between October 1993 and November 1997.

Additional Hg determinations have been made at PMRC and in the Nettle Brook watershed. During the 1994 snowmelt, meltwater was collected daily at PMRC from a 1-m^2 Teflon-coated lysimeter installed on the ground beneath the snowpack [Scherbatskoy et al., 1997]. In 1997, soil water was collected from the O-horizon and B-horizon at 2 sites along Nettle Brook using non-tension fiberglass wick collectors. In December, 1997, soil samples were collected using a soil corer at 3 depths (20, 300, and 500 mm) at PMRC and at 2 depths (60-100 mm and 220-280 mm) from soil water sampling areas in the Nettle Brook area. Soil sample collection and analysis methods are described in Rea [1998]. Stream, snowmelt, and soil water samples were collected in acid-cleaned, triple-bagged Teflon bottles. Particle-free gloves were worn during sample collection.

Additional samples have occasionally been collected from four other sites, one headwater stream and three river sites, in the Lamoille River drainage basin (Figure 1). The headwater stream, Stream 10, drains a mixed conifer and hardwoods watershed between 775 and 1180 m elevation in Mt. Mansfield State Park. The stream originates near the Nose portion of the summit ridge of Mt. Mansfield and descends steeply down the west face. Of the three river sites, one sampling site was on the lower Browns River downstream of Nettle Brook, and two were on the Lamoille River, one upstream of the Browns River in Jeffersonville and the other near Lake Champlain at the walleye spring spawning grounds in Milton (Figure 1). Samples from these sites were collected by submerging the sample bottles 30 cm below the water surface approximately 1.0 to 1.5 m from the river bank. Sample bottles were extended into the river using a custom-made pole sampler constructed with a PVC pipe end and a broom handle, with all non-PVC parts wrapped in Teflon tape.

Samples were shipped by overnight courier to the University of Michigan Air Quality Laboratory (UMAQL) in Ann Arbor, MI, where they were analyzed for total and dissolved (<0.22 µm) Hg was performed in a Class 100 ultra-clean laboratory. After arrival at the laboratory, Hg was stabilized until analysis by oxidation in a 1% BrCl solution. For total Hg determination, the oxidized Hg was reduced by NH_2OH and $SnCl_2$ to elemental Hg, bubbled out of solution in a Hg-free nitrogen gas stream, and captured onto a gold-coated bead trap. Mercury was then thermally desorbed from the trap in a Hg-free helium gas stream and quantified by cold vapor fluorescence spectrometry. Analytical and quality assurance details are given in Scherbatskoy et al. [1998]. No Hg speciation was performed. The DOC (0.7 µm filtered) and suspended sediment determinations are described by Scherbatskoy et al. [1998]. Stream Hg flux at Nettle Brook was calculated using a concentration/discharge relation as described in Scherbatskoy et al. [1998].

On August 27, 1997, six plankton samples and three water samples were collected from Mallets Bay near the mouth of the Lamoille River. Three plankton samples were collected in acid-washed, triple bagged 7-ml Teflon vials in each of two net sizes, 63 µm and 202 µm. Plankton samples were stored in a freezer for one month, then ground with acid-cleaned borosilicate glass Duall tissue grinders in a portable HEPA clean chamber at PMRC. Water samples were collected using the pole sampler described above. Particle-free gloves were worn at all times when collecting and handling samples. Water and plankton samples were sent on ice via overnight courier to the University of Wisconsin water laboratory for analysis of total Hg and Me-Hg as described in Back et al. [1995].

RESULTS

In this section we have compiled available published information as well as previously unpublished data on Hg concentrations, pools, and fluxes in the Lake Champlain basin. The Hg record at Nettle Brook for March 1994 through February 1996 as reported in Scherbatskoy *et al.* [1998] is updated to include the 1996 and 1997 snowmelts and several rain events during the summer of 1997. We also report new data on Hg concentrations in soil water, soil, and plankton.

Hg in Atmospheric Deposition

Mercury emissions to the atmosphere generally occur in three forms: gaseous elemental Hg (Hg^0 (g)), gaseous divalent Hg (Hg(II) (g)), and particulate Hg. Gaseous elemental Hg is relatively insoluble and is not susceptible to wet or dry deposition. It can travel long distances and has an atmospheric residence time of one-half to two years. Because of its long residence time, Hg^0 (g) is the most prevalent form of Hg in the atmosphere. Gaseous divalent Hg is subject to rapid dry or wet deposition, as are particulate forms of Hg. Deposition of Hg(II) (g), wet divalent Hg (Hg(II) (aq)), and particulate forms of Hg occurs close to their source, with a typical residence time in the atmosphere of hours to weeks [U. S. EPA, 1997; NESCAUM *et al.*, 1998]. When re-deposited, these forms of Hg are commonly quantified in dry deposition (vapor or particulate phase) and wet deposition (precipitation).

At PMRC, total atmospheric Hg deposition was 425 and 463 mg ha^{-1} for 2 years beginning March 1994 (Table 1) [Scherbatskoy *et al.*, 1998]. Of this total, dry deposition accounted for most of the Hg flux, with only 75 and 93 mg ha^{-1} as wet deposition. Dry deposition was calculated by the "big leaf" model given in Scherbatskoy *et al.* [1997] as amended by Lindberg [1993]. Vapor phase Hg (mostly Hg^0 (g) and some Hg(II) (g)) dominates dry deposition [Rea, 1998] and occurs at a relatively constant concentration of approximately 2 ng m^{-3} throughout the year, with a slight increase in the summer months because of greater deposition velocities [Scherbatskoy, 1997]. Wet deposition of Hg also tends to be greatest in the summer primarily due to higher Hg concentrations in precipitation. This pattern has been observed in other studies [Lindberg *et al.*, 1992; Hoyer, 1995], but no causes have been identified [Scherbatskoy *et al.*, 1997]. Wet Hg deposition at the PMRC site is similar in magnitude and seasonal pattern to sites in rural northern Michigan [Hoyer *et al.*, 1995; Keeler *et al.*, 1995; Rea, 1998].

Hg in Surface Water

Total Hg (Hg_T) concentrations in surface waters within the Lake Champlain basin ranged from 0.1 to 80 ng L^{-1} [Scherbatskoy *et al.*, 1997; 1998] (Figure 2). Median Hg_T concentrations at each site generally ranged from 2 to 5 ng L^{-1}, but all sites had occurrences of >5 ng L^{-1} Hg_T. At Nettle Brook, Hg_T concentrations ranged from 0.1 to 79.7 ng L^{-1} (n=106, median 2.1 ng L^{-1}) (Figure 3). Dissolved Hg (Hg_D) concentrations in Nettle Brook streamwater ranged from 0.1 to 3.5 ng L^{-1} (n=106, median 1.4 ng L^{-1}). The

TABLE 1. Inputs and outputs of water and mercury at Nettle Brook catchment for two study years (From Scherbatskoy et al. [1998])

	Water		Mercury					Retention
	Precip.	Stream	*Input (mg Hg ha^{-1})*			*Output (mg Hg ha^{-1})*		
Date	(mm)	flow (mm)	Wet	Dry	Total	Particulate	Dissolved	Total
3/94 - 2/95	1139	587	74.5	350	425	23.6	8.4	31.9	92%
3/95 - 2/96	1109	568	92.7	370	463	13.4	8.2	21.7	95%

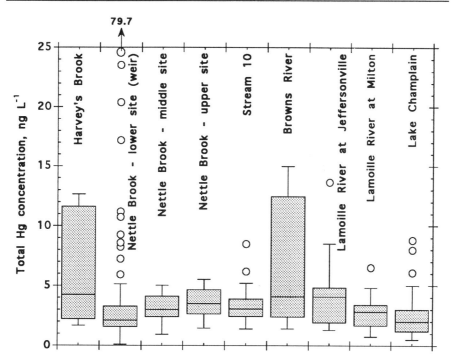

Figure 2. Boxplot indicating total Hg concentrations at all surface water sites in Lake Champlain basin with available data [Cleckner et al., 1995; Scherbatskoy et al., 1997; 1998]. Site locations are described in text and/or shown in Figure 1. For each site, stippled area spans concentration range from 25th to 75th percentile, with horizontal line denoting median; horizontal tick above and below stippled areas represents 10th and 90th percentile; points are individual values beyond 90th percentile.

low but consistent Hg$_D$ concentrations account for about 33% of the annual Hg flux at Nettle Brook [Scherbatskoy et al., 1998]. Most of the remaining 67% of the annual Hg export in streamwater occurred as particulate Hg during brief periods in extreme hydrologic events [Scherbatskoy et al., 1998]. The total Hg flux for the two years beginning March 1994 was 32 and 22 mg ha^{-1} yr^{-1} respectively.

Total Hg concentrations at Nettle Brook in 28 samples collected subsequent to the study of Scherbatskoy et al. [1998] did not exceed 10 ng L^{-1}. Some of these samples from summer storms represent the first non-snowmelt high flows sampled at Nettle Brook. Total Hg concentrations during these storms were only somewhat above typical baseflow

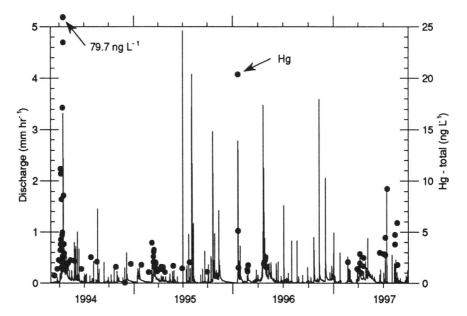

Figure 3. Discharge (mm hr^{-1}) and stream water total Hg concentration (ng L^{-1}) at the Nettle Brook weir from March 1994 through September 1997. Water was sampled for Hg analysis on a daily basis during spring snowmelt and periodically during the remainder of the year.

concentrations; the highest Hg$_T$ concentration observed in 1997 (9.3 ng L^{-1}) occurred during the event with the highest peak flow of that year (Figure 3). Although peak flows were in excess of 3.5 mm hr^{-1} in the 1996 melt and 2.0 mm hr^{-1} in the 1997 summer storm, flows at the time of sampling were considerably lower. The new data did not change the concentration-discharge relation (Figure 4) used to calculate stream Hg flux as described in Scherbatskoy *et al.* [1998].

Mercury concentration ranges and temporal dynamics at downstream sites on larger streams mimicked those at the headwater sites. Based on synoptic sampling at 3 sites downstream of Nettle Brook within the Lamoille River watershed in 1994-1997 (n=13 at each site), Hg$_T$ concentrations ranged from 0.7 to 52.6 ng L^{-1} (median 3.0 ng L^{-1}), while Hg$_D$ ranged from 0.5 to 5.1 ng L^{-1} (median 1.3 ng L^{-1}). There was significant seasonal variation in Hg concentration due to spring snowmelt or flooding summer rains. Under base flow conditions, total Hg concentrations were 2-3 ng L^{-1} (except for one outlier of 12 ng L^{-1}), with approximately 50% of the Hg$_T$ as Hg$_D$. During the flood of 7 August, 1995, however, Hg$_T$ concentration at the Milton site, furthest downstream near the mouth of the Lamoille River at Lake Champlain, attained 52.6 ng L^{-1} (Figure 5), while Hg$_D$ increased only slightly.

Total Hg in Lake Champlain lakewater ranged from ~1 ng L^{-1} at a shallow site in Mallets Bay [T. Scherbatskoy, A. Donlon, and M. Watzin, University of Vermont, unpublished data] to 8 ng L^{-1} (median ~3 ng L^{-1}) in the deep-water lake 1.6 km west of Burlington [Cleckner *et al.*, 1995] (Figure 2).

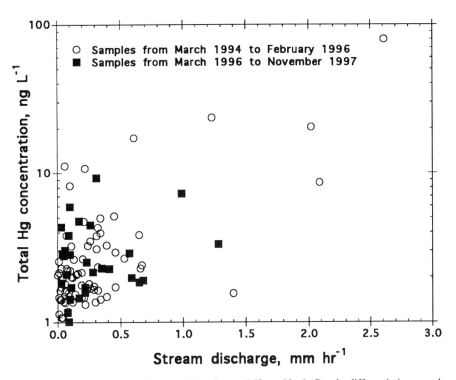

Figure 4. Concentration - discharge relation for total Hg at Nettle Brook, differentiating samples from 2 different time periods. Open circles: Scherbatskoy *et al.* [1998]; Closed squares: previously unpublished data.

Other Hg Determinations

Snowpack and snowmelt. Only one determination of Hg in the snowpack was made; 1.8 ng L^{-1} in March 1994 [Scherbatskoy *et al.*, 1997]. Mercury in snow meltwater, also in 1994, ranged from 2 to 9. ng L^{-1} and averaged 4.8 ng L^{-1} [Scherbatskoy *et al.*, 1998]. Haines *et al.* [1995] measured 3.5 ng L^{-1} in the snowpack at a Vermont site just outside the Lake Champlain basin.

Throughfall. Throughfall was collected at PMRC in Underhill Center, VT during August and September of 1994 [Rea *et al.*, 1996]. The mean volume-weighted Hg concentration was 12.0 ± 8.5 ng L^{-1} with a range of 4.5 to 35.8 ng L^{-1}. Net throughfall for 2 months was 12 mg ha^{-1} [Rea *et al.*, 1996].

Foliage. Rea [1998] found that Hg concentration consistently increased in foliage throughout the growing season and peaked in litterfall. In 1995, foliar Hg concentration increased from 3.6 ± 2.6 ng g^{-1} at budbreak to 28.8 ± 2.4 ng g^{-1} in September foliage to 47.1 ± 5.6 ng g^{-1} in October litterfall [Rea, 1998]. In 1994, Hg increased from 34.2 ± 7.2 ng g^{-1} in August foliage to 53.2 ± 11.4 ng g^{-1} in October litterfall [Rea *et al.*, 1996]. The

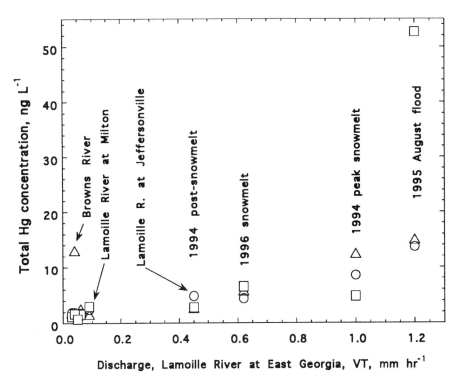

Figure 5. Total Hg concentrations at the Lamoille River and Browns River sites for base flow and selected events, plotted as a function of normalized discharge at a nearby USGS gaging station.

total flux of Hg from the canopy (throughfall + litterfall) was attributed primarily to dry deposition and was ~2 times the flux from precipitation [Rea et al., 1996].

Litterfall. Mercury in litterfall in the fall of 1996 represented an annual areal Hg flux of 158 mg ha^{-1} yr^{-1} [Rea, 1998].

Soils. Mercury concentrations in soils from PMRC [Rea, 1998], at Nettle Brook, and Stream 10 [A. Donlon, unpublished data] agreed very closely. The Oa horizon had 210 ng g^{-1} (n=3) at PMRC, 204 ng g^{-1} (n=3) at Nettle Brook, and 169 ng g^{-1} (n=1) at Stream 10. The B horizon had 70 ng g^{-1} (n=3) at PMRC, 58 ng g^{-1} (n=4) at Nettle Brook, and 75 ng g^{-1} (n=2) at Stream 10 (Figure 6).

Soil water. Between April and November, 1997, soil water was sampled at two sites along Nettle Brook and one site along Stream 10 using non-tension fiberglass wick collectors. Soil water at the middle elevation location along Nettle Brook had an average total Hg concentration of 20.1 ng L^{-1} (n=13) just below the Oa soil horizon and 3.0 ng L^{-1} (n=4) within the B horizon (Figure 6). Concentrations at an upper elevation location near the headwaters of Nettle Brook averaged 3.5 ng L^{-1} (n=9) below the Oa horizon and 1.1 ng L^{-1} (n=10) within the B horizon. The upper elevation site at Nettle Brook is located on a wet hillslope, and its consistently low Hg concentrations may reflect the chemistry of

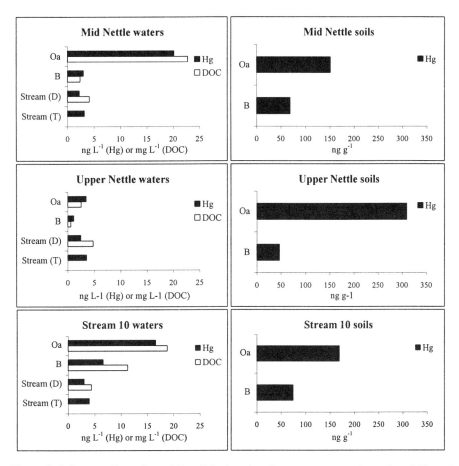

Figure 6. Soil water (from Oa and B soil horizons) and stream concentrations of total Hg and dissolved organic carbon (DOC), and Hg concentrations in soils at three locations in Underhill Center, Vermont. Stream (D) refers to dissolved Hg concentrations in stream water; stream (T) refers to total Hg concentrations in stream water.

discharging groundwater rather than soil water. At Stream 10, the average Hg concentration was 16.6 ng L^{-1} (n=6) just below the Oa soil horizon and 6.6 ng L^{-1} (n=4) within the B horizon.

Mercury and DOC concentrations were positively correlated at all sites. DOC concentrations at the middle site at Nettle Brook averaged 22.7 mg L^{-1} (n=12) and 2.39 mg L^{-1} (n=2) at the Oa and B horizons, respectively. By contrast, DOC concentrations at the upper site averaged only 2.53 mg L^{-1} (n=8) and 0.55 mg L^{-1} (n=8) at the two depths, respectively. DOC in soil water at Stream 10 was 18.8 mg L^{-1} (n=6) below the Oa horizon and 11.25 mg L^{-1} (n=3) in the B horizon.

Sediments. Two surveys of Hg in sediments were conducted in the basin in the early 1990's: a survey of Hg in stream sediments by the USGS [Colman and Clark, 1994] and a survey of Hg in lake bottom sediments [MacIntosh, 1994] (Table 2). In each survey, the

TABLE 2. Sediment Hg concentrations ($\mu g\ g^{-1}$) in Lake Champlain and inlet streams

	n	median	mean	std. dev.
Streams [Colman & Clark, 1994]	76	0.18	0.29	0.43
Lake [MacIntosh, 1994]	30	0.20	0.20	0.15

median Hg concentration was near 200 ng g^{-1}. At six sites where sampling locations approximately coincided in the 2 studies, near stream outlets to the lake, Hg concentrations agreed closely (r^2=0.67).

Plankton. In 1997 Hg and Me-Hg (dry weight) concentrations were determined in plankton from Lake Champlain waters [T. Scherbatskoy, A. Donlon, and M. Watzin, University of Vermont, unpublished data]. Mercury concentration averaged 410 ± 110 ng g^{-1} and 760 ± 20 ng g^{-1} in 3 samples each of small (between 63 and 202 µm) and large (>202 µm) plankton, respectively. Corresponding Me-Hg concentrations were 5 ± 3 ng g^{-1} for small plankton and 21 ± 3 ng g^{-1} for large plankton.

Fish. A compilation of Hg concentrations in fish tissue in the Lake Champlain basin from 1970 to present is available from the Vermont Agency of Natural Resources (VANR) and other government sources in Vermont, New York, and Québec, and the University of Vermont [State of Vermont, 1990; 1997]. The highest concentrations are typically found in walleye, but elevated Hg occurs consistently in lake trout, smallmouth bass, and northern pike, and frequently in largemouth bass, yellow perch, and pickerel. Higher Hg concentrations in fish are favored in acidic and brownwater lakes [State of Vermont, 1997]. Mercury is detectable (>50 ng g^{-1}) in virtually all fish tissue analyzed. Mercury concentration is typically greater than 500 to as much as 2000 ng g^{-1} for walleye. Mercury concentration for brown bullhead and trout other than lake trout more typically range from 100 to 200 ng g^{-1}. Mercury in walleye in the Missisquoi River averaged 700 ng g^{-1} in 1970 (n=4) and 990 ng g^{-1} in 1990 (n=10) [State of Vermont, 1990]. Because Hg concentrations in fish tissue are reported on a wet weight basis, concentrations in fish are actually considerably higher than those in sediment and plankton, as would be expected from bioaccumulation. While they continue routine monitoring, VANR is collaborating with the New Hampshire Department of Environmental Services and the U.S. EPA on a comprehensive interpretive study on fish/water/sediment Hg and Me-Hg in lakes throughout Vermont and New Hampshire, including 19 lakes within the Lake Champlain basin [N. Kamman, VANR, personal communication].

MERCURY INPUT-OUTPUT BUDGETS

Scherbatskoy *et al.* [1998] determined watershed retention rates for Hg of 92 and 95% in the Nettle Brook catchment for the two years beginning March 1994 (Figure 7). Because of the tendency for episodic transport of Hg, retention may be considerably less than this amount during snowmelt and other high flow periods, and considerably greater than this during base flow periods. The net Hg retention at Nettle Brook (92-95%) is greater than the 80% to 90% retention reported in Sweden [Aastrup *et al.*, 1991; Johanssen *et al.*, 1991]. It is comparable to the 92% retention (an underestimate because

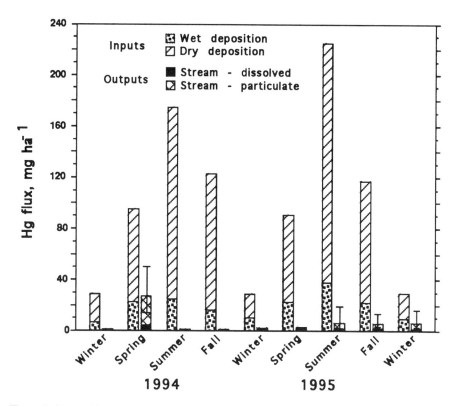

Figure 7. Seasonal input (wet and dry deposition) and stream export (dissolved and particulate) of total Hg in Nettle Brook catchment from December 1993 through February 1996, expressed as mg ha^{-1}. Modified from Scherbatskoy et al. [1998].

dry deposition was not considered) determined by Krabbenhoft et al. [1995] at a Wisconsin site.

The Hg retention percentages above are based solely on mass balance considerations. Mercury deposited to the basin that is not exported in streamflow is said to be "retained" in the basin, though in fact much of the atmospheric Hg incident to the basin may be re-volatilized to the atmosphere, as observed in catchments in Tennessee [Johnson and Lindberg, 1995] and Sweden [Bishop et al., 1998]. Because most of the Hg input is by dry deposition to foliage (Table 1), volatilization may occur directly from foliar surfaces.

A simple extrapolation of the Hg input-output budget from the 11-ha Nettle Brook catchment to the entire Lake Champlain basin may be misleading. The basin as a whole may have a Hg deposition rate somewhat lower than that of Nettle Brook because 36% of the basin area is non-forested land, where the dry deposition rate may be limited by a lower stomatal uptake potential. On the other hand, non-atmospheric anthropogenic sources of Hg may be significant at the larger scale. An estimate of Hg flux from Lake Champlain itself can be made by assuming that the median Hg concentration from a set of samples from the deep part of the lake [Cleckner et al., 1995] approximates the mean annual Hg concentration of lake water. The product of this Hg concentration (3 ng L^{-1}) and the mean annual discharge from the lake [Shanley and Denner, this volume] yields a

Hg output of ~15 mg ha^{-1} yr^{-1}. This value has a high uncertainty, but the results of this exercise suggest that the Hg retention for the basin as a whole is similar to that of Nettle Brook, i.e. > 90%.

SOURCES OF MERCURY

Atmospheric Hg is the dominant source of Hg to the Lake Champlain basin [NESCAUM et al., 1998]. Atmospheric Hg is in part natural, and in part derived from industrial sources including smelting, fossil fuel combustion, and waste incineration. There is no significant mineralogical source of Hg in the Lake Champlain basin, thus direct geologic contributions of Hg to basin waters from geologic weathering are negligible. Other potential non-atmospheric transfers of Hg to the landscape include leaching of Hg from landfills, discharge of Hg-containing industrial effluent and wastewater treatment plants to streams, and land application of Hg-containing sludge. Also, Hg may have an agricultural source because it was commonly used as an agricultural fungicide prior to the 1970's [Hem, 1985]. Industrial and agricultural sources of Hg are believed to be minor in the Lake Champlain basin.

About half of the Hg deposition in the Lake Champlain basin may be from in-basin sources. NESCAUM et al. [1998] used the Regional Lagrangian Model of Air Pollution (RELMAP) to simulate patterns of Hg transport and deposition. The results, which should be regarded as best estimates, suggested that 47% of Hg deposition in the northeast U.S. is attributable to regional sources; 30% to U.S. sources outside the region; and 23% from the global atmospheric reservoir. About 27% of the modeled Hg deposition originated from utility and non-utility boilers, and more than 50% originated from municipal waste combustion (Figure 8) [NESCAUM et al., 1998]. One of the largest waste incinerators in the basin is a medical waste combustion facility in Colchester, Vermont. In a monitoring check in June 1996, the facility was emitting Hg at the rate of 4 g hr^{-1} [Vermont DEC, 1996]. If extrapolated to an annual rate over the entire basin, this emission equates to 17 mg ha^{-1} yr^{-1}, or enough to account for 4% of the deposition in the basin.

TERRESTRIAL PROCESSES

From March 1995 to February 1996, Scherbatskoy et al. [1998] determined an annual Hg deposition to the landscape of 463 mg ha^{-1}, of which 370 mg ha^{-1} was dry deposition. Within this period (summer of 1995), Rea [1998] independently determined Hg dry deposition rates of ~200 mg ha^{-1} yr^{-1}. A comparable Hg flux of 158 mg ha^{-1} was deposited as litterfall in the Nettle Brook area in 1995 [Rea, 1998]. Rea [1998] concluded that uptake of gaseous Hg through leaf stomata, which represented most of the dry deposition flux, was able to account for the Hg flux in litterfall. A key finding in support of gaseous uptake was the progressive increase in foliar Hg through the growing season. Increases in Hg concentration from rainfall to throughfall, together with the high Hg concentrations in litterfall [Rea et al., 1996], suggest that much of the incident Hg finds its way to the forest floor. Considerable uncertainty remains about processes controlling the cycle of Hg in the canopy and forest floor.

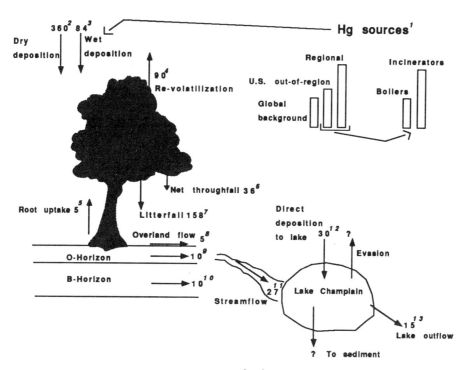

Figure 8. Schematic depicting fluxes of Hg (mg ha^{-1} yr^{-1}) in the Lake Champlain basin. Superscript numbers indicate the derivation of fluxes as follows: 1. NESCAUM *et al.*, 1998 (bars are proportional based on percentages); 2. Scherbatskoy *et al.*, 1998; 3. Scherbatskoy *et al.*, 1998; 4. Bishop *et al.*, 1998 (site in Sweden - assumed Vermont is similar); 5. Rea [1998]; 6. Rea *et al.*, 1996; extrapolated to 6 month growing season from 2 month study; 7. Rea [1998]; 8, 9, 10. Estimated from consideration of soil pools [this paper], and stream Hg dynamics [Scherbatskoy *et al.*, 1998], constrained to match stream flux; 11. Scherbatskoy *et al.*, 1998; 12. Total deposition [Scherbatskoy *et al.*, 1998] prorated to lake area; 13. Estimated from Hg concentration in lake [Cleckner et. al., 1995] and mean annual lake discharge [Shanley and Denner, this volume].

Most of the Hg input to the forest floor is inherently associated with organic matter by virtue of its incorporation in litterfall. Mercury reaching the forest floor by other means -- direct precipitation or washoff of particulate dry deposition -- also likely associates with organic matter [Aastrup *et al.*, 1991; Mierle and Ingram, 1991]. Mercury concentrations in forest floor soil solution collected during storms at Nettle Brook were an order of magnitude greater than those in streamwater (Figure 6), suggesting that forest floor Hg may be flushed during storms. The positive correlation of Hg and DOC in soil water suggests that Hg is associated with dissolved organic acids. These mobile organic ions may help transport Hg to a stream, or downward in the profile where they adsorb to larger soil particles (causing the decrease in Hg concentrations in B-horizon soil water). Bishop *et al.* [1998] found that conifers took up small but variable amounts of Hg in soil water, depending on the tree species. Applying the results of Bishop *et al.* [1998] to soil water Hg concentrations at Nettle Brook, Rea [1998] estimated that between 3 and 14% of Hg in litterfall could be accounted for by uptake of soil water Hg.

What is the fate of the large pool of Hg "retained" in the catchment, represented by the large difference between the incoming flux of Hg in atmospheric deposition and the outgoing flux in streamflow (Figure 7)? The unaccounted Hg may either exit the catchment by volatilization to the atmosphere [Siegel and Siegel, 1988; Hanson et al., 1995, Johnson and Lindberg, 1995], or it may remain in the catchment and accumulate in the soil. Volatilization in the terrestrial landscape occurs from foliar surfaces as well as from the forest floor, but its importance is poorly quantified.

Lindberg [1997] estimated that 657 mg ha^{-1} yr^{-1} Hg was volatilized from soil at a Tennessee site; Bishop et al. [1998] estimated 90 mg ha^{-1} yr^{-1} for a site in Sweden. Given the similarity in climate between Sweden and Vermont, the latter figure may more closely approximate the re-emission flux at Nettle Brook. If re-emission of Hg is of this magnitude, the input-output balance requires that a somewhat greater amount of Hg accumulate in catchment soils. Using some basic assumptions to convert the soil Hg concentrations to Hg pools (O-horizon: thickness = 8 cm; density = 1.1 g cm^{-3}; porosity = 0.6; B-horizon: thickness = 20 cm; density = 2.5 g cm^{-3}; porosity = 0.4), the existing Hg pool in the O-horizon can be accounted for by 150 years of deposition at the current rate; the combined Hg pool in the O- and B-horizons would require 600 years. It is not known how much of the soil Hg would be present naturally.

AQUATIC PROCESSES

On an annual basis, about 67% of Hg export is in the form of particulate Hg (primarily during high-flow events) and about 33% is as Hg$_D$. The tendency for event-related transport is reflected in the 3-month period of March-May, when about 84% of the annual flux of total Hg occurred with only 56% of the annual streamflow (Figure 7). The concentration of Hg$_D$ remains fairly constant regardless of discharge [Scherbatskoy et al., 1998], leading to a more "continuous" export of Hg$_D$ in contrast to the episodic export of Hg$_T$.

As summarized in Scherbatskoy et al. [1998], Hg concentrations at Nettle Brook are typical of upland forested catchments for Hg$_D$ and Hg$_T$, except at high flows. During events, Hg concentrations at Nettle Brook exceeding 20 ng L^{-1} are among the highest reported in the literature. For example, among several catchments in Sweden, maximum Hg concentrations did not exceed 15 ng L^{-1} [Johanssen et al., 1991; Lee and Iverfeldt, 1991; Hultberg et al., 1995] except for 21 ng L^{-1} in runoff from a peat bog [Westling, 1991].

What controls Hg concentrations during high-flow events? Hg concentrations in Nettle Brook tend to increase with increasing stream discharge in a complex fashion, generally in conjunction with increased sediment transport at higher flows [Scherbatskoy et al., 1998]. Total Hg concentrations attained very high values in the 1994 spring snowmelt (79.7 ng L^{-1}) and the 1996 January thaw (20.4 ng L^{-1}) [Scherbatskoy et al., 1998]; these were not matched in subsequent sampling in 1996 and 1997, despite a focus on high-flow sampling in the latter year (Figure 3). In one of the few studies that conducted event sampling for Hg, Bishop et al. [1995] also found a positive correlation between Hg and flow during snowmelt in Sweden, although their streamwater Hg concentration peaked at only 6 ng L^{-1}, compared to 80 ng L^{-1} at Nettle Brook. Bishop et al. [1995] found that 37%

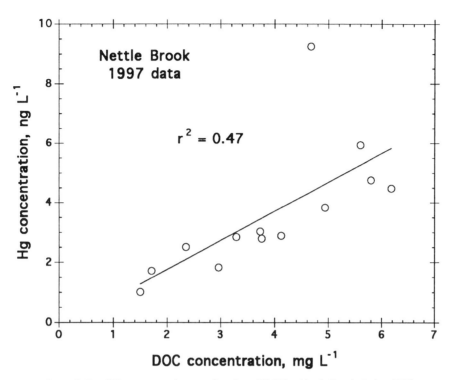

Figure 9. Total Hg concentration as a function of DOC at Nettle Brook during 1997.

of total annual Hg flux occurred in a three week spring snowmelt period, compared to 50% in the single day of peak snowmelt in 1994 at Nettle Brook.

Scherbatskoy et al. [1998] found a good correlation between particulate Hg (Hg_T - Hg_D) and the organic fraction of suspended sediment. This relation is consistent with the known affinity of Hg(II) for organic matter in forest soils [Lindqvist, 1991; Mierle and Ingram, 1991; Schuster, 1991]. Total Hg was less well-correlated with stream DOC [Scherbatskoy et al., 1998], but the most recent sampling has shown a strong relation between Hg_T and DOC (Figure 9). Driscoll et al. [1994] also found a positive correlation between Hg_T and DOC in Adirondack lakes. Likewise, in a small forested stream in Wisconsin, Krabbenhoft et al. [1995] found that Hg_T increased sharply with releases of DOC, while Me-Hg did not. Bishop et al. [1995] found that the increase in Hg concentration during the spring snowmelt was associated with total organic carbon.

A critical question is whether the occasional high episodic fluxes of Hg from upland catchments persist downstream to the larger rivers and to Lake Champlain. Synoptic sampling of some larger streams downstream of Nettle Brook [Scherbatskoy et al., 1997] suggest that Hg dynamics are similar to those in the headwater catchment. As at Nettle Brook, Hg_D is consistently 1 to 3 ng L^{-1} in the larger streams, but concentrations of Hg_T increase during large events, such as the 1994 snowmelt or the August 1995 flood (Figure 5). These episodic pulses of Hg in the larger rivers may reflect transport from steep upland headwater catchments such as Nettle Brook, or they may be caused by more local sources such as runoff from agricultural or disturbed lands. This important question is the subject of ongoing research.

The Hg output flux from Lake Champlain was estimated above as 15 mg ha^{-1} yr^{-1}. This compares to 25-30 mg ha^{-1} yr^{-1} Hg entering from inlet streams and an equal amount of Hg from direct atmospheric deposition to the lake surface. Much of the excess Hg is probably incorporated in lake sediment (some passing first through the food chain), whereas some Hg is probably evaded from the lake surface.

CONCEPTUAL MODEL OF HG CYCLING IN THE BASIN

The small flux of Hg in streamflow relative to inputs in atmospheric deposition raises several questions about the fate of the Hg remaining in the catchment. How is Hg sequestered in the terrestrial landscape, and what internal processes control its release to surface waters? How much Hg is volatilized to the atmosphere, and how much of the remaining terrestrial Hg pool is cycled within the catchment? We have presented thus far a summary of the existing state of knowledge about Hg pools and fluxes in the Lake Champlain basin (Figure 8). Although many gaps and questions remain, the following is an attempt to synthesize our existing understanding into a conceptual model of Hg cycling in the basin.

Mercury enters the basin primarily from the atmosphere [NESCAUM et al., 1998], and is therefore relatively uniformly distributed throughout the basin. Because the primary mechanism of dry deposition is by absorption through leaf stomata, total deposition is influenced by land cover type. Stomatal density in agricultural cropland at maturity may be comparable to that of a forest, but vegetation is removed by harvesting. Accordingly, deposition may be lower in agricultural and developed areas. In any case, of the Hg incident to the landscape, whether absorbed by foliage or passing to land surface, a sizeable fraction is re-volatilized to the atmosphere. The remaining Hg becomes bound by soil organic matter, the majority (in the forested landscape) as litterfall.

The pool of Hg in the forest floor establishes equilibrium with soil solution. A small amount of the dissolved Hg in soil solution is translocated to foliage by root uptake and returns to the foliage/litterfall/forest floor cycle [Bishop et al., 1998; Rea, 1998]. Some of the Hg in soil solution moves to the stream in association with DOC. Another portion of the Hg moves with water seeping deeper in the soil where it establishes equilibrium at a lower Hg concentration. Mercury reaching the groundwater supplies the low base flow Hg concentrations in streamwater. The higher stormflow Hg$_T$ concentrations result from erosion of streambanks or riparian soils and/or mobilization of organic-rich stream sediments. Because of the linkage of Hg transport and suspended sediment concentrations, erosion-prone farmlands and developed lands may be more important Hg sources than the forest, but this possibility cannot be assessed from existing data.

The low but constant dissolved Hg concentrations in streamwater suggest a constant transfer of Hg from the basin to Lake Champlain. The higher episodic transport of particulate Hg is reflected in elevated Hg concentrations downstream as well, and represents pulse Hg inputs. Reduced gradients on approach to the lake result in sediment deposition and Hg accumulation in stream and lake sediments, but these deposits are likely periodically remobilized during subsequent storms. Available data preclude determining whether farmlands and developed lands contribute to the episodic Hg pulses in downstream locations. Mercury reaching the lake, whether by stream inputs or direct deposition to the lake surface, is divided among uptake in the food chain, deposition in sediment, evasion from the lake surface, and export from the lake outlet.

Nearly all of the terrestrial Hg research in the Lake Champlain basin has been in the forested environment. While forests represent the major land cover in the basin, the linkage between sediment and Hg suggests that agricultural and urban land uses may export considerable Hg. An agricultural landscape in Minnesota [Balogh et al., 1997] and an urban landscape near Washington D. C. [Mason and Sullivan, 1998] export more Hg than a typical undisturbed forest. Our current research on Hg cycling in the Lake Champlain basin is addressing the role of the agricultural landscape.

MERCURY AND HUMAN HEALTH

Mercury bioaccumulates in fish to levels that are potentially dangerous to humans, even though the concentrations in water and sediments are very low. No specific studies have been done in the basin on Hg in the food web, but our limited data suggest that Me-Hg increases up the food chain. Total Hg concentrations increased from lake sediment (a mixture of organic detritus with inorganic matter) (200 ng g^{-1}) [MacIntosh, 1994]; to small plankton (410 ng g^{-1}; 5 ng g^{-1} Me-Hg); to large plankton (760 ng g^{-1}; 21 ng g^{-1} Me-Hg); to fish (100-2000 ng g^{-1} wet weight). The Hg concentrations in fish would be several times higher if expressed on a dry weight basis similar to the plankton. Walleye and lake trout may have some of the highest Hg contents because they subsist almost exclusively on other fish.

Methylmercury, the form of Hg toxic to humans, is generally present in waters in much lower concentrations than Hg_T, but it comprises greater than 95% of the Hg in fish [Porcella, 1994]. Fish do not methylate inorganic Hg within their tissues, but rather obtain Me-Hg from their diet, and to a lesser extent, through their gills [Weiner and Spry, 1996]. Generally less than 1% of Hg in atmospheric deposition is as Me-Hg [Porcella, 1994]. Methylmercury is formed subsequent to Hg deposition under reducing conditions such as in wetlands [Krabbenhoft et al., 1995; Branfireun et al., 1996] and in stream and lake sediments, soils, and lake water [Verta et al., 1994]. Some studies suggest that wetlands are primarily responsible for Me-Hg mobilization [Driscoll et al., 1994; Hurley et al., 1995; Branfireun et al., 1996]. In the Experimental Lakes Area of Ontario, Me-Hg ranged between 0.46% and 31% of Hg_T, but peaked as high as 92% after a flood in an experimental wetland [Kelly et al., 1995]. Mercury dynamics in wetlands have not been investigated in the Lake Champlain basin.

Methylmercury concentrations were determined on a small set of samples from Lake Champlain waters [M. Watzin, University of Vermont, unpublished data]. Methylmercury concentrations were extremely low, on the order of 0.03 ng L^{-1}, while corresponding Hg_T concentrations were near 1.2 ng L^{-1}. Methylmercury concentration was 0.23 ng L^{-1} on the Browns River. When one considers that nearly all the Hg in fish is in the Me-Hg form, and that the starting Me-Hg concentration in waters is so very low, this is strong testimony to the bioconcentration of Me-Hg.

CONCLUSIONS

The Lake Champlain basin receives about 450 mg ha^{-1} yr^{-1} Hg in atmospheric deposition. About 30% is wet deposition and 70% is dry deposition, mostly by absorption through leaf stomata. Some of the incident Hg is re-emitted to the atmosphere; most of the remainder is sequestered in the forest floor by organic matter, including a large portion in litterfall. Only a minor amount of this forest floor Hg pool reaches a surface stream; in 2 years of study, only 5 to 8% of the incident Hg was exported in streamflow. Stream Hg export occurred as a small but constant dissolved Hg flux (about one-third) and an episodic organic-particulate Hg flux (about two-thirds). High Hg concentrations in O-horizon soil water suggest that the upper soil zone is a primary pathway for Hg from the terrestrial ecosystem to the stream.

Larger streams exhibit similar Hg dynamics to the headwater Nettle Brook. However, we do not know whether Hg transported in mountain streams consistently reaches the lowlands, or whether the more proximal agricultural and developed landscapes contribute to the observed episodic Hg pulses. Dissolved Hg concentrations in Lake Champlain itself are similar to observed concentrations of dissolved Hg in streamwater. Particulate Hg is probably incorporated in lake sediments. Some Hg enters the food chain, and bioaccumulation has elevated Hg concentrations in some fish species to levels that are hazardous for human consumption. Mercury uptake and accumulation in fish is nearly exclusively as Me-Hg; human health risks should be assessed in light of methylation potential.

Acknowledgments. Support was provided from EPA and NOAA through the Cooperative Institute of Limnology and Ecosystems Research agreement no. NA37RJ0204. We wish to acknowledge the University of Vermont Proctor Maple Research Center and the assistance of Joanne Cummings and Jon Denner. This research was undertaken in cooperation with the Vermont Forest Ecosystem Monitoring Program. The manuscript benefitted greatly from reviews by two anonymous reviewers.

REFERENCES

Aastrup, M., J. Johnson, E. Bringmark, I. Bringmark and A. Iverfeldt. 1991. Occurrence and transport of mercury within a small catchment area. *Water, Air, and Soil Pollution* 56: 155-167.

Back, R. C., V. Visman, and C. J. Watras. 1995. Microhomogenization of individual zooplankton species improves mercury and methylmercury determinations. *Canadian Journal of Fisheries and Aquatic Sciences* 52: 2470-2475.

Balogh, S. E., M. D. Meyer, and D. K. Johnson. 1997. Mercury and suspended sediment loadings in the lower Minnesota River. *Environmental Science and Technology* 31: 198-202.

Bishop, K., Y. -H. Lee, J. Munthe, and E. Dambrine. 1998. Xylem sap as a pathway for total mercury and methylmercury transport from soils to tree canopy in the boreal forest. *Biogeochemistry* 40: 101-113.

Bishop, K., Y. -H.. Lee, C. Pettersson and B. Allard. 1995. Methylmercury output from the Svartberget catchment in northern Sweden during spring flood. *Water, Air, and Soil Pollution* 80: 445-454.

Branfireun, B. A., A. Heyes, and N. T. Roulet. 1996. The hydrology and methylmercury dynamics of a Precambrian Shield headwater peatland. *Water Resources Research* 32: 1785-1794.

Burke, J., M. Hoyer, G. J. Keeler, and T. Scherbatskoy. 1995. Wet deposition of mercury and ambient mercury concentrations at a site in the Lake Champlain Basin. *Water, Air, and Soil Pollution* 80: 353-362.

Cleckner, L. B., E. S. Esseks, P. G. Meier, and G. J. Keeler. 1995. Mercury concentrations in two "Great Waters." *Water, Air, and Soil Pollution* 80: 581-584.

Colman, J. A. and S. F. Clark. 1994. *Geochemical Data on Concentrations of Inorganic Constituents and Polychlorinated Biphenyl Congeners in Streambed Sediments in Tributaries to Lake Champlain in New York, Vermont, and Québec, 1992.* U.S. Geological Survey Open-File Report 94-472, 65 pp.

Driscoll, C. T., C. Yan, C. L. Schofield, R. Munson, and J. Holsapple. 1994. The mercury cycle and fish in Adirondack lakes. *Environmental Science and Technology* 28:136A-143A.

Engstrom, D. R. and E. B. Swain. 1997. Recent declines in atmospheric mercury deposition in the Upper Midwest. *Environmental Science and Technology* 31: 960-967.

Fitzgerald, W. F. 1993. Mercury as a global pollutant. *The World & I*. October: 192-198.

Fitzgerald, W. F., D. R. Engstrom, R. P. Mason, and E. A. Nater. 1998. The case for atmospheric mercury contamination in remote areas. *Environmental Science and Technology* 32: 1-7.

Haines, T., B. Mower, and R. Perry. 1995. Regional patterns of mercury content in snowpack, sphagnum, and fish in northeastern North America. Poster presented at Society of Environmental Chemistry and Toxicology Second World Congress, Vancouver, BC, Canada. Nov. 5-9, 1995.

Hanson, P. J., S. E. Lindberg, T. A. Tabberer, J. G. Owens, and K. –H. Kim. 1995. Foliar exchange of mercury vapor: evidence for a compensation point. *Water, Air, and Soil Pollution* 80: 373-382.

Hem, J. D. 1985. *Study and Interpretation of the Chemical Characteristics of Natural Water.* U.S. Geological Survey Water Supply Paper 2254, 263 pp.

Hoyer, H., J. Burke, and G. J. Keeler. 1995. Atmospheric sources, transport and deposition of mercury in Michigan: two years of event precipitation. *Water, Air, and Soil Pollution* 80: 199-208.

Hultberg, H., J. Munthe, and A. Iverfeldt. 1995. Cycling of methyl mercury and mercury -- responses in the forest roof catchment to three years of decreased atmospheric deposition. *Water, Air, and Soil Pollution* 80: 415-424.

Hurley, J. P., J. M. Benoit, C. L. Babiarz, M. M. Shafer, A. W. Andren, J. R. Sullivan, R. Hammond, and D. A. Webb. 1995. Influences of watershed characteristics on mercury levels in Wisconsin rivers. *Environmental Science and Technology* 29: 1867-1875.

Johansson, K., M. Aastrup, A. Andersson, L. Bringmark, and A. Iverfeldt. 1991. Mercury in Swedish soils and waters - assessment of critical load. *Water, Air, and Soil Pollution* 56: 267-281.

Johnson, D.W. and S. E. Lindberg, S.E. 1995. The biogeochemical cycling of Hg in forests: alternative methods for quantifying total deposition and soil emission. *Water, Air, and Soil Pollution* 80: 1069-1077.

Keeler, G. J., G. Glinsorn, and N. Pirrone. 1995. Particulate mercury in the atmosphere: its significance, transport, transformation, and sources. *Water, Air, and Soil Pollution* 80: 159-168.

Kelly, C. A., J. W. M. Rudd, V. L. St. Louis, and A. Heyes. 1995. Is total mercury concentration a good predictor of methylmercury concentration in aquatic systems? *Water, Air, and Soil Pollution* 80: 715-724.

Krabbenhoft, D. P., J. M. Benoit, C. L. Babiarz, J. P. Hurley, and A. W. Andren. 1995. Mercury cycling in the Allequash Creek watershed, northern Wisconsin. *Water, Air, and Soil Pollution* 80: 425-433.

Landis, M. S. and G. J. Keeler. 1997. Critical evaluation of a modified automatic wet-only precipitation collector for Hg and trace element determinations. *Environmental Science and Technology* 31: 2610-2615.

Lee, Y. -H. and A. Iverfeldt. 1991. Measurement of methylmercury and mercury in run-off, lake and rain waters. *Water, Air, and Soil Pollution* 56: 309-321.

Lindberg, S. E. 1993. *Journal of Geophysical Research* 97: 14677.

Lindberg, S. E. 1997. Forests and the global biogeochemical cycle of mercury: the importance of understanding air/vegetation exchange processes. In: Baeyens, W. and O. Vasiliev, eds. *Global and Regional and Mercury Cycles: sources, fluxes, and mass balances.* NATO Advanced Science Institute. Boston, Massachusetts: Kluwer Academic.

Lindberg, S. E., T. P. Meyers, G. E. Taylor, R. R. Turner, and W. H. Schroeder. 1992. Atmosphere-surface exchange of mercury in a forest: results of modeling and gradient approaches. *Journal of Geophysical Research* 97: D-2: 2519-2528.

Lindqvist, O. 1991. Mercury in the Swedish environment. Chapter 8: Mercury in terrestrial systems. *Water, Air, and Soil Pollution* 55: 73-100.

MacIntosh, A. (ed.). 1994. *Lake Champlain Sediment Toxics Assessment Program: an assessment of sediment associated contaminants in Lake Champlain, Phase I.* Lake Champlain Technical Report No. 5, Lake Champlain Basin Program, Grand Isle, Vermont.

Mason, R. P. and K. A. Sullivan. 1997. Mercury in Lake Michigan. *Environmental Science and Technology* 31: 942-947.

Mason, R. P. and K. A. Sullivan. 1998. Mercury and methylmercury transport through an urban watershed. *Water Resources Research* 32(2): 321-330.

Mierle, G. and R. Ingram. 1991. The role of humic substances in the mobilization of mercury from watersheds. *Water, Air, and Soil Pollution* 56: 349-357.

Miskimmin, B. M., J. W. M. Rudd, and C. A. Kelly. 1992. Influence of dissolved organic carbon, pH, and microbial respiration rates on mercury methylation and demethylation in lake water. *Canadian Journal of Fisheries and Aquatic Sciences.* 49: 17-22.

NESCAUM (Northeast States for Coordinated Air Use Management), NEWMOA (Northeast Waste Management Officials Association), NEIWPCC (New England Interstate Water Pollution Control Commission), and EMAN (Canadian Ecological Monitoring and Assessment Network). 1998. *Northeast States and Eastern Canadian Provinces Mercury Study: a framework for action.* February, 1998.

Nriagu, J. O. and J. M. Pacyna. 1988. Quantitative assessment of worldwide contamination of air, water, and soils by trace metals. *Nature* 333: 134-139.

Olmez, I., M. R. Ames, and G. Gullu. 1998. Canadian and U.S. sources impacting the mercury levels in fine atmospheric particulate material across New York State. *Environmental Science and Technology* 32: 3048-3054.

Porcella, D.B. 1994. Mercury in the environment: biogeochemistry. In: C. J. Watras and J. W. Huckabee (eds.), *Mercury Pollution: Integration and Synthesis.* Boca Raton: Lewis Publishers. p. 3-19.

Rea, A. W. 1998. Accumulation of atmospheric mercury in foliage in the Lake Champlain and Lake Huron watershed. Ph.D. dissertation, University of Michigan, University Microfilms, Inc.

Rea, A. W., G. J. Keeler and T. Scherbatskoy. 1996. The deposition of mercury in throughfall and litterfall in the Lake Champlain watershed: a short-term study. *Atmospheric Environment* 30: 3257-3263.

River Watch Network. 1998. *River Peoples of Color Project, Final Report.* U. S. EPA Environmental Justice Small Grants Program, 7 pp.

Scherbatskoy, T., J. M. Burke, A. W. Rea, and G. J. Keeler. 1997. Atmospheric mercury deposition and cycling in the Lake Champlain Basin of Vermont. In: J.E. Baker (ed.), *Atmospheric Deposition of Contaminants to the Great Lakes and Coastal Waters.* Society of Environmental Toxicology and Chemistry Special Publication Series. Pensacola, FL: SETAC Press, Pensacola, FL.

Scherbatskoy, T., J. B. Shanley, and G. J. Keeler. 1998. Factors controlling mercury transport in an upland forested catchment. *Water, Air, and Soil Pollution* 105: 427-438.

Schuster, E. 1991. The behavior of mercury in the soil with special emphasis on complexation and adsorption processes -- a review of the literature. *Water, Air, and Soil Pollution* 56: 667-680.

Siegel, S.M. and B. Z. Siegel. 1988. Temperature determinants in plant-soil-air mercury relationships. *Water, Air, and Soil Pollution* 40: 443-448.

Slemr, F. and E. Langer. 1992. Increase in global concentrations of mercury inferred from measurements over the Atlantic Ocean. *Nature* 355:434-437.

State of Vermont. 1990. *Fish Contaminant Monitoring Program Report*, 4 pp.

State of Vermont. 1997. *A Compendium of Fish Tissue Contaminant Data from Vermont and Adjoining Waters 1970-1990*, 56 pp.

Swain, E.B., D. R. Engstrom, M. E. Brigham, T. A. Henning, and P. L. Brezonik. 1992. Increasing rates of atmospheric mercury deposition in midcontinental North America. *Science* 257: 784-787.

United States Environmental Protection Agency. 1997. *Mercury Study Report to Congress.* Office of Air Quality Planning and Standards and Office of Research and Development. Washington, D. C. EPA-452/R-97-003. December 1997.

Vasu, A. B. and M. L. McCullough. 1994. *First Report to Congress on Deposition of Air Pollutants to the Great Waters.* U.S. EPA Office of Air Quality and Planning Standards, Research Triangle Park, North Carolina. EPA-453/R-93-055.

Vermont DEC. 1996. *June 4-11, 1996, Emission Test Report Review of SMS Incinerator.* August 28, 1996. Air Pollution Control Division.

Verta, M., T. Matilainen, P. Porvari, M. Niemi, A. Uusi-Rauva, and N. S. Bloom. 1994. Methylmercury sources in boreal lake ecosystems. In: C.J. Watras and J.W. Huckabee (eds.), *Mercury Pollution: Integration and Synthesis.* Boca Raton: Lewis Publishers, p. 119-136.

Watzin, M. C. 1992. A research and monitoring agenda for Lake Champlain: proceedings of a workshop, December 17-19, Burlington, VT. *Lake Champlain Basin Program Technical Report No. 1.* U.S. EPA, Boston, Massachusetts.

Weiner, J. G. and D. J. Spry. 1996. Toxicological significance of mercury in freshwater fish. In: Beyer, W. N., G. H. Heinz, and A. W. Redmon-Norwood (eds.), *Environmental Contaminants in Wildlife: Interpreting Tissue Concentrations.* New York: Lewis Publishers, Inc.

Westling, O. 1991. Mercury in runoff from drained and undrained peatlands in Sweden. *Water, Air, and Soil Pollution* 56: 419-426.

Wilmot, S. H. and T. Scherbatskoy. 1994. *Vermont Monitoring Cooperative Annual Report for 1993.* VMC Annual Report Number 3. Vermont Department of Forests, Parks, and Recreation, Waterbury, Vermont.

Lower Trophic Level Interactions in Pelagic Lake Champlain

S. N. Levine, M. A. Borchardt, A.D. Shambaugh, and M. Braner

ABSTRACT

Four experiments were conducted in Lake Champlain to estimate primary and bacterial productivity and identify and quantify feeding relationships among the plankton. Primary productivity was much greater in spring than in summer, but generally occurred at rates typical of a mesotrophic lake (350-1660 mg C m^{-2} d^{-1}). Bacterial productivity was minimal in spring and maximal in summer. Both macrozooplankton (Cladocera and copepods >200 μm in size) and microzooplankton (20-200 μm; rotifers and copepod nauplii) fed primarily on phytoplankton in springtime (>98% of consumed C), but in summer and fall, macrozooplankton derived 85-90% and microzooplankton 50% of their C from bacteria and heterotrophic protozoa. Experimental manipulation of nutrient and grazer levels in a 2 X 3 factorial design showed that phytoplankton growth rates are principally controlled by nutrients, while heterotrophic protozoa are more closely regulated by grazers, and bacteria are sensitive to both variables. A trophic cascade was observed involving 4 trophic levels: macrozooplankton depressed rotifers which allowed protozoa to increase and depress bacteria. Further study of Lake Champlain's food web and of controls on productivity is needed to ensure wise management of the lake.

INTRODUCTION

Aquatic food webs are composed not only of the fish, aquatic plants and benthos obvious to the common observer, but of a myriad of microbial species. Phytoplankton, the microscopic plants of the plankton, are at the base of the pelagic food web. Using light as an energy source and dissolved carbon dioxide and minerals as building materials, they produce the organic matter that is passed up food webs. Zooplankton feed on phytoplankton and in turn, are consumed by fish, thus completing the herbivore-based food chain familiar to all lake managers. Also present in lakes is a less well-known trophic pathway referred to as the "microbial loop" [Azam et al, 1983]. Based on dissolved organic carbon (DOC) "excreted" by phytoplankton (1-50% of primary productivity is immediately leaked into the water [Cole et al. 1982]) or released during cell lysis and decomposition, the loop includes DOC-consuming bacteria, bacterivorous protozoa and zooplankton, and zooplankton that prey on protozoa. Both herbivory and the microbial loop pass C, nutrients and energy to zooplankton and thus support the pelagic fishery.

Another largely unappreciated aspect of the pelagic food web is nutrient recycling from upper to lower trophic levels. Primary production is highly dependent on the P and N waste products of zooplankton and fish [Lehman, 1980; Elser and Goldman, 1991]; thus the paradox of high primary productivity during periods of intense zooplankton grazing [Elser and MacKay, 1989]. Bacteria also may regenerate limiting nutrients, or consume them in competition with phytoplankton, depending on the C:N and C:P ratios of available substrates [Caron, 1991].

Because of the many trophic and competitive linkages between food web components, the potential exists for the propagation of disturbances up, down and across food webs. Many such "trophic cascades" have been described in the literature, both at the top of the food web and within the microbial loop [e.g., Berquist et al., 1985; Riemann 1985; Scavia and Fahenstiel 1987; Carpenter, 1988; Carpenter and Kitchell, 1993]. It follows that management programs designed to meet specific goals (e.g., enhanced sports fisheries through fish stocking, or reduced algal turbidity through phosphorus control) frequently yield indirect effects on nontargeted populations. Holling [1986] refers to ecological effects far removed from the initial impact of a disturbance (or from the intent of a management program) as "ecological surprise". Surprise can be minimized through careful analysis of food web structure and function, followed by predictive modeling. However, few lakes have been studied to the extent that realistic modeling can occur.

The purpose of the study described here was to initiate analysis of the microbial portion of the pelagic food web of Lake Champlain (VT-NY-Quebec), and thus contribute to the development of a Lake Champlain food web model. The bioenergetics of major piscivores in the lake have been studied to contribute to this effort [LaBar, 1993; LaBar and Parrish, 1995], but analysis of the remainder of the food web has been restricted to estimates of organism densities. For phytoplankton and zooplankton, there has been a regular summer monitoring program in place since 1991 [McIntosh et al., 1993; Shambaugh et al. 1999; C. Siegfried, NY State Museum and S. Quinn NY Department of Conservation, pers. comm.]. For bacteria and protozoa, however, the current study provides all the information available on non-sewage-associated organisms.

Food web models require flux as well as standing stock measurements. Levels of primary and bacterial productivity should be known, as well as the rates of C and nutrient transfer along food web pathways. Furthermore, there must be an understanding of how environmental factors affect process rates. Our study focused on obtaining data in three arenas: phytoplankton and bacterial productivity, C transfer between trophic levels, and the potential for top-down versus bottom-up trophic cascades. We were able to accomplish the first two goals using a methodology developed by Lehman and Sandgren [1985]. In this method, prey populations are exposed to multiple levels of grazer density and their growth rates determined. Thereafter, prey growth rates are regressed against grazer biomass to yield an equation whose slope is an estimate of grazer clearance rate and whose y-intercept approximates the prey growth rate in the absence of grazer mortality. The method assumes that cell division rates are similar across experimental systems, and that grazing is the only source of mortality in the experimental systems. The method has been used successfully in the past to estimate zooplankton grazing on phytoplankton [e.g., Elser and Goldman, 1991; Elser, 1992; Cyr and Pace, 1993], but its use for measuring zooplankton grazing on bacteria and protozoa is new. To obtain information on the relative importance of top-down versus bottom-up controls on community structure, we performed an experiment in which zooplankton and nutrient levels were manipulated simultaneously in a 2 X 3 factorial design. Resulting changes in phytoplankton, bacterial and protozoan growth rates were interpreted using 2-way ANOVA.

Some of the results of our study have been presented in earlier papers; the impact of zooplankton grazing on phytoplankton species composition is considered in Levine et al. [1999], and the structure and function of the microbial loop in Borchardt et al. [1999]. Levine et al. [1997] describe the nutrient status of phytoplankton in the lake during the grazing studies. The current paper synthesizes all aspects of the study for the first time, thus allowing an image of Lake Champlain's microbial food web to emerge.

MATERIALS AND METHODS

Study Site

Lake Champlain has been described in other papers of this volume [Manley et al., 1999; Smeltzer, 1999]. Our study took place at the broadest expanse of the lake, about 1 km northwest of Juniper Island (44°27' N, 73°17' W) and at a water depth of 70 m. The sampling site was chosen because it is within 0.5 km of the principal sampling station of the New York and Vermont Departments of Environmental Conservation (Station 19), and thus is better characterized than any other location on the lake.

Grazing Rates

We estimated grazing rates on three occasions (25-27 July and 27-29 September 1994, and 8-11 May 1995) using a modified version of the Lehman-Sandgren method described above. Three different grazer size classes were manipulated, macrozooplankton (animals > 200 μm, including cladocerans, adult and copepodite copepods, and the predaceous rotifer *Asplanchna*), microzooplankton (20-200 μm; mostly rotifers and copepod nauplii) and nanograzers (1-20 μm; protozoa). For the first two size classes, three levels were created in duplicate containers (10 L clear plastic cubitainers): one about 4 times ambient (called "high"), one at the ambient density ("ambient"), and one in which as many grazers as possible were removed through sieving ("low"). For the nanograzer experiment, we were unable to produce a "high" treatment, as we could not collect the animals except through filtration, a process which killed them.

Water and organisms were collected over the depth of the lake's mixed layer, plus 2 m into the metalimnion (except in spring, when we terminated water collection at 20 rather than 70 m), and mixed in a large tank prior to dispersal into experimental units. Figure 1 shows the sequence of manipulations followed in setting up the study. Macrozooplankton levels were reduced by sieving water through a 220 μm sieve or a 202 μm net, and augmented by adding animals collected with a Wisconsin net (202 μm mesh) hauled up through the depth of the mixed layer. The ambient treatment for microzooplankton was the same as the low macrozooplankton treatment. The low microzooplankton treatment was created by passing some of the 220 μm sievate through a 20 μm mesh sieve, and the high treatment by adding animals collected on this sieve back to a portion of the 220 μm

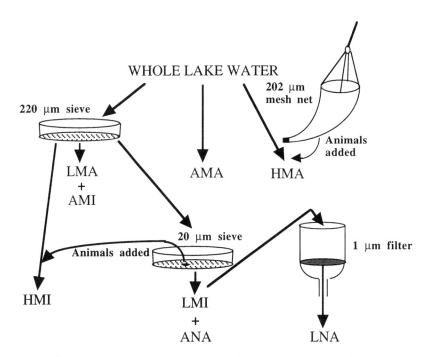

Figure 1. The procedure used to create gradients in macrozooplankton (>200 μm), microzooplankton (20-200 μm), and nanograzers (<20 μm; protozoa) in Lehman-Sandgren grazing experiments. HMA, AMA, and LMA refer to high, ambient and low densities of macrozooplankton; HMI, AMI, and LMI, to high, ambient and low densities of microzooplankton; and AN and LN, to ambient and low levels of nanograzers.

sievate. The low microzooplankton treatment doubled as the ambient nanograzer treatment, and the low nanograzer treatment was created by collecting water passed through Whatman GFF filters (nominal pore size about 1 μm).

Nutrients (0.3 μM KH_2PO_4, 5 μM NH_4Cl and 5.7 μM dextrose) were added to the cubitainers to ensure nutrient-saturated algal and bacterial growth and thus eliminate the possibility that growth rates would be affected by zooplankton excretion [Elser, 1992]. Because grazer abundance was unequal in the cubitainers, the level of nutrient recycling was as well. The cubitainers were incubated in Burlington Harbor for 2-3 days at one-half the Secchi depth (1.5-2 m), and sampled for prey over a time course (generally on 4 occasions; duplicate samples).

Phytoplankton were preserved in acid Lugol's solution and counted under an inverted microscope following the Utermöhl technique [Wetzel and Likens, 1991]. Cell biovolume was determined for each species present by measuring cell dimensions and applying geometric formulae. To calculate biomass and C content, we used the conversion factors reported by Reynolds [1984] for large worldwide data sets, 0.47 pg dwt μm^{-3} and 0.5 μg C per μg dwt. The appropriate conversion factors for individual species vary; thus our estimates may be off by a few percentage points. Bacteria were preserved in 3% formalin and enumerated with flow cytometry [Borchardt et al., 1999]. We did not measure bacterial volumes, biomass or C content directly, but used values for bacteria in Lake Michigan [Scavia and Laird 1987], which has a pelagic environment and trophic state similar to Lake Champlain. Our samples for heterotrophic protozoa were treated sequentially with alkaline Lugol's solution, 3% formalin, sodium thiosulfate, and DAPI stain, and then filtered onto black Nucleopore filters, mounted on slides (under immersion oil) and counted and measured under an epifluorescence microscope [Sherr et al. 1993]. Biovolume, biomass and C content were determined as for phytoplankton, but the conversion factors used were 0.40 pg dwt·µm^{-3} and 0.5 µg C per µg dwt [Borsheim and Bratbak 1987]. Growth rates were estimated from the regression of prey biomass (log-transformed) over time.

At the end of the experiment, the entire contents of the cubitainers was passed through a 63 µm (July) or a 20 µm (other experiments) sieve to collect zooplankton. The animals were rinsed into a bottle where they were anaesthetized with carbonated water and preserved with a solution of 10% formalin, sucrose and Rose Bengal [Haney and Hall, 1975]. Zooplankton in the reservoir water used to create treatments were sampled by passing 10 L of the water through the above sieves. Animals were counted and measured under the inverted microscope. Macrozooplankton biomass was estimated from biomass-length relationships and rotifer biomass from shape dimensions and geometry [Ruttner-Kolisko, 1977; McCauley, 1984]. Cyclopoid copepods older than copepodite stage III were not included in the estimate of macrograzer biomass, as these animals are principally carnivorous. Greater detail on treatment set up, the sampling protocol, organism preservation, counting and measurement, and data analysis are available in Levine et al. [1999] and Borchardt et al. [1999].

Productivity Measurements

The y-intercepts of the Lehman-Sandgren regression lines provided estimates of bacterial and algal growth rates under the conditions of the experiment (2 m incubation depth, and nutrient enrichment). Multiplying growth rates by algal or bacterial biomass and C content yielded estimates of net primary and bacterial productivity.

To assess primary productivity within the mixed layer of the lake (outside the cubitainers), we used the ^{14}C technique in a laboratory incubator [Fee et al., 1992]. Radiolabelled dissolved inorganic carbon (DIC) was removed from incubated samples through acidification and bubbling [Schindler et al. 1972]; thus both DOC excreted from algae and newly-produced algal C were included in the gross primary production estimates. Ambient light conditions were accounted for by measuring solar irradiance with a LiCor quantum meter and probe, and light extinction coefficients were calculated from change in light intensity with depth at the sampling site as measured with a submersible quantum probe. The numerical model of Fee [1990] was employed to estimate daily primary production from solar irradiance, light extinction rate, the ^{14}C-determined productivity-light relationships (Figure 2), and ^{12}C-DIC concentration, as estimated from sample alkalinity and pH [Wetzel and Likens, 1991]. Chlorophyll a concentrations in the samples were measured by filtering 1 L of lakewater through Whatman GFF filters, extracting the collected pigment in hot ethanol [Sartory and Grobbelar, 1984], and measuring pigment concentration with a spectrophotometer (with phaeophytin correction) [Lorenzen, 1967].

Controls on Food webs

The nutrients versus grazers experiment was conducted from 31 July- 4 August 1995. Macrozooplankton were manipulated to three levels as in the grazing studies, but

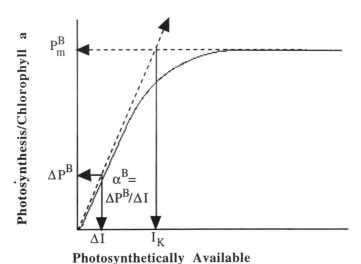

Figure 2. Diagram showing aspects of the photosynthesis-light response curve used in assessing phytoplankton light status (modified from Fee et al. 1992). P_m^B is the maximum rate of photosynthesis attainable per µg of chlorophyll a, α is the slope of the light-limited part of the response curve, and I_k (= P_m^B/α^B) is the light intensity (PAR) above which photosynthesis is light limited.

treatments were done in triplicate rather than in duplicate. Half of the cubitainers at each grazing level were fertilized with P, N and dextrose as in the grazing experiment; half were left unfertilized. Chlorophyll a (a proxy for phytoplankton in this experiment), bacteria, and protozoa were sampled 3 times, at 0, 23, and 92 h, using the methodology employed during the grazing experiment. Only initial and final phytoplankton samples were collected and only those of the fertilized cubitainers were counted. Two-way analysis of variance was used to detect significant treatment effects on prey populations. From the data on fertilized systems, we obtained a fourth assessment of rates of macrozooplankton grazing.

RESULTS AND DISCUSSION

Experimental Conditions

The three grazing studies were conducted during different seasons to maximize the diversity of patterns observed. Surface water temperatures ranged from 5°C in May, to 22°C in July 1994, to 17°C in September, while mixing depths during these same experiments were 70 m (the lake bottom), 10 m and 18 m. Nutrient concentrations were greatest in May, probably because the lake was in overturn and spring runoff was high, and lowest in July, while the lake was well-stratified (Table 1). The September experiment took place during lake cooling and thermocline erosion. Physicochemical conditions during the nutrients vs. grazers experiment (July 1995) were similar to those during the July 1994 grazing experiment (Table 1).

Our May grazing experiment took place during Lake Champlain's annual diatom bloom [Meyer and Gruendling 1979; McIntosh et al. 1993]. Phytoplankton biomass was greater than 10 times that present during the other experiments, and diatoms accounted for 96% of phytoplankton biomass. One species, *Aulacoseira* sp. (formerly *Melosira*) was particularly abundant, making up 67% of phytoplankton biomass on its own. Diatoms were also common during the September experiment, but mixed in with green algae and cryptophytes. The phytoplankton communities during the two July experiments were composed principally of green algae and cryptophytes. However, cyanobacteria were also common in 1995.

Zooplankton biomass varied greatly between experiments. During the September 1994 and July 1995 experiments, animals were so abundant that zooplankton biomass exceeded phytoplankton biomass (Figure 3). By contrast, the phytoplankton: zooplankton biomass ratio during the other two studies was about 5:1. Zooplankton species composition was highly diverse during the May diatom bloom; cladocerans (especially *Daphnia retrocurva*, and *Daphnia galeata mendotae*), cyclopoid copepods (*Diacyclops thomasi*, *Acanthocyclops robustus* and *Tropocyclops prasinus prasinus*), calanoid copepods (mostly *Diaptomus sicilis* and *Diaptomus minutus*), and microzooplankton all were present and in nearly equal proportions (Figure 3). By contrast, the July 1994 community was almost entirely Cladocera (*D. galeata mendotae*, *D. retrocurva*, and *Eubosmina coregoni*), and the September community largely cyclopoid copepods (*A. robustus*, *Mesocyclops edax*, and *D. thomasi*). Both Cladocera (especially *Daphnia galeata mendotae*) and cyclopoid copepods (dominant, *Mesocyclops edax*) were common during the July 1995 experiment. Phytoplankton biomass during this experiment was unusually low.

The bacterivorous protozoa of Lake Champlain consist of both heterotrophic and mixotrophic species. The latter organisms are photosynthetic, but supplement their nutrient supply through phagotrophy. *Chroomonas* sp., the most abundant phytoplankter in Lake Champlain [Shambaugh et al., 1999], is mixotrophic. The heterotrophic protozoa that we counted in samples were principally small species, 5-12 µm in length (Figure 3). Protozoan biomass was greater in spring and fall than in summer, and always much less than phytoplankton or zooplankton biomass.

Bacterial densities were relatively uniform over the experiments, ranging from $1-2.5 \times 10^9$ cells L^{-1}. The highest densities were measured in July, when protozoa were most scarce and temperatures warmest (Figure 3).

Phytoplankton Productivity

The primary production rates that we measured for Lake Champlain using the ^{14}C technique (Table 2) were within the range of values characteristic of deeper mesotrophic lakes [Wetzel 1983]. For theoretical cloudless conditions (best for comparisons), we estimated areal production rates of 1660 and 349 mg C m^{-2} d^{-1} in May and September, respectively. Chlorophyll-*a*-specific primary production was much lower in May than in September (3 vs. 10 mg C mg chl a^{-1} d^{-1}), but light penetration was deeper and phytoplankton biomass greater. Thus volumetric production rates (mixed-layer averages) were similar during the two months, 19 and 24 mg C m^{-3} d^{-1}, respectively.

The parameters of the light response curves for photosynthesis provide information on the light status of phytoplankton. I_k, the light intensity at which photosynthesis saturates, is reduced by light stress, whereas, α, the rate of change in chlorophyll-a specific photosynthesis with changing light intensity at low light levels, increases (Figure 2). Light limitation of photosynthesis is expected when the ratio of the average light intensity in the mixed layer (I_{ave}) to I_k is < 1.0 [Fee et al. 1992]. Both our May and

TABLE 1. Physicochemical conditions at the Lake Champlain sampling site during the three grazing studies and the nutrients vs. grazers experiment. The chemistry data are from the Vermont Department of Environmental Conservation, and are for samples taken within a few days of our experiments. All values pertain to the mixed layer.

Parameter	Grazing Studies			Nutrients vs. Grazers
	July 1994	Sept. 1994	May 1995	July 1995
Mixed layer depth (m)	10	18	70	12
Secchi depth (m)	5	6.5	-	6
Temperature (°C)	22	17	5	22
TP ($\mu g\ L^{-1}$)	10	9	11	9
TN ($\mu g\ L^{-1}$)	406	532	686	322

September analyses of primary production showed α and I_k values typical of phytoplankton receiving low levels of light [Fee et al. 1992]. Given the depth of the mixed layer during the two experiments (70 and 18 m), these findings were not surprising. Phytoplankton probably spent a good deal of each day in the dark. $I_{ave}:I_k$ ratios were just 0.1 and 0.3 (May and September, respectively). Obviously, light is a variable which will require much more attention if the phytoplankton dynamics of Lake Champlain are to be understood.

The Lehman-Sandgren method yielded estimates of net primary production (NPP) for all four studies, but only for 2 m depth (where light was abundant) and with N, P and dextrose enrichment. Because the grazing curves had considerable scatter, the accuracy of the estimates was probably considerably less than that of the ^{14}C method. In May, the estimate of net primary production in the cubitainers was 65% of that for gross primary production (GPP) at the experimental incubation depth (Table 3). In September, the same comparison yielded an estimate of 23%. Because algal respiration is on the order of 10% of gross primary productivity [Reynolds 1984], we suspect that a fairly substantial proportion of photosynthate (as much as 25-65%) was excreted during the experiments. Other experimenters have reported similarly high phytoplankton excretion rates [Cole et al. 1982]. For the two July studies, when GPP was not estimated, NPP rates in the cubitainers were 15 and 5 $\mu g\ C\ L^{-1}\ d^{-1}$ (1994 and 1995, respectively).

Bacterial Productivity

The Lehman-Sandgren regressions for protozoan grazing on bacteria were used to arrive at estimates of net bacterial production rates for the three grazing experiments (Table 3). For July 1995, when only macrozooplankton grazing was examined, we used the y-intercept of the macrozooplankton curve for the productivity estimate. The error involved was probably small as the y-intercepts of our zooplankton and protozoan grazing curves were generally within 20% of one another. Bacterial productivity was much greater in July (13-15 $\mu g\ C\ L^{-1}\ d^{-1}$; 87- 300% of NPP) than in May and September (2 and 4 $\mu g\ C\ L^{-1}\ d^{-1}$; 2 and 31% NPP), probably because bacterial growth is highly temperature sensitive (more so than phytoplankton growth) [Scavia and Laird 1987].

Zooplankton Harvesting of Phytoplankton

Macrozooplankton consumed from 0-11% of the phytoplankton standing stock daily, while microzooplankton depleted another 4-15% (Table 4). Macrozooplankton biomass

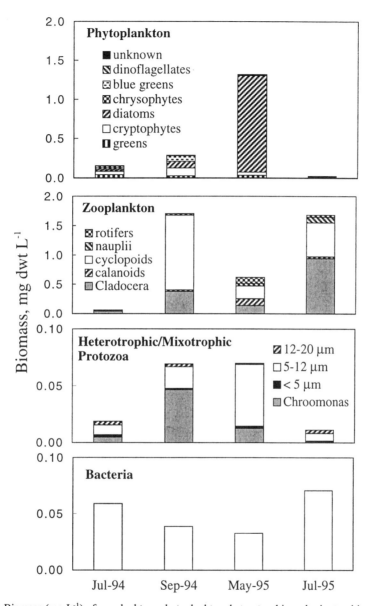

Figure 3. Biomass (mg L⁻¹) of zooplankton, phytoplankton, heterotrophic and mixotrophic protozoa, and bacteria in Lake Champlain during the four experiments. The taxonomic composition of zooplankton and phytoplankton and the size distribution of protozoa are also shown. All samples were taken during daytime and over the depth of the mixed layer (plus 2 m into the metalimnion). Note the smaller scale for protozoa and bacteria.

TABLE 2. Light and photosynthesis parameters for Lake Champlain during the grazing experiments conducted during September 1994 and May 1995. I_{ave} is the average light intensity (PAR) over the mixing depth; P_m^B is the rate of photosynthesis per mg of chlorophyll a at light saturation (the plateau of the curve); α^B is the rate at which photosynthesis (per mg of chlorophyll a) increases with light intensity at low light intensities (the slope of the linear portion of a plot of photosynthesis per mg chlorophyll a vs. light intensity); and I_k (= P_m^B/α^B) is the light intensity below which photosynthesis is primarily light limited. Specific and volumetric productivity values are averages for the mixed layer.

Parameter	Sept. 1994	May 1995
Light extinction coefficient (m^{-1})	0.76	0.41
Mixed layer depth (m)	18	70
I_{ave} (Ei m^{-2} h^{-1}), cloudless	0.13	0.11
Chlorophyll a (mg m^{-3})	1.9	7.4
P_m^B (mg C mg chl a^{-1} h^{-1})	7.9	2.7
α^B (mg C mg chl a^{-1} Ei m^{-1})	10.2	5.5
I_k (Ei m^{-2} h^{-1})	0.49	0.77
I_{ave}: I_k (cloudless)	0.3	0.1
Specif. Prim. Prod. (mg C mg chl a^{-1} d^{-1})		
Cloudless	10.2	3.2
Actual Light	7.0	2.1
Areal Prim. Prod. (mg C m^{-2} d^{-1})		
Cloudless	349	1660
Actual Light	236	1093
Volumetric Prim. Prod. (mg C m^{-3} d^{-1})		
Cloudless	19	24
Actual Light	13	16
Vol. Prim. Prod. at 2 m (mg C $m^{-3}d^{-1}$) (Actual light)	56	184

exceeded microzooplankton biomass during all 3 grazing studies so that one would expect these animals to dominate grazing. However, microzooplankton attained substantially greater biomass-specific clearance rates, which made up for their small representation in the plankton. Both groups processed considerably more phytoplankton C during the May diatom bloom (combined rate, 165 µg C L^{-1} d^{-1}) than during the other studies (<5 µg C L^{-1} d^{-1}).

Zooplankton grazing rates were estimated for separate phytoplankton taxa as well as for the entire assemblage. The results are presented in detail in Levine et al. [1999], and will not be repeated here, apart from stating the major conclusion of the work: phytoplankton species vary widely in their vulnerability to grazing. During our grazing study, many green algal species, dinoflagellates, and cryptophytes lost 20-30% of their biomass daily to grazers. By contrast, many cyanobacterial and diatom species had loss rates of <1% per day.

Zooplankton and Protozoan Harvesting of Bacteria

A variety of zooplankton are capable of concentrating bacteria from lakewater and using them as food. Cladocera, and particularly large daphnids, appear to be well-equipped for this process [Peterson et al. 1978], as are a number of the filter-feeding

TABLE 3. Carbon fluxes (µg C L^{-1} d^{-1}) in pelagic Lake Champlain as estimated from growth and clearance rates obtained in Lehman-Sandgren grazing experiments. All estimates are for a 2 m incubation depth, for N-P-and-dextrose-enriched conditions, and for plankton obtained over the depth of the mixed layer (plus 2 m into the metalimnion). A "+" indicates that prey responded positively to grazer density; prohibiting determination of grazing rates (which were probably minimal).

	Experiment			
	July 1994	Sept. 1994	May 1995	July 1995
Gross primary production (GPP), 2 m[a]		56	184	
Net primary production (NPP)[b]	15	13	119	5
NPP as Percentage of GPP		*23*	*65*	
Bacterial production	13	4.0	1.9	15
Percentage of GPP		*7*	*1*	
Percentage of NPP	*87*	*31*	*2*	*300*
Grazing on phytoplankton:				
by macrozooplankton	0.31	0.00	66	1.16
Percentage of NPP	*2.1*	*0.0*	*55*	*23*
by microzooplankton	3.06	4.29	99	-
Percentage of NPP	*20*	*33*	*83*	-
Grazing on bacteria:				
by macrozooplankton	2.03	+	0.21	4.48
Percentage of Bact. Prod.	*16*	-	*11*	*30*
by microzooplankton	3.08	+	0.34	-
Percentage of Bact. Prod.	*24*	-	*18*	-
by protozoa	+	0.4	+	+
Percentage of Bact. Prod.	-	*10*	-	
Grazing on heterotrophic protozoa				
by macrozooplankton	1.65	0.24	0.3	1.66
Percentage of Bact. Prod.	*13*	*6*	*16*	*11*
by microzooplankton	+	0.01	2.07	-
Percentage of Bact. Prod.	-	*0.2*	*109*	-

[a]Includes excreted C, and is before respiratory loss
[b]Carbon used in realized growth.

rotifers [Nauwerck, 1959]. Calanoid copepods consume some bacteria, but inefficiently [Belyatskaya, 1958], and cyclopoid copepods, which are raptorial carnivores, probably are unable to capture cells of this size [Nauwerck, 1963]. We estimated macrozooplankton grazing on bacteria during 3 of our experiments; microzooplankton grazing estimates were obtained in 2 experiments (Tables 3, 4). We did not obtain usable estimates of grazing during the September experiment because bacteria responded positively to additions of either animal group. Since dextrose and nutrients were added to the experimental cubitainers, the bacterial response was probably not due to nutrient recycling by grazers but to indirect predator-prey interactions. Macrozooplankton grazing on protozoa was relatively intense (see below) and may have depressed protozoan grazing to the extent that bacterial populations were able to increase. The estimates of macrozooplankton bacterivory that we obtained ranged from 0.2-4.5 µg C L^{-1} d^{-1}, and

TABLE 4. Clearance rates of grazers (CR, ml d^{-1} µg dwt grazer^{-1}) during the four experiments and the percentage of prey biomass lost per day (% Lost) as a result of grazing. These estimates exclude the biomass of cyclopoid copepods older than copepodite stage III as these animals are carnivorous and feed little on the prey represented. + indicates that prey increased with increasing grazers, prohibiting estimation of grazing. * indicates r^2 value >0.2 for the clearance rate curve.

	Experiment							
	July 1994		Sept. 1994		May 1995		July 1995	
Grazer-Prey Pair	CR	% Lost	CR	% Lost	CR	% Lost	CR	% Lost
Macroz.-phytop.	0.06	0.4	-0.02	0	0.26	10	0.17*	11
Microz.-phytop.	1.47	1.4	0.49	1.7	0.76	11	-	-
Total zooplankton-phytop.		1.8		1.7		21		-
Macroz.-bacteria	1.16*	8.9	+	-	0.05	1.7	0.3*	16.4
Microz.-bacteria	8.18*	13.5	+	-	0.15*	2.7	-	-
Total zooplankton-bacteria		22.4		-		4.4		>16.4
Protoz.-bacteria	+	-	4.9*	0.3	+	-		
Macroz.-protozoa	3.1*	23.9	0.07*	2.1	0.03	1.1	0.6*^{a}	32.7
Microz.-protozoa	+	-	0.03	0.1	0.4*	7.2	-	-
Total zooplankton-protozoa		23.9^b		2.2		8.3		>32.7

^{a}This rate is for the last 3 days of the experiment; during the first day, protozoan growth was positively related to macrozooplankton concentration.
^{b}While microzooplankton grazing could not be estimated, the positive slope of the Lehman-Sandgren regression suggests it was very low.

thus involved 11-30% of bacterial productivity. For the standing stocks present, removal rates were 2-16% per day.

The microzooplankton (20-200 µm) fraction in our experiments included a mixture of rotifers, nauplii (previously shown to have very low bacterial use rates [Pedrós-Alió and Brock, 1983]), and a few particularly large heterotrophic protozoa (e.g., ciliates). In general, metazoan animals greatly outnumbered protozoa in this size fraction. For the May and July experiments, we obtained estimates of microzooplankton grazing on bacteria similar in magnitude to those obtained for macrozooplankton grazing, 0.3 and 3 µg C L^{-1} d^{-1}, the equivalent of 3 and 14% of bacterial biomass per day, and 18 and 24% of bacterial productivity. In July 1995, we manipulated only macrozooplankton, and therefore have no estimates of grazing by smaller animals.

We were able to estimate nanograzer (protozoan) bacterivory only once. For two experiments, bacteria grew somewhat faster at high than at low protozoan densities. Clearance rates were probably very low at these times, and their determination made difficult by the fact that we had just 4 data points to work with (duplicates for 2 treatments). During the September experiment, protozoa removed just 0.3% of the bacterial standing stock per day. The C flux from bacteria to protozoa was 0.4 µg C L^{-1} d^{-1}, or 10% of bacterial productivity.

Zooplankton Predation on Protozoa

Macrozooplankton grazing on heterotrophic protozoa was measured on two occasions (July and September 1994) and micrograzing on one (May 1995). The other experiments yielded positive response curves. Macrograzers removed 27 and 0.6% of protozoan biomass per day during July and September, respectively (Table 4). Micrograzers removed about 5% per day in May. While biomass removal rates were not large, the C flux reaching zooplankton via protozoa was a substantial portion of the C influx to the microbial loop at the time (>100% of bacterial productivity; Table 3). We conclude that predation on protozoa is at least at times a significant linkage between microbial C processing and the traditional food chain.

Grazing vs. Nutrients as Controls on Microbial Communities

The results of our nutrients versus grazers experiment are presented in detail in Levine et al. [1999] and Borchardt et al. [1999] and will be repeated here only to the extent necessary to highlight major findings. While the experiment was designed to use 2-way ANOVA as an instrument for weighing the relative importance of grazing mortality and resources (as induced by nutrient levels) on the growth rates of major microbial groups, the manipulations yielded numerous "ripple" effects that impacted microbial populations indirectly and thus complicated interpretation of the results. Our ANOVA was divided into two parts, 0-23 and 23-92 hours, as major changes in the dynamics of the food web occurred part-way through the study.

During the first 23 h of in-situ incubations, chlorophyll a levels declined slightly in the cubitainers with ambient nutrient concentrations and increased slightly in the fertilized systems (Table 5). ANOVA indicated that phytoplankton growth rates (as calculated from the changes in chlorophyll a concentrations) were significantly different at the two nutrient levels. Grazer level, on the other hand, did not affect phytoplankton growth during this initial period of algal response.

Bacteria experienced positive growth in all of the experimental cubitainers during the first 23 h of incubation (Table 5). Growth rates were augmented both by nutrient addition and the addition of high densities of macrograzers ($p < .001$, for both treatments). Bacterivory by macrozooplankton was substantial (Table 3), but apparently did not overwhelm growth-inducing indirect effects brought on by the presence of the animals. The protozoa data provide a possible explanation. The densities of these animals increased in cubitainers with low and ambient levels of macrozooplankton (LMA and AMA), but declined in those with a high macrozooplankton level (HMA). The diminished protozoan populations may have allowed for greater bacterial survival. It is also likely that zooplankton contributed to the supply of DOC and inorganic nutrients essential for bacterial growth, but our nutrient additions to half of the cubitainers should have eliminated this effect in them. Protozoan growth rates were not dependent on nutrient concentrations during any part of the experiment. In a nutrients vs. grazers experiment very similar to ours in design, Pace and Funke [1991] also found that protozoan growth was under the control of grazers, while bacterial growth appeared to be principally regulated by nutrient availability.

During the last 3 days of the experiment, phytoplankton growth rates were, on average, twice as high in cubitainers enriched with nutrients as in those without ($P = 0.03$), confirming the importance of nutrient availability to the dynamics of this group (Table 5). Contrary to expectations, however, phytoplankton grew just as quickly in the HMA treatment without fertilization. Apparently, the zooplankton biomass in these cubitainers (which was mostly cyclopoid copepods) was sufficient to supply the phytoplankton with as much nutrient as we provided other cubitainers through fertilization.

Bacteria declined in abundance during the last 3 days of the experiment. Strangely, those treatments that during the first 23 h of the study had shown the highest bacterial

TABLE 5. Treatment averages for the growth rates of phytoplankton (based on chlorophyll a), bacteria and protozoa during the nutrients vs. grazers experiment. P values are for 2-way ANOVA. Separate analyses are provided for 0-23 and 23-92 h.

Parameter	Treatment	Level	0-23 h r (day^{-1})	p	23-92 h r (day^{-1})	p
Phytoplankton	Nutrients	Ambient	-.328	.012	.160	.032
		Enriched	.169		.328	
	Grazers	Low	.054	.152	.188	.011
		Ambient	-.299		.124	
		High	.007		.421	
	Interaction			.184		.118
Bacteria	Nutrients	Ambient	.171	<.001	-.244	<.001
		Enriched	.276		-.362	
	Grazers	Low	.152	<.001	-.341	<.001
		Ambient	.091		-.169	
		High	.429		-.398	
	Interaction			<.001		<.001
Protozoa	Nutrients	Ambient	.085	.990	.672	.145
		Enriched	.088		.507	
	Grazers	Low	.121	.085	.341	<.001
		Ambient	.518		.235	
		High	-.380		1.193	
	Interaction			.116		.037

growth rates (those fertilized or with high macrozooplankton biomass) now showed particularly high rates of bacterial decline. The dynamics of protozoa in the cubitainers also changed between the two time periods. The depressed populations in HMA treatments rebounded, and grew at a faster rate than in other treatments. The lowest growth rates for protozoa were now in the ambient treatments.

In theory, our macrozooplankton manipulations should have resulted in a treatment scheme in which both cyclopoid copepods and Cladocera increased predictably along the LMA to AMA to HMA gradient, and microzooplankton maintained uniform densities across all treatments. Initially, such conditions existed [Borchardt et al., 1999]. However, their maintenance depended on both a lack of trophic interactions between zooplankton groups and a slow response of zooplankton to changes in resource availability. We conducted ANOVA on final biomass values to determine whether nutrient and macrograzer levels were in fact irrelevant to nauplii and rotifer populations. As Table 6 makes clear, nauplii biomass did not change with treatment, while rotifer biomass did. Being small and parthenogenic, rotifers reproduce rapidly and thus are better able to track changes in algal availability. Consequently, mean rotifer biomass was significantly greater in our fertilized than unfertilized cubitainers after 4 days of incubation in situ ($P < 0.001$). Rotifers were also influenced in a negative fashion by grazer level ($P < 0.001$). Hence, they were most abundant in fertilized LMA treatments.

Cladocera also suffered losses in the HMA systems (Table 6). By the experiment's end, cladoceran abundance in the HMA treatment was only one-third that in the AMA

TABLE 6. Treatment averages for the final biomasses of different zooplankton groups during the nutrients vs. grazers experiments. P values are for 2-way ANOVA (rotifers and nauplii only; copepods and cladocerans were manipulated as part of the treatment and thus cannot be analyzed in ANOVA).

Parameter	Treatment	Level	Biomass ($\mu g\ L^{-1}$)	P	Parameter	Treatment	Level	Biomass ($\mu g\ L^{-1}$)
Rotifers	Nutrients	Ambient	64	<.001	Cladocerans	Nutrients	Ambient	536
		Enriched	347				Enriched	462
	Grazers	Low	383	<.001		Grazers	Low	32
		Ambient	179				Ambient	1122
		High	54				High	343
	Interaction			<.001				
Nauplii	Nutrients	Ambient	37	0.23	Cyclopoids	Nutrients	Ambient	2941
		Enriched	32				Enriched	2936
	Grazers	Low	32	0.39		Grazers	Low	217
		Ambient	38				Ambient	656
		High	34				High	7942
	Interaction			0.76				

treatment (although still 10 times that in the LMA treatment). Food availability for Cladocera was greatest in the HMA treatment. Therefore, the poor survival of the animals in this treatment probably was related to grazing losses to cyclopoid copepods.

Undoubtedly, alterations in zooplankton populations were responsible for some of the changes observed among microbial populations. For example, the exceptionally low protozoan growth rates in AMA cubitainers could easily be a consequence of the exceptionally high cladoceran levels in these systems. Cladocerans are known to feed efficiently on heterotrophic protozoa [Pace and Funke, 1991]. These cubitainers also had bacterial growth rates that were substantially less negative than in the other treatments. This makes sense in light of the depression of protozoan grazers that occurred. A very different situation was observed in the HMA treatments, where rotifer numbers were brought down to very low levels by macrozooplankton grazing, and cladoceran numbers were also below ambient levels. Here protozoan growth rates were very high, and bacteria responded with exceptionally negative growth rates (i.e., with mortality rates greatly in excess of reproductive capabilities).

The Big Picture

The results of our study suggest a highly dynamic and interactive food web in pelagic Lake Champlain (Figure 4). Primary production rates probably depend on light intensity and thus mixing depth during much of the year (certainly from October-May), while during summer stratification, nutrients also are important. In our nutrients vs. grazers experiment, nutrient enrichment consistently stimulated phytoplankton growth. Nutrient limitation experiments that we conducted in concert with the grazing studies (the results of which are presented in Levine et al. [1997]) provided more details on the phytoplankton-nutrient dynamic. In May, additions of N, P and Si to cubitainers elicited no phytoplankton growth, whether added alone or in combination; this response suggested limitation by a factor other than nutrients (e.g., by light or grazing). In July and September, combined N and P addition greatly stimulated phytoplankton growth, but

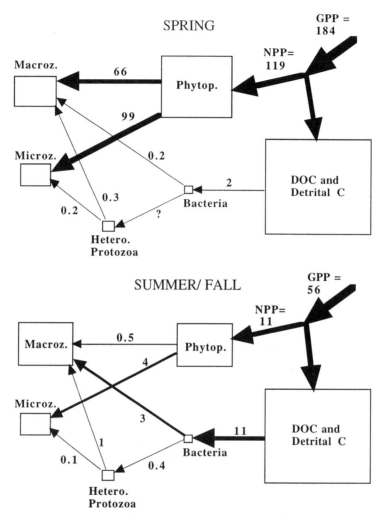

Figure 4. Summary of measured C flows between plankton groups. GPP is gross primary production (^{14}C-determined); NPP is net primary production (from the y-intercepts of the grazing curves). The difference represents phytoplankton "excretion" and respiration. The summer/fall data are the averages for 3 experiments. The spring data are for May 1995. Note that the arrow scales are different for the two diagrams.

singular additions did not. These results suggest a phytoplankton community fairly well-balanced in its N and P use relative to supply rates [Goldman, 1980].

The outcome of our nutrients vs. grazers study pinpointed an important source of nutrients for phytoplankton, zooplankton recycling. High zooplankton levels (mostly cyclopoids) stimulated rather than depressed phytoplankton growth rates. Clearly the phytoplankton-zooplankton link in the food web is a complex one, with both groups dependent on one another for a supply of nutrients, and with the species composition of both the animals and algae influencing the dynamics of the interaction [Sterner, 1990; Elser and Goldman, 1991; Elser and Frees, 1995].

Our study suggested that grazing is a major source of phytoplankton mortality in Lake Champlain in springtime (when loss rates were about 25% daily), but not during lake stratification. Although some phytoplankton species (certain green algae, dinoflagellates, and *Cryptomonas*) lost 20% or more of their biomass daily during the summer, most dominants (e.g., cyanobacteria) sustained grazing losses of <5% day^{-1} [Levine et al., 1999], and the daily loss of total phytoplankton biomass ranged from just 3 to 11%. It may be that, except for short-term events in which large populations of particularly vulnerable phytoplankton species are caught off-guard by burgeoning zooplankton populations (unlikely as phytoplankton grow more quickly than zooplankton), the principal response of phytoplankton to grazers is change in community composition. Our findings of low grazing rates were not unusual; Cyr and Pace (1992) measured zooplankton grazing rates on phytoplankton of <10% day^{-1} in two-thirds of 30 phytoplankton communities that they examined, and rates of < 5% day^{-1} in over half. Literature reports of grazing rates in excess of 50% day^{-1} generally involve measurements made using cultured (and usually highly edible) algae as an added food item, rather than natural populations (Cyr and Pace 1992).

Where the unconsumed phytoplankton C goes in lakes is not entirely known. A good deal probably sediments. Another portion is released into the water column as DOC during cell lysis and decomposition. DOC is, of course, the beginning substrate of the microbial loop. Our study suggested that recovery of Lake Champlain's DOC by bacteria is meager in springtime, but substantial in summer. During our two July experiments, bacterial production rates were 38 and 150% of NPP (compared with 2% in May). Studies of C flow through the microbial loops of other lakes also have documented bacterial processing of more than half of primary productivity [e.g., Cole et al., 1988; Weisse et al. 1990].

Several researchers have suggested that protozoa are the principal bacterivores in lakes, [e.g., Fenchel, 1986; Sanders et al., 1989; Beringer et al., 1991], while others have assigned the role to zooplankton, in particular daphnids [e.g., Wylie and Currie, 1991; Pace and Cole, 1994]. The relative importance of the two groups may depend on lake trophic status, with protozoa most important in oligotrophic systems [Stockner and Porter, 1988]. For Lake Champlain, we found evidence of substantial protozoan feeding on bacteria during our nutrients vs. grazers experiment, but not during any of the three grazing experiments. In two of the latter experiments, bacteria responded positively to protozoan presence, prohibiting the calculation of grazing rates, while in the third, only 10% of bacterial productivity was consumed by protozoa.

The role of zooplankton as bacterivores undoubtedly depends on the species composition of the zooplankton community, on zooplankton densities, and on the existence of alternative food sources [Pace and Cole, 1994]. When Cladocera are present to consume bacteria directly, C can be efficiently passed up the food web without high respiratory costs. Since Cladocera are prime food for planktivorous fish, their key role in C transport is obvious. A macrozooplankton community dominated by copepods, on the other hand, requires an additional step in the microbial loop; bacteria must be consumed by protozoa or by rotifers before the C is in packages large enough for macrozooplankton consumption. In our study, >30% of bacterial productivity was consumed per day, mostly by zooplankton. In addition, rates of macrozooplankton grazing on protozoa and rotifers were high. Thus there appears to be significant linkage between the microbial loop and the herbivore-based food chain in Lake Champlain, a situation which permits higher zooplankton and fish productivity.

A different perspective on C dynamics is gained by focusing on predators rather than on prey when evaluating grazing dynamics. Macrozooplankton are the linchpin in the transfer of energy and nutrients from lower trophic levels to fish, thus they deserve special attention. In May 1995, essentially all (99%) of the non-detrital C reaching macrozooplankton was obtained through herbivory (66 µg C L^{-1} d^{-1} vs. 0.2 and 0.3 µg C L^{-1} d^{-1} from bacteria and protozoa). The diatom bloom was in full swing at the time, and bacterial productivity was held back by low temperatures; thus predominance of the traditional grazing chain was not surprising. In July, however, a very different feeding environment existed. Herbivory provided just 0.3-1.2 µg C L^{-1} d^{-1} to macrozooplankton while feeding on bacteria and protozoa contributed 2-5 and about 2 µg C L^{-1} d^{-1} (85-90% of the total).

For microzooplankton, the seasonal pattern in herbivory vs. dependency on the microbial loop was similar; 98% of C was phytoplankton-derived in May, whereas, in July, bacteria and phytoplankton provided similar amounts of C (feeding on protozoa could not be measured as protozoa responded positively to microzooplankton density). Our general impression is that the Lake Champlain food web switches from one driven by new productivity early in the growth season to one reliant on nutrient and organic matter recycling during summer and early fall. More studies must be conducted to confirm this.

Our findings also illustrated the large number of trophic linkages in Lake Champlain and the ease with which they can lead to indirect effects. The alternating cascade of impacts that we observed on rotifers, protozoa and bacteria in response to macrozooplankton enhancement showed that "trophic ripples" can be far-reaching, and their total effects difficult to predict. We are left wondering whether the extensive piscivorous fish stocking that goes on in the lake might at times depress planktivorous fish populations to the extent that macrozooplankton increase to densities capable of inducing the "domino" effects that we observed in our experimental systems. Trout-affected bacteria seems counterintuitive, but the linkages exist for such an impact to occur. Of course multiple linkages can mean a dampening of effects as well [Mills et al., 1987; Schindler, 1987].

It is also worth noting that feeding relationships were not the only important interactions between microbial groups that we observed in the lake. Nutrients play a major role in regulating phytoplankton and bacterial populations and grazer densities simultaneously affect nutrient recycling rates and mortality. Thus in our nutrients vs. grazers study, phytoplankton responded positively to zooplankton additions at ambient nutrient concentrations, due to the nutrients provided by zooplankton excreta.

In closing, it should be pointed out that the C flux data presented here are sensitive to the conversion factors used to translate biovolume measurements to C mass. The values that we used for bacteria and protozoa were for systems similar to Lake Champlain in trophic state and community structure , while the factors for phytoplankton were averages for very large data sets. Because the C content of organisms increases to as much as 70% of dry weight during nutrient limitation (vs. the normal 50%), and declines to as low as 35% dwt during light or C limitation, our flux estimates may be off by as much as 20% (perhaps more if biovolume to dry weight conversions are in error). However, nutrient deficiencies among phytoplankton tend to be passed along to zooplankton and bacteria [Tezuka 1989; Sterner, 1990]. Therefore, we expect that inaccuracies in our flux estimates were in the same direction for all plankton groups, and that the basic C flow structure revealed was correct.

Studies of the plankton of Lake Champlain always have emphasized macrozooplankton and phytoplankton. We will judge the current paper a success if its impact is to convince researchers that bacteria, heterotrophic protozoa, and rotifers are important players in the food web and deserve considerable attention. Future research on the lake should include measurement of primary and bacterial productivity over an entire annual cycle, assessment of the role of light in regulating primary productivity, further analysis of the roles of invertebrate predators (especially cyclopoid copepods, but also mysids) in the food web, and additional measurements of grazing rates so as to better characterize ranges of normal activity.

Acknowledgements. We thank H. McKinney, S. Pomeroy and S. Spencer for technical assistance and the Vermont Department of Environmental Conservation for their free sharing of data. This project was funded by the U.S. Environmental Protection Agency through the New England Interstate Water Pollution Control Commission (LC-RC92-6-NYRFP).

REFERENCES

Azam, F., T. Fenchel, J. G. Fields, J. S. Gray, L. A. Meyer-Reil, and F. Thingstad, The ecological role of water-column microbes in the sea. *Mar. Ecol. Prog. Ser.*, 10, 257-263, 1983.

Belyatskaya, I. S., Seasonal changes of the total numbers and biomass of bacteria in waters of three lakes of different types, *Microbiology*, 27, 112-117, 1958.

Beringer, U.-G., B. J. Finlay, P. Kuoppo-Leinikki, Protozoan control of bacterial abundance in fresh water, *Limnol. Oceanogr.*, 30, 139-147, 1991.

Berquist, A. M., S. R. Carpenter and J. C. Latino, Shifts in phytoplankton size structure and community composition during grazing by contrasting zooplankton assemblages, *Limnol. Oceanogr.*, 30, 1037-1045, 1985.

Borchardt, M., S. Levine, A. Shambaugh and M. Braner, Importance of grazers in controlling bacterioplankton and heterotrophic protozoa in Lake Champlain, *J. Great Lakes Res.* (in review), 1999.

Borsheim, K.Y., and G. Bratbak, Cell volume to cell carbon conversion factors for a bacterivorous *Monas* sp. enriched from seawater, Mar. Ecol. Prog. Ser., 36, 171-175, 1987.

Caron, D. A., Evolving role of protozoa in aquatic nutrient cycles, in *Protozoa and Their Role in Marine Processes*, edited by P. C. Reid, pp. 387-415, Springer-Verlag, Berlin, Heidelberg, 1991.

Carpenter, S. R., *Complex interactions in lake communities*, Springer-Verlag, New York, 1988.

Carpenter, S. R. and J. F. Kitchell, *The Trophic Cascade in Lakes*, Cambridge University Press, New York, 1993.

Cole, J. J., S. Findlay, and M. L. Pace, Bacterial production in fresh and saltwater ecosystems: a cross-system overview, *Mar. Ecol. Prog. Ser.*, 43, 1-10, 1988.

Cole, J. J., G. E. Likens and D. L. Strayer, Photosynthetically produced dissolved organic carbon: An important source for planktonic bacteria, *Limnol. Oceanogr.*, 27, 1080-1090, 1982.

Cyr, H. and M. L. Pace, Grazing by zooplankton and its relationship to community structure, *Can. J. Fish. Aquat. Sci.*, 49, 1455-1465, 1992.

Elser, J. J., Phytoplankton dynamics and the role of grazers in Castle Lake, California, *Ecology*, 73, 887-902, 1992.

Elser, J. J., and D. L. Frees, Microconsumer grazing and sources of limiting nutrients for phytoplankton growth: Application and complications of a nutrient-deletion/dilution gradient technique, *Limnol. Oceanogr.*, 40, 1-16, 1995.

Elser, J. J. and C. R. Goldman, Zooplankton effects on phytoplankton in lakes of contrasting trophic status, *Limnol. Oceanogr.*, 36, 64-90, 1991.

Elser, J. J. and N. A. MacKay, Experimental evaluation of effects of zooplankton biomass and size distribution on algal biomass and productivity in three nutrient-limited lakes, *Arch. Hydrobiol.*, 114, 481-496, 1989.

Fee, E. J., Computer programs for calculating in situ phytoplankton photosynthesis, 1740, *Can. Tech. Rep. Fish. Aquat. Sci.*, v+27, 1990.

Fee, E. J., J. A. Shearer, E. R. BeBruyn, and E.U. Schindler, Effects of lake size on phytoplankton photosynthesis, *Can. J. Fish. Aquat. Sci.*, 49, 2445-2459, 1992.

Fenchel, T., The ecology of heterotrophic microflagellates, in *Advances in Microbial Ecology*, edited by K.C. Marshall, Plenum, New York, pp. 57-97, 1986.

Goldman, J.C., Physiological processes, nutrient availability and the concept of relative growth rate in marine phytoplankton ecology, in *Primary Production in the Sea*, edited by P. Falkowski, Plenum, New York, pp. 179-194, 1980.

Haney, J.F., and D. J. Hall, Sugar-coated *Daphnia*: A preservation technique for Cladocera. *Limnol. Oceanogr.* 18, 331-333, 1975.

Holling, C. S., Resilience of ecosystems; local surprise and global change, in *Sustainable Development of the Biosphere*, edited by W.C. Clark and R.E. Munn, Cambridge Univ. Press, Cambridge, England, pp. 292-317, 1986.

LaBar, G. W, Use of bioenergetics models to predict the effect of increased lake trout predation on rainbow smelt following sea lamprey control, *Trans. Amer. Fish. Soc.*, 122, 942-950, 1993.

LaBar, G. W., and D. L. Parrish, *Bioenergetics modeling for lake trout and other top predators in Lake Champlain., Final Report to the Lake Champlain Management Conference,* Lake Champlain Basin Program, 1995.

Lehman, J. T., Release and cycling of nutrients between planktonic algae and herbivores, *Limnol. and Oceanogr.*, 25, 620-632, 1980.

Lehman, J. T. and C. D. Sandgren, Species-specific rates of growth and grazing loss among freshwater algae, *Limnol. Oceanogr.*, 30, 34-46, 1985.

Levine, S. N., A. d. Shambaugh, S. E. Pomeroy, and M. Braner, Phosphorus, nitrogen, and silica as controls on phytoplankton biomass and species composition in Lake Champlain (USA-Canada), *J. Great Lakes Res.*, 23, 131-148, 1997.

Levine, S. N., M. Borchardt, M. Braner and A. d. Shambaugh, The impact of zooplankton grazing on phytoplankton species composition and biomass in Lake Champlain (USA-Canada), *J. Great Lakes Res.*, 25, 61-77, 1999.

Lorenzen, C. J., Determination of chlorophyll and pheopigments: Spectrophotometric equations, *Limnol. Oceanogr.*, 12, 343-346, 1967.

Manley, T., K. L. Hunkins, J. Saylor, and P. L. Manley, Aspects of summertime and wintertime hydrodynamics in Lake Champlain, this volume, 1999

McCauley, E., The estimation of the abundance and biomass of zooplankton in samples., in *A manual of methods for the assessment of secondary productivity in freshwaters*, edited by D. Downing and F. Rigler, pp. 228-265, Blackwell, Oxford, 1984.

McIntosh, A., E. Brown, A. Duchovnay, A. Shambaugh, and A. Williams, *1992 Lake Champlain Biomonitoring Program*, Vermont Water Resources and Lake Studies Center, University of Vermont, Burlington, VT, 1993.

Mills, E. L., and J. L. Forney, Trophic dynamics and development of freshwater pelagic food webs, in *Complex Interactions in Lake Communities*, edited by S.R. Carpenter, Springer-Verlag, New York, pp. 13-30, 1988.

Myer, G. E., and G. K. Gruendling, *Limnology of Lake Champlain,* New England River Basin Commission, Burlington, VT, 1979.

Nauwerck, A., Die Beziehungen zwischen Zooplankton und Phytoplankton im See Erken, *Symbolae Botanica Upsaliensis*, 17, 1963.

Nauwerck, A., Zur Bestimmung der Filtrierrate limnishen Planktontiere, *Arch. Hydrobiol.* Supplements, 25, 83-101, 1959.

Pace, M. L. and J. J. Cole, Comparative and experimental approaches to top-down and bottom-up regulation of bacteria, *Microbial Ecol.*, 28, 181-193, 1994.

Pace, M. L. and E. Funke, Regulation of planktonic microbial communities by nutrients and herbivores, *Ecology*, 72, 904-914, 1991.

Pedrós-Alió, C. and T. D. Brock, The impact of zooplankton feeding on the epilimnetic bacteria of a eutrophic lake, *Freshwater Biol.*, 13, 227-239, 1983.

Peterson, B. J., J. E. Hobbie and J. F. Haney, Daphnia grazing on natural bacteria, *Limnol. Oceanogr.*, 23, 1039-1044, 1978.

Reynolds, C. S., *The ecology of freshwater phytoplankton.*, Cambridge University Press, Cambridge, UK, 1984.

Riemann, B., Potential importance of fish predation and zooplankton grazing on natural populations of freshwater bacteria., *Appl. Environ. Microbiol.*, 50, 187-193, 1985.

Ruttner-Kolisko, A., Suggestions for biomass calculation of plankton rotifers, *Arch. Hydrobiol. Beih. Ergebn. Limnol.*, 8, 71-76, 1977.

Sanders, R. W., K. G. Porter, S. J. Bannett, and A. E. Debase, Seasonal patterns of bacterivory by flagellates, ciliates, rotifers, and cladocerans in a freshwater planktonic community, *Limnol. Oceanogr.*, 34, 673-684, 1989.

Sartory, D. P. and J. U. Grobbelaar, Extraction of chlorophyll a from freshwater phytoplankton for spectrophotometric analysis, *Hydrobiol.*, 114, 177-187, 1984.

Scavia, D. and G. L. Fahnenstiel, Dynamics of Lake Michigan phytoplankton: Mechanisms controlling epilimnetic communities., *J. Great Lakes Res.*, 13, 103-120, 1987.

Scavia, D. and G. A. Laird, Bacterioplankton in Lake Michigan: Dynamics, controls, and significance to carbon flux, *Limnol. Oceanogr.*, 32, 1017-1032, 1987.

Schindler, D. W., Detecting ecosystem responses to anthropogenic stress, *Can. J. Fish. Aquat. Sci.*, 44, Supplement No. 1, 6-25, 1987.

Schindler, D. W., R. V. Schmidt, and R. A. Reid, Acidification and bubbling as an alternative to filtration in determining phytoplankton production by the ^{14}C method, *J. Fish. Res. Board Can.*, 29,1627-1631, 1972.

Shambaugh, A., A. Duchovnay, and A. McIntosh, A plankton survey of Lake Champlain, this volume, 1999

Sherr, E. B., D. A. Caron, and B. F. Sherr, Staining of heterotrophic protists for visualization via epifluorescent microscopy, in *Handbook of Methods in Aquatic Microbial Ecology*, edited by P.F. Kemp, B. F. Sherr, E. B. Sherr, and J. J. Cole, Lewis Publishers, Boca Raton, LA, pp. 213-227, 1993.

Smeltzer, E., Phosphorus management in Lake Champlain, this volume, 1999

Sterner, R. W., The ratio of nitrogen to phosphorus resupplied by herbivores: Zooplankton and the algal competitive arena, *Amer. Nat.*, 136, 209-229, 1990.

Tezuka, Y. 1989, The C:N:P ratio of phytoplankton determines the relative amounts of dissolved inorganic nitrogen and phosphorus released during aerobic decomposition, *Hydrobiol.* 173, 55-62, 1989.

Weisse, T., H. Muller, R. M. Pinto-Cowlho, A. Schweizer, D. Springmann, and G. Baldringer, Response of the microbial loop to the phytoplankton spring bloom in a large prealpine lake, *Limnol. Oceanogr.* 35, 781-794, 1990.

Wetzel, R. G., and G. E. Likens, *Limnological Analyses,* Springer-Verlag, New York, 1991.

Wylie, J. L. and D. L. Currie, The relative importance of bacteria and algae as food sources for crustacean zooplankton, *Limnol. Oceanogr.*, 36, 708-728.

A Survey of Lake Champlain's Plankton

Angela Shambaugh, Alan Duchovnay, and
Alan McIntosh

ABSTRACT

An evaluation of phytoplankton and macrozooplankton communities was undertaken in Lake Champlain in 1991 and 1992. Samples collected prior to, during and after stratification indicated that *Chroomonas* spp. were numerically dominant in essentially all phytoplankton samples. The most abundant macrozooplankton were *Bosmina* spp., *Daphnia retrocurva*, and *Diacyclops thomasi*. Comparisons with available historical data suggest there may have been changes in algal community structure from dominance by diatoms and Cyano-bacteria to increased densities of cryptophyte flagellates, and to a larger proportion of clad-ocerans in the macrozooplankton community. While phosphorus reduction has been a high priority of lake management in recent years, it is unclear whether this has contributed to the changes in plankton assemblages. This study also documented the first occurrence of *Thermocyclops crassus* in North America, one of the limited occurrences of the calanoid copepods *Skistodiaptomus pygmaeus* and *S. oregonensis* in the same body of water, and the occurrence of *Eubosmina coregoni* outside the Great Lakes area.

INTRODUCTION

Data on the plankton communities in Lake Champlain have been collected as far back as 1929 [Muenscher 1929] and more recently in the mid 1970's [Myer and Gruendling 1979]. The study described here was undertaken in 1991 and 1992 to evaluate the current composition of plankton in Lake Champlain, compare it to available past data, and assess possible changes in community structure. To our knowledge, this is the first comprehensive lakewide listing of plankton species completed for Lake Champlain since 1976.

METHODOLOGY

Twenty-three stations, chosen to provide broad coverage, were located throughout the five major sub-basins of Lake Champlain [Figure 1], and ranged in depth from 4 to 125 m. Samples were collected between May and October, four sets in 1991 and four in 1992, representing conditions prior to, during and after stratification. Three samples were collected within a predetermined circle encompassing each station. A random number generator was used to determine sampling sites within a circle, and new sites were generated for each collection. Secchi depths, temperature and conductivity data were collected from each station. Sample data within the circles were averaged; however, for clarity, error bars have not been included in the summary figures. Coefficients of variation typically were less than 35% for the phytoplankton (N = 3), occasionally 50 to 60%. Coefficients of variation for the zooplankton during the final two collections of 1991 ranged from 11% to 1.27% (N = 3).

Phytoplankton. Whole water grab samples were collected in triplicate at a depth of 1 m with an horizontally oriented 1 L Van Dorn bottle and preserved with acid Lugols solution [APHA 1989]. Phytoplankton enumeration to lowest feasible taxon followed protocols outlined for Utermöhl settling chambers [APHA 1989], utilizing the field counts. Counts were made at 200x with an Olympus CK2 inverted microscope. Enumeration continued until 100 individuals of the most abundant taxon had been counted, or a minimum of 100 fields had been observed. Identification of diatoms to species requires special preparation and was done infrequently. Densities are reported as natural units per L (i.e. number of single cells, colonies or filaments), an approach which gives equal weighting to taxa ranging in size from small flagellates to multi-celled colonies.

Macrozooplankton. Samples were obtained in triplicate by vertical tows of a 202 µm mesh plankton net, diameter 0.5 m. A reducing collar was added after the initial sample collection. Tow depths were to within 3 m of the bottom at shallow stations and a maximum of 50 m at deeper sites. Samples were anaesthetized with carbonated water and preserved with 10% buffered sucrose formalin to a final concentration of 5% [Hall 1973]. Sample volumes were calculated using standard geometric formulae for determining the volume of a cylinder. Samples containing fewer than 200 organisms were analyzed in their entirety. Those with larger numbers of individuals were subsampled. Individuals were identified to the lowest feasible taxon utilizing an Olympus CK 2 inverted microscope. When higher resolution was needed, individuals were examined with a standard compound microscope. Densities were determined following Standard Methods [APHA 1989] and reported as number of individuals per L.

Samples were collected during daylight hours; therefore, macrozooplankton exhibiting diurnal migrations, such as *Mysis relicta,* were unlikely to be collected. Rotifers, a numerically large portion of the zooplankton population in the lake, were not enumerated in this study (hence the designation macrozooplankton). Recent information on these organisms can be found in Levine et al [1999a and 1999b].

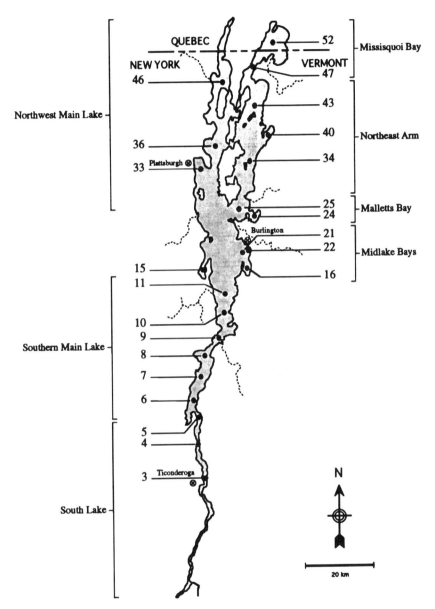

Figure 1. Location of plankton sampling locations on Lake Champlain. Major sub-basins in the lake and the stations located within them are identified. Stations were sampled as part of a previous project and not renumbered for this study, hence they do not follow in sequence.

RESULTS

Phytoplankton

Phytoplankton densities ranged from a high of 7,930,000/L recorded in Missisquoi Bay in 1992 to a low of 230,000/L in 1991 at a station located in the southern portion of the main lake [Figure 2]. Highest densities, found most consistently in shallow Missisquoi Bay and the equally shallow south lake, occurred in May collections. Lowest densities were found primarily at the midlake stations during May and June collections both years, but were not limited to these stations.

A total of 216 taxa was observed in the 1991 samples, and an additional 10 taxa were noted in 1992 [Table 1]. Seven divisions were represented: Bacillariophyta, Chlorophyta, Chrysophyta, Cryptophyta, Cyanobacteria, Euglenophyta and Pyrrophyta. Samples collected in 1991 were dominated by the Cryptophyta, primarily *Chroomonas* spp. These organisms were predominant in all but 13 samples. *Chroomonas* was also the most common organism noted in the 1992 samples, with the exception of May, when communities at 14 of the 23 stations were dominated by a tiny unidentified flagellate. While many taxa were widely distributed throughout the lake during the collection period, only two, *Chroomonas* and *Cryptomonas*, were identified in every sample.

With the exception of the Pyrrophyta and Euglenophyta, which never comprised more than 5% of the total density, divisions other than the Cryptophyta occasionally represented up to 60% of the algae in an individual sample, but rarely dominated the phytoplankton community [Table 2]. In May and June 1991, locally abundant taxa included diatoms and indeterminates. In July/August and September/October, Cyanobacteria were present in greater densities. In 1992, locally abundant divisions included chrysophytes, diatoms and indeterminates in May, with diatoms decreasing in June. Cyanobacteria were abundant in July/August, while indeterminates and diatoms increased in September/October.

Macrozooplankton

Maximum zooplankton densities were observed most consistently in the southern portions of Lake Champlain [Figure 3]. Lowest densities were observed primarily in May samples collected from the southern main lake and mid-lake bays. Greatest observed densities at an individual station occurred most frequently during June. Thirty-three taxa, including copepods (calanoid, cyclopoid and harpacticoid species) and a variety of cladocerans, were observed in 1991 samples, and 45 in 1992 [Table 3]. The most abundant macrozooplankton in both years were *Daphnia retrocurva, Bosmina* spp, *Diacyclops thomasi, Mesocyclops edax, Daphnia galeata mendotae,* and *Eubosmina coregoni.*

In 1991, May samples were dominated by cladocerans in the southern portions of the lake and cyclopoid copepods in the northern reaches [Figure 4]. Mid-lake stations at this time were almost devoid of macrozooplankton ($<20/m^3$). In June, cladocerans predominated at most sites, with a large percentage of the remaining macrozooplankton represented by cyclopoid copepods. Cladocerans were abundant in the south lake during July/August, a period when macrozooplankton populations in the rest of the lake were low. Higher

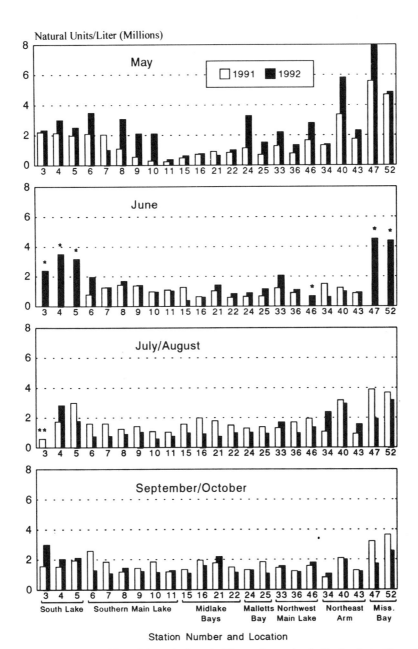

Figure 2. Mean total densities of phytoplankton (millions of natural units/liter) at Lake Champlain sampling locations in 1991 and 1992. N = 3 at all stations. * and ** indicate stations for which data were not available in 1991 and 1992, respectively.

Table 1. Phytoplankton species of Lake Champlain, 1991 and 1992, by location: south lake (1), southern main lake (2), midlake bays (3), Malletts Bay (4), northwest main lake (5), Northeast Arm (6), Missisquoi Bay (7). * indicates taxa observed only in 1992.

Taxon	Location	Taxon	Location
Chlorophyta			
Actinastrum spp.	5 6	*Oedogonium* spp.	3
Ankistrodesmus spp.	3 4 5 6	*Oocystis* spp.	1 2 3 4 5 6 7
A. convolutus Corda	1 2 3 4 5 6 7	*Pandorina morum* (Muell) Bory	1 2 3 4 5
A. falcatus (Corda) Ralf	1 2 3 4 5 6 7	*Pediastrum Boryanum*	
Ankyra spp.	1 2 3 4 5 6 7	(Turp.) Meneghini	2 4 5 6 7
Arthrodesmus spp.	1 2 3 4 5 6	*P. duplex* Meyen	1 2 3 4 5 6 7
Botryococcus spp.	1 2 3 4 5 6	*P. simplex* (Meyen) Lemmermann	1 2 3
Closteriopsis spp.	1 2 3 5	*P. simplex* var. *duodenarium*	
Closterium spp.	1 2 3 5 6 7	(Bailey)	1 2 3 5
Coelastrum spp.	1 2 5 6 7	*P. tetras* (Ehrenb.) Ralfs	5
C. microporum Naegeli	1 2 3 4 5 6	*Quadrigula* spp.	1 2 3 4 5 6 7
C. reticulatum (Dang) Senn	2	*Q. closteroides* (Bohlin) Printz	2 3 4 5 6 7
Cosmarium spp.	1 2 3 4 5 6	*Scenedesmus* spp.	1 2 3 4 5 6 7
Crucigenia spp.	1 2 4 5 6 7	*S. abundans* (Kirch) Chodat	1 2 6 7
C. fenestrata Schmidle	1 2 4 5 7	*S. arcuatus* Lemmermann	1 2 3 4 6 7
C. Lauterbornii Schmidle	1 2 5 6 7	*S. bijuga* (Turp.) Lagerheim	1 2 3 4 5 6 7
**C. quadrata* Morren	3	*S. dimorphus* (Turp.) Kuetzing	1 2 3 5 6 7
C. rectangularis (A. Braun)		*S. quadricauda* (Turp.)	
Gay	7	de Brebisson	1 2 3 4 5 6 7
C. tetrapedia (Kirch)		*Schroederia setigera* (Schroed.)	
West & West	5	Lemmermann	1 2 3 4 5 6 7
**Dictyosphaerium* Ehren-		*Selenastrum* spp.	1 2 3 4 5 6 7
bergianum Naegli	1 2 3 6 7	*S. minutum* (Naeg.) Collins	1 2 3 4 5 6 7
D. pulchellum Wood	1 2 3 5 6 7	*Sphaerocystis Schroeteri* Chodat	1 2 3 4 5 6 7
Elakatothrix gelatinosa		*Spirogyra* spp.	2 6
Wille	1 3 4 5	*Spondylosium* spp.	2 5
E. viridis (Snow) Printz	1 2 3 4 5 6 7	*Staurastrum* spp.	1 2 3 4 5 6 7
Eudorina spp.	1 2 3 4 5 6	*Staurodesmus* spp.	2 3 4 5
Gloeocystis spp.	1 2 3 4 5 6 7	*Tetraedron* spp.	1 3 6 7
G. gigas (Kuetz)		*T. caudatum* (Corda) Hansgirg	1 3 4 5 6
Lagerheim	1 2 3 4 5 6 7	*T. lobulatum* (Naeg.) Hansgirg	6
Gonium pectorale Mueller	1 6	*T. minimum* (A. Braun) Hansgirg	1 2 3 5 6 7
Kirchneriella spp.	2 7	*T. regulare*	1 5 6 7
K. contorta (Schmidle)Boelin	7	*T. regulare* var. *bifurcatum*	1
Lagerheimia longiseta		*Tetraspora* spp.	1 2 3 4 5 6 7
(Lemm.) Printz	1 6 7	*T. lacustris* Lemmermann	1 2 3 4 5 6
Micractinium pusillum		*T. lamellosa* Prescott	2 3 7
Fresenius	1 2 3 5 6 7	*Treubaria* spp.	1 3 5
Mougeotia spp.	1 2 3 4 5 6 7	*Ulothrix* spp.	5
Nephrocytium spp	1 2 3 4 5 6	*Volvox* spp.	5
N. limneticum (G.M. Smith)		*V. aureus* Ehrenberg	1
G.M. Smith	1 2 3 4 5 6 7	*V. globator* Linnaeus	3

Table 1. Phytoplankton species of Lake Champlain, continued.

Taxon	Location	Taxon	Location
Chlorophyta (continued)		Chrysophyta	
Westella botryoides		*Dinobryon* spp.	1 2 3 4 5 6 7
(W. West) de Willemann	7	*D. bavaricum* Imhof	1 2 4 5 6 7
Xanthidium spp.	6	*D. cylindricum* Imhof	2 3 5
Indeterminate	1 2 3 4 5 6 7	*D. divergens* Imhof	1 2 3 4 5 6 7
		D. sertularia Ehrenberg	1 2 3 4 5 6 7
Cryptophyta		*D. sociale* Ehrenberg	1 2 3 4 5 6 7
[1]*Chroomonas* spp.	1 2 3 4 5 6 7	*Mallomonas* spp.	1 2 3 4 5 6 7
[2]*Cryptomonas* spp.	1 2 3 4 5 6 7	*M. pseudocoronata* Prescott	1 2 3 4 5 6 7
		Synura uvella Ehrenberg	1 2 3 4 5 6 7
Bacillariophyta[3]		*Ureglenopsis* spp.	1 2 3 4 5 6 7
Asterionella formosa	1 2 3 4 5 6 7	Indeterminate	1 2 3 4 5 6 7
Attheya spp.	1 2 3 5 6		
[4]*Aulicoseira* spp.	1 2 3 4 5 6 7	Cyanobacteria	
[4]*A. granulata*	1 2 5 6 7	*Anabaena* spp.	1 2 3 4 5 6 7
[4]*A. islandica*	2 3 4 5	*A. circinalis* Rabenhorst	2 7
[4]*A. italica*	1 2 3	*A. flos-aquae* (Lyngb.)	
Cocconeis spp.	1 2 3 4 5 6 7	de Brebisson	1 2 3 4 5 6 7
[5]*Cyclotella* spp.	1 2 3 4 5 6 7	*A. helicoidea* Bernard	2
C. meneghiniana	1	**A. planctonica* Brunnhaler	3 6
Cymbella prostrata var.		*Aphanizomenon flos-aquae*	
auerwaldii	3	(L.) Ralfs	1 2 3 4 5 6 7
Diatoma spp.	3	*Aphanocapsa* spp.	1 2 4 5 6 7
Entomoneis spp.	1 2 3 4 5 6	*A. elachista* West & West	6
Fragilaria spp.	1 2 3 4 5 6 7	*Aphanothece* spp.	1 2 3 4 5
F. crotonensis	1 2 3 4 5 6 7	*Chroococcus* spp.	1 2 3 4 5 6 7
Nitzschia spp.	1	*C. limneticus* Lemmerman	1 2 3 4 5 6 7
N. acicularis	1	*C. Prescottii*	1 2 3 4 5 6 7
Rhizosolenia spp.	1 2 3 4 5	*Coelosphaerium* spp.	1 2 3 4 5 6 7
Stephanodiscus spp.	1 2 3 4 5 6 7	**Gloeotrichia echinulata*	
S. niagare	1 2 3 4 7	(J.E. Smith) P. Lichter	6
Synedra spp.	1 2 3 4 5 6 7	*Gomphosphaeria lacustris*	1 2 3 4 5 6 7
S. delicatissima	1 2 3 5	*Merismopedia* spp.	3 4 7
S. ulna	3 4 5	*Microcystis* spp.	1 2 3 4 5 6 7
Tabellaria spp.	1 2 3 4 5 6 7	*M. aeruginosa*	1 2 3 4 5 6 7
T. fenestrata	1 2 3 4 5	*Oscillatoria* spp.	2 3 4 5 7
Indeterminate	1 2 3 4 5 6 7	Indeterminate	1 2 3 4 5 6 7

[1]*Chroomonas acuta* and *C. minutus* identified after sample analysis.
[2]*Cryptomonas ovata* and *C. erosa* identified after sample analysis.
[3]Individual diatoms not identified to species in all samples.
[4]formerly known as *Melosira*.
[5]identified after sample analysis.

Table 1. Phytoplankton species of Lake Champlain, continued.

Taxon	Location	Taxon	Location
Euglenophyta		Pyrrophyta	
Euglena spp.	1 2 3 5 6 7	*Glenodinium* spp.	1 2 3 4 5 6 7
Phacus spp.	5	*G. armatum* Levander	2 3 5
Trachelomonas spp.	1 2 3 4 5 6 7	*Gymnodinium fuscum* (Ehrenb.) Steind	1 2 3 4 5 6 7
Pyrrophyta		*Peridinium* spp.	1 2 3 4 5 66 7
Ceratium hirundinella (O.F. Muell) Dujardin	1 2 3 4 5 6 7		

cladoceran densities were noted in the northern and southern portions of the lake during the September/October sampling. The first confirmed North American observation of the cyclopoid copepod *Thermocyclops crassus*, an organism widespread in Africa and Eurasia, was made in Missisquoi Bay in 1991 [Duchovnay et al 1992]. In 1992, it was found in a single sample from Lake Champlain, although it is likely that the species was present in other samples as unidentifiable immature individuals.

In May 1992, the cyclopoid species *D. thomasi* predominated in the samples, except at the south lake stations, where *Bosmina* spp. (Cladocera) were the most abundant macrozooplankton. In the June and July/August collections, copepods represented a smaller portion of the population than the cladocerans. This was due not to a decrease in copepod abundance but to an increase in cladoceran density. In September/October, cladocerans and copepods were similar in abundance at most stations, with cladocerans predominant in northern and southern stations and copepods in deep waters. *Skistodiaptomus pygmaeus*, reported primarily east of the Connecticut River [Stemberger 1995], was identified for the first time in Lake Champlain in the 1992 samples.

DISCUSSION

We have presented an overview of spatial and temporal variability in plankton community composition in Lake Champlain during 1991 and 1992. The data, summarized in two technical reports [VWRLSC 1992 and 1993], provide a substantial basis for future reference. Such data will be especially valuable as they describe plankton communities present before the invasion of zebra mussels, known to have significantly affected composition and densities of plankton in the Great Lakes since their introduction in the late 1980's [Nicholls and Hopkins 1993, MacIsaac 1996].

Phytoplankton. There have been relatively few published studies describing Lake Champlain's phytoplankton. Muenscher [1929] quantified phytoplankton in whole water samples from several depths at 16 stations, ten shallow and six deep, from June to August 1929. Data were reported as nannoplankton and densities by genus as individuals and colonies per liter. The samples were dominated by diatoms (*Cyclotella, Melosira, Synedra, Stephanodiscus, Navicula*). Blue-green algae (*Microcystis, Aphanothece, Aphanocapsa, Chroococcus, Coelosphaerium, Merismopedia, Gomphosphaeria*) were locally abundant in

Table 2. Relative abundance, as a percentage, of phytoplankton divisions in Lake Champlain, 1991 and 1992. Relative abundance is calculated from total number of natural units/L. P - observed outside the designated counting area. NP - not observed in the sample.

Division	May		June*		July/Aug		Sept/Oct	
	Range	Mean	Range	Mean	Range	Mean	Range	Mean
Cryptophyta								
1991	15 - 77	53	43 - 88	66	14 - 78	45	42 - 78	69
1992	13 -54	35	55 - 89	73	56 - 90	70	53 - 83	69
Cyanobacteria								
1991	NP - 1	0.5	P - 8	3	2 - 61	26	P - 14	6
1992	NP - 0.7	0.1	NP - 14	1	0.6 - 35	5	P - 11	3
Chrysophyta								
1991	P - 36	6	2 - 14	6	1 - 17	8	1 - 17	7
1992	2 - 15	5	0.5 - 23	5	2 - 27	7	1 - 12	6
Bacillariophyta								
1991	4 - 26	14	P - 29	15	P - 8	2	2 - 34	8
1992	0.9 - 26	10	0.3 - 13	6	0.4 - 8	3	1 - 30	7
Chlorophyta								
1991	1 - 10	3	P - 20	3	2 - 22	6	1 - 12	5
1992	0.3 - 19	3	0.5 - 10	3	2 - 15	8	0.8 - 11	6
Pyrrophyta								
1991	P - 4	2	P	–	P - 2	0.4	NP - 1	0.2
1992	P - 4	1	P -0.7	0.1	P - 2	0.5	NP - 0.5	0.2
Euglenophyta								
1991	NP - P	–	NP - P	–	NP - P	–	NP - 5	0.4
1992	NP - 0.3	<0.1	NP - 0.1	<0.01	NP- 0.1	<0.01	NP - 0.9	0.1
Unknown								
1991	NP - 58	21	P - 31	6	P -43	13	1 - 11	4
1992	16 - 70	45	0.7 - 31	12	NP - 20	7	0.8 - 30	9

* stations 3, 4, 5, 46, 52, and 470 not collected in 1991

Burlington and Missisquoi Bays, and near Plattsburgh in August. Green algae (*Oocystis, Sphaerocystis, Dictyosphaerium, Scenedesmus*) were described as "unevenly distributed" and highly variable in density. The Cryptophyta, typically small and light-colored flagellates, were not documented and likely not counted.

Gruendling [summarized in Myer and Gruendling 1979] sampled 20 stations from February 1970 to October 1974. Discrete wholewater samples were taken over 10 m and combined prior to analysis [Gruendling, personal communication]. Data, reported as cell counts and biomass by species, documented diatoms, Cyanobacteria and cryptophytes as the abundant taxa in Lake Champlain. Cell densities were dominated by diatoms (*Fragilaria,*

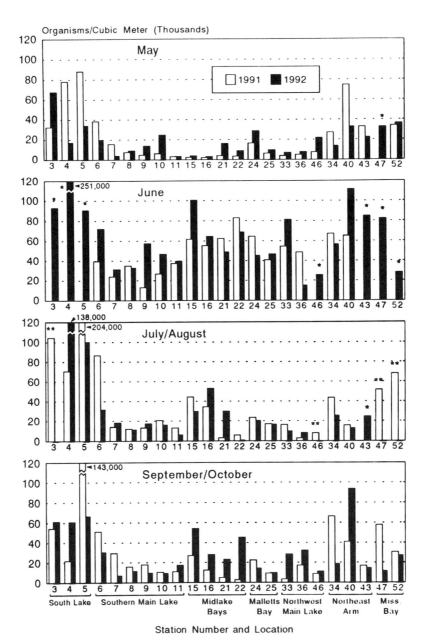

Figure 3. Mean total densities of macrozooplankton (thousands/m³) at Lake Champlain sampling locations in 1991 and 1992. N = 1 for May and June 1991; N = 3 for July/August and September/October 1992 and for station 43 in May 1992. N = 2 for all other locations in 1992. * and ** indicate stations for which data were not available in 1991 and 1992, respectively.

Table 3. Macrozooplankton in Lake Champlain, 1991 and 1992 by location: south lake (1), southern main lake (2), midlake bays (3), Malletts Bay (4), northwest main lake (5), Northeast Arm (6), Missisquoi Bay (7).

Copepoda		Cladocera	
Calenoida			
Epischura lacustris	1 2 3 4 5 6 7	**Acroperus harpae*	1 3
Leptodiaptomus minutus	1 2 3 4 5 6 7	*Alona* spp.	1 2 3
Limnocalanus macrurus	1 2 3 4 5 6 7	*Alonella* spp.	5
Senecella calanoides	2 3 5	[1]*Bosmina* spp.	1 2 3 4 5 6 7
Skistodiaptomus		*Camptocercus* spp.	3 5
oregonenesis	1 2 3 4 5 6 7	**Ceriodaphnia* spp.	1 2 3 4 5 6 7
S. pygmaeus	1 2 3 4 5 6 7	*C. lacustris*	2 5 7
S. sicilis	1 2 3 4 5 6 7	*Chydorus* spp.	1 2 3 4 5 6 7
		Daphnia galeata mendotae	1 2 3 4 5 6 7
Cyclopoida		*D. longiremis*	4
Acanthocyclops robustus	1 2 3 4 5 6 7	*D. retrocurva*	1 2 3 4 5 6 7
Cyclops scutifer	1 2 3 7	*Diaphanosoma birgei*	1 2 3 4 5 6 7
Diacyclops thomasi	1 2 3 4 5 6 7	*Eubosmina coregoni*	1 2 3 4 5 6 7
Eucyclops serrulatus	1	*Eurycerces lamellatus*	1
*Macrocyclops alibidus	4	*Holopedium gibberum*	1 2 3 4 5 6 7
Macrocyclops ater	1	*Leptdora kindtii* 1 2 3 4 5	
Mesocyclops edax	1 2 3 4 5 6 7	**Pleuroxis* spp.	6
Thermocyclops crassus	6 7	**Sida crystalline*	5
Tropocyclops spp.	1 2 3 4 5 6 7		
Harpacticoida			
Indeterminate	1 2 3 5 6 7		

*found in 1992 samples only
[1]identified as *B. longirostis*, however De Melo and Hebert (1994) suggest this species does not occur outside the Great Lakes.

Melosira, Asterionella, Stephanodiscus) throughout most of the year, with some Cyanobacteria (*Anabaena, Aphanizomenon, Coelosphaerium, Microcystis*) abundant in late summer and fall. Cryptophytes (*Cryptomonas, Rhodomonas*) were abundant throughout the year, especially in Malletts Bay and the Northeast Arm.

The present survey documents a lakewide phytoplankton community dominated by small cryptophyte flagellates (*Chroomonas*) in 1991 and 1992. The two taxa representing this division were found in every sample collected during the two-year survey. Diatoms and Cyanobacteria were strong seasonal components, with diatoms abundant during May and June, and Cyanobacteria in August and September.

In 1992, the Long-Term Water Quality and Biological Monitoring Project, a joint collaboration of the New York and Vermont Departments of Environmental Conservation, began to collect basic limnological and biological data to assist in the detection of long-term changes in Lake Champlain [LCBP 1998]. Included in the program was phytoplankton and zooplankton sampling throughout the lake basin. At a main lake index station, vertical variability was explored by the collection of discrete samples at 10 depths and integrated

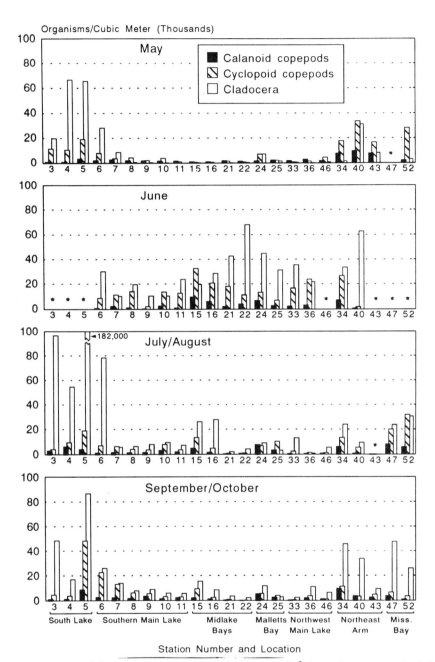

Figure 4. Mean total densities of macrozooplankton (thousands/m³) by major groups at the Lake Champlain sampling locations in 1991. N = 1 for May and June; N = 3 for July/August and September/October. * indicates stations for which data were not available.

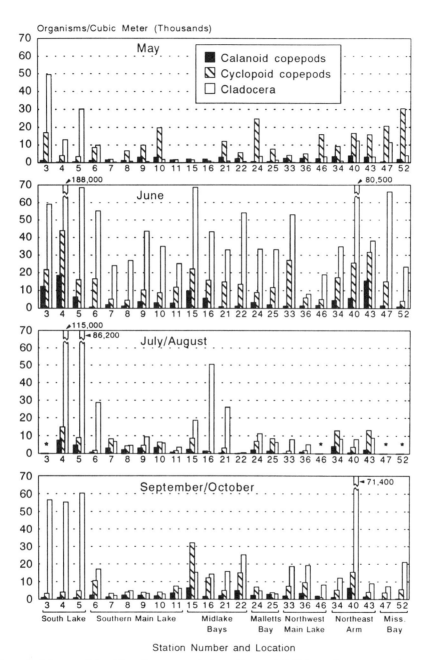

Figure 5. Mean total densities of macrozooplankton (thousands/m³) by major groups at the Lake Champlain sampling locations in 1992. N = 3 for station 43 in May; N = 2 for all other locations. * indicates stations for which data were not available.

samples over two times the Secchi depth biweekly from May to October, with results reported as cell densities per liter. Preliminary data from the main lake index site in 1992 and 1993 [NYDEC, personal communication] documented May samples dominated by diatoms (primarily *Melosira*) and green algae (primarily *Mougeotia*). In June, cryptophytes (*Cryptomonas*), diatoms (*Fragilaria*), and Cyanobacteria (*Coelosphaerium*) were the dominant groups. The cyanobacterium *Coelosphaerium* was the most abundant taxon in the July samples in both years, with the green alga *Mougeotia* also common in 1992. *Coelosphaerium* remained dominant in September and October.

Levine et al [1997] evaluated phytoplankton samples near Juniper Island in the main lake during 1994 and 1995 as part of an investigation of lower food web dynamics. Water was collected from 1 m utilizing an 8 L Van Dorn bottle, and multiple hauls were combined before analysis. Densities were reported as cell counts per liter. In June 1994 and May 1995, diatoms dominated the total cell densities. Cryptophytes were common in May, abundant in June and represented approximately 40% of the total cells in July 1994. Green algae and Cyanobacteria each comprised nearly 20% of the total cells in July. The September 1994 community was dominated by cryptophytes and diatoms. Cyanobacteria represented approximately 20% of the total densities.

Apparent in this brief synopsis is the variety of sampling and reporting methodologies used in the various studies. The present survey and Muenscher report data on a natural unit basis, while the others have utilized cell counts and biomass. The number of filaments and colonies are counted when utilizing the natural unit method, rather than the individual cells comprising them. This more conservative method results in lower overall phytoplankton density estimates when compared with cell counts, and reduces the likelihood that composition will be skewed by a few large, multi-celled taxa in a population of single-celled algae. However, the procedure gives equal weighting to organisms as diverse as small flagellates and multi-celled colonies, and does not lend itself well to assessments of productivity and food webs. Evaluating the phytoplankton community on a biomass basis results in a similar dilemma, as a single diatom cell or cyanobacterial colony contains biomass equivalent to several hundred small flagellates, an important difference when considering food web interactions.

Detailed comparisons were hampered by the differences in methodology; however, some general trends can be discerned. Both Muenscher and Gruendling characterized the phytoplankton community as one dominated by diatoms and seasonally abundant Cyanobacteria. Gruendling noted, in addition, that cryptophyte taxa (a division not reported by Muenscher, probably because these small, lightly colored flagellates are easily missed) were abundant throughout the year. In a discussion of long-term changes from the 1930's to the 1970's, Gruendling indicated that there was evidence for increased cyano-bacterial densities, and changes in diatom composition from species less tolerant of eutrophic conditions to more tolerant species [Myer and Gruendling 1979]. Data from the 1990's have suggested that phytoplankton communities are undergoing another change, as cryptophyte flagellates appear to have increased in density and distribution throughout the lake, while diatoms and Cyanobacteria continue to dominate cell densities seasonally. The data now being collected as part of the Long-Term Water Quality and Biological Monitoring Project should allow better assessment of this change once they are analyzed and made available.

A major focus of Lake Champlain management since the 1970's has been phosphorus reduction, with point source loading in Vermont estimated to have decreased 40% by 1991, and an additional 21% lakewide estimated for the period 1991 - 1995 [LCMC 1996, LCBP 1998]. While it may seem logical to look for corresponding changes in phytoplankton with reductions in phosphorus loading to Lake Champlain, Levine et al [1997] concluded that phytoplankton in the lake are not severely phosphorus-limited, responding more to combined P and N supplementation than to P alone. Phosphorus reductions in the Great Lakes resulted in overall lower phytoplankton biomass [Nicholls and Hopkins 1993], although changes in species composition did not always occur [Nicholls and Hurley 1989]. Great Lakes phytoplankton communities are typically numerically dominated by diatoms, especially in the spring [Pappas and Stoermer 1995, Nicholls and Hopkins 1993]. Munawar and Munawar [1986] considered diatoms to be the most abundant taxonomic group in the Great Lakes. Cyanobacteria and cryptophytes represent seasonally significant components of the phytoplankton community, with cryptophytes often numerically dominant in Lake Superior and portions of Lake Erie [Munawar and Munawar 1986]. Phytoplankton communities in Lake Champlain in the 1990's do not differ greatly from those of the Great Lakes, although apparent increasing densities of cryptophytes are a recent occurrence, the underlying cause of which is unknown.

Macrozooplankton. The past studies on Lake Champlain summarized above also included a zooplankton component. Methodologies were similar, with zooplankton collected as integrated net samples and summarized as number of individuals per cubic meter. Muenscher [1929] enumerated zooplankton found in net samples collected in 1929 to genus only. The Cladocera (*Bosmina, Daphnia, Diaphanosoma*) were noted as most abundant and widely distributed. *Cyclops* and *Diaptomus* represented the only common copepods in these samples.

A summary of studies conducted between the 1930's and 1970's [Myer and Gruendling 1979] identified the abundant copepods in Lake Champlain as the calanoids *Diaptomus sicilis*, *Diaptomus oregonensis* and *Limnocalanus macrurus* and the cyclopoids *Mesocyclops edax* and *Cyclops bicuspidatus thomasi*. The abundant cladocerans were *Daphnia retrocurva, D. galeata mendotae, Eubosmina coregoni,* and *Bosmina longirostris*. The copepods represented the dominant group in the main lake, while the cladocerans were more abundant in other parts of the lake.

The plankton survey presented in this paper identified the most abundant taxa in 1991 and 1992 to be cyclopoid copepods (*Mesocyclops edax* and *Diacyclops thomasi*) and cladocerans (*Daphnia retrocurva, Bosmina* spp., *D. galeata mendotae,* and *Eubosmina coregoni*). Cladocerans were dominant in most of the lake samples collected between June and October. Cyclopoid copepods comprised a large part of the community and were seasonally dominant at some stations, especially in May.

The preliminary data available through the Long-Term Water Quality and Biological Monitoring Project [NYDEC, personal communication] at the main lake index site for 1992 indicated that cyclopoid copepods, primarily nauplii, dominated throughout the year, with *Daphnia* abundant in May and calanoid copepod nauplii abundant in July and August.

Levine et al [1999a and 1999b] identified the community near Juniper Island in the main lake during May 1995 as a mixture of cladocerans (*Daphnia retrocurva, D. galeata mendotae*), cyclopoid copepods (*Diacyclops thomasi, Acanthocyclops robustus, Tropocyclops* spp.), and calanoid copepods (*Leptodiaptomus sicilis, L. minutus*) in nearly equal proportions. July 1994 samples were almost entirely comprised of cladocerans, while the September 1994 community was dominated by cyclopoid copepods.

In comparing the 1970's data to those of Muenscher, Gruendling noted that there appeared to have been no significant changes in the crustacean zooplankton since the earlier report, with main lake crustacean zooplankton populations described as 80 - 100% copepods in the spring, around 50% in the summer, and 70% in the fall [Myer and Gruendling 1979]. In the 1990's, there may be indications that cladocerans have become more important in the main lake population. This survey documented a macrozooplankton population in the southern main lake that was roughly equal in copepod and cladoceran densities in all four collections during 1991, and in two of the collections during 1992. May cyclopoid densities in 1992 were two to three times higher than cladoceran densities, while in June 1992, cladocerans were present in densities two to three times higher than the copepods. The Long-Term Water Quality and Biological Monitoring Project index site yielded data similar to Gruendling's summary in 1992 [NYDEC personal communication], with copepods dominating in spring and fall, and cladocerans representing approximately half the population during the summer. The July 1994 population evaluated by Levine et al [1999a and 1999b] was largely cladocerans, similar to the 1992 data noted here, while September was dominated by copepods and the May 1995 sample was about two-thirds copepods.

Phosphorus load reductions are unlikely to have a direct role in increasing cladoceran densities, as herbivorous zooplankton community structure is influenced more by planktivory than lake productivity [Hessen et al 1995, Almond et al 1996, Proulx et al 1996]. In the St. Lawrence River drainage, the majority of lakes have zooplankton compositions resembling those of Lake Champlain. Almond et al [1996] found the presence of mysids and planktivorous fish strongly influenced zooplankton community structure in these lakes. Predation by the animals reduced biomass of both small and large bodied cladocerans and copepods. Sampling variability and timing of individual collections among the Lake Champlain studies seem more likely to account for any differences in macrozooplankton communities. Data from the Long-Term Biological Monitoring Project, with its higher sampling frequency at selected sites, may address this issue.

Zooplankton composition has been strongly linked to post-glacial hydrology [Stemberger 1995], and Lake Champlain contains an assemblage of glacial relict copepods similar to that in the Great Lakes (*Skistodiaptomus oregonensis, Diacyclops thomasi*) as well as taxa widely distributed across the Northeast (*Epischura lacustris, Leptodiaptomus minutus, Tropocyclops spp.*) The co-occurrence of *S. oregonensis* and *S. pygmaeus* in Lake Champlain is unusual, marking the boundary between two distinct post-glacial drainage basins. The lake also supports a population of *Eubosmina coregoni*, a cladoceran introduced from Europe in the 1960's, and thought to be found only in the Great Lakes area [De Melo and Herbert 1994]. The occurrence of *Thermocyclops crassus*, documented in Africa and Eurasia, adds to the unusual zooplankton diversity in the lake. The significance of this to lake biology is unknown, an intriguing question for future research.

Lake Champlain is experiencing major perturbations at this time: the increasing population of zebra mussels, implementation of extensive phosphorus reduction measures, and pressure from an increasing human population in the basin. To sort through these perturbations, to make decisions regarding future management, and to evaluate the long-term overall health of Lake Champlain, a monitoring program is necessary. The Long-Term Water Quality and Biological Monitoring Project, implemented in 1992, is comprehensive and can supply this vital information. Every effort must be made to continue this program and make the resulting water quality and biological data available to managers and researchers alike.

Acknowledgments. Funding for this study was provided by the Linthilhac Foundation of Burlington VT. Samples were collected by the crew of the university research vessel MELOSIRA - Captain Richard Furbush, assistant Sally Keefer, and others. The comments of Dr. Suzanne Levine (University of Vermont) and an anonymous reviewer were beneficial in shaping this manuscript.

REFERENCES

Almond, M., E. Bentzen, and W. Taylor, Size structure and species composition of plankton communities in deep Ontario lakes with and without *Mysis relicta* and planktivorous fish, *Can. J. Fish. Aquat. Sci.*, 53, 315-325, 1996.

APHA, Standard Methods for the Examination of Water and Wastewater, 17th edition, American Public Health Association, Washington D.C., 1989.

De Melo, R. and P. Hebert, A taxonomic reevaluation of North American Bosminidae, *Can. J. Zool.*, 72, 1808-1825, 1994.

Duchovnay, A., J.W. Reid, and A. McIntosh, *Thermocyclops crassus* (Crustacea: Copepoda) present in North America: a new record for Lake Champlain, *J. Great Lakes Res.*, 18(3), 415-419, 1992.

Hall, D. J., Sugar-coated *Daphnia:* a preservation technique for Cladocera, *Limnol. Ocean.*, 18(2), 331-333, 1973.

Hessen, D., B. Faafeng, and T. Andersen, Replacement of herbivore zooplankton species along gradients of ecosystem productivity and fish predation pressure, *Can. J. Fish. Aquat. Sci.*, 52, 733-742, 1995.

LCBP, Long-term water quality and biological monitoring project for Lake Champlain: cumulative report for 1992 - 1996, Technical Report No. 26, Lake Champlain Basin Program, 1998.

LCMC, Opportunities for action: an evolving plan for the future of the Lake Champlain Basin, Lake Champlain Management Conference, draft 1996.

Levine, S., A. Shambaugh, S. Pomeroy, and M. Braner, Phosphorus, nitrogen, and silica as controls on phytoplankton biomass and species composition in Lake Champlain (USA - Canada), *J. Great Lakes Res.*, 23(2), 131-148, 1997.

Levine, S., M. Borchardt, M. Braner, and A. Shambaugh, Impact of zooplankton grazing on phytoplankton species composition and biomass in Lake Champlain (USA - Canada), *J. Great Lakes Res.*, In Press, 1999a.

Levine, S., M. Borchardt, M. Braner, and A. Shambaugh, Lower trophic level interactions in pelagic Lake Champlain, in this volume, American Geophysical Union, 1999b.

MacIsaac, H., Potential abiotic and biotic impacts of zebra mussels on the inland waters of North America, *Amer. Zool.*, 36, 287-299, 1996.

Munawar, M. and I. Munawar, The seasonality of phytoplankton in the North American Great Lakes, a comparative synthesis, *Hydrobiologia*, 138, 85-115, 1986.

Muenscher, W.G., Plankton studies in the Lake Champlain watershed, in *A biological survey of the Lake Champlain watershed,* NY State Conservation Dept., Supplement to the 19th Annual Report, 1929.

Myer, G.E. and G.K. Gruendling, *Limnology of Lake Champlain,* Lake Champlain Basin Study, New England River Basins Commission, Burlington VT, 1979.

Nicholls, K. and G. Hopkins, Recent changes in Lake Erie (north shore) phytoplankton: cumulative impacts of phosphorus loading reductions and the zebra mussel introduction, *J. Great Lakes Res.*, 19(4), 637-647, 1993.

Nicholls, K, and D. Hurley, Recent changes in the phytoplankton of the Bay of Quinte, Lake Ontario: the relative importance of fish, nutrients, and other factors, *Can. J. Fish. Aquat. Sci.*, 46, 770-779, 1989.

Pappas, J. and E. Stoermer, Effects of inorganic nitrogen enrichment on Lake Huron phytoplankton: an experimental study, *J. Great Lakes Res.*, 21(2), 178-191, 1995.

Proulx, M., F. Pick, A. Mazumder, P. Hamilton, and D. Lean, Effects of nutrients and planktivorous fish on the phytoplankton of shallow and deep aquatic systems, *Ecol.*, 77(5), 1556-1572, 1996.

Stemberger, R., Pleistocene refuge areas and postglacial dispersal of copepods of the northeastern United States, *Can. J. Fish Aquat. Sci.*, 52, 2197-2210, 1995.

VWRLSC, 1991 Lake Champlain Biomonitoring Program, Vermont Water Resources and Lake Studies Center, School of Natural Resources, University of Vermont, Burlington VT., 54 p., 1992.

VWRLSC, 1992 Lake Champlain Biomonitoring Program, Vermont Water Resources and Lakes Study Center, School of Natural Resources, University of Vermont, Burlington VT., 61 p., 1993.

Analysis of Fish DNA Integrity as an Indicator of Environmental Stress

Glenn A. Bauer, Brian T. Dwyer, Brennan J. Leddy,
Ryan P. Maynard, and Leah M. Moyer

ABSTRACT

Through the use of an alkaline unwinding assay, the integrity of DNA molecules can be determined. Over the course of the past three years we have been investigating the use of DNA integrity as a measure of environmental stress on fish (i.e. as a biomarker). Our first study consisted of collecting rock bass (*Ambloplites rupestris*) from Burlington Harbor and the Inland Sea, followed by measuring the integrity of their liver DNA. The results from this study indicated that fish from a contaminated site (Burlington Harbor) had a higher level of DNA integrity (fewer nicks) than fish from a less-contaminated site (the Inland Sea). We followed-up on this observation with a controlled experiment in which fallfish (*Semotilus corporalis*) were raised in aquaria containing either clean, commercially obtained sand or sludge dredged from Burlington Harbor. Once again, we found that fish from the clean tanks had more nicks in their DNA than the fish containing the Burlington Harbor sludge. These results are in contrast to those found by other workers, where contaminants cause a transient increase in the number of nicks in DNA. We propose that fish from contaminated sites have their DNA repair systems working at a higher rate than those fish from relatively less contaminated sites. Further work to explore this question will require molecular analysis of DNA repair genes to see if their gene expression is higher in fish from a contaminated site when compared to fish from a less-contaminated site. We have also explored the feasibility of using DNA obtained from blood samples so fish would no longer need to be sacrificed in order to obtain DNA integrity measurements. DNA from blood appears to be less sensitive to the effects of environmental contaminants as compared to DNA from liver. Because our results are in disagreement with previously published work, we cannot conclude, without further investigation, that DNA integrity is a reliable biomarker for environmental stress.

INTRODUCTION

The impact of chemical pollution on aquatic environments has generated considerable interest in the context of biochemical contamination and its subsequent effect on fish [Maccubbin, 1994; Stein et al., 1992]. One common chemical contaminant in aquatic environments is polycyclic aromatic hydrocarbons (PAHs), which are formed by the inefficient burning of fossil fuels and produced naturally by bacteria and higher plants [Harvey, 1982]. Malins [1985] has shown a positive correlation between liver neoplasms in English sole (*Parophrys vetulus*) and sediment concentrations of the aromatic hydrocarbons benzo[a]pyrene and benz[a]anthracene. One contributing factor to the carcinogenicity of PAHs is the formation of unstable adducts resulting in the nicking of the DNA strand [Phillips, 1983]. The effects of contamination on organisms in aquatic environments can be quantified by measurement of biological markers such as DNA integrity, hematological parameters, liver somatic index, oxidative stress, or indices of biotransformation [Stein et al., 1992; McCarthy and Shugart, 1990]. The determination and assessment of DNA damage has been proposed as a biological marker of environmental pollution. Previous research found that fish living in environments contaminated with pollutants have a lower percentage of double-stranded DNA, indicating a greater number of strand-breaks or nicks.

A relationship between DNA integrity and contamination was observed in a field study of fish collected from freshwater and marine creeks contaminated with polychlorinated biphenyls (PCBs), PAHs, and cadmium in East Tennessee and Mississippi Sound [Everaarts et al., 1993]. The DNA isolated from liver tissue of redbreast sunfish (*Lepomis auritus*) from the contaminated freshwater creeks had significantly more strand-breaks that that of sunfish from the pristine reference sites. Likewise, hardhead catfish (*Arius felis*) from the pristine reference creeks had a higher percentage of double-stranded DNA than those from the contaminated marine sites. Similar results were found in a 28-day laboratory study of the DNA integrity of channel catfish (*Ictalurus punctatus*) exposed to sediments obtained from Black Rock Harbor and from a reference site [DiGiulio et al., 1993]. Black Rock Harbor is known to be polluted with PCBs and PAHs, as well as other organic contaminants. Following alkaline unwinding assays, a greater fraction of double-stranded DNA was found in the reference fish than in those exposed to the contaminated sediment at all time points, indicating more strand breaks in the DNA of catfish exposed to the Black Rock Harbor sediments. It is also important to recognize that the fraction of double-stranded DNA decreased over the first three time points (namely 2,7 and 14 days), but then increased at the conclusion of the experiment (day 28), indicating that the catfish was ultimately able to restore its DNA integrity.

We chose to investigate the efficacy of using DNA integrity measurements as a means of detecting environmental stress in fish from Lake Champlain. In order to validate the procedure, we chose the Inland Sea as our reference site and Burlington Harbor as our contaminated site. High levels of silver, mercury, PAHs, and PCBs have been reported within inner Burlington Harbor. In addition to this field study, fish were also raised in controlled environments exposed to either clean, commercially obtained sand or sludge from Burlington Harbor. Finally, we looked at both liver and blood as a source of the DNA to be used in the integrity measurements. It was our hope that if blood proved suitable, we could conduct the DNA analysis without sacrificing the fish.

MATERIALS AND METHODS

Field Study

Rock bass (*Ambloplites rupestris*) were collected from two sites in Lake Champlain, Vermont. Burlington Harbor was our contaminated site and the Inland Sea was our reference site. All fish were collected by electrofishing and euthanized with tricane methansulfonate (MS222; from Sigma). Total and standard lengths, and sex were taken of all fish; age of fish was determined by analysis of the annual rings on the fish scales. Blood was obtained from the dorsal aorta, collecting 1 ml of blood into an eppendorf tube containing 200 µl of 15% EDTA. Livers were removed, weighed, checked for physical abnormalities, and stored at -20°C. A total of forty-six fish were sampled from Burlington Harbor and eighteen were sampled from the Inland Sea.

Controlled Aquaria Experiment

Young fallfish (*Semotilus corporalis*) were separated into sixteen 10 gallon aquaria and raised at 15°C. Half of the tanks contained clean, commercially obtained sand and the other half held sludge dredged from the bottom of Burlington Harbor. The fallfish were divided evenly among the tanks and randomly sampled at 0, 2, 4, 6, and 12 weeks. The fish were euthanized with MS-222 and their weight and standard length recorded. After removal of the spleen (used for another project studying macrophage aggregation), the fish were stored at -20°C.

Extraction of DNA from Tissue

A modified version of Shugart's (1988) procedure was used. Liver samples (200-400 mg) from the field study fish were homogenized (6-8 strokes) in 1 ml of 1 M NH_4OH and 0.2% Triton X-100. In the controlled experiment, the entire fish was homogenized due to their small size. The homogenate was transferred to a centrifuge tube with 2 ml of distilled water and 6 ml of chloroform/isoamyl alcohol/phenol (24:1:25), mixed by inversion, and allowed to stand for ten minutes. The phases were separated by centrifugation at 10,000xG for twenty minutes. Next, 1.25 ml of the aqueous phase were placed on a Sephadex G-50 column (Sigma; 1 cm id, 3.5 ml settled bed volume) equilibrated in G-50 buffer (150 mM NaCl, 10 mM Tris-Cl (pH 7.4) 1 mM $MgCl_2$, 0.5 mM EDTA). The sample was allowed to flow into the column and the eluate discarded. The DNA was eluted by adding 1.25 ml of fresh G-50 buffer to the column and stored at -20°C. When whole fish were homogenized, if the weight of the fish exceeded 1 gm, the sample was run through two consecutive G-50 columns.

Extraction of DNA from Blood

Three hundred microliters of blood was centrifuged and prepared according to the directions of the Wizard® genomic DNA preparation kit from Promega. Samples were thawed only once after their initial freezing.

Alkaline Unwinding Assay

Fifty microliters of 0.05 N NaOH were added to 100 µl of DNA in G-50 buffer on ice. The double-stranded sample was left on ice, while the experimental sample was incubated for 5 minutes and the single-stranded sample incubated for 20 minutes at 80°C. After the incubation period, each sample was rapidly neutralized by the addition of 50 µl of 0.05 N HCl. Five microliters of 0.2% SDS and 2 mM EDTA were immediately added to each sample and then each sample was passed forcefully through a 20 gauge needle several times. Three milliliters of 0.2 M potassium phosphate buffer (pH 6.9) and 3 µl of Hoechst dye 33258 (1 mg/ml) were added to each sample and then incubated in the dark for 15 minutes. Fluorescence was measured using a Hoefer DyNAQuant 200 fluorometer (Ex: 365 nm; Em: 460 nm). The fraction of ds DNA present in the sample after alkaline unwinding was calculated using equation 1.

$$F(x) = \frac{X_{sample} - X_{ssDNA}}{X_{dsDNA} - X_{ssDNA}} \quad (1)$$

$F(x)$ is the fraction of the experimental sample that is still double-stranded after alkaline unwinding. X_{sample} is the fluorescence of the experimental sample after 5 minutes of alkaline unwinding. X_{ssDNA} is the fluorescence of a single-stranded DNA sample (20 minutes of alkaline unwinding). X_{dsDNA} is the fluorescence of double-stranded DNA (0 minutes of alkaline unwinding).

Table 1. DNA integrity measurements using liver and blood samples of four year old rock bass (*Ambloplites rupestris*) taken from two locations in Lake Champlain.

Site	Frequency of ds DNA[a]	
	Liver DNA	Blood DNA
Burlington Harbor	0.387 ± 0.032 (n = 10)	0.53 ± 0.082 (n = 9)
Inland Sea	0.196 ± 0.076 (n = 6)	0.41 ± 0.081 (n = 12)

[a] frequency of ds DNA after alkaline unwinding; values given are averages ± standard error about the mean.

Statistical Analysis

All statistical analyses were carried out using Microsoft Excel.

RESULTS

As can be seen from Table 1, both DNA sources (liver and blood) showed a higher level of ds DNA in the samples from Burlington Harbor compared to the Inland Sea samples. Liver DNA from Burlington Harbor fish contained 38.7% ds DNA after alkaline unwinding whereas liver DNA from the Inland Sea had 19.6% ds DNA. Another way of stating this is that in their native state, DNA from Inland Sea contained more nicks than DNA from Burlington Harbor. This same trend was also observed in the blood DNA samples from these two sources. From the blood DNA analysis, Burlington Harbor DNA had 53% ds DNA and Inland Sea DNA isolated from blood had 41% ds DNA. There was a noticeable change in the difference between the ds DNA values at the two sites, depending upon the source of the DNA used in the analysis. Liver DNA differed by 19.1% in the amount of ds DNA between Burlington Harbor and the inland Sea (t-test, $P<.05$). The blood DNA differed by only 12% in the amount of ds DNA between the two sites (t-test, $P\cong.10$).

Table 2 and Figure 1 show the results from the tank study where young fallfish were raised under two different conditions: clean sand and Burlington Harbor sludge. The level of ds DNA remained relatively steady in the fish raised in the environment containing the clean sand. For the fish raised in aquaria containing sludge from Burlington Harbor, there was a noticeable increase in the amount of ds DNA detected after 2 - 4 weeks of growth, but then returned to values indistinguishable from those of the fish in the clean sand tanks.

Table 2. DNA integrity measurements taken from aquaria-raised fish (*Semotilus corporalis*) over a period of 8 weeks. Fish were grown in either clean or contaminated conditions. For all data points, n = 10.

Time	Frequency of ds DNA[a]	
	Sand environment[b]	sludge environment[c]
0 weeks	0.59 +/- 0.10	
2 weeks	0.56 ± 0.09	0.62 ± 0.12
4 weeks	0.51 ± 0.11	0.88 ± 0.05
8 weeks	0.54 ± 0.22	0.73 ± 0.08

[a] frequency of ds DNA after alkaline unwinding; values given are averages ± standard error about the mean.
[b] clean sand purchased locally.
[c] sludge dredged from the bottom of Burlington Harbor.

DISCUSSION

The ds DNA values from both the liver and blood DNA analysis is contradictory to other reported results. Shugart (1988) clearly saw a transient decrease in the amount of ds DNA (*i.e.* more nicks) in analyzing fish that were exposed to benzo[a]pyrene over a 40 day period. Our data shows just the opposite. Fish from our contaminated site (Burlington Harbor) contained more ds DNA (less nicks) than the DNA from fish in our reference site (Inland Sea). It was our hope that DNA from blood would serve as a suitable source for the analysis of DNA integrity, since this meant that we would no longer have to sacrifice fish; we would just need a small aliquot (300 µl) of blood. Although the blood data were qualitatively similar to that from liver, there appeared to be a loss in sensitivity as seen in the smaller difference in the ds DNA values between the reference and contaminated sites. Because we have supporting data from two different DNA sources (liver and blood), we are confident that the numbers are reliable. Since our field study results were contrary to those previously reported (Shugart, 1988; Everaarts *et al.*, 1993), we chose to conduct a tank study where the fish were raised under controlled conditions while being exposed to contaminants. Fish were raised in either aquaria containing clean, commercially obtained sand or sludge from Burlington Harbor. Once again, we saw that the fish grown in contaminated conditions exhibited an increase in the amount of ds DNA present after alkaline unwinding. This means that the DNA in these fish contained less nicks that their counterparts grown under non-contaminated conditions.

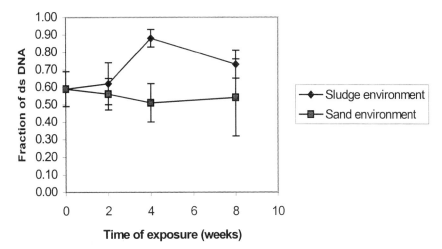

Figure 1. Fraction of ds DNA after alkaline unwinding of DNA samples from aquaria raised fish (*Semotilus corporalis*) over time. Fish were raised in aquaria containing either clean, commercially obtained sand or sludge from Burlington Harbor. DNA was extracted from whole fish. Values are averages ± standard error about the mean. For all data points, n = 10.

We propose that one explanation for these results may lie in the level of gene expression of DNA repair genes in the fish. Fish living in contaminated conditions may have responded to this chronic attack on the integrity of their DNA by increasing the expression of genes responsible for DNA repair. Presumably this benefits the fish. It does, however raise the question of what, if any, long term effect this will have on the health of the fish. In an effort to confirm our hypothesis, we are attempting to measure the gene expression of certain DNA repair genes in the fish to see if their expression truly has been increased. Because our results are in disagreement with previously published work, we cannot conclude, without further investigation, that DNA integrity is a reliable biomarker for environmental stress.

Acknowledgements. This work was funded by an NSF EPSCoR Small College Development grant, the VT EPSCoR Outreach program, and the Saint Michael's College Faculty Development fund. We would like to thank Doug Facey for helpful discussion of various aspects of this work.

REFERENCES

DiGiulio, R.T., Habig, C., and Gallagher, E.P., Effects of Black Rock Harbor sediments on indices of biotransformation, oxidative stress, and DNA integrity in channel catfish, *Aquat. Toxicol.*, 26, 1-22, 1993.

Everaarts, J.M., Shugart, L.R., Gustin, M.K., Hawkins, W.E., and Walker, W.W., Biological markers in fish: DNA integrity, hematological parameters and liver somatic index, *Marine Environ. Res.*, 35, 101-107, 1993.

Harvey, R.G., Polycyclic hydrocarbons and cancer, *American Sci.*, 70, 368-393, 1982.

Maccubbin, A.E., DNA adduct analysis in fish: laboratory and field studies, in *Aquatic Toxicology: Molecular, Biochemical, and Cellular Perspectives,* edited by D.C. Malins and G.K. Ostrander, pp267-293, Lewis Publishers, Boca Raton, FL, 1994.

Malins, D.C., Toxic chemicals in sediment and biota from a creosote-polluted harbor: relationship with hepatic neoplasms and other hepatic lesions in English sole (Parophrys vetulus). *Carcinogenesis*, 6, 1463-1469, 1985.

McCarthy, J.F., and Shugart, L.R., Biomarkers of Environmental Contamination, pp3-12, CRC Press, Boca Raton, FL, 1990.

Phillips, D.H., Fifty years of benzo[a]pyrene, *Nature*, 303, 468-474, 1983.

Shugart, L.R., Quantitation of chemically induced damage to DNA of aquatic organisms by alkaline unwinding assay, *Aquat. Toxicol.*, 13, 43-52, 1988.

Stein, J.E., Collier, T.K., Reichert, W.L., Casillas, E., Horn, T., and Varanasi, U., Bioindicators of contaminants exposure and sublethal effects: studies with benthic fish in Puget Sound, Washington, *Environ. Toxicol. Chem.*, 11, 701-714, 1992.

Physiological Indicators of Stress Among Fishes From Contaminated Areas of Lake Champlain

Douglas E. Facey, Cynthia Leclerc, Diana Dunbar,
Denise Arruda, Lori Pyzocha, and Vicki Blazer.

ABSTRACT

From 1992 to 1997, we conducted a series of investigations to determine whether fishes from contaminated areas of Lake Champlain showed evidence of physiological stress. We compared the number of and percent area occupied by macrophage aggregates in spleens of yellow perch (*Perca flavescens*), brown bullhead (*Ameiurus nebulosus*), and rock bass (*Ambloplites rupestris*) from known contaminated sites to reference sites with lower contaminant levels. Yellow perch had no significant differences in macrophage aggregate parameters among sites. Brown bullhead from inner Cumberland Bay did have significantly higher macrophage aggregate area than similarly aged fish from a reference site. Brown bullhead from the contaminated site also had high rates of external lesions, barbel deformities, and a higher incidence of liver lesions associated with exposure to contaminants. Rock bass collected from Burlington Harbor in 1992 had significantly higher splenic macrophage aggregate area and significantly larger livers than similarly aged fish from reference sites. Rock bass collected in 1997 from Burlington Harbor had significantly less splenic area occupied by macrophage aggregates and smaller livers than had been found in 1992. There was no significant difference in macrophage aggregate area between the 1997 Burlington Harbor fish and similarly aged fish captured from any of the reference areas. Comparison of gonads of rock bass from Burlington Harbor and the Inland Sea in 1997 also showed no significant differences in fecundity, egg diameter, gonad weight or histology of the testes.

INTRODUCTION

Concern regarding the effects of contaminants on aquatic ecosystems and fish health has generated considerable interest in finding suitable biomarkers in fishes to help evaluate water quality [eg., Adams 1990; McCarthy and Shugart 1990a; Huggett et al. 1992; Peakall 1992; Couch and Fournie 1993]. The use of biomarkers is based on the concept that the expression of stress at any level of biological organization most likely is preceded by changes at a lower level. Therefore, sublethal stress is first expressed at the molecular and biochemical levels before changes can be seen in cells, tissues, and organs. Alteration of structure and function of cells, tissues, and organs can interfere with integrated physiological functions. This subsequently may affect growth or reproduction, thereby impacting populations, communities, and ecosystems.

Biomarkers generally are considered to be those biological indicators that are expressed at the cellular and subcellular level [McCarthy and Shugart 1990b], making them more sensitive than indicators at the population and community level. Biomarkers can be extremely valuable in environmental monitoring because they can provide valuable information that would not be available from chemical analyses of water or sediments. Although chemical analyses are necessary to document the presence of contaminants, they do not provide information on the biological availability of the contaminants or their effects on exposed organisms. Chemical analyses also may not detect contaminants that are readily taken up, metabolized, and eliminated by organisms. Biomarkers, however, may identify physiological stress resulting from exposure. In addition, they can provide information regarding the synergistic effects of toxins that may occur together in the environment. Biomarkers explain the link between contaminants and their effects on populations, and may also provide a means to assess improvements in environmental quality.

Wolke et al. [1985] suggested that macrophage aggregates in livers and spleens might be good biomarkers of fish exposure to environmental contamination after noticing that winter flounder (*Pseudopleuronectes americanus*) from polluted coastal areas had significantly more and larger macrophage aggregates in these organs than flounder from clean reference sites. Several other studies have found similar results in largemouth bass (*Micropterus salmoides*) [Blazer et al. 1987]; whitemouth croaker (*Micropogonias furneri*) [Macchi et al. 1992]; roundnose grenadier (*Coryphaenoides rupestris*) [Lindesjoo et al. 1996]; and white sucker (*Catostomus commersoni*) [Couillard and Hodson 1996].

Macrophages are phagocytic cells and play an important role in vertebrate immune systems. They are involved in both the nonspecific and specific immune responses, and are capable of degrading and clearing some foreign substances from the body [Thaler et al. 1977]. Macrophages often are concentrated in the liver and spleen where they can detect and engulf recognized non-self particles such as a viruses, bacteria, or protozoans [Roitt 1977]. Macrophages are also important in destroying fragments of cells that have been killed by toxins, parasites, or other stressors. Hence, fishes collected from polluted environments tend to exhibit an increase in the area, density, and frequency of macrophage aggregates.

Macrophage aggregates accumulate as a fish ages [Brown and George 1985; Blazer et al. 1987]. Therefore, macrophage aggregates provide some indication of past exposure to environmental stressors. Because macrophage aggregate number and area increase with fish age, however, it is important to use fish of similar age when comparing contaminated and reference sites.

Liver size also can be an indicator of fish exposure to some types of contamination. Because the liver is so important in detoxification, exposure to contaminants can lead to an increase in liver size due to an increase in the size of liver cells (hypertrophy) or an increase in the number of cells (hyperplasia) [Goede and Barton 1990]. Studies evaluating the relative liver size of fishes from contaminated and reference sites often utilize the Hepatosomatic Index (HSI), which expresses liver size as a percentage of total body weight.

Toxins in the environment may also interfere with fish reproductive systems. For example, female mosquito fish (*Gambusia affinis*) exposed to kraft mill effluent exhibited male characterisitics [Howell et al. 1980; Bortone and Davis 1994]. Munkittrick et al. [1992] noted testicular atrophy in males and abnormal oocytes in female whitefish (*Coregonus clupeaformis*) exposed to kraft mill effluent. Both male and female yellow perch (*Perca flavescens*) from environments with elevated levels of cadmium, zinc, PAHs and PCBs, had smaller gonads than fish from less contaminated sites [Hontela et al. 1995]. Exposure to DDT derivatives and PCBs resulted in the feminization of male spotted seatrout (*Cynocion nebulosus*) and rainbow trout (*Oncorhynchus mykiss*) [Sumpter and Jobling 1995].

In a series of studies from 1992 to 1997 we evaluated macrophage aggregate area, compared liver size and, in 1997, compared gonads of fishes from contaminated and references sites in Lake Champlain and Sunset Lake, a small, isolated lake with no visible sources of toxic pollution. Although numerous studies have addressed many aspects of the Lake Champlain ecosystem, including toxic pollutants in the lake, none had attempted to use fish health indicators to identify the presence, magnitude, or potential danger of contamination. The underlying question in our investigation was whether fishes collected from contaminated areas of Lake Champlain would have a larger area of splenic macrophage aggregates, proportionately larger livers, or show some abnormalities in their reproductive systems when compared to fishes from reference sites.

METHODS

Sampling locations were selected based on the results of sediment analyses completed in 1991 by Dr. Alan McIntosh (Water Resources and Lake Studies Center, University of Vermont). This survey of 30 sites in Lake Champlain revealed that although much of the lake sediments appeared relatively clean, several sites were contaminated with PCBs, PAHs and some heavy metals, including lead and cadmium. Three of the more contaminated sites identified included (1) the inner portion Burlington Harbor, which is partially separated from the main lake by a breakwater and had elevated levels of some metals, PAHs, and some PCBs; (2) inner Cumberland Bay, especially the area adjacent to Wilcox Dock, where PCB levels were extremely high; and (3) the area south of the mouth of the LaChute River, which used to receive effluent from the old International Paper plant in Ticonderoga, New York, and which has elevated levels of PAHs and PCBs (Figure 1). Several cleaner areas of Lake Champlain were selected as reference sites: the mouth of Shelburne Bay, shallow sections of outer Malletts Bay, and the west Milton shore of the Inland Sea. We also collected some fish from Sunset Lake, a small, isolated lake

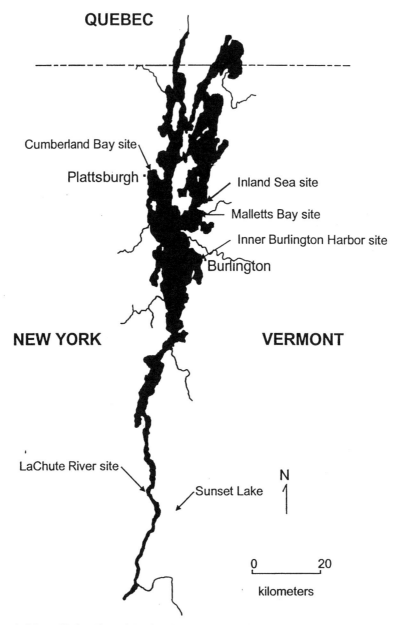

Figure 1. Map of Lake Champlain showing sites from which rock bass were collected. One additional reference site, Sunset Lake, is located about 8 km east of southern Lake Champlain, near the town of Orwell, Vermont.

approximately 8 km east of southern Lake Champlain with no apparent inputs of industrial or municipal pollution.

Fishes were collected by electrofishing, except in Sunset Lake fish where we used angling. Species were selected for this study based primarily on their availability at both contaminated and reference sites. We selected yellow perch (*Perca flavescens*), brown bullhead (*Ameiurus nebulosus*), and rock bass (*Ambloplites rupestris*). We used both male and female fish, but did not try to compare the sexes within a species due to small sample size.

After capture, fishes were transported live back to the lab. Within 24 hours they were killed with an overdose of anesthetic (tricaine methansulfonate, MS 222). Within one hour after death fishes were measured, weighed, and evaluated for external lesions. Livers were removed and weighed for later determination of the hepatosomatic index (HSI = weight of liver x 100 /total fish weight). During the first year of the study, tissue samples of liver, spleen, head kidney, and hind kidney were removed and fixed in 10% buffered formalin. In subsequent years only spleens were removed and preserved, until 1997 when we also weighed and preserved gonads of rock bass.

Tissues were embedded in paraffin, sectioned (12 microns), and stained with hematoxylin and eosin. For macrophage aggregate comparisons, fishes of the same age were compared among sites because macrophage aggregates accumulate with age [Brown and George 1985, Blazer et al. 1987]. Age was determined by reading annual rings on scales or spines. The mean area of spleen tissue occupied by macrophage aggregates and the mean hepatosomatic index were compared among sites using a Tukey's multiple means comparison test.

In the most recent phase of the study, gonads of rock bass from Burlington Harbor were compared to those from the Inland Sea. The number of eggs, egg diameter, and gonad weight of females were measured, and the testes of males were weighed and histologically examined for evidence of intersex development. Mean egg number, diameter, and gonad weight were compared using a student's t-test.

RESULTS

During the initial year of the study we showed that macrophage aggregate parameters from spleens did show significant differences between contaminated and reference sites, whereas there were no significant differences in macrophage aggregate parameters in livers or head kidneys [C. Leclerc, unpublished manuscript]. Therefore, only spleens were used in subsequent years.

Results from yellow perch and brown bullhead have been published elsewhere [Blazer et al. 1994]. There were no significant differences in macrophage aggregate parameters or hepatosomatic index among 2 year old yellow perch from Shelburne Bay, Cumberland Bay, and Burlington Harbor, or between 3 year old yellow perch from Malletts Bay and the mouth of the LaChute River. The data from the 2 year olds did indicate that yellow perch from Burlington Harbor had more abundant macrophage aggregates and larger spleen area occupied by macrophage aggregates, but the difference was not statistically significant due to small sample size and high variability. Brown bullhead from inner Cumberland Bay did have a significantly higher number and greater area of macrophage aggregates than

TABLE 1. Mean (± one standard deviation) percent area of spleen tissue displaced by macrophage aggregates (MA area) and mean hepatosomatic index (HSI) in 3 and 4 year old rock bass from Burlington Harbor and three reference sites. Data were analyzed using Tukey's multiple means comparison test. Arcsin tranformation of the data, to account for the fact that they are presented as percent values, did not change the outcome of the Tukey's test.

	3 year olds			4 year olds		
	MA area	HSI	n	MA area	HSI	n
Burlington Harbor 1992	3.77 (± 2.49)	1.66 (± 0.33)	6	11.44 (± 9.87)	1.37 (± 0.27)	8
Malletts Bay 1992	1.36 [a] (± 1.01)	1.06 [a] (± 0.19)	8	2.98 [b] (± 2.44)	1.09 [a] (±0.19)	11
Sunset Lake 1994	n/a	n/a	0	2.02 [b] (± 1.82)	0.80 [b, c] (± 0.19)	10
Burlington Harbor 1997	0.39 [b] (± 0.53)	1.08 [a] (± 0.36)	6	1.72 [b] (± 0.89)	1.12 (± 0.13)	7
Inland Sea 1997	1.09 [a] (± 0.36)	1.81 (± 0.48)	6	2.13 (± 0.65)	1.18 (± 0.06)	2

[a] values significantly lower than those from Burlington Harbor 1992 ($P<0.05$)
[b] values highly significantly lower than those from Burlington Harbor 1992 ($P < 0.01$)
[c] mean HSI for Sunset Lake 1994 fish was significantly higher than mean HSI from Burlington Harbor 1992 ($P<0.01$), Malletts Bay 1992 ($P<0.05$), and Burlington Harbor 1997 ($P<0.05$).

bullhead from a reference area in Malletts Bay. In addition, six of the nine brown bullheads from Cumberland Bay had external lesions and barbel deformities, abnormalities that were not seen in the reference population. Subsequent histopathological evaluation of sections of these livers showed several fish with regions of a type of abnormal cell growth often associated with exposure to contaminants. Of the nine fish examined, four showed ceroid-containing cells throughout the parenchyma; four showed bile duct proliferation; three exhibited altered cell foci considered to be preneoplastic; and one exhibited a cholangiocellular carcinoma. None of the fourteen fish examined from the reference site exhibited any of these abnormalities.

Both three and four year old rock bass from Burlington Harbor in 1992 had a significantly larger area of macrophage aggregates in their spleens than did fish from Malletts Bay (1992) and Sunset Lake (1994) (Table 1). Rock bass captured from Burlington Harbor in 1992 usually had a larger number of macrophage aggregates in their spleens than did fish from Malletts Bay (1992) and Sunset Lake (1994), but this difference was not statistically significant. Hepatosomatic index (HSI) also showed significant differences, with 1992 Burlington Harbor fish having the largest livers and Sunset Lake fish the smallest (Table 1). Our results also showed a rather dramatic difference in macrophage aggregate area between age 3 and age 4 fish in both contaminated and reference sites.

TABLE 2. Comparison of gonads between rock bass captured from the Inland Sea and Burlington Harbor in 1997. P values represent the results of a student's t test comparing the two sites.

	Inland Sea	Burlington Harbor	P =
Females			
N	9	13	
Age (yr)	4.1 ($\pm$ 1.6)	4.1 ($\pm$ 1.2)	0.95
Std length (cm)	14.7 ($\pm$ 2.3)	13.1 ($\pm$ 1.5)	0.07
Mean gonad wt (%)	5.7 ($\pm$ 1.9)	7.9 ($\pm$ 3.9)	0.12
Mean number of eggs	4058 ($\pm$ 3544)	3618 ($\pm$ 1954)	0.70
Mean egg diam (mm)	24.7 ($\pm$ 5.0)	27.3 ($\pm$ 8.4)	0.33
Males			
N	7	16	
Age (yr)	3.6 ($\pm$ 1.4)	4.7 ($\pm$ 1.0)	0.02
Std length (cm)	15 ($\pm$ 2.9)	14 ($\pm$ 1.8)	0.22
Mean gonad wt (%)	0.65 ($\pm$ 0.3)	0.77 ($\pm$ 0.4)	0.41

Both three and four year old rock bass from Burlington Harbor in 1997 had significantly smaller macrophage aggregate area than did Burlington Harbor fish from 1992 (Table 1). Macrophage aggregate area of three and four year old Burlington Harbor rock bass from 1997 was not significantly different than that of fish from reference sites. HSI of Burlington Harbor rock bass from 1997 was not different than HSI of rock bass from Malletts Bay 1992, but was larger than HSI of rock bass from Sunset Lake 1994.

There were no significant differences in gonad parameters between male and female rock bass in 1997 from Burlington Harbor and the Inland Sea (Table 2). Histological observations of male gonads showed no evidence of intersex development or feminization similar to that noted by Bortone and Davis [1994] and Sumpter and Jobling [1995].

DISCUSSION

Our results showed that in 1992, fishes from contaminated areas of Lake Champlain showed evidence of physiological stress. The results from our comparison of splenic macrophage aggregates support several other studies which also have shown that fishes in contaminated areas have more numerous and/or larger macrophage aggregate area than fishes in reference areas. These studies support the belief that macrophage aggregates might be useful as monitors of exposure of wild fishes to contaminants. Wolke et al. [1985] found that splenic macrophage aggregates in winter flounder from clean sites were significantly smaller and less numerous than was the case in winter flounder from polluted sites. There was no difference in the number of liver macrophage aggregates, but they were significantly larger in the flounder from the polluted sites. Macchi et al. [1992] found similar results when comparing splenic macrophage aggregates in whitemouth croaker. Thermally stressed female largemouth bass had more numerous and a larger percent area of both liver and spleen macrophage aggregates than non-stressed fish [Blazer et al. 1987]. Stressed males also had a significantly larger number of liver and splenic macrophage aggregates, but the difference in percent area was not great enough to be statisitically significant. Similar results were reported in roundnose grenadier [Lindesjoo et al. 1996] and white sucker [Couillard and Hodson 1996].

The reason for the apparent increase in macrophage aggregate area in physiologically stressed fish has not been completely explained. Macrophages are an important part of both the specific and nonspecific immune responses, and their production may become activated under stressful conditions. They are important in the phagocytosis of cellular debris, and may be needed to clean up cell and tissue damage resulting from environmental stressors. They may also be needed in higher numbers because some environmental toxins diminish their effectiveness. Macrophages extracted from estuarine fishes from areas polluted with high levels of PAHs showed generally decreased phagocytic and chemotactic abilities than macrophages taken from fishes in cleaner reference sites [Weeks et al. 1986; Seeley and Weeks-Perkins 1991].

Whatever the reason, there is considerable evidence that macrophage aggregates are good indicators of fish exposure to environmental contaminants. Our own results showed that fish collected in 1992 from the contaminated areas of inner Cumberland Bay and inner Burlington Harbor had more spleen tissue occupied by macrophage aggregates than did similar aged conspecifics from less contaminated reference sites. Our results for yellow perch, however, did not show significant differences among sites. This probably is due to the tendency of yellow perch to move around the open lake quite a bit, whereas rock bass tend to remain within a smaller area and rarely move away from structure, such as rock piles, and brown bullhead are constantly in contact with the substrate. Therefore, species specific behavior will tend to make some fishes more vulnerable to the effects of sediment contamination.

Our data also showed that rock bass captured in 1997 from Burlington Harbor were not significantly different than rock bass from reference sites, and that the area of spleen tissue displaced by macrophage aggregates was less than it had been in 1992. Although this is only one small study with a limited sample size, it does suggest that the environmental quality of inner Burlington Harbor may be improving. This is consistent with expectations based on the completion in 1994 of a substantial upgrade of the main Burlington sewage treatment plant, including an extension of the discharge pipe so that effluent is no longer released into the inner harbor. However, we must note that the rock bass captured in 1992 and 1997 were not caught in the same location within Burlington Harbor. The 1992 fish were caught near the southern end of the breakwater, which is closer to areas of higher sediment toxicity (off the mouth of the former sewage treatment discharge pipe). In 1997, due to crowding by recreational boaters, we captured fish about 700 meters north of this location. After the collections and data analyses were done, it was pointed out that this region has lower levels of sediment contamination. It is possible, therefore, that the difference in macrophage aggregate parameters was due to different levels of exposure due to location, and not due to improvement of habitat quality. A follow-up study of rock bass captured during the same year from different sites within Burlington Harbor would help answer this question.

The results of the hepatosomatic index comparisons also led us to similar conclusions. Rock bass from the most contaminated area (Burlington Harbor in 1992) had the largest livers. When fishes are exposed to some contaminants, particularly heavy metals and organic compounds, the liver must work harder to detoxify these compounds. This can lead to an increase in liver size due to hypertrophy, hyperplasia, or both [Goede and Barton 1990]. In some cases, prolonged exposure to contaminants can cause some liver cells to become damaged or die, which then may result in the regeneration of additional cells. If exposure is severe enough, centers of abnormal cells can develop within the liver tissue, some of which may lead to tumors [Hinton and Lauren 1990]. Several cases of preneoplastic altered foci within the liver parenchyma were seen in some of the brown bullhead from inner Cumberland Bay.

In 1997 we decided to look at the gonads of rock bass from Burlington Harbor because of concern about possible effects of PCBs, PAHs, and heavy metals on their reproduction. Other studies have shown that exposure to heavy metals or organic contaminants can interfere with proper sexual development, and presumably impact reproduction and recruitment [Howell et al. 1980; Bortone and Davis 1994; Munkittrick et al. 1992; Hontela et al. 1995; Sumpter and Jobling 1995]. The 1992-1994 phases of our study had shown that rock bass from Burlington Harbor exhibited evidence of physiological stress associated with exposure to environmental contaminants. If exposure to these toxins also affected the reproductive capacity of these fish, it is possible that rock bass populations could become affected. In 1997, however, there was no evidence that rock bass in Burlington Harbor were experiencing any detrimental effects on their reproductive system that might be caused by environmental contamination. The results of the gonad evaluations were consistent with the 1997 macrophage aggregate data that also showed no significant difference from reference populations. Perhaps if a study of reproductive systems had been done in 1992, when macrophage aggregate results showed definite signs of physiological stress, the results might have been different. However, our 1997 data indicate no differences in splenic macrophage aggregate area or gonad parameters between Burlington Harbor rock bass and rock bass from the reference site.

In conclusion, we have shown (1) that splenic macophage aggregate area is a good biomarker of exposure to environmental contamination in some fishes, and (2) that some fishes from contaminated areas of Lake Champlain did show evidence of physiological stress.

Acknowledgments. Special thanks to Madeleine Lyttle and the U. S. Fish and Wildlife Service (Essex Junction, Vermont) for the use of their electrofishing boat and to Kristen Bartlett and Darlene Bowling for many hours of excellent laboratory work. Funding was provided by the U. S. Geological Survey, Vermont EPSCoR, and St. Michael's College.

REFERENCES

Adams, S.M, *Biological indicators of stress in fish* American Fisheries Society Symposium 8. Bethesda, MD, 1990.

Blazer, V. S., D. E. Facey, J. W. Fournie, L. A. Courtney, and J. E. Summers, Macrophage aggregates as indicators of environmental stress, pp. 169-185 in J. S. Stolon and T. C. Fletcher, editors, *Modulators of Fish Immune Responses, Vol. 1*, SOS Publications, Fair Haven, N.J., 1994.

Blazer, V. S., R. E. Wolke, J. Brown, and C. A. Powell, Piscine macrophage aggregate parameters as health monitors: effects of age, sex, relative weight, season and site quality in largemouth bass (*Micropterus salmoides*), *Aquatic Toxicology*, 10: 199-215, 1987.

Bortone, S. A., and W. P. Davis, Fish Intersexuality as Indicators of Environmental Stress. *BioScience*, 44(3): 165-172, 1994.

Brown, C.L. and C.J. George, Age-dependent accumulation of macrophage aggregates in the yellow perch, *Perca flavescens* (Mitchill), *Journal of Fish Diseases* 8: 135-138, 1985.

Couch, J. A. and J. W. Fournie, *Pathobiology of marine and extuarine organisms*, CRC Press, Inc. Boca Raton, FL. 552 p., 1993.

Couillard, C. M., and P. V. Hodson, Pigmented macrophage aggregates: a toxic response in fish exposed to bleached-kraft mill effluent? *Environmental Toxicology and Chemistry* 15: 1844-1854, 1996.

Goede, R. W., and B. A. Barton, Organismic indices and an autopy-based assessment as indicators of health and condition of fish, *American Fisheries Society Symposium*, 8: 93-108, 1990.

Hinton, D. E., and D. J. Lauren, Liver structural alterations accompanying chronic toxicity in fishes: potential biomarkers of exposure, pp. 17-57 in J.F. McCarthy and L.R. Shugart, editors, *Biomarkers of Environmental Contamination*, Lewis Publishers, CRC Press, Inc., Boca Raton, FL, 1990.

Hontela, A., P. Dumont, D. Duclos, and R. Fortin, Endocrine and Metabolic Dysfunction in Yellow Perch, *Perca flavescens*, Exposed to Organic Contaminants and Heavy Metals in the St. Lawrence River, *Environmental Toxicology and Chemistry*, 14(4): 725-731, 1995.

Howell, W. M., A. D. Black, and T. Denton, Abnormal expression of secondary sex character in a population of mosquitofish, *Gambusia affinis holbrook*: evidence for environmentally induced masculinization, *Copeia*, 1980: 676-681, 1980.

Huggett, R. J., R. A. Kimerle, P. M. Mehrle, Jr., and H. L. Bergman, *Biomarkers: Biochemical, physiological, and histological markers of anthropogenic stress*, Society of Environmental Toxicology and Chemistry, Special Publication 6, Lewis Publishers, Ann Arbor, MI, 347 p., 1992.

Lindesjoo, E., A.-M. Husoy, I. Petterson, and L. Forlin, Histopathological and immunohistochemical studies in roundnose grenadier (*Coryphaenoides rupestris*) in the Skagerrak, North Sea *Marine Environmental Research*, 42: 229-233, 1996.

Macchi, G. J., L. A. Romano, and H. E. Christiansen, Melano-macrophage centres in whitemouth croaker, *Micropogonias furneri*, as biological indicators of environmental changes, *Journal of Fish Biology* 40: 971-973, 1992.

McCarthy, J.F., and L.R. Shugart, *Biomarkers of Environmental Contamination*, Lewis Publishers, CRC Press, Inc., Boca Raton, FL, 1990a.

McCarthy, J.F., and L.R. Shugart, Biological Markers of Environmental Contamination, pp.3-14 in J.F. McCarthy and L.R. Shugart, editor, *Biomarkers of Environmental Contamination* Lewis Publishers, CRC Press, Inc., Boca Raton, FL, 1990b.

Munkittrick, K. R., M. E. McMaster, C. B. Portt, G. J. Van der Kraak, I. R. Smith, and D. G. Dixon, Changes in maturity, plasma sex steroid levels, hepatic mixed-function oxygenase activity, and the presence of external lesions in lake whitefish exposed to bleached kraft mill effluent, *Canadian Journal of Fisheries and Aquatic Sciences*, 49: 1560-1569, 1992.

Peakall, D., *Animal Biomarkers as Pollution Indicators*, Chapman & Hall, New York, 1992.

Roitt, I.M., *Essential Immunology*, Oxford: Blackwell Scientific Publications, Boston, MA, 95-99, 1977.

Seeley, K. R., and B. A. Weeks-Perkins, Altered phagocytic activity of macrophages in oyster toadfish from a highly polluted subestuary, *Journal of Aquatic Animal Health*, 3: 224-227, 1991.

Sumpter, J. P., and S. Jobling, Vitellogenesis as a Biomarker for Estrogenic Contamination of the Aquatic Environment, *Environmental Health Perspectives*, 103 (7): 173-178, 1995.

Thaler, M.S., R.D. Klausner, & H.J. Cohen, *Medical Immunology*, J.B. Lippincott Company, Philadelphia, 1977.

Weeks, B. A., J. E. Warinner, P. L. Mason, and D. S. McGinnis, Influence of toxic chemicals on the chemotactic response of fish macrophages, *Journal of Fish Biology*, 28: 653-658, 1986.

Wolke, R.E., R.A. Murchelano, C.D. Dickstein, and C.J. George, Preliminary evaluation of the use of macrophage aggregates (MA) as fish health monitors, *Bulletin of Environmental Contamination and Toxicology*, 35:222-227, 1985.

High Rates of Brown-headed Cowbird Occurrence in Champlain Valley Forests: Conservation Implications for Migratory Songbirds

Steven D. Faccio and Christopher C. Rimmer

ABSTRACT

Using breeding bird census data collected during the Vermont Forest Bird Monitoring Program (FBMP) from 1989 - 1997, we evaluated the number of Brown-headed Cowbird (*Molothrus ater*) occurrences by study site and physiographic region. Data were collected at 17 study sites located in large tracts (≥40.5 ha) of mature, forested habitats in 6 different physiographic regions of the state. Most cowbirds (95.0%) occurred at the 5 FBMP sites located in the Champlain Lowlands, suggesting that even relatively large forested tracts in this region may be subject to high rates of cowbird parasitism. Other pressures associated with forest fragmentation may also be negatively affecting the breeding bird populations of the Champlain Basin, including high rates of nest predation, and reduced food supply. In addition, we hypothesize that the highly fragmented forests in this region may function as population sinks, with reproductive rates below the levels necessary to compensate for adult mortality. At the same time, the large interior forests of nearby areas may support source populations that produce enough surplus birds to sustain the losses of the sinks. Further research is needed to investigate this and to develop effective conservation measures.

INTRODUCTION

Small, isolated forest patches have a high amount of forest edge relative to interior area, resulting in an increase in "edge effects" [Temple 1986]. Numerous recent studies have demonstrated that edge effects reduce the quality of habitat for many Neotropical migrant songbirds, resulting in lower productivity [Faaborg et al., 1993; Donovan et al., 1995; Robinson et al., 1995; Burke and Nol 1998]. These negative effects, which may be contributing to the decline of some songbird populations, include high rates of nest predation [Wilcove, 1985; Small and Hunter, 1988; Yahner and Scott, 1988] and brood parasitism by Brown-headed Cowbirds (*Molothrus ater*) [Mayfield, 1977; Brittingham and Temple, 1983; Robinson et al., 1993], reduced pairing success [Gibbs and Faaborg, 1990; Villard et al., 1993] and low food abundance [Burke and Nol, 1998].

Brood parasitism by the Brown-headed Cowbird has become one of the major threats to breeding populations of Neotropical migrants [Robinson et al., 1993] for the following reasons: (1) female cowbirds are capable of laying more than 40 eggs per season on average; (2) cowbirds often remove one host egg from all parasitized nests; (3) cowbird eggs have a shorter incubation period than most host eggs, giving cowbird nestlings a head start; and (4) cowbird nestlings grow faster, beg more loudly and have larger gapes than host nestlings [Robinson et al., 1993].

In comparison to other edge effects, brood parasitism is a particularly severe problem since most migratory forest birds lack any compensating defenses against it. When nests are depredated, many songbirds re-nest in a new area, which over time may result in a systematic departure from areas of persistent nest predation [Robinson, 1997]. Similarly, female Ovenbirds (*Seiurus aurocapillus*) may avoid small fragments as breeding sites because they contain reduced food supplies [Burke and Nol, 1998]. When nests are parasitized by cowbirds however, host species behave as though their breeding efforts were successful, even if they fledged only cowbirds [Robinson, 1997].

Brown-headed Cowbirds spatially partition their breeding and feeding activities [Darley, 1982; Airola, 1986]. Cowbird feeding areas are located in a variety of open habitats, but are concentrated in areas associated with livestock [Darley, 1982; Airola, 1986]. Breeding areas are concentrated in host-rich zones such as clear-cuts [Verner and Ritter, 1983], riparian zones [Airola, 1986], and edges between forests and meadows [Gates and Gysel, 1978]. As a result, landscapes that support both breeding and feeding areas in close proximity are likely to have high densities of cowbirds and increased rates of parasitism throughout forest patches, while areas with low cowbird densities may result in low parasitism rates with no apparent edge effects [Hoover, 1992]. In the agricultural Midwest, parasitism rates have been reported at over 80% for many species [Robinson, 1997]. In Illinois forests, where high nest parasitism and predation rates have resulted in extremely low reproductive success for many migratory birds, Brawn and Robinson [1996] suggest that much of the state may be a population sink, producing virtually no surplus young. This lack of reproductive success is sustained by source populations presumably from large, contiguous forests of neighboring Midwestern states that have much lower rates of nest predation and parasitism [Brawn and Robinson, 1996].

It is not clear how far edge effects extend into a forest patch, but the distance probably varies with the regional landscape [Freemark et al., 1993]. While edge vegetation may extend less than 30 m into a forest [Wilcove et al., 1986; Saunders et al., 1991], studies of edge-related nest predation rates have ranged from 50 to 100 m [Gates and Gysel, 1978; Burger, 1988], to 600 m into a forest [Wilcove et al., 1986]. In landscapes dominated by agriculture, edge effects may occur at a much larger fragment size than in forest-dominated landscapes [Sabine et al., 1996].

In 1989 the Vermont Institute of Natural Science (VINS) initiated the Vermont Forest Bird Monitoring Program (FBMP). The program was designed to detect changes in breeding populations over time for all forest-dwelling songbirds, and to collect habitat-specific baseline data on species composition and relative abundance at undisturbed, interior forest sites in Vermont. Using data collected at 17 FBMP study sites throughout Vermont between 1989-1997, we evaluated the occurrences of Brown-headed Cowbird by study site and physiographic region. In this paper, we present evidence that one physiographic region, the Champlain Lowlands, may be subject to high rates of cowbird parasitism. In addition, we hypothesize that because of the highly fragmented landscape in this region, the forests may function as population sinks, with reproductive rates below the levels necessary to compensate for adult mortality.

METHODS

Site and Observer Selection

Several criteria were used to select FBMP study sites. Minimum requirements were that sites: 1) consist of mature, contiguous stands of homogeneous forest type, 2) be 40.5 ha or larger in order to avoid edge effects, and 3) be permanently protected from silvicultural practices, development, and other large-scale human-induced habitat changes. These criteria ensured that the study sites were located in relatively undisturbed, interior forest habitats that will remain relatively stable over the long-term, and in which bird populations should be little affected by the problems associated with forest fragmentation.

At each site, 5 point counts (stations) were established at least 100 m from the nearest forest edge and 200 m apart. Each station was clearly marked and labeled with survey flagging and an aluminum tree tag.

Volunteer observers were encouraged to participate in the project if they possessed a high degree of competency in both aural and visual bird identification, and could make a multi-year commitment.

Survey Methods

Survey methods consisted of unlimited distance point counts, based on the approach described by Blondel et al. [1981] and used in the Ontario FBMP [Welsh, 1995]. The count procedure was as follows:

1) Counts began shortly after dawn on days where weather conditions were unlikely to reduce count numbers (i.e., calm winds and very light or no rain). Censusing began shortly (<1min.) after arriving at a station.

2) Observers recorded all birds seen and heard during a 10-min sampling period, which was divided into 3 time intervals: 3, 2, and 5 mins. Observers noted in which time

interval each bird was first encountered and were careful to record individuals only once. To reduce duplicate records, individual birds were mapped on standardized field cards and known or presumed movements noted. Different symbols were used to record the status of birds encountered (i.e., singing male, pair observed, calling bird, etc.).

3) Each site was sampled twice during the breeding season; once during early June (ca. 2-12 June) and once during late June (ca. 14-25 June). Observers were encouraged to space their visits 7-10 days apart. All stations for each site visit were censused in a single morning.

4) Following each survey, field data were transcribed by the observer from the field cards onto data coding sheets. The level of breeding evidence determined whether a bird was assumed to indicate a pair or a single bird: Singing males, observed pairs, occupied nests, and family groups were considered to represent a pair and counted as 2 birds. All other individuals seen or heard were counted as singles.

In 1989, 17 study sites (85 stations) were established in 6 of Vermont's 7 physiographic regions and censused by VINS staff and volunteers (Figure 1). Six sites were located in northern hardwood forests, 6 were in transition hardwood forests dominated by oak (*Quercus sp*), hickory (*Carya sp*), maple (*Acer sp*), and American beech (*Fagus grandifolia*), 3 were in lowland spruce (*Picea sp*)-balsam fir (*Abies balsamea*) forests, and 2 were in montane red spruce (*Picea rubens*)-balsam fir forests.

In determining cowbird occurrence at each study site, we used the higher of the 2 values recorded for that species as the station estimate for each year.

RESULTS & DISCUSSION

To determine the extent to which Brown-headed Cowbirds penetrated into interior FBMP forest sites, we examined occurrence data for this species by physiographic region (Table 1). Ten of 17 study sites recorded at least 1 occurrence of Brown-headed Cowbird between 1989-1997, and the species was recorded on 28.5% of all FBMP site visits, with a total of 141 individual occurrences. While these numbers were higher than anticipated for this species, most cowbirds (95.0%) were found at 5 FBMP sites located in the Champlain Lowlands, a physiographic region of the state that is largely agricultural with a highly fragmented forest landscape (Figure 2).

Our results corroborate those of Coker and Capen [1995], who found that cowbirds in Vermont were most likely to occur in forested habitats if they were near large cleared areas (>9.8 ha) with high concentrations of livestock. Conversely, Coker and Capen [1995] found that small forest openings (<4 ha) located at least 7 km from livestock areas were unlikely to attract cowbirds. Airola [1986] demonstrated that distance from a cowbird feeding area was an indicator of the rate of cowbird parasitism; sites that were closer to feeding areas had higher parasitism rates.

Since the Champlain Valley is largely agricultural with a high density of dairy farms, we suspect that many bird species breeding at FBMP sites in this area may experience relatively high rates of parasitism by Brown-headed Cowbirds, while those at study sites in the other physiographic regions probably have much lower rates. Other pressures associated with forest fragmentation may also negatively affect breeding bird populations

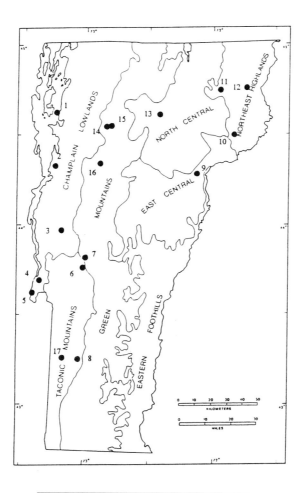

SITE #	SITE NAME	TOWN
1	Sandbar	Milton
2	Pease Mt.	Charlotte
3	Cornwall Swamp	Cornwall
4	Shaw Mt.	West Haven
5	Bald Mt.	West Haven
6	Sugar Hollow	Pittsford
7	The Cape	Chittenden
8	Dorset Bat Cave	East Dorset
9	Roy Mt.	Barnet
10	Concord Woods	Concord
11	May Pond	Barton
12	Moose Bog	Ferdinand
13	Bear Swamp	Wolcott
14	Underhill State Park	Underhill
15	Mt. Mansfield	Stowe
16	Camel's Hump	Huntington
17	Merck Forest	Rupert

Figure 1. Distribution of Vermont Forest Bird Monitoring sites by physiographic region, 1989-1997.

Table 1. Brown-headed Cowbird occurrences by FBMP site and physiographic region, 1989-1997.

Site Name	Physiographic Region	1989	1990	1991	1992	1993	1994	1995	1996	1997	Total
Bald Mt.	Champlain Lowlands	3	3	6	5	2	2	2	2	1	26
Pease Mt.	Champlain Lowlands	0	0	0	3	4	4	4	4	4	23
Cornwall Swamp	Champlain Lowlands	3	1	1	0	0	4	5	3	1	18
Sandbar	Champlain Lowlands	5	3	3	0	3	6	0	6	6	32
Shaw Mt.	Champlain Lowlands	3	9	9	9	5	0	0	0	no data	35
Bear Swamp	N. Central	no data	0	0	0	2	0	0	0	0	2
Roy Mt.	E. Central	1	0	0	0	0	0	0	0	0	1
Dorset Bat Cave	Taconics	0	0	1	0	0	0	0	0	0	1
Merck Forest	Taconics	0	0	0	0	0	1	0	0	0	1
Sugar Hollow	Taconics	0	1	0	0	0	0	0	0	1	2
	Total Count	15	17	20	17	16	17	11	15	13	141

of the Champlain Valley, including high rates of nest predation, low pairing success, and reduced food abundance.

Nest predation rates increase around forest edges due to a number of mammalian and avian predators which are more abundant in these ecotones. These include the Raccoon (*Procyon lotor*), Opossum (*Didelphis marsupialis*), Blue Jay (*Cyanocitta cristata*), and American Crow (*Corvus brachyrhynchos*) [Robbins, 1979; Whitcomb et al., 1981].

Studies of one Neotropical migrant, the Ovenbird (*Seirus aurocapillus*), have demonstrated that many males occupying forest fragments and edges are unmated [Gibbs and Faaborg, 1990; Burke and Nol, 1998]. This low pairing success has often been attributed to predator avoidance by females. However, Burke and Nol [1998] have suggested that female Ovenbirds avoid fragments because of a reduced abundance of soil-dwelling arthropods, their primary food, resulting from the drier, more dessicated leaf litter found in small forest tracts. It is possible however, that foliage-dwelling arthropods may be more abundant around edges where foliage is more dense, or that fragments in areas with rich soils may support suitable prey biomass and excellent foraging conditions. This may encourage some species to select edges and fragments making them susceptible to ecological traps.

As a result, we theorize that even the relatively large forest fragments of the Champlain Valley may act as population sinks for some migratory species, with reproductive rates that are below the levels necessary to compensate for adult mortality [e.g., see Brawn and Robinson, 1996; Brittingham and Temple, 1983]. At the same time, the large, contiguous forests of Vermont and the Adirondaks in New York, may support source populations which produce enough surplus birds to sustain the losses of the sinks.

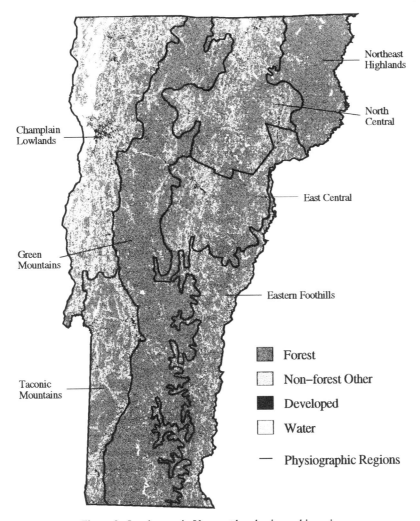

Figure 2. Land cover in Vermont by physiographic region.

FUTURE DIRECTIONS AND IMPLICATIONS

Further research is needed to determine if the forests of the Champlain Valley are population sinks, to what extent they might be impacting source populations, and what conservation measures, if any, would be most effective. Specifically, data are needed on the rates of productivity, nest predation, and brood parasitism, as well as on demographic traits such as adult survivorship, sex ratios, and pairing success. In order to identify source populations, data on dispersal, either from radio-tracking or recoveries of previously banded birds, are required. Dispersal data however, will be challenging to obtain due to extremely low band recovery rates and the potentially long dispersal distances of many Neotropical migrants. Currently, a University of Vermont graduate student is investigating the productivity and demographics of Black-throated Blue Warbler populations in Champlain Valley forest fragments [D. Capen pers. comm.]. This study may provide more detailed information regarding source/sink dynamics in the Champlain Basin.

If the forest fragments of the Champlain Valley are identified as population sinks, development of a regional conservation strategy that would benefit forest-dwelling birds and other organisms might be warranted. Such an approach should seek to maintain the source populations while minimizing the "drain" of the sinks by: (1) consolidating forested tracts into a few large reserves when possible; (2) allowing some adjacent forest patches to re-connect, forming core areas that are greater than 20 ha in size [Burke and Nol, 1998]; and (3) preventing further fragmentation of the largest woodlots in the Champlain Valley such as Snake Mt. and Cornwall Swamp.

Acknowledgments. We would like to thank the many organizations and agencies that have supported, assisted, and/or cooperated with the FBMP. They are: the Vermont Chapter of The Nature Conservancy, Vermont Fish and Wildlife Department, Vermont Department of Forests, Parks, and Recreation, the Center for Northern Studies, the University of Vermont, Merck Forest and Farmland, Green Mountain National Forest, and the Canadian Wildlife Service (CWS). Funding has been generously provided by the Merck Family Fund, Norcross Wildlife Foundation, the Sudbury Foundation, Vermont Forest Ecosystem Monitoring, and members and friends of VINS. Special thanks to Ernie Buford and the University of Vermont Spatial Analysis Lab for providing the map used in Figure 2. Finally, the FBMP would not have been possible without the efforts of the many skilled birders who monitored one or more FBMP sites during this study.

REFERENCES

Airola, D. A. 1986. Brown-headed Cowbird parasitism and habitat disturbance in the Sierra Nevada. *J. Wildl. Manage.* 50: 571-575.

Blondel, J., C. Ferry, and B. Frochot. 1981. Point counts with unlimited distance. Pp. 414-420, *In* Ralph, C. John; Scott, J. Michael (Eds.). *Estimating numbers of terrestrial birds. Studies in Avian Biology* 6: 630pp.

Brawn, J. D., and S. K. Robinson. 1996. Source-sink population dynamics may complicate the interpretation of long-term census data. *Ecology* 77: 3-12.

Brittingham, M. C., and S. A. Temple. 1983. Have cowbirds caused forest songbirds to decline? *BioScience* 33: 31-35.

Burger, L.D. 1988. Relations between forest and prairie fragmentation and depredation of artificial nests in Missouri. M.A. Thesis, University of Missouri-Columbia, Columbia, MO.

Burke, D.M., and E. Nol. 1998. Influence of food abundance, nest-site habitat, and forest fragmentation on breeding Ovenbirds. *The Auk* 115: 96-104.

Coker, D. R., and D. E. Capen. 1995. Landscape-level habitat use by Brown-headed Cowbirds in Vermont. *J. Wildl. Manage.* 59: 631-637.

Darley, J. A. 1982. Territoriality and mating behavior of the male Brown-headed Cowbird. *Condor* 84: 15-21.

Donovan, T. M., F. R. Thompson, III, J. Faaborg, and J. R. Probost. 1995. Reproductive success of migratory birds in habitat sources and sinks. *Conservation Biology* 9: 1380-1395.

Faaborg, J., M. Brittingham, T. Donovan, and J. Blake. 1993. Habitat fragmentation in the Temperate Zone: a perspective for managers. Pp. 331-338, *In* D. M. Finch and P. W. Stangel, (Eds.). Status and management of Neotropical migratory birds. General technical report RM-229,. U.S. Forest Service, Rocky Mountain Forest and Range Experiment Station, Fort Collins, Colorado. 422pp.

Freemark, K., J. Probst, J. B. Dunning, and S. Hejl. 1993. Adding a landscape ecology perspective to conservation and management planning. Pp. 346-352, *In* D. M. Finch and P. W. Stangel, (Eds.). Status and management of Neotropical migratory birds. General technical report RM-229,. U.S. Forest Service, Rocky Mountain Forest and Range Experiment Station, Fort Collins, Colorado. 422pp.

Gates, J. E., and L. W. Gysel. 1978. Avian nest dispersion and fledging success in field-forest ecotones. *Ecology* 59:871-883.

Gibbs, J. P., and J. Faaborg 1990. Estimating the viability of Ovenbird and Kentucky Warbler populations in forest fragments. *Conservation Biology* 4: 193-196.

Hoover, J. P. 1992. Factors influencing Wood Thrush nesting success in a fragmented forest. M. S. Thesis, The Pennsylvania State University, University Park, PA.

Mayfield, H. F. 1977. Brown-headed Cowbird: Agent of extermination? *American Birds* 31: 107-113.

Robbins, C. S. 1979. Effects of forest fragmentation on bird populations. Pp. 198-212, *In* R. M. DeGraaf and K. E. Evans, (Eds.). Management of north-central and northeastern forests for nongame birds. U.S. Department of Agriculture ForestService, General Technical Report, NC-51.

Robinson, S. K., J. A. Grybowski, S. I. Rothstein, M. C. Brittingham, L. J. Petit, and F. R. Thompson III. 1993. Management implications of cowbird parasitism for Neotropical migrant songbirds. Pp. 93-102, *In* D. M. Finch and P. W. Stangel, (Eds.). Status and management of Neotropical migratory birds. General technical report RM-229,. U.S. ForestService, Rocky Mountain Forest and Range Experiment Station, Fort Collins, Colorado. 422pp.

Robinson, S. K., F. R. Thompson, III, T. M. Donovan, D. Whitehead, and J. Faaborg. 1995. Regional forest fragmentation and the nesting success of migratory birds. *Science* 267: 1987-1990.

Robinson, S. K. 1997. The case of the missing songbirds. *Consequences* 3(1).

Sabine, D. L., A. H. Boer, and W. B. Ballard. 1996. Impacts of habitat fragmentation on pairing success of male Ovenbirds in southern New Brunswick. *Canadian Field-Naturalist* 110: 688-693.

Saunders, D. A., R. J. Hobbs, and C. R. Margules. 1991. Biological consequences of ecosystem fragmentation: a review. *Conservation Biology* 5: 18-32.

Small, M. F., and M. L. Hunter. 1988. Forest Fragmentation and avian nest predation in forested landscapes. *Oecologia* 76: 62-64.

Temple, S. A. 1986. Predicting impacts of habitat fragmentation on forest birds: A comparison of two methods. Pp. 301-304 *In* Verner, J., M. L. Morrison, and C. J. Ralph, (Eds.). *Wildlife 2000: Modelling habitat relationships of terrestrial vertebrates.* Univ. of Wisconsin Press, Madison.

Verner, J., and L. V. Ritter. 1983. Current status of the Brown-headed Cowbird in the Sierra National Forest. *The Auk* 100: 355-368.

Villard, M.-A., P. R. Martin, and C. G. Drummond. 1993. Habitat fragmentation and pairing success in the Ovenbird. *The Auk* 110: 759-768.

Welsh, D. A. 1995. An overview of the Forest Bird Monitoring Program in Ontario, Canada. Pp. 93-97, *In* C. J. Ralph, J. R. Sauer, and S. Droege, (Eds.). Monitoring bird populations by point counts. General technical report PSW-GTR-149. Pacific Southwest Research Station, Forest Service, U.S. Dept. of Agriculture, Albany, California. 181pp.

Whitcomb, R. F., C. S. Robbins, J. F. Lynch, B. L. Whitcomb, K. Klimkiewicz, and D. Bystrak. 1981. Effects of forest fragmentation on avifauna of the eastern deciduous forest. Pp. 125-205, *In* R.L. Burgess and D. M. Sharpe, (Eds.). *Forest island dynamics in man-dominated landscapes.* New York: Springer-Verlag.

Wilcove, D. S. 1985. Nest predation in forest tracts and the decline of migratory songbirds. *Ecology* 66: 1211-1314.

Wilcove, D.S., C.H. McClellan, and A.P. Dobson. 1986. Habitat fragmentation in the temperate zone. Pp. 237-256, *In* M.E. Soule, (Ed.). *Conservation biology: the science of scarcity and diversity.* Sinaur Associates, Sunderland, MA.

Yahner, R., and D. P. Scott. 1988. Effects of forest fragmentation on depradation of artificial nests. *J. Wildl. Manage.* 52: 158-161.

Research Management of the Common Tern on Lake Champlain, 1987-1997: A Case Study

Mark S. LaBarr and Christopher C. Rimmer

ABSTRACT

The common tern (*Sterna hirundo*) is a colonial nesting waterbird that nests annually on Lake Champlain, Vermont. Common terns were listed as a Vermont endangered species in 1989 due to declining population levels. This decline was due to decreasing numbers of adult breeders and extremely low reproductive success during the 1970's and 1980's. Reasons for the decline included the direct and indirect effects of nocturnal avian predation, inter-specific competition for nesting space with ring-billed gulls (*Larus delawarensis*), and human disturbance. Intensive monitoring and management began in 1987 and resulted in increased numbers of breeding pairs and higher productivity. Chick shelters, control of ring-billed gull populations, a bait/barrier system against ants, warning-sign buoys, and lake-user education successfully reduced predation, over-crowding, and human disturbance. Population levels increased from a low of about 50 breeding pairs in 1988 to approximately 170 breeding pairs in 1997. Annual productivity increased from 0.3 chicks/pair to 0.7 chicks/pair during the same period. Studies of color-banded adult terns and chicks conducted between 1987-1996 revealed high site fidelity of adult breeders, recruitment of locally-fledged chicks into the breeding population, and immigration of terns from sites as distant as the Connecticut coast and Lake Ontario.

INTRODUCTION

Nesting common terns *(Sterna hirundo)* have been documented on 1-6 small rocky islands (< 0.5 hectares each) in northern Lake Champlain, Vermont since at least 1892

[Chapman 1904]. Although historical information is incomplete, surveys conducted in the late 1960's indicated that Lake Champlain supported approximately 300-400 pairs of breeding common terns [Spear 1966]. Since then, breeding terns on Lake Champlain have experienced problems similar to those in other colonies [e.g., Morris and Hunter 1976, Nisbet 1975, Courtney and Blokpoel 1983, Nisbet and Welton 1984]. Reduced numbers of adult breeders and poor reproductive success during the 1970's and 1980's threatened the viability of this species on Lake Champlain.

Lake Champlain's common tern population has been monitored annually since 1980. Survey data between 1980-1988 showed a steady decline in the number of breeding adults. As a result, common terns were legally designated a Vermont state endangered species in 1989. Predation by great horned owls *(Bubo virginianus)* was suspected of causing large numbers of nest failures and resulting in regular nocturnal desertions of incubating terns. Predation by black-crowned night-herons *(Nycticorax nycticorax)* and tiny thief ants *(Solenopsis molesta)*, competition for nesting space with ring-billed gulls *(Larus delawarensis)*, and human disturbance also contributed to this population's chronic low breeding success (LaBarr and Rimmer 1994). Intensive monitoring and management of this population began in 1987 in an attempt to reverse the continuing decline. Chick shelters, control of ring-billed gull populations, warning-sign buoys, and a bait/barrier system against ants were used to reduce predation, over-crowding, and human disturbance. This paper discusses the reasons for the decline, the current status of the population and monitoring and management techniques that have been employed since 1987.

METHODS

Population Monitoring

Routine surveys of common terns on Lake Champlain were conducted between 1987-1997 on six islands; Hen, Gull Rock, Popasquash, Grammas, Rock and Savage. All six islands are located east of Grand Isle and North Hero, Vermont. Islands with active common tern colonies were censused every two to six days. Islands with no tern activity were censused less regularly. Surveys were conducted between late April and early September in all years. Islands were visited less frequently between 1980-1986 (three to four times a season) and not at all in 1982. Although this data is not as complete as in later years (i.e. fewer surveys), we believe it still provides a relatively accurate assessment of adult numbers and reproductive success for Lake Champlain common terns.

Adult terns were censused from the boat, and nests and chicks on foot, carefully minimizing length of colony visits and degree of disturbance. Colonies were entered at two to six day intervals and only during favorable weather conditions. Landing parties included biologists and trained volunteers. In 1987-1997, nests were individually marked, and nest locations were plotted on colony maps. Data collected during colony visits included numbers of adults and chicks, number and status of nests, evidence of predation or other disturbance, extent of inter- and intra-specific behavioral interactions, and numbers of ring-billed gull adults, nests and chicks. Behavioral watches at various times of day were conducted from the boat for all islands and from a blind on Popasquash Island.

We conducted 116 dusk watches between 1987-1996 at active colonies to monitor the extent and effects of nocturnal desertion. Watches began approximately 1.5 hours before sunset and ended about 1 hour after sunset. Some of these watches were followed by pre-dawn watches to determine whether birds remained in the colony and, if not, at what time and from what direction they returned.

Management

We employed various management techniques to reduce predation, crowding by gulls, and human disturbance. Signed buoys were placed around active tern colonies at an average distance of 15 m, to mark the island and immediate lake area as "restricted". These consisted of two metal signs mounted on buoys or inflated tires, anchored to the lake bottom. Buoys notified boaters that the islands were restricted and that disturbances of nesting terns were punishable by a fine under state law. We removed buoys following the completion of tern breeding activities.

Shelters to protect chicks from predation were constructed from natural materials on the nesting islands (i.e., rocks, driftwood) or from lumber, plywood and/or cinder blocks. The majority of shelters were patterned after the "teepee" design described by Burness and Morris [1991]. Chick shelters were placed 10-25 cm from nests just prior to hatching or next to nests with newly hatched chicks.

We controlled the ring-billed gull breeding populations on Popasquash Island in 1987-1997 and on Rock Island in 1987-1994. We selectively destroyed gull nests from within and around tern nesting areas. Nests were removed if discovered within a tern nesting area or if their proximity to nests provoked aggressive interspecific behavior by nesting terns. Gull eggs were removed from the island and nest material was scattered.

An ant-specific bait system was combined with physical barriers on Popasquash Island in 1989-1997 to reduce predation of tern chicks. The bait system consisted of household ant traps filled with "Drax Ant Kill Gel " (a 5% boric acid based bait) and placed within tern nesting areas. Traps were deployed from late May through July.

The barrier system consisted of circles of plastic vegetation matting approximately 13 cm in diameter placed beneath selected tern nests prior to hatching. The barriers were permeable to water but prevented ants from accessing chicks from below ground.

We controlled encroaching late season vegetation by physically removing, by hand, vegetation around active tern nests. The growth of dense vegetation around and above the nest restricted the access of incubating terns to their eggs. Vegetation removal allowed terns to access nests that would have otherwise been impenetrable to incubating adults.

Education

Lake users and the general public were educated about common tern biology and conservation throughout each field season. Conversations with anglers and pleasure boaters during tern watches, as well as newspaper/newsletter articles, informational posters, and slide lectures, were used to convey information to the public.

Banding

A long-term banding project was initiated in 1987 to monitor reproductive success, site fidelity, recruitment, and immigration. Incubating adult terns were captured in a drop trap triggered from a blind and in Potter traps. Adults received a U.S. Fish and Wildlife

TABLE 1. Peak adult and nest counts, total numbers of nests and breeding population estimates of Common Terns in Vermont, 1987-97.

	Peak Adult Numbers	Peak Nest Numbers	Total Nest Numbers	Estimated Number of Breeding Pairs
1987	135	53	168	55
1988	115	48	128	50
1989	105	49	108	50
1990	130	61	115	65
1991	130	64	143	65
1992	140	76	139	80
1993	180	91	226	120
1994	195	109	216	130
1995	210	127	267	150
1996	200	109	233	135
1997	220	158	258	170

Service (USFWS) stainless steel band and a unique combination of 3 Darvic colored leg bands. We marked all banded adults with picric acid (a harmless temporary dye) in a pattern specific each colony. Chicks received a USFWS stainless steel band and a single yellow Darvic colored leg band.

RESULTS

Population levels and breeding success

The number of common terns observed and the number of nesting pairs on Lake Champlain increased steadily between 1989 and 1997 (Table 1). Although the use of individual nesting islands fluctuated from year to year, Popasquash Island supported the highest number of terns in all years (Table 2). A drop in adult numbers was observed in 1996. We suspect this was due to in part to roosting double-crested cormorants *(Phalacrocorax auritus)* on Popasquash Island in May of that year. This was the first and only year in which cormorants roosted on Popasquash Island in large numbers (>100). We believe the roosting cormorants displaced terns from traditional nesting areas, resulting in the decline in adult numbers.

Overall breeding success (number of chicks fledged) and productivity (chicks fledged/pair) also increased between 1989 and 1997 (Fig. 1, Fig.2). The number of pairs

TABLE 2. Peak nest numbers, estimated breeding pairs, hatching success (percentage of nests hatching at least one egg) and number of chicks fledged by Common Terns on Vermont tern nesting islands, 1987-1997.

	1987	1988	1989	1990	1991	1992	1993	1994	1995	1996	1997
Popasquash I.											
Peak Nest Numbers	34	24	36	52	58	72	86	96	127	91	154
Estimated Pairs	35	30	40	55	60	75	90	110	150	130	160
Rock I.											
Peak Nest Numbers	20	23	13	13	7	5	3	7	0	0	13
Estimated Pairs	25	25	15	15	10	6	7	8	0	0	13
Grammas I.											
Peak Nest Numbers	0	2	0	0	0	0	20	5	0	5	0
Estimated Pairs	0	2	0	0	0	0	20	6	0	5	0
Hen I.											
Peak Nest Numbers	15	13	18	5	0	11	7	0	0	12	0
Estimated Pairs	15	15	20	6	0	11	10	0	0	12	0
Savage I.											
Peak Nest Numbers	0	0	0	0	15	1	2	0	0	0	0
Estimated Pairs	0	0	0	0	47	0	0	0	0	0	0
Gull Rock I.											
Peak Nest Numbers	4	11	0	0	0	0	0	0	0	0	0
Estimated Pairs	4	15	0	0	0	0	0	0	0	0	0

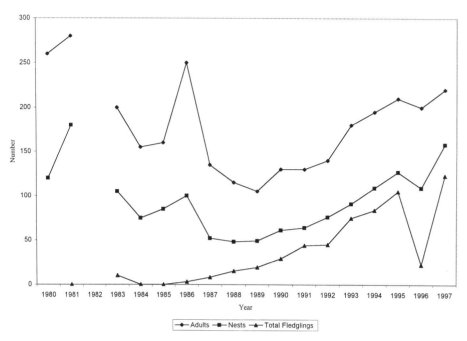

Figure 1. Population levels and reproductive success (peak counts) of common terns on Lake Champlain, 1980-1997.

successfully fledging young and the number of fledglings rose steadily in all years except 1996. Again we believe the decline in 1996 was due to the impacts of inter-species competition with cormorants coupled with increased predation. Higher adult numbers, productivity and reproductive output were observed in 1997 when cormorants were present in lower numbers on Popasquash Island and predation rates declined.

Causes of Breeding Failure

Nocturnal avian predation caused nocturnal desertions and most nest failures between 1988-1997 (LaBarr and Rimmer 1997). Great horned owls visited islands with breeding terns in each year, as evidenced by the presence of contour feathers, owl pellets, and fresh prey remains. Remains of ring-billed gull adults and juveniles were located at each island in at least one year and on Popasquash Island in every year. Gulls appeared to have been taken in a manner described by Nisbet [1975] for great horned owls. We observed physical evidence of owls taking adult terns on 2 occasions, on Popasquash Island in 1994 and on Grammas Island in 1996.

Black-crowned night herons were observed on numerous occasions on Popasquash, Rock and Hen islands. Black-crowned night herons have been documented taking eggs and chicks in other tern colonies (Nisbet and Welton 1984, Morris et.al.1991). We were unable to obtain direct evidence of predation by night herons; however, their presence at the nesting islands in all years suggests that they were responsible for both egg and chick

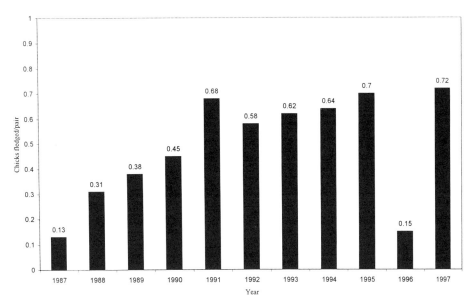

Figure 2. Lake Champlain common tern productivity, 1987-1997.

predation. Although younger chicks may have been predated by both owls and Black-crowned Night Herons, disappearances of older tern and gull chicks probably resulted solely from owl predation.

Nocturnal desertion of terns and gulls from the nesting islands was documented during 104 of the 116 (89.7%) dusk watches conducted during 1988-1996. Although ring-billed gulls remained down on a small proportion of the monitored nights, we suspect most terns abandoned their nests. This was based on the observed exodus of large numbers of terns at dusk in each year and extended incubation lengths of up to 35 days. Nisbet (1975) showed a direct relationship between nocturnal abandonment of nests and increased incubation lengths beyond the normal 21-23 days documented for common terns.

Tiny thief ants were first documented on Popasquash Island in 1988 and were responsible for chick mortality at this site in all years. Ants attacked newly hatched chicks from below, swarming them until dead. Attacks by ants resulted in both individual chick death and the failure of entire clutches. Adult terns incubated chicks throughout the process.

Management

Chick shelters limited predation in all years. Wooden and rock shelters provided chicks with important means of concealment throughout the season. This was especially critical for early hatching nests, which produced chicks during periods of limited natural vegetative shelter. Although an exact number is not known, a large proportion of the chicks that fledged between 1988-1997 were observed using shelters.

The protein-based bait/barrier system successfully limited ant predation on Popasquash Island. Ant predation dropped from 18 nests and 34 chicks in 1988 to 1 nest and 2 chicks

TABLE 3. Peak number of Ring-billed Gull nests on Vermont tern nesting islands, 1987-1997.

	Peak Nest Numbers										
	1987	1988	1989	1990	1991	1992	1993	1994	1995	1996	1997
Popasquash I.	116	190	189	220	180	271	41	34	316	156	203
Rock I.	107	108	3	20	88	34	0	42	0	13	0
Totals	223	298	192	240	268	305	41	76	316	169	203

in 1997. We believe the boric acid-based bait was the primary factor responsible for this decline. Although barriers physically limit access of ants to tern chicks, they were often displaced or compromised by vegetation.

The control of vegetation was effective at providing access to some nests by adult terns when vegetation encroached on the nest area during incubation. Although we obtained no quantitative data, regular removal of vegetation allowed access by adults to a number of nests, in each year, that produced young. Nests, however, where encroaching vegetation was not controlled due to logistical constraints (i.e. time) were usually abandoned [LaBarr and Rimmer pers. obs.].

The combination of buoys and lake user education limited human disturbance at colony sites. Anglers comprised the majority of boaters around the nesting islands. Most of the boats observed near the islands remained outside the restricted areas, rarely disturbing the nesting terns. Those that did pass within the buoys left the area after being informed about common terns and the reason for the island closures. Compliance increased as the buoys became a fixture around the islands during progressive summers.

Ring-billed Gull Management

The number of ring-billed gulls nesting on Popasquash and Rock islands fluctuated from year to year (Table 3). In years where the number of nesting gulls was high, terns had limited access to traditional nesting sites as gull nesting expanded into tern nesting areas. Removal of gull eggs and nests early in the nesting season made these traditional sites available for terns. In years when gull nesting was low (i.e. 1993-1994) terns were not affected by nesting gulls. The number of nests removed in any given year ranged from 40 nests to about 150.

Banding

A total of 167 incubating terns were trapped, color-banded and marked in the Lake Champlain colonies between 1987-1996. The adult banding program was discontinued in 1997. Twenty-six adult terns originally banded as chicks in the Lake Champlain colonies (recruits) were retrapped between 1990-96, representing an average annual recruitment rate of 12%. Eighteen adult terns originally banded in colonies outside Lake Champlain were recaptured between 1990-1996. Of these, 5 were banded on Lake Oneida, 2 on the St. Lawrence River, 1 on Lake Ontario and 1 at Faulkner Island, Connecticut. One hundred and thirty (78.8%) of the 165 adult terns banded between 1987 and 1995 were observed nesting on Lake Champlain during at least 2 breeding seasons, while 106 (74.6% of 142 possible) were observed nesting during at least 3 breeding seasons. Seventeen of the 22 (80.9%) recruits retrapped between 1990-1995 were observed breeding on Lake Champlain in at least one subsequent year. Eleven (73.3%) of the 15 immigrant terns banded between 1990-1995 have also nested on Lake Champlain in one subsequent year. These data indicate high adult, natal, and immigrant site fidelity to the Lake Champlain nesting colonies.

DISCUSSION

Morris et al. [1991] suggested that sustained management efforts may be required to prevent extirpation of declining common tern populations on the Great Lakes. Increases in the breeding population and reproductive success of Lake Champlain common terns since 1987 suggest that management has had positive effects. Management techniques have increased chick survival, maintained available nesting space, and limited human disturbance. Chick shelters, control of ring-billed gull nesting populations and warning sign buoys have had the greatest impacts. However, control of tiny thief ant populations and continued efforts to educate lake users have also contributed to recent gains in productivity.

Productivity for this population still falls below the 1.1 fledglings/pair suggested by Nisbet [1978] and DiConstanzo [1980] as necessary for a self-sustaining population. Although this seems problematic, as young birds are needed to replace older adult breeders, we believe that immigration offsets this population's sub-optimal productivity. The combination of higher reproductive output, immigration of breeding terns into this population, and high natal and adult site fidelity suggest that this population should continue to grow if management strategies are maintained at their current levels.

The decline in adult numbers and reproductive success in 1996 indicates that Lake Champlain common terns are still highly susceptible to the adverse effects of overcrowding and predation. The growing number of double-crested cormorants on Lake Champlain [Fowle et. al., in review] and their potential impacts on Common Tern nesting islands [LaBarr and Rimmer 1996] suggests that cormorants could be a problem for nesting terns in the future. Of immediate concern is the loss of Popasquash Island to nesting terns due to competition for space with cormorants. This island supports the largest number of breeding terns and has the highest reproductive success of any of the Lake Champlain colonies. It is uncertain whether management techniques that are effective in reducing gull populations (i.e. removal of nests and eggs) will be as effective with cormorants.

Shifts in current predation patterns of owls and night herons could also negatively affect tern productivity. Increases in predation rates of eggs, chicks or both could result in a significant drop in productivity. This was observed in 1996 when predation of tern chicks increased. Although sporadic increases in annual predation rates may be offset by higher productivity in other years (i.e. 1997), sustained increases could lead to a long-term decline in adult numbers.

Common tern conservation efforts on Lake Champlain produced a steady increase in adult numbers and an accompanying rise in productivity between 1987-1997. However, 1996 brought a return to the extremely low reproductive success and population declines of the 1970's and early 1980's. Continued increases in the adult breeding population and its productivity, combined with enhanced efforts to control predation, overcrowding and to protect critical nesting sites will be necessary to maintain the viability of the Lake Champlain common tern colonies.

Acknowledgments: Funding for this project was provided by the Vermont Fish and Wildlife Department's Nongame and Natural Heritage Program, through the Nongame Wildlife Fund, the Lake Champlain Citizens Advisory Committees Partnership Program, The H.E. Thompson Foundation and several private contributors. We thank The Frank M. Chapman Memorial Fund and the Eastern Bird Banding Association for support of our research efforts. The Vermont Department of Forests, Parks and Recreation deserves special thanks for its logistical support. We thank the Green Mountain Audubon Society, The Nature Conservancy, the Lake Champlain Land Trust and the Vermont State Police for their efforts to protect tern nesting islands. We appreciate field assistance from Vermont Institute of Natural Science staff, Green Mountain Audubon Society Conservation Interns, and our many volunteers.

REFERENCES

Burness, G.P. and R.D. Morris. 1991. *Shelters decrease avian predation on chicks at a Common Tern colony.* J. Field Ornithology 62(2): 186-189.

Chapman, F.M. 1904. *Handbook of the birds of eastern North America. (6th ed.).* D. Appleton, New York, NY.

Courtney, P. and H. Blokpoel. 1983. *Distribution and numbers of Common Terns on the lower Great Lakes during 1900-80: a review.* Colonial Waterbirds 6:107-120.

DiConstanzo, J. 1980. *Population dynamics of a Common Tern colony.* Journal of Field Ornithology 51:229-243.

Fowle, M. R., D.E. Capen and N.J. Buckley. In review. *Population growth and dynamics of double-crested cormorants in Lake Champlain.* Northeast Wildlife.

LaBarr, M.S. and C.C. Rimmer. 1996. *The 1996 breeding status of Common Terns on Lake Champlain.* Unpubl. report, Vermont Institute of Natural Science, Woodstock, VT.

Morris, R.D. and R.A. Hunter. 1976. *Factors influencing desertion of colony sites by Common Terns.* Can. Field Nat. 90:137-143.

Morris, R.D., H. Blokpoel and G.D. Tessier. 1991. *Management efforts for the conservation of Common Tern colonies: Two case histories.* Biol. Cons. 60:7-14.

Nisbet, I.C.T. 1975 *Selective effects of predation in a tern colony.* Condor 77:221-226.

Nisbet, I.C.T. 1978. *Population models for Common Terns in Massachusetts: Present numbers and historic changes.* Bird-Banding 44:241-245.

Nisbet, I.C.T., and W.J. Welton. 1984. *Seasonal variations in breeding success of Common Terns: consequences of predation.* Condor 86:53-60.

Spear, R.N. 1966. *History of the common tern in Vermont to June 1966.* Report to Vermont Fish and Wildlife Dept., Waterbury, VT.

Lake Champlain Basin Education and Outreach Programs

Thomas R. Hudspeth and Patricia Straughan

ABSTRACT

A much richer array of education and outreach programs related to the Lake Champlain Basin exists in 1998 than in 1989. One of the major players has been the Champlain Basin Education Initiative, with its summer institutes and one-day workshops and the work of its various partners. Another major player has been the University of Vermont, with its "place-based" environmental education programs linking university students with Burlington middle schools and its summer courses for middle and elementary school teachers and associated academic summer camp programs for Burlington youth. These efforts are similar in that they: target K-12 teachers and their students; depend on grants; emphasize partnerships; are interdisciplinary, hands-on, field-based; and offer activities aligned with school standards in the two states in the Basin. However, much more needs to be done, including: greater coordination between educators and researchers; greater coordination among water quality monitoring programs; and rigorously evaluating existing education efforts. The advent of the Sea Grant Program in Vermont offers promise for increased attention to education and outreach efforts in the Lake Champlain Basin.

INTRODUCTION

For a Lake Champlain Symposium at Basin Harbor Club in 1989, Vermont Agency of Natural Resources compiled "Lake Champlain Educational Resources." This pamphlet listed primarily slide programs, pamphlets and other publications, exhibits, self-guided nature trails, and venues for field trips offered by non-governmental organizations and government agencies in the Basin. The sole academic curriculum listed was the *Human Impact on the Lake Champlain Basin* program at Colchester High School. It was a year-long course taught by science, social science, English, and computer teachers. The students went on field trips, heard presentations by guest speakers, and worked on projects which they presented at a fair at the end of the school year to which parents, community members, guest speakers, and others were invited.

While excellent education and outreach programs and curriculum materials existed for Chesapeake Bay, the Great Lakes, and other basins in the United States in 1989, relatively little was available for the Lake Champlain Basin at that time. However, in the 1990's efforts on a variety of fronts have resulted in a much richer array of education and outreach programs related to the Lake Champlain Basin. Most of these developments have taken place since Vermont, New York, and Quebec signed a Memorandum of Understanding in 1988 and since the Lake Champlain Special Designation Act was signed into law in 1990, although some build on previous research [Hudspeth, 1988a; Hudspeth, 1988b]. This paper describes these efforts and provides recommendations for building on these initiatives in the future.

CHAMPLAIN BASIN EDUCATION INITIATIVE

The Champlain Basin Education Initiative (CBEI) was formalized in 1992 with funding from the Lake Champlain Basin Program to foster cooperation among environmental education organizations in the Basin. The original partners in the consortium were:
- Shelburne Farms (VT)
- Lake Champlain Basin Program
- Adirondack Visitor Interpretive Center (NY)
- Adirondack Teacher Center (NY)
- Center for Educational Leadership, McGill University (QU)

New partners added in 1997 were:
- Green Mountain Audubon Nature Center (VT)
- Lake Champlain Basin Science Center (VT)
- National Wildlife Federation
- Amy Demarest, author of *This Lake Alive!: An Interdisciplinary Handbook for Teaching and Learning about the Lake Champlain Basin* [1997].

Each of the individual partners in CBEI offers its own programs for school groups and the general public, including activities related to Lake Champlain. By combining forces in CBEI, they have been able to provide richer, more varied offerings.

The goal of CBEI is to equip middle-level teachers with the skills necessary to educate peers and students to be wise decision-makers in the Basin. In 1992-93, a series of one-day workshops were held at various venues in the Basin. They combined educational activities with presentations by resource specialists, and included an assessment of teachers' needs for education programs and materials related to Lake Champlain. CBEI used the results of the needs assessment and focus groups together with the priorities of the Lake Champlain Management Conference and the Lake Champlain Basin Program to develop and offer week-long Educational Leadership Institutes at Basin Harbor Club in the summers of 1993, 1994, and 1995. These institutes constituted a total immersion in lake studies, offering hands-on field experiences with experts (water quality monitoring in rivers and the lake, geological studies, cultural history and drama presentations and song-writing, etc.); a resource fair with more experts, government agencies, and businesses; technology training; networking opportunities; and assistance and funding possibilities for teachers to develop their own programs. Over 90 teachers attended. In turn, these teachers reached over 2000 students in their classes. Each teacher brought a school administrator and local community member to sessions on fundraising and showcasing Lake-related programs; this strategy was aimed at helping to gain local support for the teachers when they developed their own model programs.

Since 1996 CBEI has had decreased funding available to offer the week-long summer institutes and so has returned to a series of one-day workshops held at various venues in the Basin, focusing on such topics as Monsters in Lake Champlain (toxics and invasive species), Legends and Lore of the Lake (geology, history, and drama), and Four Lakes in One (seasonal changes). *This Lake Alive!* [1997] has been disseminated to the 90 teachers who have participated in these workshops. More workshops are planned for 1998-99. Also, CBEI co-sponsored the Framing the Basin Institute for middle school teachers in June, 1998, following the summer institute model and incorporating standards-based curriculum development.

"PLACE-BASED" ENVIRONMENTAL EDUCATION PROGRAMS LINKING UNIVERSITY OF VERMONT (UVM) STUDENTS WITH AREA MIDDLE SCHOOLS AND UTILIZING THE LAKE CHAMPLAIN BASIN AS AN INTEGRATING THEME IN SCIENCE, SOCIAL STUDIES, LANGUAGE ARTS, AND MATHEMATICS

With funding from the Geraldine Dodge Foundation and The Orion Society, the 22 students in Tom Hudspeth's Environmental Education class at the University of Vermont (UVM) engaged "Watershed Partnerships" with two teachers in the sixth grade Endeavor Team and with four teachers in the seventh and eighth grade Roots and Branches Team at Edmunds Middle School during the fall semester of 1995 [Hudspeth, 1996a]. And the 26 students in his class engaged in "Watershed Partnerships" with two teachers at Edmunds Middle School in Burlington, two teachers at Hunt Middle School in Burlington, one two-teacher team at Burlington High School, and one two-teacher team at South Burlington High School during the fall semester of 1996 [Hudspeth, 1997]. The UVM students developed and field-tested interdisciplinary environmental education curriculum units on a wide variety of topics related to Lake Champlain (e.g., geology, wetlands, history, Native Americans, zebra mussels, agriculture, etc.). The grant allowed for field trips, guest speakers, and purchase of supplies and equipment, such as for water testing. The six participating teachers each year included science and mathematics instructors and–through team arrangements at the middle school level–social studies, language arts, art, computer, and industrial arts/technology teachers. These partnership programs were examples of place-based environmental education.

Place-based environmental education

Each locale, each watershed, each bioregion has its own distinctive character, or what Alan Gussow [1971] calls its unique "sense of place." Natural history and cultural features contribute to the character of a place and help to define that place. Natural history includes physical properties such as geology, soils, weather patterns, climate; and biological properties such as native plants and animals which have adapted to survive there. Cultural features include human settlement patterns, indigenous traditions, more recent human historical experience, ways people make their living from the land, and stories.

Place-based environmental education (EE)–also termed "locally-based EE" or "locally-focused EE" uses place as a unifying theme for EE. Through what John Elder [1998] terms the "localized practice of attentiveness," students get to know the place where they live. Via direct, firsthand experience, they examine the natural features and cultural stories of their own place that make it unique and special.

The students are exposed to the wonders and beauty of nearby nature, and are provided direct experience in the natural world that nourishes and encourages what Rachel Carson [1965] calls their "sense of wonder." As students follow a local stream, track animals, make maps, and explore the climate, water systems, soil types, native plants of their immediate environment, nature becomes less abstract and more concrete to them, and they are reminded that people are an integral part of nature and that we are all linked and connected. Equally important, the students trace the human history and modern activities and unique stories that their places have to tell. They make connections rooted in their own rich experiences, bringing alive the natural and historical place where they live and becoming grounded in the process.

By its very nature, place-based EE is interdisciplinary, integrating writing and the arts with the natural sciences and social sciences. Stories of the interrelationship of social, economic, political, historical, and ecological issues are woven together to reveal a tapestry of the interdependence of cultural and natural systems in their place. Such an inductive and comprehensive study of the environment unifies students' experience–rather than simply adding a new environmental studies unit–and provides wholeness in learning.

By paying close attention to the character of their own homes, the students discover their sense of place, their connection with their surroundings... before turning their attention to the pressing environmental problems of the day. From studying events in their local watershed or bioregion, they gain a context for extending beyond to explore larger environments and problems faced by those environments.

Heightened awareness of the local place from diligent study and exploration and celebration of the unique, dynamic characteristics of that place leads to increased attachment to, love for, concern for, devotion to, care for, respect for, and sense of stewardship for the local place. The students develop what Aldo Leopold [1987] terms an "environmental ethic," an ethic of environmental stewardship. They feel a part of their local landscape and a sense of connectedness, of belonging to the community as citizens. Such a sense of community encourages the students to cultivate and sustain the unique qualities of their place, and provides an impetus for activism when the integrity of that environment becomes threatened.

Hudspeth [1998] describes the culminating event in the fall, 1995, "place-based" partnership between his UVM students and the middle school teachers and students:

"There was standing room only at Memorial Auditorium in Burlington, Vermont, on that crisp December evening. The Governor of Vermont was there, as was the Mayor of Burlington, 200 sixth-grade students from the Endeavor Team and seventh and eighth-grade students from the Roots and Branches Team at Edmunds Middle School, their teachers and parents and siblings, countless interested community members, and the 26 University of Vermont (UVM) students in my Environmental Education (EE) course who had worked with the middle schoolers several hours per week throughout the fall semester. A reporter and photographer covered the event for the *Burlington Free Press*, and camera persons from two of the local television stations shot the proud students standing beside and explaining their research projects and documentation of their field trip experiences. These included: working models of a living machine and a tertiary wastewater treatment plant; their displays on such topics as native plants and animals, the geology of Button Bay, aquatic nuisances in the Lake Champlain Basin (zebra mussels, Eurasian milfoil, water chestnuts, purple loosestrife, etc.), non-point source water pollution, an E.P.A. Superfund site called the Burlington Barge Canal, wetlands in the Basin, acid rain, shipwrecks in Lake Champlain, and Champ; their Hyperstudio presentation on the geology, plants and animals, and human history of Redstone Quarry; a videotape of their trip on Lake Champlain aboard the UVM research vessel *Melosira* on which they had conducted tests of turbidity, conductivity, temperature, and dissolved oxygen and carried out plankton tows and bottom grabs; graphs comparing the quality of water they had sampled from headwater streams on the summit of Mount Mansfield while on an overnight trip, from the LaPlatte River while on an all day canoeing trip, and from the Winooski River and Lake Champlain; maps they had made; artwork, poems, essays, photographs, and field journals based on their field trips; plaster casts of animal tracks they had followed; their own creation myths and stories they had made up after listening to an Abenaki storyteller; photographs of salmon they had raised from eggs in their classroom and released into rivers; activity sheets from trips to the Lake Champlain

Basin Science Center and to the Lake Champlain Maritime Museum; a videotape of a role-playing simulation (mock New England town meeting) dealing with a proposed development adjacent to a significant wetland; and much more. The event was the Lake Champlain Basin Community Exposition, and it culminated the semester-long Watershed Partnership between my EE students and six Burlington middle school teachers and their students."

UNIVERSITY OF VERMONT (UVM) SUMMER COURSES FOR MIDDLE SCHOOL AND ELEMENTARY SCHOOL TEACHERS IN BURLINGTON FOCUSING ON THE LAKE CHAMPLAIN BASIN, LINKED WITH ACADEMIC SUMMER CAMP PROGRAMS FOR ELEMENTARY AND MIDDLE SCHOOL YOUTH IN BURLINGTON

Since 1991, UVM's Lake Champlain Summer Institute has provided participants the opportunity to explore the aquatic environments of the Lake Champlain Basin by offering such courses as: Ecology of a Large Lake, Field Methods in Water Resources, Aquatic Botany of the Lake Champlain Basin, Ecology of Vermont Wetlands, Natural Areas of the Lake Champlain Basin, Lake Champlain: A Living System, Exploring Historic Lake Champlain, Nautical Archaeology Field School, Exhibit Design for the Lake Champlain Basin Science Center, Aquatic and Wetlands Education for Teachers, Applied Wetlands Studies and Living Machines, Lake Champlain Basin Interdisciplinary Curriculum for Middle School Educators, and Constructivist Biology in the Lake Champlain Basin. The last two courses, offered specifically for Burlington teachers, are described below.

Lake Champlain Basin Interdisciplinary Curriculum for Middle School Educators

With funding from the Higher Education Cooperative for Instructional Development, Eisenhower Professional Development Program, Tom Hudspeth [1996b] cooperated with Edmunds Middle School teachers Kim Frashure and Matt Chandler in the summer of 1996 to offer Lake Champlain Basin Interdisciplinary Curriculum for Middle School Educators.

The course focused on the use of integrating themes and environmental education in interdisciplinary middle school curriculum development, using Lake Champlain Basin and rivers/wetlands/watersheds within the basin as examples. Participating teachers were sent textbooks and an extensive binder of required readings in advance of the course. A week-long teacher institute, based at the Lake Champlain Basin Science Center, with extensive field trips, was offered in June. Topics included:
- One day on Lake Champlain aboard UVM research vessel Melosira, where researchers discussed current research projects and offered suggestions about ways middle school students could replicate their research and assist with monitoring.
- Exploration of bogs, marshes, and swamps in the basin, including a canoe trip exploring the LaPlatte River/McCabe Creek marshes
- Water quality monitoring for biological, chemical, and physical parameters–using GLOBE and GREEN-compatible protocols–at Stevens Brook, Browns River, Winooski River, LaPlatte River, and Lake Champlain
- Using a variety of art media (including presentation and demonstration by artist Ginny Mullen) and journaling
- Role-playing simulation dealing with wetlands
- Computer and modem technology allowing the Burlington middle school students to share data they collect that relate to Lake Champlain, the rivers that flow into the lake, and the wetlands associated with the lake

- Stories about shipwrecks in Lake Champlain and Native American tales (told from memory by the teachers during the lunch break each day of the teacher institute)
- Existing curricula and resources related to Lake Champlain Basin (including presentation by Amy Demarest, Milton Middle School social studies teacher, on *This Lake Alive!*)
- Vermont's Framework of Standards and Learning Opportunities (including presentation by Bill Romond, Colchester high School science teacher and technology coordinator and Vermont Institute for Science, Mathematics, and Technology associate)

Course participants designed team-taught interdisciplinary curriculum units–linked with Vermont's Framework of Standards and Learning Opportunities and integrating science, mathematics, computers, social studies, language arts, art–on the Lake Champlain Basin or a river/wetland/watershed within the basin. During the week following the teacher institute, they field-tested activities from their unit with middle-school aged campers at Champlain Quest, an academic summer camp program for middle school youth in Burlington also held at the Lake Champlain Basin Science Center. The grant paid the teachers a stipend for being camp counselors at Champlain Quest. The teachers further refined their units on Lake Champlain Basin with their own classes during the fall. Tom Hudspeth provided follow-up for the teachers in the fall, and students in his UVM Environmental Education class worked with them. Finally, the teachers participated in a day-long summary retreat in November at which they shared their experiences, assessed their activities, and decided on joint future activities. The summer course was an extended-credit offering, with grades submitted in November after the concluding discussion. The teachers offered Champlain Quest summer camp in 1997 and 1998 as well.

Constructivist Biology in the Lake Champlain Basin

With funding from Title II of the Elementary and Secondary Education Act, Tom Hudspeth cooperated with Burlington elementary school teachers Colleen Cowell and Anne Tewksbury Frye in the summer of 1998 to offer Constructivist Biology in the Lake Champlain Basin, a summer course for elementary school teachers in Burlington focusing on the Lake Champlain Basin and constructivist science approaches. The course was modeled after the middle-school educators course, with a week-long teacher institute followed by an academic summer camp program for elementary youth in Burlington, but with several changes. Participating teachers were sent *This Lake Alive!*, some textbooks and photocopied handouts on constructivist science teaching, and an extensive binder of required readings on Lake Champlain in advance of the course. Each day for one hour in the week-long teacher institute, Colleen and Anne presented lectures, videotapes, and discussion on constructivist approaches to science teaching. And each day for one hour Charlie Cavanaugh made presentations in the Computer Lab on use of computers in elementary science programs. Georgine Gregory made several presentations and demonstrations on using a variety of art media. Erick Tichonuk led a field trip to the Lake Champlain Maritime Museum, focusing on elementary field trip programs offered by the Museum. Natural Resources graduate student Marc Companion led a field trip to the Bartlett's Bay Living Machine (an artificial wetland inside a greenhouse for treating waste water) in South Burlington. Tom Hudspeth covered wetlands, research aboard the *Melosira*, and water quality monitoring (focusing on macroinvertebrates, pH, dissolved oxygen, temperature, and turbidity) as in the middle school course, but offered the investigations as ways of engaging in place-based environmental education to better understand the "sense of place" of local watersheds.

In teams of two with high school aides, the 20 elementary school teachers carried out the academic summer camp for 180 elementary youth in Burlington during one of the rainiest June weeks on record. As with the middle school educators, these 20 elementary school teachers participated in a day-long summary retreat in November after they had implemented their units on Lake Champlain Basin with their own classes during the fall.

SUMMARY

Since 1989, the Lake Champlain Basin and its environmental problems have received increased public attention. Media coverage has focused on issues such as non-point source water pollution from agricultural runoff, invasion of the lake by zebra mussels and Eurasian milfoil and water chestnuts, and other exotics. U.S. Senator Patrick Leahy (D-VT) has been successful in securing funding for the Lake Champlain Basin Program through the Lake Champlain Basin Special Designation Act and also for the Lake Champlain Basin Science Center in Burlington.

A much richer array of education and outreach programs related to the Lake Champlain Basin exists in 1998 than in 1989. One of the major players has been CBEI, with its summer institutes and one-day workshops and the work of its various partners. Another major player has been UVM, with its programs linking university students with Burlington middle schools and its summer courses for teachers and associated academic summer camp programs for youth. CBEI's and UVM's efforts are similar in that they:
- target teachers in public schools and, through them, K-12 students;
- depend on aggressive grantsmanship for funding;
- emphasize partnerships, collaboration, networking, cooperation among many players from many sectors (formal education, non-formal education, non-governmental organizations, government, business, etc.);
- exhibit many of the attributes of sound environmental education: interdisciplinary, hands-on, field-based focus;
- offer learning activities aligned with school standards in Vermont and New York.

However, much more needs to be done to carry out the "Action Plan for Educating and Involving the Public" *Opportunities for Action* (1994), and many other players need to be involved. Closer coordination between educators in the Basin (CBEI, UVM, teachers, etc.) and researchers in the Basin (as represented by the Lake Champlain Research Consortium) is needed. Greater efforts must be made to incorporate research findings by science and social science investigators into education and outreach programs offered by CBEI, UVM, and other players in the Basin.

Also needed is greater coordination among the numerous water quality monitoring programs throughout the Basin (GLOBE, GREEN, The Vermont Rivers Program, River Watch Network, Vermont Lay Monitoring Program, etc.). Rigorous evaluation of existing education and outreach programs in the Basin is also needed before groups head out in new directions.

Amending the National Sea Grant Program Act in 1998 to include Lake Champlain, thus giving educators access to funding for education, outreach, and technology transfer just as it gives scientists access to research funds, is very promising. The advent of the Sea Grant Program in Vermont offers hope for stepping up education and outreach efforts in the Lake Champlain Basin in the future. Vermont will be able to learn from the efforts of Sea Grant educators from throughout the United States who have been involved in related work for 30 years. Hopefully, a Lake Champlain Education Consortium coordinating all the formal and non-formal educators in the Basin (CBEI, UVM, teachers, etc.) and closely linked with researchers in the Basin in the Lake Champlain Research Consortium will evolve as the Sea Grant program in Vermont develops.

REFERENCES

Carson, R., *A Sense of Wonder*, Harper and Row, New York City, NY., 1965.

Demarest, A. B., *This Lake Alive!: An Interdisciplinary Handbook for Teaching and Learning about the Lake Champlain Basin*, Shelburne Farms, Shelburne, VT., 1997.

Elder, J., in *Stories in the Land: A Place-based Environmental Education Anthology*, edited by The Orion Society, The Orion Society, Great Barrington, MA., 1998.

Gussow, A., *A Sense of Place: The Artist and the American Land*, Friends of the Earth/Seabury Press, San Francisco, CA., 1971.

Hudspeth, T. R., Lake Champlain Basin Bioregion research database: Final report to Vermont Community Foundation, 1988a.

Hudspeth, T. R., Lake Champlain Basin Bioregion: Final report to the UVM University Committee on Research and Scholarship, 1988b.

Hudspeth, T. R., Partnerships between students in the UVM environmental education class and Edmunds Middle School students exploring the Lake Champlain Basin: Final report to the Orion Society and Geraldine Dodge Foundation, 1996a.

Hudspeth, T. R., Integration of the middle school curriculum in Burlington, VT., around the theme "Lake Champlain Basin": Final report to the Higher Education Cooperative for Instructional Development, Eisenhower Professional Development Program, 1996b.

Hudspeth, T. R., "Partnerships between students in the UVM environmental education class and Burlington's middle schools and high school": Final report to the Orion Society and Geraldine Dodge Foundation, 1997.

Hudspeth, T.R., Getting to know the Lake Champlain Bioregion, in *Stories in the Land: A Place-based Environmental Education Anthology*, edited by The Orion Society, pp. 93-103, The Orion Society, Great Barrington, MA., 1998.

Lake Champlain Basin Program, *Opportunities for Action: An Evolving Plan for the Future of the Lake Champlain Basin*, 1994.

Leopold, A., *A Sand County Almanac, and Sketches Here and There*, Oxford University Press, New York City, NY., 1987.

Vermont Agency of Natural Resources, Lake Champlain educational resources (pamphlet), Vermont Agency of Natural Resources, Waterbury, VT., 1989.

Lake Champlain Cultural and Social Resource Management in the 1990's: You Can't Get Where You're Going Until You Know Where You've Been

Susan Bulmer, Art Cohn, and Ann Cousins

ABSTRACT

The Lake Champlain Basin Program included social-economic considerations in its watershed planning process by involving cultural resource managers, recreation planners, sociologists and economists in research, demonstration, and implementation projects. As a result the Pollution Prevention, Control and Restoration Plan for Lake Champlain includes an historic context based on the lake's human connection as well as thoughtful consideration of long-term social-economic impacts for residents and visitors to the region. A summary of the process and findings are presented.

INTRODUCTION

On November 16, 1990, the Lake Champlain Special Designation Act was signed into law, establishing a thirty-one member Management Conference to develop a comprehensive pollution prevention, control and restoration plan for Lake Champlain. The Statement of Legislative Intent that accompanied the law called on the Management Conference to institute a multi-disciplinary research program unlike any previous federal watershed program. While the overriding concern was water quality, the Statement challenged the Management Conference to consider the breadth of interests, including cultural heritage and social resources. Paramount to the Plan was to be a human connection.

As the Management Conference organized to define its mission, cultural resource managers and recreation planners were invited to the table with biologists, chemists, fisheries specialists, farmers and others to talk about the future of Lake Champlain. Cultural and recreation management research projects received a share of the Basin Program's federal dollars, and *Opportunities for Action: an Evolving Plan for the Future of the Lake Champlain Basin* (Lake Champlain Management Conference, 1996), included a chapter on "Recreational and Cultural Resources" along with "Water Quality and the Health of the Lake," and "Living Natural Resources."

The Lake Champlain Basin Program helped define and fuel a social-economic re-awakening for watershed planning.
- In 1991-93 the *Archeology on the Farm* project documented the value of working closely and cooperatively with farmers, farm-oriented government agencies, and local organizations to protect important heritage sites on farms, without impeding farming activities.
- In 1992, the landmark *Mount Independence-Fort Ticonderoga Management Survey* demonstrated the richness and fragile nature of the lake's submerged resources.
- The 1993 *Recreation Survey* highlighted the challenge to balance increasing recreational access with mechanisms to protect against over-use and user conflicts.
- In 1993 the *Cultural Resources Planning Needs Assessment* evaluated current protective mechanisms, resource inventories and scholarly research to determine scholarship, protection, and *public* interpretation gaps in the Basin's 10,000-year human history.
- In 1993, the *Lake Champlain Economic Database Project* developed a socio-economic profile, database, and description of the tourism economy for the Lake Champlain Basin while assessing the potential applications of economic instruments for environmental protection in the Lake Champlain Basin.
- In 1995, the Lake Champlain Maritime Museum, with funding from the Lake Champlain Basin Program developed new scholarship about the *Impact of Zebra Mussels on Shipwrecks*, leading to an expedited timeline for a multi-year survey of the lake bottom.

The Lake Champlain Basin Program's cultural and social research brought to the forefront the lake's human history and human use. The research helped pique new interests and inspire public-private, bi-state partnerships to improve access and appreciation for the human side of the lake. Implementation projects now underway building from Basin Program research include the Lake Champlain Heritage Corridor study, Lake Champlain Byways, Lake Champlain Bikeways, the Paddlers' Trail, the Lake Champlain Underwater Historic Preserve program, and various local bicycle and heritage trail initiatives in communities around the lake.

In the area of submerged cultural resources, the Lake Champlain Basin Program provided structure and funding to greatly enhance our approach to these finite, publicly owned resources. In 1992, the Lake Champlain Basin Program, reacting to the arrest of a diver pilfering Revolutionary War artifacts from the waters off Mount Independence in Orwell, Vermont, commissioned the Lake Champlain Maritime Museum to perform a survey and management assessment in the greater Fort Ticonderoga/Mount Independence region of the lake. This survey was carried out in partnerships with the University of Vermont, Middlebury College, the Institute of Nautical Archaeology at Texas A&M

University, and the Vermont and New York State Historic Preservation Offices. It located a number of 19th century commercial vessels, two 19th century railroad drawboats, re-located and documented the Revolutionary War "Great Bridge" that spanned Mount Independent and Fort Ticonderoga in 1777, and clarified the size and complexity of a submerged artifact collection located off Mount Independence. The resulting management recommendations led, in the next season, to the recovery from the lake of over 1,000 Revolutionary War artifacts, including a rare cannon. Funded by the State of Vermont, recovery included conservation and public interpretation at the Lake Champlain Maritime Museum, and a total of five technical reports produced by this project. The Mount Independence State Historic Site continues to exhibit and interpret this exciting chapter of Lake Champlain history.

When zebra mussels were discovered in Lake Champlain, the Lake Champlain Basin Program reacted by asking the Maritime Museum to investigate the implication of this new non-native invader to the Lake's extraordinary collection of shipwrecks. The resulting report, *Zebra Mussels and Their Impact on Historic Shipwrecks* (Cohn, 1995) served as a catalyst for action. The report concluded that zebra and quagga mussels would, in all probability, encrust most of the lake's submerged cultural resources. Because relatively a small portion of the bottom of the lake had been inventoried, the need to accelerate the survey to inventory yet-unknown cultural sites was imperative. This led, in 1996, to the beginning of a landmark survey project designed to find and document the lake's historic legacy before encrustation. Now in its third year, this project has examined almost one hundred square miles of lake bottom and located and documented more than twenty new historic sites. Additionally, building off the innovative format of the Mount Independence Project, the survey is simultaneously capturing geological data. This will result in a bank of information for researchers looking at the geology, bathymetry and hydrology of Lake Champlain.

The documentation process includes physical recordation, archival research, and assessment for public access through the Underwater Historic Preserve Program. The Federal Abandoned Shipwreck Act of 1987 mandates that states take responsibility for managing shipwreck sites within their jurisdiction and provide appropriate public access. Vermont has had an underwater historic preserve program since 1985, and New York instituted a Submerged Cultural Preserve program at Lake George in 1993. In an effort to expand and build bi-state cooperation, the Lake Champlain Basin Program provided funding to perform an underwater historic preserve feasibility study of the *Champlain II*, in Westport, New York. In 1998, with funding from the Lake Champlain Basin Program, New York Department of Environmental Conservation opened the shipwreck as New York's first underwater historic preserve site on Lake Champlain. That year the Basin Program also assisted the State of Vermont with funding to open its sixth underwater preserve, the *O.J. Walker*, located in Burlington harbor.

Expanding the underwater historic preserve program is particularly attractive in light of the emerging national tourism industry based on history and recreation, a high interest for visitors to the Lake Champlain Basin (Schaefer, pg 24). Tourism and recreation opportunities bring economic benefit and contribute to a high quality of life for residents. Despite the recognized positive impacts, research has also shown that Lake Champlain is under-realized in terms of its tourism identity. Political boundaries have encouraged promotion on a local and state basis while discouraging regional promotion of the entire Basin. Tourist maps tend to stop at State or County borders (Schaefer, 1997). Sub-regions have been exclusively promoted as destinations, such as the "Green Mountain

State," the "Adirondacks," and the "Richelieu Valley" in Quebec (Holmes and Associates, 1993). *The 1996 Lake Champlain Basin Cultural Heritage Tourism Survey and Marketing Plan* recommends a marketing approach based on cooperative public-private partnerships with a centralized point of responsibility promotion (Schaefer, 1997). At the time of this writing, there are two studies looking at the feasibility of a unified tourism program: one, through federal Heritage Corridor designation; the other through the Scenic Byways program.

In 1993, results and recommendations were released from a series of *Recreational User Survey* questionnaires (Dziekan, 1995) and the *Economic Database Project* (Holmes and Associates, 1993), part of which addressed the role the tourism industry plays in the Basin. Combining the results from these and various other studies has given researchers, planners, and managers a better picture of recreation and heritage tourism for the region.

The Recreational User Survey consisted of a series of questionnaires developed for eight targeted recreational user groups:
- public access users,
- shoreline property owners,
- canal lock users,
- snowmobilers,
- ice anglers,
- marina users,
- SCUBA divers, and
- park users.

Over the course of two years, 12,000 questionnaires were distributed to diverse recreationists with a general response rate of approximately 28 percent (Dziekan, 1995 and Smith, 1996). The data generated from the User Survey was used as an advisory piece of information to supplement the overall recreation planning effort. Even with its limitations, the results provide some interesting insights. Results suggest that public access to the lake, or improved public access, is needed in Essex, Plattsburgh/Valcour Island, Westport, Ticonderoga, and Point Au Roche in New York. For Vermont, areas with access needs include Burlington, Malletts Bay, SE Inland, Keeler Bay, Woods Island, and St. Albans Bay. Areas of high use concentrations on the Lake included, Burlington, Shelburne Bay and Malletts Bay in Vermont, and Plattsburgh/Valcour Island, Point Au Roche area, and Chazy Landing in New York. Non-boating activity and facility needs, in priority order, were 1) bike trails, 2) beach and swimming areas, 3) hiking trails, 4) campgrounds, and 5) picnic areas. All user groups surveyed tended to agree with the statement that "pollution is a severe problem on Lake Champlain." This is further supported by open-ended comments indicating that the majority had a concern over the water quality of the lake. All user groups agreed that the government should purchase more land for recreation, public access, and for preserving the open space landscape around Lake Champlain.

Trip expenditures reported varied greatly among user groups, with the majority of recreationists' expenditures at the Lake, rather than in-transit or at home in preparation for the trip. On average, recreational user groups spent between $33.00 - $465.00 for either one-day or multi-day trips. The canal lock users and marina users tended to spend more, per trip average, than park users, snowmobilers, SCUBA divers, public access site users, and ice anglers because their trips were longer in duration than the day users.

The tourism component of the Economic Database Project (Holmes and Associates, 1993) attempted to assess the overall economic impact of tourism in the Lake Champlain Basin through review and research of existing studies and information. Using the best available information, Holmes and Associates collected data relative to three types of tourists: international, U.S., and internal visitors (residents from the Champlain-Adirondack region).

Overall, the following data was found:
- It is estimated that as many as 6 million visitors and 600,000 residents of the Basin enjoy some form of recreation in the Lake Champlain area annually.
- During 1991/92 (federal fiscal year) a total of 7.9 million non-U.S. citizens entered the United States through the fourteen ports of entry to the Champlain Basin. An average of 541,396 non-U.S. citizens has been crossing into the Lake Champlain Basin per month during the previous five years. Boat border crossings at Rouses Point have annually averaged 26,343 people traveling in 7,318 boats.
- Total tourism-related expenditures in the Basin were estimated at $2.2 billion in 1990, with approximately 71% in the Vermont portion of the Basin ($1.6 billion) and 29% in the New York portion ($638 million).
- Tourism-related expenditures of internal visitors accounted for an estimated $968 million in 1990, which is 44 percent of the total tourist expenditures in the Basin.
- Approximately 40 percent, or $880 million, of the 1990 Basin tourism expenditures occurred in shoreline towns.

Other studies suggest the prominent role of recreation in generating tourism dollars. In 1991, an estimated 32,500 bicycle tourists spent $13.1 million at a rate of $115 per person per day in the state of Vermont (Burgess, 1992). In the Adirondack North Country Region cyclists averaged $45 per person per day as reported in a 1993 bicycling survey (Holmes and Associates, 1994). In 1988, angling related expenditures for Lake Champlain reached an excess of $61 million, about $25 million of it in shoreline communities (Gilbert, 1990). Based on data obtained from a 1993 study on the economic and social values of Vermont State Parks, annual park-related dollars spent within the Basin area of Vermont were estimated at $42.2 million. Of that amount $15.2 million accounts for annual expenditures at parks on Lake Champlain (University of Vermont, 1995). Lake Champlain tourism-related employment is estimated at 16,400 jobs in New York and Vermont, equaling approximately 6.3 percent of all employment. In 1990, the 64 marinas on Lake Champlain employed about 344 people with total annual payroll of $3.4 million (Farnum, 1995).

History, scenery and cultural events add to the Lake's tourism identity. While national surveys suggest about 37-38% of all tourists are interested in historic sites, surveys of visitors and those inquiring about the Champlain Valley put that number at 57-60% (Schaefer, 1997). In 1996, the Lake Champlain Basin Program contracted with MarketReach, Inc. to conduct a cultural heritage tourism survey. The survey was designed to capture visitor characteristics, visit expectation, visitor satisfaction, and other key elements that would allow historic sites and tourism agencies to market to a broader base of potential visitors. Overall, 2,565 surveys were completed. In general, visitors to the region liked what they saw. Over 90% stated that their expectations were "Met" or "Exceeded" (Schaefer, 1997). While several of the larger, well-known historic sites target geographic markets and receive significant visitors, most historic sites operate in relative isolation, many with no marketing budgets. They rely on regional tourism campaigns and word of mouth. This pattern is echoed in the visiting pattern of most tourists, who stated that they come to the region because they had a general idea about the destination, and once here, plan their itinerary, relying on signage, word of mouth, and brochures. Suprisingly, the sites themselves have yet to fully appreciate this pattern. There is little cross-promotion of sites. Only 6 percent of visitors learned about the site they visited at another historic site (Schaefer, 1997). This research underscores the opportunity to develop theme-based itineraries as a means to attract visitors and to improve the quality of the visitor experience. The Lake Champlain valley provides an ideal setting for not only single-theme itineraries, but also an opportunity to focus on regional tours that link an array of natural and cultural resources to create trails and networks of unrelated sites. Whether for walkers, bicyclists, motorists, divers, or boaters, these trails link sites and provide the visitor with easy access.

In Opportunities for Action (Lake Champlain Management Conference, 1996), the development and promotion of heritage tourism bridges the goals and objectives of the Recreation and Cultural Heritage Programs of the Lake Champlain Basin Program. Through combined efforts, a number of implementation projects are underway in response to the various studies described above. In all of these programs, the Lake is the centerpiece.

Lake Champlain Bikeways

Lake Champlain Bikeways is a public/private initiative to create a network of bicycle routes on existing roads around Lake Champlain in New York, Vermont, and Quebec. A Steering Committee was formed to promote bicycling throughout the region for the purposes of increasing opportunities for bicycle recreation and transportation, improving the quality of life, enhancing the economic vitality, and raising public awareness and appreciation of inherent scenic, historic, cultural, natural, and recreational resources throughout the area. A primary route of 350 miles has been established and ten theme loops are linked to the principle loop to enhance bicycling opportunities. Major efforts to promote the region for bicycling have met with great success. To date, Lake Champlain Bikeways has supported a variety of products: 1) Basin maps depicting the recreational and cultural resources of the region including the principal Lake Champlain Bikeways route; 2) a Directory of Bicycle-Friendly Accommodations and Services in the Champlain Valley; 3) many local theme loop guidebooks; 4) a website for Bikeways; and 5) a decal that can be displayed at bicycle-friendly businesses and establishments. In addition, the Steering Committee has sent feature releases to a multitude of national, regional and local publications resulting in numerous articles in magazines, newspapers, and travel section publications since 1996. The onslaught of about 1000 inquiries in 1996 and 1997 prompted a Bicycling Marketing Study to determine the type of person requesting information and if they actually visited the Basin for a vacation and bicycling. Of those surveyed, almost 30 percent decided to visit the area with 18 percent visiting for the first time. Other preliminary results indicated the Champlain valley was their primary destination, and bicycling was the primary reason for their trip. Their trip was not part of an organized group bicycle tour, but they did bicycle the recommended Lake Champlain Bikeways principal route. While in the Champlain Valley, they also participated in hiking, walking, water sports; visiting historic sites; nature interpretation and viewing; and other activities. They had a good bicycling experience and would recommend the Champlain valley as a primary destination for bicycling.

Lake Champlain Byways

An outgrowth of the Lake Champlain Bikeways initiative is the Lake Champlain Byways Program. The states of Vermont and New York received Federal Highway Administration funding through the National Scenic Byways Program to develop a corridor management plan for the Basin, building upon the international flavor, cultural resources, and Lake Champlain Bikeways initiative. The corridor management plan is being developed at the grass-roots level with local citizen groups in each of the seven counties surrounding the Lake. Each local group is looking at a new approach to economic development through the recognition of local heritage, natural, cultural, and social resources. A shared identity for the entire Lake Champlain region is the basis for an action plan that will establish a vision and methods to balance economic development and tourism with stewardship of the resources. Several unifying themes, such as "The Lives of Lake Champlain: Exploration, Military, Commercial, Recreational," have been developed to tell the story of the region. The ultimate outcome of the program is an increased quality of life for residents through economic development while providing the tourist with enhanced experiences while visiting the area.

Lake Champlain Paddlers' Trail

The Lake Champlain Paddlers' Trail was initiated by the Lake Champlain Kayak Club, the Lake Champlain Committee, and interested paddlers, as a water trail linking together primitive camping sites along the shores of the Lake to provide a multi-day paddling experience. Camping sites have been established at a number of state parks and other public lands, and contacts have been made with numerous private landowners to develop additional sites on private land. Stewardship of the resources is a primary management goal of the organizers of the trail, and site management plans have been developed for each site. As kayaking and canoeing have become more and more popular in Vermont, New York and the United States, it is anticipated that there will be tremendous interest in the trail once fully developed and promoted.

Lake Champlain Underwater Historic Preserve

The Vermont Division for Historic Preservation established the Vermont Underwater Historic Preserve Program in 1985. Through the efforts of the Lake Champlain Basin Program, the program was expanded in 1998 to New York as a one-year pilot bi-state program. An ad-hoc Lake Champlain Underwater Historic Preserve Advisory Committee provides direction and makes recommendations to the Vermont Division for Historic Preservation and the New York Department of Environmental Conservation on behalf of the program. The preserve includes seven historic sites, marked with special buoys. Staff and volunteer Preserve Monitors assist divers and assure that each diver has registered for the program (no cost), and has received a booklet that tells the story of each site and highlights safety considerations of each of the historic sites.

The Lake Champlain Basin Program nurtured each of these programs. Each provides a unique opportunity to experience the lake. The Plan for the future of the lake recognizes the importance of looking beyond the science of clean water. The lure of the lake is more than natural beauty. It is a combination people, places, experiences and opportunities. The success of the restoration plan for Lake Champlain is dependent on individual's sense of respect, pride and commitment. The research and implementation projects reinforce the human connection to the lake--past, present and future. Results quantify the lake's value and potential for tourism and recreation. And, through an exploration of historic context, the research kindles appreciation, nurtures a sense of stewardship, and builds public commitment to plan for the future of Lake Champlain.

REFERENCES

1992 Fort Ticonderoga-Mount Independence Submerged Cultural Resources Survey. Lake Champlain Basin Program Demonstration Report Series, Grand Isle, VT.
- A) Executive Summary, Art Cohn, 1995.
- B) Submerged Cultural Resources Survey, Art Cohn, 1995
- C) The Great Bridge From Fort Ticonderoga to Mount Independence, Art Cohn, 1995.
- D) Geophysical Reconnaissance in Mount Independence Area: Larrabee's Point to Chipman's Point, Pat Manley, Roger Flood, Todd Hannahs, 1995.
- E) Ticonderoga's Floating Drawbridge, Peter Barranco, 1995.
- F) Bottom Morphology and Boundary Currents of Lake Champlain, Hollister Hodson, 1995.

Baldwin, Elizabeth, Arthur B. Cohn, Kevin J. Crisman and Scott McLaughlin, *Underwater Historic Preserve Feasibility Study of the Lake Champlain Steamboat "Champlain II"*, Lake Champlain Maritime Museum: Ferrisburg, VT, 1996.

Cohn, Art, *Zebra Mussels and Their Impact on Historic Shipwrecks*, Lake Champlain Basin Program Technical Report No. 15, Grand Isle, VT, 1996.

Dziekan, K., *Lake Champlain Recreation User Surveys*, New York Office of Parks, Recreation and Historic Preservation, Albany, N.Y. and Vermont Department of Forests, Parks and Recreation, Waterbury, VT 1995.

Farnum, G., *Lake Champlain Outdoor Recreation Facilities Inventory -Near Shore Sites*, Vermont Department of Forests, Parks and Recreation, Waterbury, VT and New York Office of Parks, Recreation and Historic Preservation, Albany, NY, 1995.

Gilbert, A., *Resident and Non-resident Hunting, Fishing, and Trapping Expenditures in Vermont*. Unpublished. Forest Parks and Recreation, Waterbury, VT, 1990.

Holmes and Associates, *Lake Champlain Economic Database Project, Socio-Economic Profile, Database and Description of the Tourism Economy for the Lake Champlain Basin*. Lake Champlain Basin, Lake Champlain Basin Program Technical Report No. 4B, Grand Isle, VT, 1993.

Lake Champlain Management Conference, *Opportunities for Action: An Evolving Plan for the Future of the Lake Champlain Basin*, Lake Champlain Basin Program, Grand Isle, VT, 1996.

McLaughlin, Scott A., Anne W. Lessmann, and Arthur B. Cohn, *Lake Champlain Underwater Cultural Resources Survey*, Lake Champlain Maritime Museum: Ferrisburg, VT, 1998.

Rossen, Jack, *The Archeology on the Farm Project: Improving Cultural Resource Protection on Agricultural Lands, A Vermont Example*, Demonstration Report #3, Lake Champlain Basin Program, Grand Isle, VT, 1994.

Schaefer, David and K. Celeste Gaspari: MarketReach, *Lake Champlain Basin Cultural Heritage Tourism Survey and Marketing Plan*, Lake Champlain Basin Program Grand Isle, VT, 1996.

Smith, M., *Lake Champlain Recreation: Assessment Report*. Vermont Department of Forests, Parks and Recreation, Waterbury, VT. and New York Office of Parks, Recreation and Historic Preservation, Albany, N.Y, 1996.

Economic Analysis of Lake Champlain Protection and Restoration

Timothy Holmes, Anthony Artuso, and Douglas Thomas

ABSTRACT

The Lake Champlain Management Conference (LCMC) was charged by Congress in 1990 with creating a comprehensive plan for protecting and enhancing Lake Champlain and its watershed area. Since the beginning of the five-year planning effort, the LCMC was interested in integrating a vital economy with protection and enhancement of Lake Champlain. The goal was to promote economic strategies that were compatible with water quality objectives, and to tailor pollution prevention and control strategies for economic efficiency as well as environmental effectiveness. It is in the realm of economic impacts, and the associated issues of cost effectiveness and the equitable distribution of costs, that this research provided information to the LCMC and the public. This and prior research demonstrates the significant economic value of Lake Champlain as a tourism and recreation resource. The research findings also provide evidence substantiating the economic value to basin residents in protecting and restoring Lake Champlain. The economic analysis and resulting economic models and databases contribute economic tools that can help decision-makers and the public to develop and maintain a lake management program that is efficient, cost effective and equitable.

INTRODUCTION

The Lake Champlain Management Conference is a 31-member board that represented a broad spectrum of businesses, organizations and government in the Lake Champlain basin. The Lake Champlain Basin Program (LCBP) is the organization established to coordinate the work of the LCMC in carrying out the activities envisioned in the Lake Champlain Special Designation Act of 1990. The primary goal of the Act was to create a plan for protecting the future of Lake Champlain and its surrounding watershed. The resulting plan, published in 1996, is entitled "Opportunities for Action: An Evolving Plan for the Future of Lake Champlain." The Plan identified three actions of highest priority and five other broad priority areas. The highest priority actions are:

1. Reducing phosphorus in targeted watersheds of the lake.
2. Preventing and controlling persistent toxic contaminants found lake wide or in localized areas of the lake.
3. Developing and implementing a comprehensive management program for nuisance nonnative aquatic species.

The other five priority areas are protecting human health, managing fish and wildlife, protecting wetlands, managing recreation, and protecting cultural heritage resources.

This article summarizes the economic impacts, both costs and benefits, of addressing all of the identified priority issues in the Lake Champlain basin. The first section discusses the primary sources of the lake's water quality problems and includes a detailed description of the phosphorus control model. Based on linear programming techniques, this model enabled the study team to model various methods by which a desired level of phosphorous reduction could be achieved. The second section discusses the human activities that influence phosphorus levels in the lake and their economic characteristics. The third section illustrates how the current recreational use of Lake Champlain was valued and incorporated into the economic modeling effort. The fourth section is a detailed description of the IMPLAN modeling effort, with tables illustrating summary findings. The fifth section is a discussion on steps taken to understand perceptions within the business community on the economic implications of the Plan. The article concludes with recommendations for further research and a concluding description of costs, benefits and the cost of no action.

MODELING THE COSTS OF PHOSPHORUS CONTROL

The Lake Champlain phosphorus control model developed as part of this project has four integrated components. The first section of the model is a detailed compilation of actual, permitted and projected point source flows and loadings, given current policies and assuming implementation of controls to achieve phosphorus discharge concentrations of either 0.8 mg/l and 0.5 mg/l at facilities with permitted flows greater than .2 mgd. Also included in this component of the model is the capital and operating cost of upgrading each treatment plant with a permitted flow of greater than 0.2 mgd to meet either a 0.8 mg/l and 0.5 mg/l discharge standard. This point source section of the model updates the information developed for the Diagnostic-Feasibility (D/F) Study with new estimates of capital and operating costs of point source upgrades [VT Natural Resources & NY Environmental Conservation 1994]. These estimates include the additional operating and maintenance costs required for sludge handling. The point source component of the model allows the user to calculate the costs and discharge reductions that would result from implementation of phosphorus control measures at any combination of facilities.

The economic phosphorus control model developed for this study, also incorporates projections of point source flows through the year 2010 derived from town level population projections. Projections of treatment facility flows were calculated by identifying the town(s) served by each treatment facility and increasing the current flow by the percentage change in population. If the population served by the treatment facility was less than 100% of the population of the town(s) it serves, the current flow was increased to incorporate the projected population change in the service area and to reflect the assumption that by 2010 sewer service would be extended to the other sections of the service area until 100% of the permitted capacity of the treatment facility is utilized. This methodology may tend to underestimate future point source loads since it does not assume any increase in permitted point source flows and caps the projected flow of any treatment facility at its permitted flow.

The second component of the model incorporates information on nonpoint source phosphorus loadings and control costs for each lake segment watershed. Starting with the total nonpoint source loads estimated in the Diagnostic Feasibility (D/F) Study, the relative phosphorus contributions of agriculture, urban and forest lands were estimated on the basis of current and projected land use data and estimated phosphorus exports rates of each land use.

The third component of the model incorporates the costs and effectiveness of agricultural and urban Best Management Practices (BMP's). The analysis included development of new estimates on the total number of animal units in each lake segment watershed, prevalence of existing best management practices, and costs of BMP's and the potential phosphorus reductions. The urban BMP's included non-structural measures and minor capital improvements, as well as more capital intensive measures such as construction of detention basins and stormwater treatment facilities. The costs of nonstructural urban BMPs incorporated into the model ranged from $24 to 72$ per hectare, structural urban BMP costs were assumed to range from $687 to $6,132 per hectare of urban land.

The final component of the phosphorus control model integrates the data on point and nonpoint source loading and control costs with the "BATHTUB" hydrodynamic model of Lake Champlain developed for the D/F study. The BATHTUB model permits simultaneous computation of phosphorus concentrations in each lake segment given point and nonpoint source loadings in each watershed. By combining the essential aspects of the BATHTUB model with data on phosphorus reduction costs in each watershed, the model can be used to determine how point and nonpoint source phosphorus loadings can be reduced at least cost in order to achieve a given set of in-lake phosphorus concentration standards.

The Economic Analysis Report presents several scenarios for Plan implementation, based on four alternatives for controlling phosphorus in the basin. In addition to the no action scenario, four phosphorus reduction strategies were evaluated using the phosphorus control model, as shown in Figure 1. The first reduction strategy (Policy 2) examines the cost of implementing a point source phosphorus standard of 0.8 milligrams per liter (mg/l) for all facilities with permitted flow in excess of 0.2 million gallons per day (mgd) and then seeks to target nonpoint source controls in a cost-effective manner to achieve the in-lake phosphorus criteria.

The second phosphorus reduction strategy (Policy 3) includes a point source phosphorus standard of 0.5 mg/l for all facilities with permitted flow in excess of 0.2 mgd and then seeks to target nonpoint source controls in a cost-effective manner to achieve the in-lake phosphorus criteria. The third phosphorus reduction strategy (Least Cost) allows for complete flexibility in targeting both point and nonpoint source controls to achieve the in-lake criteria most cost effectively. The final phosphorus reduction strategy examined (Plan) involves no point source controls in New York and targeted point source controls in Vermont together with cost effective targeting of NPS controls in both New York and Vermont.

The third phosphorus reduction strategy (Least Cost) was found to be most cost effective at meeting the phosphorus targeting goals. Under this point source and nonpoint

400 Economic Analysis of Protection and Restoration

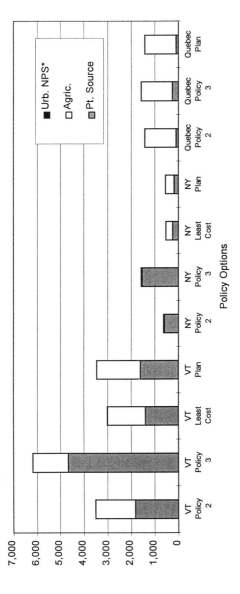

Figure 1. Summary of Phosphorus Control Costs
(annually, in $000's of 1995 dollars)

source (NPS) targeting scenario, the total present value costs decline to $109.8 million per year which is equivalent to $7.1 million on an annualized basis. Only a few treatment facilities in New York would be targeted for phosphorus control measures under this cost-effective scenario. The initial run of the model found that targeting the following facilities was most cost effective: Great Meadows Correctional Center, Port Henry, Peru, and Lake Placid. However, more detailed cost estimates might indicate that targeting other New York facilities makes more sense. The findings of the model need to be considered in light of the specific situation and needs at each facility.

The Least Cost strategy includes continued implementation of agricultural NPS controls in the South Lake A, South Lake B, Otter Creek and Main Lake watersheds in both New York and Vermont. Approximately 25% of the costs of this phosphorus control strategy would be for point source controls and 75% would be for nonpoint source controls. Of the $7.1 million in total annualized control costs for this strategy, approximately $4.5 million or 63% would be for point and nonpoint source controls in Vermont, slightly less than $1 million or 13% would be for phosphorus controls in New York, and $1.7 million or 24% would be for phosphorus control measures in Quebec.

All of the phosphorus control strategies outlined above, with the exception of the 0.5 mg/l point source standard (Policy 3), rely heavily on nonpoint source controls. Even with a less stringent in-lake phosphorus target for Missisquoi Bay, 30-50% of the cost of the above scenarios is attributable to that watershed. In addition, phosphorus control costs in Missisquoi Bay are highly sensitive to changes in the in-lake phosphorus concentration target. If the in-lake target for Missisquoi Bay is made even 5% more stringent than the 28.5 ug/l included in the above analyses, the present value of phosphorus control costs increase by more than $100 million. This is because in order to achieve in-lake phosphorus concentrations lower than 28.5 ug/l in Missisquoi Bay, relatively expensive urban BMP's must be implemented on a high percentage of urban land in that watershed. However, due to the lack of detailed land use and agricultural data for the Quebec portion of the Missisquoi Bay watershed, results of the model for this watershed should be interpreted only as general approximations. More detailed modeling of this watershed in cooperation with the Province of Quebec is needed before a cost-effective phosphorus control strategy can be reliably developed for Missisquoi Bay.

The present value of the capital and operating cost of achieving the phosphorus loading targets outlined in the Phosphorus Agreement (illustrated as "Plan" in Figure 1) is estimated to be approximately $83.5 million which is equivalent to $5.4 million on an annualized basis.

REFINING AND UPDATING THE COSTS OF PHOSPHORUS CONTROL

Wastewater Treatment Facilities

Wastewater treatment facilities were recognized as the primary point-source of phosphorus in the Lake Champlain Basin. The study team initiated a detailed cost evaluation of upgrading treatment plants to include additional sludge handling facilities and sludge disposal, and to develop accurate cost estimates for upgrading phosphorus removal capabilities. All costs were preliminary and assumed chemical addition for phosphorus removal. Some of the treatment facilities within the basin already have chemical storage and addition facilities; however, since phosphorus limits are not in their permit, these chemical facilities are currently not being used. Treatment facilities such as Saranac Lake, Peru, Crown Point, and Willsboro, New York, which were recently upgraded or newly constructed include chemical addition facilities since the design was prepared in anticipation of phosphorus limits ultimately being imposed. Most treatment facilities within the basin, however, will require new chemical addition facilities. The method of sludge handling at each facility and the existing sludge handling facilities at each plant was taken into consideration when estimating the costs for treatment plant upgrades.

On the New York side of Lake Champlain, the following treatment plants were targeted to obtain more accurate cost estimates for upgrading the facilities:

Champlain	Lake Placid	Rouses Point
Dannemora	Peru	Saranac Lake
Granville	Plattsburgh	Ticonderoga
Keeseville	Port Henry	Whitehall

The researchers met with New York DEC facilities engineers for the above plants to obtain as much information as possible regarding these plants. The following information was obtained for each of the plants:

1. Process descriptions of the plants, including the plant flow schematic.
2. Current phosphorus effluent levels for each plant for the past two years.
3. The current method of sludge handling and the sludge disposal.
4. General operating conditions of the plants.

After reviewing New York DEC files and speaking with the facilities engineers, several treatment facilities were visited to obtain first hand information regarding the treatment plants. Cost estimates for upgrading the treatment plants were then developed for each treatment plant. Cost estimates include equipment costs, contractor installation and markup costs, engineering design costs, and construction observation costs. Besides actual equipment costs, costs of buildings required to house new equipment and yard piping and repiping costs were also estimated. Equipment costs were obtained from vendors when possible. Operation and maintenance costs were also estimated. Costs for disposing of increased sludge volumes were based on average actual costs that municipalities in the basin are currently spending on sludge disposal. All costs were compared with recent costs from similar construction projects and were adjusted, as needed, for each treatment plant.

Only five Vermont treatment plants listed in the Diagnostic Feasibility Study required a more detailed cost estimate for upgrading. These plants include Brandon, Fair Haven, Richmond, Northfield, and West Rutland.

The State of Vermont made a commitment to finance necessary municipal wastewater treatment plant up-grades at the state level. The financing for wastewater treatment facilities in New York was undetermined at the time of this research. In an economic review of local fiscal capacity in the New York portion of the basin, the picture that emerges is one of a region with income levels that are substantially below the state norm, with the resulting limitations on local government spending that these income levels imply. Consistent with this picture is the region's relatively high level of dependence on resources supplied from outside in the form of aid payments from other governments. Despite the region's ability to generate tax revenue from non-residents in the form of property taxes on seasonal residences, its ability to tax residents is still limited by their relatively low incomes. Subsequent to this study, and in part because of it, New York State created a bond program in 1997 aimed specifically at wastewater treatment plants in the Lake Champlain basin and in other priority watersheds.

Agriculture

Agricultural activities were recognized as one of the primary nonpoint sources of phosphorus in the basin and the major agricultural activity in the basin is dairy farming. To put the analysis in perspective it should be recognized that it will get increasingly difficult for dairy farms with less than 100 cows to remain economically viable in the Northeast unless they can continue to lower their costs. In order to stay in business the small dairy farmer will have to carefully assess the economic aspects of all phases of his operation. The economic analysis report focuses on only one of the many aspects of a modern dairy farming operation, manure management. The analysis offered farmers and decision makers one of the more complete economic analyses of manure management available for any watershed area in the nation.

Manure can be a valuable source of nutrients for cropland if managed correctly and if there is need for its nutrient value. However, if mismanaged, manure can be a source of phosphorus pollution in lakes and streams. To achieve a reduction in nonpoint pollution from agriculture in the basin will require some farms to change their manure management practices. Ecologically sound manure management practices include diversion of barnyard run off, safe storage, maintenance of grass cover between water ways and cultivated fields, and elimination of manure spreading on frozen ground. Each of these practices is designed to eliminate or reduce the amount of manure runoff into adjoining waterways. All these practices are currently being used in various locations around the basin.

The elimination of manure application on frozen ground is perceived by some farmers as a major change to their past farming practices. The elimination of daily spreading in the winter requires farmers to store their manure and apply it during periods where the potential for runoff is lower. The economics of moving away from daily spreading is a major topic of concern, especially for the small and medium sized dairy farmers. The costs to dairy farmers of alternative manure management practices to daily spreading are examined in the report. One of the conclusions from the analysis is that smaller farmers may see a positive economic impact on their operation by contracting out their manure spreading.

In addition to cost estimates, the crop nutrient (fertilizer) value of manure in each alternative management practice was estimated. The nutrient values are used in conjunction with the cost estimates to project net benefits or costs of alternative manure management practices relative to daily spreading. The sensitivity of each alternative's annual cost and net benefit to changes in government cost sharing of capital expenditures associated with building a storage structure was also examined.

Urban Runoff

Another point-source issue that can be evaluated with the aid of the economic model is the implications of developing stronger phosphorus control requirements for new urban and suburban developments. It is almost always less expensive and less difficult politically to require phosphorus control measures in new urban/suburban development rather than retrofitting existing developments. If it is assumed that new urban/suburban developments are required to install structural control measures (e.g. detention ponds) that reduce phosphorus export rates by 50%, then projected nonpoint source loadings would decline by more than 11 metric tons relative to a no action scenario. The result of this requirement would be to reduce the phosphorus export from new urban developments to a level roughly equivalent to average agricultural export rates, thereby virtually eliminating any growth in nonpoint source phosphorus loadings due to land use change.

Non-Phosphorus Actions

In addition to high rates of phosphorous, Lake Champlain's water quality is also adversely affected by toxic pollutants and nuisance aquatics. Furthermore, the lake's environmental and public health, recreational capacity and aesthetics could be raised by targeting improvements in water quality and further development of the region's recreational and cultural infrastructure. Accordingly, the assessment of the economic impacts of improved water quality in Lake Champlain considered seven non-phosphorus action categories, including toxics, human health, fish & wildlife, wetlands, nuisance aquatics, recreation and cultural heritage.

In total, Vermont's annual expenditures on all the non-phosphorous actions would be $4.8 million during each of the first five years of implementation, and New York's expenditures would be $2.0 million. Those figures are based on a 71% / 29% split of the estimated costs. The split is based on each state's share of total commerce in the Lake Champlain basin. Based on past federal funding practices, it is further projected that federal funding would provide for approximately 64% of those expenditures, reducing Vermont's share to $1.7 million and New York's to $.70 million. Under a "No Action" scenario federal funds would likely be less available for the non-phosphorus activities, increasing the local costs for any actions that the state's may decided to pursue on their own.

MEASURING THE ECONOMIC VALUE OF IMPROVED WATER QUALITY

The direct benefits of water quality improvement in Lake Champlain, are quite diverse. They include recreational and/or health benefits for anglers, swimmers and boaters; aesthetic benefits for both local residents and visitors; and the bequest and existence values that residents within the basin and elsewhere might place on a cleaner lake. These benefits are quite difficult to quantify even based upon studies designed for and conducted within the Lake Champlain basin. While at the same time benefit estimates developed in other regions can only provide a preliminary indication of the range of potential benefits of water quality improvements.

Benefit estimates gleaned from over 50 studies on the economic benefit of water quality improvement indicate the benefits of water quality improvement on Lake Champlain could range from $13 million to an upper bound of $63 million per year [Holmes & Associates and Anthony Artuso 1995a & 1995b]. That is the estimated dollar value of benefits lake users would receive from improved water quality, in addition to any increase in direct expenditures resulting from water quality improvement.

Water quality improvement or avoidance of water quality degradation would provide benefits to lakeshore property owners. Research by Feather [1992] found significant property value impacts due to water quality changes up to a mile from the lakes included in his study. The several hedonic valuation studies indicated property value impacts in the range of 10% due to changes in water quality similar to those that might result in Lake Champlain from implementation of the Plan. The total assessed value of real estate in Vermont lakeshore towns in 1991 was approximately $7.2 billion and for New York was $1.6 billion for a total assessed value of $8.6 billion [Holmes & Associates 1993]. Using a conservative estimate that 5% of this assessed value of real estate in lakeshore towns is attributable to lakefront property, then a 10% change in value in this lakefront property due to water quality changes (positive or negative) would equate to an economic impact of $43 million. Since real estate prices capitalize the use value of a property over time, this benefit (or avoided cost) of water quality improvement would be a one time rather than an annual benefit.

Estimating Current Recreational Expenditures

In order to model the economic impact of projected water quality improvement resulting from implementation of the Plan, a baseline for current Lake Champlain dependent economic expenditures had to be established. The study team took the conservative approach of estimating only water-based and Lake Champlain dependent recreational activity expenditures that occurred within three miles of Lake Champlain at recreation facilities inventoried for the Lake Champlain Basin Program [1995].

Water-based and Lake Champlain dependent recreational activity resulted in approximately $107 million in direct expenditures in the vicinity of Lake Champlain during the summer season. The study team based that estimate on two approximate measures of recreational activity and expenditure. The number of lake users was based on average attendance levels at approximately 500 recreation sites within three miles of the lake, while the average expenditures per user were based on reported expenditures in the Lake Champlain area by families and groups engaged in various recreational activities including boating, swimming, camping, fishing, etc. The expenditure data were derived primarily from surveys of visitors carried out in the Lake Champlain basin in 1992 and 1993 [Lake Champlain Basin Program 1996c].

In addition to lake related direct expenditures by visitors, the study team estimated that seasonal residents who have summer homes around the lake spent approximately $16 million in the area on food, entertainment, arts & crafts, and other non-durable items during 1994. Thus, the combined direct expenditures on Lake Champlain related tourism and recreation by both visitors and seasonal residents were estimated at $123 million. Approximately $77 million (63%) of the total expenditures occur in Vermont lake shore communities, and $46 million occurs in New York lake shore communities. That was believed to be a conservative estimate of Lake Champlain's current recreation value as it reflects only summer season use and is based on a 50% or less occupancy rate at public and commercial recreation sites within three miles of the lake.

The prime beneficiaries of Lake Champlain dependent economic activity are Chittenden and Grand Isle counties in Vermont, and Clinton and Essex counties in New York. As shown in Table 1, approximately $29 million of the Vermont lake related expenditures occur in Chittenden County and $21 million in Grand Isle County. In New York, Lake Champlain visitors and seasonal residents spend approximately $21 million in Clinton County and $21 million in Essex County. Those would also be the counties receiving the majority of possible increases in recreational expenditures should the Plan be successfully implemented.

The values displayed in Table 1 reflect direct expenditures as determined by the study team. IMPLAN analysis was used to determine the total direct, indirect, and multiplier effects of those expenditures. The analysis found a total value added by tourism expenditures along the lakeshore of $97 million in Vermont and $57 million in New York, for

Table 1. Summary of Direct Lake Champlain Related Visitor and Seasonal Resident Expenditures, by County

	Visitor Expenditures[1]	Seasonal Household Expenditures[2]	Total
New York Counties	$38,936,709	$6,906,462	$45,843,171
Clinton County	$20,112,116	$1,246,628	$21,358,744
Essex County	$17,502,051	$3,525,564	$21,027,615
Washington County	$1,322,542	$2,134,270	$3,456,812
Vermont Counties	$67,796,000	$9,353,282	$77,149,282
Grand Isle County	$18,043,990	$3,405,902	$21,449,892
Franklin County	$8,185,683	$2,037,826	$10,223,509
Chittenden County	$27,298,869	$2,055,686	$29,354,555
Addison County	$12,678,772	$1,564,536	$14,243,308
Rutland County	$1,588,685	$289,332	$1,878,017
Totals	$106,732,709	$16,259,744	$122,992,453

1. Visitor expenditure data based on 1992 and 1993 dollars.
2. Seasonal resident expenditure data based on 1994 dollars.

a total economic impact of Lake Champlain related tourism expenditures of $154 million in 1992.

MODELING THE ECONOMIC IMPACTS OF PLAN IMPLEMENTATION

To understand the economic impacts of both the cost of Plan implementation and the benefits of improved water quality on the Lake Champlain basin's economy, the direct economic effects must be traced through the economy. Direct economic effects may be defined as those changes in final demands that are immediately stimulated by a given event. For instance, the event of a large pre-Christmas snow storm has several direct economic effects on final demand in the tourist industry including increased ski lift ticket sales, hotel reservations and restaurant meals all of which can lead to the creation of new jobs and wages. These initial, or direct changes in final demand then lead to many more subsequent indirect changes in the demand for goods and services produced in the economy. The restaurants order more food products, consume more cooking gas, require increased linen service, etc. In turn, each of these impacts can also create new wages and jobs. These new wages and jobs provide more wealth to households, which induce them to consume more, subsequently creating more demand for goods and services. Thus, the total economic impact of an event is typically much larger than the size of its direct effects alone.

To measure how the direct effects of an event such as improved water quality reverberate through the economy, economists typically use Input-Output models that incorporate all of the direct, indirect and induced economic effects. This research used a computer generated input-output model of the Lake Champlain basin's economy based upon the IMPLAN model (IMpact analysis for PLANing). The IMPLAN input-output model was originally designed by the US Forest Service and is currently maintained by the Minnesota IMPLAN Group, Inc. (MIG) in Stillwater, Minnesota.. IMPLAN is widely accepted as a useful tool for analyzing the economic impact of natural resource based events.

IMPLAN is grounded in the inter-industry relationships between 528 different national production industries. In addition to the inter-industry relationships that embody the production structure of the economy, IMPLAN incorporates an extensive county-specific database on government expenditures, inventory purchases, gross private capital formation, Commodity Credit Corporation inventories, federal government sales, employment and earnings, population, exports and imports.

The model used was constructed on the basis of 1992 production relationships. The program includes industry-specific deflators that translate current expenditures to this 1992 base year, and all dollar reports are in 1992 dollars. The model also includes industry specific "regional purchase coefficients" (RPCs), derived through multivariate regression techniques, to estimate the proportion of increased demand for an industry's products that will be supplied from local producers.

The study team tailored the IMPLAN modeling to treat state and local government expenditures as costs by computing the economic impact those funds would have had if they were spent by consumers. The reasoning is that expenditures financed by regional tax sources represent a redistribution from households (tax payers) to state and local government. Subsequently they imply a redistribution in the region's consumption patterns. For purposes of modeling the subsequent increase in regional government spending and taxes, households' personal consumption expenditures (PCE) were reduced by an equivalent amount. Implicit in this approach is that no existing government expenditures would be replaced by these new initiatives.

Economic Benefit and Cost Projections

The projected economic impacts of Plan implementation presented in this report are based on all of the non-phosphorus action items being implemented and one of the phosphorus control scenarios being implemented. Table 2 provides a summary of the economic impacts in Vermont of each of four scenarios. The New York IMPLAN modeling was done only on Scenario 3, as shown in Table 3.

The most significant result from the preliminary IMPLAN analyses is that all four proposed scenarios would lead to increased employment over both the short and long term (Phase 1 and 2). Phase 1 economic activity represents the construction of phosphorus control facilities (e.g., waste water treatment plant improvements, manure confinement), operation and maintenance (O&M) costs beginning in Year 2, and initiating implementation of the non-phosphorus programs. Phase 1 accounts for Years 1 - 5 of Plan implementation. The Phase 1 economic impacts are very stimulative to the basin economy because of the construction involved, but short-lived as they subside once the recommended capital improvements are put into place.

Phase 2 economic activity describes the long-term economic impacts in two sub-phases. Phase 2A represents the O&M economic activity related to phosphorus control and continued implementation of the non-phosphorus programs. The duration of Phase 2A is roughly Years 6 - 10. Phase 2B represents the economic value to the basin of improved water quality in Lake Champlain. The economic impact is modeled as a potential increase of 1%, 5%, or 10% in tourism expenditures over present levels. Those possible increases in recreation use related to water quality improvement are expected to begin occurring in Year 11 and continue indefinitely. The assumption in Phase 2B is that a

Table 2. Vermont Summary of Lake Champlain Plan Economic Impacts as Determined by the IMPLAN Analysis (annually, in millions of 1992 dollars)

Phosphorus Scenario	Type of Economic Impact	PHASE 1: Implementation and Construction[1]					PHASE 2A[2] Years 6 - 10	PHASE 2B Years 11 - 30		
		Year 1	Year 2	Year 3	Year 4	Year 5		+ 10%	+ 5%	+ 1%
								(percentage tourism increase)		
Scenario 1 (.8 mg/l)	Employment [persons][4]	233	227	220	214	208	114	428	242	94
	Employee Compensation[5]	6.2	6.1	6.0	6.0	5.9	4.4	8.0	5.2	3.0
	Total Value Added[6]	8.0	7.8	7.7	7.6	7.4	4.6	12.2	7.3	3.4
Scenario 2 (.5 mg/l)	Employment [persons][4]	207	200	193	186	179	113	425	239	91
	Employee Compensation[5]	5.8	5.7	5.6	5.5	5.4	4.4	7.9	5.2	3.0
	Total Value Added[6]	7.1	7.0	6.8	6.7	6.6	4.7	12.1	7.3	3.4
Scenario 3 (targeted point source controls)	Employment [persons][4]	231	227	223	220	216	127	441	255	107
	Employee Compensation[5]	6.2	6.1	6.1	6.0	6.0	4.5	8.1	5.4	3.2
	Total Value Added[6]	7.9	7.8	7.8	7.7	7.6	4.9	12.4	7.6	3.7
Scenario 4 (no New York point source controls)	Employment [persons][4]	241	236	232	228	224	124	438	253	104
	Employee Compensation[5]	6.3	6.3	6.2	6.2	6.1	4.5	8.1	5.3	3.1
	Total Value Added[6]	8.2	8.1	8.0	8.0	7.8	4.8	12.4	7.5	3.7

See notes under Table 3.

Table 3. New York Summary of Lake Champlain Plan Economic Impacts as Determined by the IMPLAN Analysis of Scenario 3 (annually, in millions of 1992 dollars)

Phosphorus Scenario	Type of Economic Impact	PHASE 1: Implementation and Construction[1]					PHASE 2A[2] Years 6-10	PHASE 2B Years 11 - 30		
		Year 1	Year 2	Year 3	Year 4	Year 5		+10%	+5%	+1%
								(percentage tourism increase)		
Scenario 3 (targeted point source controls)	Employment [persons][4]	48	49	56	59	53	35	223	141	74
	Employee Compensation[5]	1.6	1.7	1.9	2.0	1.8	1.2	4.4	3.1	2.1
	Total Value Added[6]	1.9	1.9	2.1	2.2	2.1	1.4	7.0	4.6	2.7

1. Phase 1 is the phosphorus control construction during years 1-5. O & M costs assumed to begin in year 2. Non-phosphorus programs begin Year 1.
2. Phase 2A represents years 6 to 10 when the non-phosphorus programs continue, as do O & M costs. Annual O & M costs for years 6-10 represent annual costs for the life of the project.
3. Phase 2B is the water quality improvement stage beginning approximately 11 years after implementation, projected as a possible 10%, 5%, or 1% increase in tourism expenditures annually over current levels.
4. Employment is the net change in number of jobs in the New York or Vermont portion of the basin. The employment increase accounts for reduced employment that would result from lower consumer spending to pay for the Plan.
5. Employee compensation is total wages paid to employees working within the New York or Vermont portion of the basin.
6. Total Value Added is the net economic impact in the New York or Vermont portion of the basin from federal, state, local, and private expenditures related to the Plan. It included employee compensation as well as other types of income and tax revenues.

The economic analysis is preliminary because the benefits do not account for property value increases, potential non-use benefits (e.g., option, existence, bequest values), and other benefits related to improved water quality in Lake Champlain. In addition, the analysis assumes a certain level of federal funding.

cleaner lake will yield some level of economic benefit in the basin. Although the economic impact is assumed to be positive, at this time the exact level of recreational response to an incremental improvement in water quality is unknown. The range of 1% to 10% is assumed to bracket the likely increases in recreational activity resulting from Plan implementation.

The preliminary economic assessments for Scenario 3 (corresponding to "Least Cost" in Figure 1) demonstrates that environmental planning approaches which employ least cost methods may offer the greatest economic benefits and the smallest economic costs as measured by employment and value added. Additional major findings from the IMPLAN modeling are identified below.

As computed by the IMPLAN model, recreational activity in the vicinity of Lake Champlain resulted in 5,654 jobs in 1992, with a total value added in Vermont and New York of $154 million. The resulting total economic impact in Vermont -- direct, indirect and induced impacts -- of the $77 million in Lake Champlain dependent expenditures was estimated at 3,707 jobs and $97 million dollars in value added for the basin's economy. In the New York portion of the basin, the impact of Lake Champlain recreation was estimated at 1,947 jobs and $57 million in value added.

Preliminary results using IMPLAN indicate that in Vermont Scenario 3 could produce an additional 127 new jobs beginning in Year 6, and $4.9 million in value added annually. In New York, there would be 35 new jobs beginning in Year 6. Value added in the New York basin economy would total $2.1 million annually, of which $1.8 million would be employee compensation. In addition, possible increased recreation activity related to water quality improvement could result in 91 to 425 new jobs in Vermont, and $3.4 to $12.1 million in value added. The lower number is attributable to a 1% increase in recreational activity around the lake, while the higher number is tied to a 10% increase in recreational activity attributable to the Lake Champlain restoration and protection effort. The middle-range estimate, based on a 5% increase in recreational activity over present levels, yields an annual economic impact in Vermont of approximately 255 jobs and $7.6 million in value added. The positive economic impacts in New York would include 74 to 223 new jobs and $2.7 to $7.0 million in value added annually..

In addition to the long term impacts, short term economic impacts could produce approximately 230 jobs and $7.6 million in value added in Vermont during each of the first five years of Plan implementation. In New York the short term impact would be approximately 50 jobs and $2 million in value added annually during the first five years of implementation. These impacts result from construction and engineering expenditures related to phosphorus control, and from employment and research related to the non-phosphorus plan items. The construction phase is projected to end in year 5. In subsequent years a portion of the positive economic impact is partially off-set by operation and maintenance costs for the improvements at waste water treatment facilities.

Scenario 3 results in 231 new jobs in Vermont in Year 1, declining annually to 216 jobs in Year 5. The Plan recommendations result in 127 jobs annually in Years 6 - 30. The net benefit in Vermont is a $7.9 million in Year 1, approximately $7.7 million in Years 2 - 5, and $4.9 million in Years 6 - 10. In Phase 2B, beginning in Year 11, a 1% increase in tourism activity yields a net annual benefit of $3.7 million. A 5% increase in tourism activity results in a $7.6 million positive flow annually in Vermont, beginning in Year 11. Of the four scenarios analyzed in this research, Scenario 3 appears to the most cost effective in Vermont and results in a positive flow of capital into the basin throughout the projected period of implementation.

In New York, Scenario 3 results in 48 new high pay construction and engineering jobs in Year 1, increasing to 53 jobs in Year 5 and resulting in 35 new jobs annually in Years

6 - 10. The net benefit in the New York portion of the basin is $1.9 million in Year 1, approximately $2.1 million in Years 2 - 5, and $1.4 million in Years 6 - 10. In Phase 2B, beginning in Year 11, a 1% increase in tourism activity yields a net annual benefit of $2.7 million. A 5% increase in tourism activity results in a $4.6 million positive flow annually in New York, beginning in Year 11. Given the comparative findings of the four scenarios in Vermont, Scenario 3 appears to the most cost effective approach in New York and it results in a positive flow of capital into the basin throughout the projected period of implementation.

It is important to recognize the scope of the data and IMPLAN when interpreting these findings. The actual long term employment outcomes will depend on the level of water quality improvement and the subsequent reaction in the tourism industry. The smaller the resulting increase in tourism, the smaller the increase in employment and other economic benefits. A significant amount of federal funding is also included in these preliminary projections. Without this funding, the projected employment outcomes would be smaller. All federal funds have been treated entirely as money derived from outside sources. Of special significance is the impact of O&M expenditures in the phosphorous reduction strategies. The preliminary IMPLAN results indicate that O&M expenses have a negative net impact on job creation. Thus, if future tourism increases are not realized and/or federal funding does not occur, the economic benefits of these proposed initiatives will be reduced significantly. However, the livelihood of funding from a variety of federal agencies is significantly enhanced by the comprehensive planning effort completed to date.

UNDERSTANDING THE PERCEIVED ECONOMIC IMPLICATIONS OF PLAN IMPLEMENTATION

The prior discussion focused almost exclusively on quantifying the costs and benefits related to Plan implementation, however, economic analysis involves more than quantifying monetary transactions. Additional issues that the study team attempted to address include the following:

What are major economic trends in the basin related to Lake Champlain?
Are there other, indirect benefits and costs of improving Lake Champlain?
How will the economic benefits actually materialize in the local economy?
Where in the basin will the costs and benefits occur?

In order to gain a better understanding of how particular types of business perceive the Plan affecting their day-to-day operations, the study team convened two economic focus groups during our preliminary economic analysis work and a third economic focus group session during the economic analysis of the draft plan. In addition, the study team sent out a survey to recreation businesses around the lake and to local governments in the basin, asking for their perceptions of Plan-related economic issues.

Economic Focus Group Sessions

The economic focus group sessions were morning sessions held at three different locations on the lake shore. The original list of invitees to the focus group session totaled

156 people residing and working around the basin. Twenty individuals attended the first session, nine participated in the second, and five participated in the third. A total of 26 different individuals representing economic interests around the basin became involved in these formal discussions on the economic issues involved in the Lake Champlain planning effort. Those in attendance represented a very wide cross-section of economic interests around the lake, including: marinas, the paper industry, City of Burlington, Plattsburgh Chamber of Commerce, local government, agriculture, forestry, recreation, banking, the charter boat sports fishing industry, and watershed associations. The meetings were facilitated by a professional facilitator who was assisted by the study team and Lake Champlain Basin Program staff.

Some participants felt that the economic benefits have to be area-specific and should not reflect just the value of the lake to the greater basin population, many of whom may not receive any direct economic benefit from the lake. Others felt that since this is a Plan for the future, a wide variety of possible present and future benefits should be considered. One aspect of the basin-wide benefits of a clean lake was expressed in terms of the lake as an asset to local industry in attracting higher caliber employees.

Some felt that primary, secondary, and tertiary costs should be quantified for specific areas around the lake and that the estimated benefits should only be accounted for in relation to those specific areas. Others pointed out that recreational benefits of cleaner water could occur throughout the lake, so it would be difficult to reconcile costs and benefits for a particular bay or other location on the lake.

During the second session, the main discussion centered on specific measures to boost the economy while protecting Lake Champlain, and at least 11 distinct proposals were offered and discussed. Most seemed to be heartily supported by the group present, although there was no attempt at a group consensus. Some of the main themes included the following:

1. Innovation, ideas, creativity -- all need to be encouraged in the private sector and supported by government.
2. Pollution prevention is key to cleaning up the lake, and prevention is tied to the encouragement of innovation.
3. There is a role for government in protecting local economies while preventing pollution of Lake Champlain.
4. On-going Lake Champlain planning efforts must facilitate and accommodate the participation of local economic interests.

The third economic focus group session focused on reviewing preliminary findings of the economic analysis work, and was held on April 11, 1996. The study team was able to incorporate the edits and suggestions into the final draft of the report.

Questionnaire Surveys

The recreation site survey carried out for this research asked recreation providers to evaluate possible factors affecting recreation activity near their facilities. Respondents were asked to rate on a scale of +3 to -3 how much of an influence each of the items were on encouraging or discouraging visits to their area of the lake.

The Canadian exchange rate was the most negative item discouraging visits to the area, a factor few have any control over. Also recognized as negative factors, although near the neutral category, were some of the Lake Champlain environmental factors, namely aquatic weed growth, fish consumption advisories, and zebra mussels. There appears to be the perception among the recreation providers that these issues are adversely affecting tourism around Lake Champlain, but not to a great degree.

The local government survey listed 12 community and environmental programs that roughly parallel the major recommendations in the Lake Champlain Plan. The survey respondents were asked how their community might set funding priorities should federal or state funds become available for the given program, and then to select the highest and second highest priority from their perspective.

The highest priority item for the 17 local governments responding to this section of the survey was Safe Drinking Water Act compliance, selected by nine respondents, or 53% of the governments responding. The next highest priority was testing of private septic systems in shoreline areas, selected by three respondents. When asked about the 2nd highest priority, four local governments selected developing management programs for nuisance aquatics such as zebra mussels, lamprey eel, water chestnut, etc. Another four selected advertising and promoting the region's recreational and cultural attractions. Two selected toxic pollution control programs and two others selected wastewater treatment plant improvements to control phosphorus as priority issues.

It seems evident that local governments recognized benefits in many of the Lake Champlain priority items. Each item listed was mentioned by at least one of the 17 local governments as a priority item. As a group, local governments were most supportive of funding that would aid in Safe Drinking Water Act compliance. It is interesting to note that it is not simply the need for financial assistance that was driving their support for this recommendation. The comments illuminated their interest in protecting drinking water resources because their local tourism economies depend on the perception of a safe, healthy environment, and because drinking water "affects all life."

In summary of the community outreach aspect of the study team's work, the following benefits and costs were identified as important to local economic interests around the basin:

Benefits. The value of a cleaner lake to future generations.
    The value of a cleaner lake to residents who rarely or never use the lake.
    The influence of a cleaner lake in attracting high tech and other industries to the basin.
    The impact of reduced fish consumption advisories on fishing and other tourism activity.
    The impact of improved sewage treatment facilities on a small community's ability to attract and accommodate new industry and development.
    The influence of a cleaner lake on property values and the resulting benefit to local taxing authorities.
    The value of wetlands in the basin for flood storage, water quality, and erosion control.
    The biological value and resulting economic value of wetlands for fish and wildlife habitat and ecological diversity.

Costs. The influence of nutrient management costs on farmers' ability to compete in national markets.
    The influence of local tax increases related to wastewater treatment plant improvements on a small community's ability to attract new industries and development.
    Costs to local industries related to the control or removal of toxic waste sites.

Possible influences on the quality of jobs in the basin: gaining lower paying service sector jobs while losing higher paying manufacturing jobs. Possible environmental costs of increased tourism.

RECOMMENDATIONS FOR FURTHER RESEARCH

There are a number of recommendations for further socio-economic research in the Lake Champlain Basin, organized under four broad categories.

Cost Efficient and Equitable Phosphorus Control

There is a need for further research on the economics and institutional arrangements involved in controlling non point sources (NPS) pollution. As development in the Lake Champlain region continues, this research will be key to maintaining water quality. More field studies of agriculture and urban best management practices (BMP's) in the Lake Champlain watershed are needed, as well as further research on economic incentives, zoning and community planning approaches, transferable P reduction credits, and the institutional arrangements for administering these. The question to be explored is how we can design market-based incentives to more effectively reduce the overall level of phosphorus loading at the lowest possible cost, and how those incentives can be used to economically allocate the reductions among the point, non point agricultural and other non point sources.

We need better data on point and non point control costs and the impacts of reducing phosphorus loading from each. This means better technical and cost data on specific phosphorus producing activities, on particular phosphorus reduction techniques, and a better understanding of who is paying for the cost of phosphorus reduction.

Continued studies of the hydrodynamics and phosphorus (P) cycling within the Lake are needed. Where to concentrate P reductions depends on where the P ends up. The "Bathtub" model that was the hydrodynamic foundation for the economic phosphorus reduction model deserves further verification and refinement. That research would then feed back into refinement of P reduction strategies. A GIS-based model developed for the LaPlatte sub-basin could be expanded to the entire watershed and would provide the basis for monitoring and refinement of the water quality improvement aspects of the Plan.

Valuing a Cleaner Lake Champlain

Continued research is needed on the relationship between environmental quality and economic activity in the basin. What does a cleaner lake mean in dollar terms? Prior research started to consider this question, but due to limitations in funding and tight deadlines, it relied primarily on the compilation of existing studies and data. An integrated and on-going research program that refines the economic benefits that result from a cleaner lake is needed that involves: new and sophisticated recreational user surveys, contingent value studies, analysis of tourism trends and expenditure patterns, hedonic price studies of property values, interviews with new businesses to understand the reasons for locating in the area, etc.

Economic benefit studies are important because the costs of pollution control strategies need to be balanced against the benefits. The costs are specific to treatment plants, farms and other sources of pollutants. The benefits are much more diffuse across urban areas, industry groupings, economic activities and sectors of the economy. Benefit studies for Lake Champlain have focused mainly on fishing and boating activities. Other benefits need to be identified and explored such as swimming, beach values, quality of life, economic renewal of lakeshore communities, benefits to lakeshore property owners, etc. As tourism becomes more important to the economy and the lake becomes a central focus of that tourism economy in the basin, better information on the economic impacts of tourists will help target resources and enhance the value of the lake.

Involving the Public and Incorporating Local Economic Concerns

How can the business community, the agricultural community and other interests become more involved in the co-management of Lake Champlain? Natural resource management has gone through a series of steps since passage of the National Environmental Policy Act in 1970, evolving from total agency control, to gathering public opinion, to involving the public in prioritizing issues, to now moving towards co-management with landowners, users and other stakeholders. The dialogue among interested Lake Champlain residents begun by the Lake Champlain Management Conference, and enhanced through the economic analysis research, needs to be continued in order to maintain the cooperative approach exemplified by the Lake Champlain "Opportunities for Action".

The best application of Lake Champlain research funds and pollution control measures is usually determined within the scientific research community. However, they are also public issues comprised of social, political and economic considerations, especially when costs result in the private sector or for local governments. Prior economic analysis incorporated community involvement and facilitation exercises that identified public and economic issues of concern. A framework is needed for incorporating public concerns, economic considerations and input from the business community, local governments and others into the on-going research agenda for Lake Champlain.

The Lake Champlain Socioeconomic Database

Who will be responsible for updating and maintaining the socioeconomic data compiled for the basin to date? The 2000 Census will provide us with much new information on the basin, but the Census data are not easily transformed from the two states to a basin wide area, not to mention census data for the Quebec portion of the basin. State and federal agencies collect other data as well, but it is time consuming to transform those data to basin-specific information.

CONCLUSION

Regardless of whether it is market or nonmarket benefits and costs that are being considered, the public's understanding of benefits and costs of Lake Champlain activities would be enhanced by clear definitions of the terms. The study team, developed the following working definition of costs and benefits for use in the on-going analysis and dialogue on the economic aspects of the Lake Champlain protection and restoration.

In analyzing environmental protection and restoration programs the definitions of costs and benefits should be constructed both comprehensively and carefully. First, it is essential to be clear about the geographic perspective from which costs and benefits are defined. Federal funding for Lake Champlain environmental protection programs would be viewed as a cost at the national level, but should be counted as a benefit if the analysis is being conducted from a regional perspective. In addition, it is essential to use a consistent baseline in measuring costs and benefits. For example, if costs of environmental protection efforts are defined in relation to a no action alternative, then benefits must also be calculated in relation to the environmental and economic conditions that would have existed if no action were taken. The following definitions are intended to define the categories of potential costs and benefits of the Lake Champlain Pollution Control and Environmental Restoration Plan from a national perspective in relation to a no action alternative.

Costs

Direct costs of environmental protection or restoration efforts include capital, operating maintenance and administrative expenditures by both the public and private sector. Indirect costs include any reduction in profits in excess of the direct costs of pollution control as well as a portion of lost wages due to any increase in unemployment. Only a portion of lost wages represents a true cost since unemployed workers will find other productive, although not necessarily equally valuable, uses for their time such as continuing their education or providing additional care to family members. In estimating indirect costs or multiplier effects it is important to recognize that environmental protection efforts may reduce output or employment in one industry while increasing it in others. These partially or fully offsetting gains in other sectors of the economy must therefore be subtracted from indirect costs or included as benefits.

Benefits

The direct benefits or avoided costs of environmental restoration and protection efforts can include increased recreational enjoyment and aesthetic appreciation, reduced public health risks, and increased profits or consumer benefits from direct commercial uses of environmental resources (e.g. sale of fish caught on the lake). Direct recreational and aesthetic benefits of environmental programs are often capitalized in the form of increases (or avoided reductions) in property values in the affected area. Benefits of pollution control or remediation efforts that enhance public health can be quantified in the form of reduced medical costs and increased productivity. However, the benefits of water pollution control efforts that protect or restore public water supplies should be measured by the avoided or reduced costs of water treatment.

In addition to these direct use benefits of environmental protection, there are potential non-use benefits related to changes in option, existence and bequest values. Option value is simply the value to the individual of preserving the opportunity to use a clean environment and is therefore closely related to, but nevertheless conceptually distinct from, direct use benefits. Bequest values are based on the satisfaction that individuals derive from knowing their children, or future generations in general, will be able to enjoy a clean environment. Existence value is any additional satisfaction, apart from direct use, option, or bequest values that individuals receive simply from knowing that an important ecosystem, natural area, or endangered species has been protected.

Environmental protection efforts will also generate beneficial multiplier effects. Purchases of pollution control equipment and operation of pollution control programs will increase output and employment in certain sectors of the economy. Multiplier effects should also be taken into account if environmental programs are expected to lead to increases, or avoid reductions, in tourism and recreational expenditures. The benefit of these multiplier effects should be measured by the increase in before-tax profits of local businesses, as well as the portion of increased wages in excess of the value of leisure time. As noted above, beneficial multiplier effects for some sectors of the economy must be considered in relation to negative multiplier effects that environmental regulation and pollution control costs may create for other sectors.

Cost of No Action

The public choice of continuing to protect and improve the water quality of Lake Champlain, versus taking no action in terms of reducing nutrient pollution, could be viewed in private sector terms as protecting an investment and expanding the basin's economic potential. On the cost side, taking no action could be viewed as deferring or externalizing pollution control costs to a later date and a future group of residents. Costs of phosphorus control will likely increase in the future, while the current costs to phosphorus dischargers could be considered as artificially low in terms of the actual costs for achieving the in-lake phosphorous levels recommended by the Plan.

The relatively clean water of Lake Champlain is a valuable, limited commodity that serves multiple uses including recreation, drinking water, industrial uses and aesthetic enjoyment. As evidenced in lake and coastal areas around the country, waterfront areas are valuable assets, however, their value hinges on the quality of the water along shoreline areas. The cost of cleaning and purifying water increases as the water quality decreases. By doing nothing and refraining from making any public investment in Lake Champlain, its economic and social values will deteriorate as with any neglected public or private property, and the costs of future clean-up will be greater than at present.

According to data compiled for this research, a conservative estimate is that Lake Champlain annually generates at least $123 million directly in the basin economy in the form of recreational activity. In addition, this research has identified at least three other economic benefit categories related to the Plan:

1. Other direct benefits to users as extrapolated from benefit estimate studies are in the range of $13 to $63 million annually.
2. There is also a one time property value benefit, or avoided cost, of perhaps $43 million.
3. Estimates of increased economic activity as generated from the IMPLAN analysis (Vermont & New York combined) indicate $9.7 million in total valued added annually related to Plan implementation in Years 1 - 5, $6.3 million in Years 6 - 10, and $6.4 to $19.4 million annually in Years 11 - 30.

The economic analysis indicates that the "Opportunities for Action" priority actions for improving water quality would have positive economic impacts within the Lake Champlain basin. These impacts are not large in terms of the entire basin economy but may be sizable if concentrated in a few local economies. Given the array of economic benefits discussed above, doing nothing to protect and enhance Lake Champlain does not appear to be economically in the best interest of basin residents and businesses. The issue then becomes how to ensure efficient and equitable distribution of the costs and benefits throughout the Plan implementation process. The economic analysis provides guidance in that direction.

REFERENCES

Feather, T., *Valuation of Lake Resources Through Hedonic Pricing,* US Army Corps of Engineers, Fort Belvoir, Va., 1992.

Holmes & Associates and Anthony Artuso,*Economic Analysis of the Draft Final Plan for the Lake Champlain Management Conference,* Lake Champlain Basin Program, Grand Isle, Vt., 1996.

Holmes & Associates and Anthony Artuso,*Preliminary Economic Analysis Of The Draft Plan For The Lake Champlain Basin Program, Part 2*, Lake Champlain Basin Technical Report No. 14B, prepared by Timothy P. Holmes, Anthony Artuso, Tommy L. Brown, Lake Champlain Basin Program, Grand Isle, Vt., 1995a.

Holmes & Associates and Anthony Artuso,*Preliminary Economic Analysis Of The Draft Plan For The Lake Champlain Basin Program*, Lake Champlain Basin Technical Report No. 12B, prepared by Timothy P. Holmes, Anthony Artuso, Tommy L. Brown, Robert L. Bancroft, and James F. Dunne for the Lake Champlain Management Conference, Lake Champlain Basin Program, Grand Isle Vt., 1995b.

Holmes & Associates, *Socio-Economic Profile, Database and Description of the Tourism Economy for the Lake Champlain Basin*, Lake Champlain Basin Program, Grand Isle, Vt., 1993.

Lake Champlain Basin Program, *Opportunities for Action: An Evolving Plan for the Future of the Lake Champlain Basin*, draft Final Plan, Lake Champlain Basin Program, Grand Isle, Vt., 1996a.

Lake Champlain Basin Program, *Background Technical Information for Opportunities for Action,* Lake Champlain Basin Program, Grand Isle, Vt., 1996b.

Lake Champlain Basin Program, *Lake Champlain Recreation Assessment Report*, prepared by the Vermont Agency of Natural Resources and the NYS Office of Parks, Recreation and Historic Preservation, Lake Champlain Basin Program, Grand Isle, Vt., 1996c.

Lake Champlain Basin Program. *Lake Champlain Outdoor Recreation Facilities Inventory.* Prepared by the Vermont Department of Forests, Parks and Recreation in cooperation with the NYS Office of Parks, Recreation and Historic Preservation. Lake Champlain Basin Program, Grand Isle, Vt., 1995.

Vermont Natural Resources and New York Environmental Conservation Departments,*A Phosphorus Budget, Model and Load Reduction Strategy for Lake Champlain: Lake Champlain Diagnostic Feasibility Study Final Report*, Vermont Department of Natural Resources, Waterbury Vt., 1994.

Watershed Management at a Crossroads: Lessons Learned and New Challenges Following Seven Years of Cooperation Through the Lake Champlain Basin Program

Lee Steppacher and Eric Perkins

ABSTRACT

The Lake Champlain Special Designation Act created a program of partnerships. Local, state and federal agencies alongside non-governmental organizations, academicians, individuals and businesses from two U.S. states and one Canadian province work together to make the Lake Champlain Basin Program (LCBP) a successful model for watershed management. Each partner contributes expertise, perspective and funds to better understand and manage Lake Champlain and its resources. A broad based coordinating committee addresses a multitude of issues including water quality, fish and wildlife, and recreational and cultural heritage resources. After seven years, the participating entities have learned many things about cooperative watershed planning and management. For example, partnerships are essential and can be fostered through effective communication, local level implementation builds a sustainable program, pollution prevention is an important complement to command and control programs, broad stakeholder involvement is required for meaningful consensus, economic issues should be fully integrated into the planning process, planning must be based on sound science, and public involvement is essential for long term commitment to management decisions. As the LCBP focuses on implementation of the plan that was completed in 1996, new challenges are presented. These include: how to maintain and broaden public and financial support for management activities, how to evaluate progress, and how to address emerging issues without undermining past accomplishments.

Lake Champlain in Transition: From Research Toward Restoration
Water Science and Application Volume 1, Pages 419-433
This paper not subject to U.S. copyright
Published in 1999 by the American Geophysical Union

BACKGROUND

Lake Champlain gained national recognition in 1990 with passage of the Lake Champlain Special Designation Act. The congressional delegations and the people of Vermont and New York recognized not only the unsurpassed beauty of the lake, the fragile ecosystem, the vital communities and unique cultural history of the lake and its basin, but also the potential threats to the lake's integrity. The Special Designation Act created a framework to study and understand the systems of the lake and its basin in order to develop a comprehensive plan to protect and restore lake resources into the future.

This is not the first time a large planning effort has been undertaken on Lake Champlain. A number of interstate and international efforts have been initiated to enhance coordination and better manage lake resources since the beginning of the century. The most recent planning effort (prior to the LCBP) was the "Level B Study" conducted by the New England River Basin Commission in the late seventies. While none of the previous efforts sustained themselves over the long term, the lessons learned informed the drafters of the Special Designation Act and resulted in a stronger piece of legislation. (Refer to Yellow Wood Associates [1993] for more information on earlier management efforts.)

The Statement of Legislative Intent accompanying the Special Designation Act called for the planning process to be comprehensive in scope and inclusive of a diversity of stakeholders around the basin. Without using the term, the Legislative Intent recognized the importance of ecosystem management and the need to examine the interrelationships of biological communities with each other and with the physical and chemical environment. It also spoke of the need to include the human component.

The Special Designation Act created a 31 member board, The Lake Champlain Management Conference (LCMC), to develop a comprehensive pollution prevention, control and restoration plan. The LCMC was responsible for overseeing all the activities of the Lake Champlain Basin Program (LCBP) that would lead to development of the plan, including support of relevant research, demonstration projects, monitoring programs and education and outreach efforts. The Environmental Protection Agency was given the responsibility of administering the LCBP, in cooperation with the Vermont Agency of Natural Resources and the New York State Department of Environmental Conservation. Other federal agencies, including the Department of Agriculture, the Fish and Wildlife Service, the National Park Service and the Geological Survey, were asked to coordinate their activities with those of the LCBP. The LCMC was directed to coordinate its research program with the Lake Champlain Research Consortium.

The Special Designation Act specifically outlined the membership of the 31 member LCMC, ensuring that it included broad representation from both NY and VT, as well as from federal agencies. Citizens, academicians, business people, environmental advocates, and farmers worked together with state legislators and government regulators. Representatives from the New York and Vermont Lake Champlain Citizens Advisory Committees were also included, emphasizing the importance of citizen input to the process of plan development. The amount and breadth of participation was greater than that of any past effort on Lake Champlain and most other watershed programs nationally.

It took time and effort for the LCMC to develop into a cohesive, functional group. Trust needed to be built among individuals, and operating guidelines needed to be formulated. These guidelines emphasized the desire to conduct business in a spirit of cooperation, working toward a common goal. In this light, the LCMC aspired to make all decisions by consensus. Voting, when needed, was for the purpose of having a record of minority opinions and to ensure accountability.

An organizational structure developed around the LCMC to support the variety of tasks required for plan development. The Special Designation Act recognized that the LCMC was responsible for policy making, and created a Technical Advisory Committee (TAC) to provide technical input and guidance. The TAC included a wide range of interdisciplinary scientists, resource managers, and business people representing key technical concerns such as water quality, toxic contamination, fish and wildlife, recreation and cultural resources, economics and data management. An Education and Outreach Committee comprised of LCMC members and educators within the basin was charged with developing an outreach program to inform the public on key issues facing the lake, soliciting input from the public and supporting educational programs within the schools. A Plan Formulation Team (PFT), a subgroup of the LCMC, oversaw the details of developing the plan and presented its findings to the LCMC. A vast array of individuals and organizations participated in the LCBP, and the partnerships created have continued beyond completion of the plan. Indeed, these partnerships are largely the defining characteristic of the LCBP today.

While the Special Designation Act required a comprehensive plan, the LCMC had to define the scope of the effort for it to be manageable. Ultimately, the LCMC decided to embrace a phased approach to planning, focusing on issues facing the lake and its immediate environs first. This was done with the understanding that over the coming years, the LCBP would begin to address issues farther out into the watershed. The LCMC also chose not to address issues that were viewed more as statewide issues than Lake Champlain Basin issues. The management of deer herds, for example, while an important issue in the basin, was considered more of a statewide issue. The LCMC chose to direct its funding and energy toward issues that were specific to the lake and its basin, and encouraged state managers to implement appropriate management programs for statewide issues. Additionally, the LCMC viewed a few major issues, such as the Pine Street Barge Canal Superfund Site in Burlington, as potential energy and money sinks which could divert resources away from other needed projects. The LCMC recognized the importance of these issues to the lake but believed that sufficient attention was being placed on these through separate processes, and that LCMC involvement in these would detract from the overall planning focus.

From 1991 to 1998 federal funding for the LCBP has surpassed $29 million. The Special Designation Act authorized $2 million annually for 5 years to the EPA to administer the program, as well as $2 million for the US Department of Agriculture and $1 million to the Department of Interior to support specific activities based on agencies' particular expertise. While these exact amounts were not realized each year, significant funding was always appropriated and has continued beyond the 5 years identified in the Act (see Table 1). Additionally, the EPA and USDA funds required a local or state match. EPA funds were matched at a 25% rate, and after 8 years this match totaled approximately 4.6 million, while match for USDA funds from farmers and the states exceeded $3 million during this period. This required investment by state and local entities strengthens commitments from all partners, decreases reliance on 'free' federal dollars, and increases overall sustainability of the Program. State and local contributions to LCBP activities have greatly exceeded the match requirements over the years.

The first five years of activity under the Special Designation Act culminated in the approval of the final plan, *Opportunities for Action*, in October 1996. The plan addresses a wide variety of issues, ranging from the improvement of water quality, to the protection of the basin's living natural resources, to the preservation of the region's rich cultural heritage. From these, the plan identifies three priorities for action: the reduction of phosphorus pollution in targeted watersheds of the lake, the prevention and control of persistent toxic contaminants found lakewide or in localized areas of the lake, and the management of nuisance nonnative aquatic plants and animals.

After seven years, the participating entities have learned many things about cooperative watershed planning and management. These will be discussed in the remainder of this paper. In the years since *Opportunities for Action* was approved in 1996, these lessons continue to inform implementation activities. New challenges are also presented including how to maintain public and financial support, how to evaluate progress and how to address emerging issues.

LESSONS LEARNED

Partnerships are essential and can be fostered through effective communication

Each member of the LCMC and the TAC had an interest in the future of Lake Champlain which inspired him or her to participate in the planning process. While some were professional managers of resource programs, others depended on the lake for their livelihood, or recreated there. Each of these interests was distinct, and the challenge was to develop a set of shared goals for the future of the lake. Prior to the official convening of the LCMC in June 1991, a workshop was held in which future LCMC members developed a shared vision for the lake. This was the first step in the development of partnerships and mutual understanding among LCMC members.

Throughout the following three years, the LCMC fully discussed each issue to be addressed in the plan. A process was designed to foster discussion among key players on all aspects of each issue. The TAC was responsible for drafting an information paper which outlined the technical underpinnings of the issues and current management efforts. The policy makers on the PFT and LCMC used this as a basis for discussions regarding future management options. The LCMC met for a full day each month and participated in extensive debate on each issue. At these meetings, time was also reserved for members of the public to express their views.

While this process was at time tedious and seemingly inefficient, it set the ground work for creation of important partnerships. In some instances, this was the first time that policy makers from Vermont and New York discussed respective resource management programs together and recognized the commonality among them. For managers and scientists, this process enhanc-ed understanding of the myriad issues facing the lake and their interrelated nature.

On the LCMC this information exchange allowed for far greater understanding of the issues and more meaningful debate. Coalitions were built around particular issues, but were very fluid and often reformed around other issues. For example, at one time all local government representatives united in opposition to more stringent phosphorus controls that might increase the financial burden to municipalities, on another occasion the academic and environmental organizations joined forces to encourage government agencies to support

TABLE 1. Federal Lake Champlain Basin Program Funding[1], 1991-1998

Funding source	Amount (in millions of $)	Description of use
U.S. Environmental Protection Agency	12.75	Support development and implementation of LCBP plan through program coordination, research, demonstration projects, education and outreach.
National Park Service	1.65	Cultural and recreational planning and demonstration projects.
U.S. Fish & Wildlife Service	3.88	Lamprey control, habitat restoration, wetland restoration, fish & wildlife management.
U.S. Geological Survey	1.38	Stream gauging, GIS data development, research.
National Oceanic & Atmospheric Administration	0.90	Air mercury monitoring, lake hydrodynamic measurements.
National Marine Fisheries Service	0.10	Lamprey control.
International Fisheries Commission	0.20	Lamprey control.
U.S. Department of Agriculture	9.50+	Agricultural best management practices (BMPs) on farms

[1]These federal funds were specifically earmarked to be used to support LCBP activities. Many of these agencies expend additional funds in the Basin through other programs. State and local entities also contribute substantially to the LCBP. The EPA funds require a 25% match ($4.6 million to date) and the USDA cost share program requires at least a 25% state/local/farmer share.

more research, and another coalition formed when Vermonters unified to strongly encourage New York to support water chestnut control efforts. Slowly trust and understanding developed among individual members which led to the possibility of building consensus.

The role of the TAC chair in fostering partnerships between the technical community and policy makers cannot be overlooked. With a seat on the LCMC, the TAC chair was depended on to communicate technical issues to a nontechnical audience and explain technical implications of various management scenarios. Stories are still told of the TAC chair performing a small dance to help explain the dynamics of the lake's internal seiche. This type of exchange strengthened overall group rapport and made it easier for members to find common ground.

Local-level implementation builds a sustainable program

Citizens know and care for their local watershed area in a way that most state and federal managers cannot. In a basin as large and diverse as Lake Champlain's, citizen-based groups such as watershed associations can provide a much needed link between local community goals and basin-wide planning efforts. The LCBP has supported local projects to promote "hands-on" activities and to increase citizen awareness of the connections between their community and the lake.

Grant programs tailored for small applicants, such as the Lake Champlain Partnership Program sponsored by the citizens advisory committees, have proven to be especially effective public involvement tools. With funding from the EPA and the National Park Service, the Partnership Program awarded very small grants ($500-$5000) to grassroots efforts that demonstrated practical ways to address economic and conservation challenges and produce tangible benefits. Each year approximately 20 projects were funded throughout the basin including projects as divergent as building a canoe launch on the Saranac River, restoring Carillon Park in Ticonderoga and studying cormorant feeding habits. The popularity of this program is reconfirmed each year when far more applications are received than can be funded.

Since 1994, over $1.1 million has been distributed to local entities through grant programs supporting pollution prevention and demonstration projects, education, and local implementation efforts. Projects are tailored to address the issues identified in the plan at the local level. Over time, projects throughout the basin, in lakeshore communities as well as farther into the watershed, have been supported. Each project puts additional citizens in touch with the LCBP and builds awareness of the plan and basin-wide issues.

These projects have also provided an opportunity for the LCBP to help link local groups to technical assistance available through state and federal agencies. Watershed associations completing streambank restoration projects with LCBP funds, for example, typically receive supplies and technical assistance from the US Fish and Wildlife Service's Partners for Wildlife program, and additional technical assistance from the state Department of Environmental Conservation and the USDA Natural Resources Conservation Service.

Pollution prevention is an important complement to command and control programs

The traditional approach to solving industrial pollution problems is to regulate treatment or removal of pollutants at their point of discharge. While this "command and control" approach continues to play an important role in the protection of water quality (for example, chemical removal of phosphorus from municipal treatment plant discharges is a key component of the Lake Champlain phosphorus strategy) it can be expensive and sometimes only transfers the problem from water to land or air. These concerns led the LCMC to place great emphasis on the concept of pollution prevention. Pollution prevention encompasses many different activities aimed at reducing or eliminating the generation of pollutants at the source. Pollution prevention has the exciting potential for both protecting the environment and strengthening the economy through more efficient manufacturing and raw material use. Pollution prevention can be pursued by a diverse constituency of individuals, government entities, and the business community.

The LCMC initiated a grant program for pollution prevention projects in 1994. One of the grants awarded through this program enabled Retired Engineers And Professionals (REAP), a nonprofit organization comprised of retired industry professionals, to provide education and technical assistance to vehicle service facilities. As is the case with many businesses in the basin, vehicle service facilities are small (typically employing less than 10 people) and staff seldom have the opportunity to leave the premises to take advantage of pollution prevention seminars offered off site. The REAP project was designed to provide direct onsite assistance to help operators implement site specific pollution prevention measures. REAP provided assistance to 20 facilities through this project [Retired Engineers and Professionals, 1996]. Illustrating that pollution prevention may also be practiced by a municipality, REAP was funded in 1996 to work with a municipality to address the source of heavy metals found in sewage sludge from the wastewater treatment plant. Through this project REAP is helping the municipality identify the source of the metal, and is designing an appropriate plan to reduce, recycle or eliminate the metal from the waste stream [Retired Engineers and Professionals, 1997].

The Lake Champlain Committee (a nonprofit lake advocacy organization established in 1963) in cooperation with the Burlington Department of Health received a grant in 1994 to promote and demonstrate an array of pollution prevention activities for individuals and businesses. This project demonstrated alternatives to pesticide applications along a railroad right of way, at a golf course and on individual lawns, organized a drive-out day to reduce car use, produced public service announcements regarding wood burning stoves, promoted the use of alternative cleaning products in area businesses and created a toxic reduction tool kit to expand these types of activities into the community.

The success of these projects and the enthusiasm with which they were received has reinforced the LCBP's commitment to pollution prevention. While not yet a realistic substitute for end-of-the-pipe control programs, pollution prevention has been shown to be an important complement to these traditional approaches. These projects also helped to strengthen the tenant held by the LCMC that everyone within the basin is responsible for helping to protect the basin's resources, and that working partnerships are critical to this process. The task now is to build on these demonstrations and to institutionalize them throughout the basin.

Broad stakeholder involvement is essential for meaningful consensus

Consensus within the LCMC did not necessarily equate to consensus among the public, as was discovered in 1994 when a draft version of the plan *Opportunities for Action* was completed and released to the public. At this time, a series of 'listening sessions' were held throughout the basin to solicit input on the plan. While there were many people who were very supportive of the efforts of the LCMC, there was, in particular, one well organized group in NY which was strongly outspoken against the plan and much of what the LCBP represented. This outpouring was partially based on misinformation and/or lack of information. There was a fear that the LCBP would become a regulatory organization even though the plan stated that this was not the case. More importantly, concern was expressed about possible economic implications of the plan for local individuals and businesses in NY. This protest was effective in that it was organized, received media attention, gained the attention of federal congressional representatives, and could have derailed further planning efforts.

Soon after the listening sessions were completed, two representatives of those opposed to the plan were appointed to the LCMC. Another process of reviewing, refining and setting priorities began based on public comment. The two new members speaking for their constituents expressed numerous concerns and painfully convinced LCMC members to consider economic impacts more directly. As a result, additional economic studies were undertaken, business leaders and economists were brought together to discuss impacts of the plan, and the LCMC embraced an overarching tenant that the plan would not impose any new financial responsibilities on local governments. As difficult as it was to rethink and refine the plan, by listening and responding to the concerns of the business community, new partnerships were formed. When the final draft plan was presented to the public, the business community in NY, and the majority of those who opposed the earlier draft, were supportive of the effort. This experience demonstrated that meaningful consensus can only be achieved when the full spectrum of interests is represented at the table.

Economics should be fully integrated into the planning process

In the early stages of the planning process, economic considerations were treated as somewhat of an afterthought. While the original intent was to integrate economic considerations into each section of the plan as sections were developed, this generally did not happen. Following significant public concern regarding the lack of attention to economics in the 1994 draft plan, the LCBP contracted with outside consultants to analyze the economic implications of all plan recommendations in 1995. The consultants initially prepared a "preliminary" analysis [Holmes and Associates and Artuso, 1995a and 1995b], drawing heavily on relevant contingency valuation studies from other parts of the country. A subsequent analysis of the final plan [Holmes and Associates and Artuso, 1996] included an input/output model that used local data to predict the overall impact of plan implementation on the regional economy. The model's middle-range estimate, for example, predicted that plan implementation would result in 329-478 additional jobs and $10.3 to $14.6 million in value added to the basin economy- beginning in year six of the implementation process. The detailed analysis of expected costs together with the model's economic benefit estimates played a significant role in obtaining final approval of the plan in October 1996.

Holmes and Artuso [1995] also developed an optimization procedure to determine the cost-effectiveness of various strategies for attaining in-lake phosphorus goals. Linked to the lake phosphorus model [VT DEC and NYSDEC, 1997], the optimization procedure is a powerful decision-making tool that enabled the LCMC and other committees to evaluate the cost-effectiveness of a multitude of combinations of point and nonpoint source phosphorus reductions predicted to attain in-lake phosphorus goals. The procedure ultimately formed the technical basis for the landmark Lake Champlain phosphorus agreement included in *Opportunities for Action*. The phosphorus agreement provided a timetable for the states to achieve specific phosphorus reduction targets from point and nonpoint sources for 19 subwatersheds within the Lake Champlain Basin.

At least two main lessons can be learned from these experiences: 1) economic analyses of plan recommendations are critical to gaining public support and ultimate consensus on recommended actions, and 2) economic analyses can be more effective if they are used early on in the development of recommendations rather than simply to evaluate the impact of previously developed recommendations.

Planning must be based on sound science

The Special Designation Act called for the Management Conference to establish a "multi-disciplinary environmental research program for Lake Champlain" to be planned and conducted jointly with the Lake Champlain Research Consortium. The LCRC and LCMC organized a major research and monitoring workshop in 1991 to identify basin research needs [Watzin, 1992]. Much of this research agenda has now been completed, and this work combined with other research already in progress provided a strong technical basis for the recommendations in the plan. Both the phosphorus and toxics sections of the plan, for example, relied heavily on a series of research efforts.

The Lake Champlain Diagnostic-Feasibility Study [VTDEC and NYSDEC, 1997] is a fundamental underpinning of the plan's phosphorus reduction strategy. This bi-state research effort, initiated in 1989, measured phosphorus loadings to the lake from all tributaries, wastewater discharges, and other major sources, and developed a whole-lake phosphorus mass balance model for Lake Champlain. The final phase of the study incorporated the optimization procedure developed by Holmes and Artuso [1995] to derive load reduction targets and costs by lake segment watershed (watersheds corresponding to each of the lake's thirteen segments). This study answered a number of key questions including: How much phosphorus is entering the lake each year and where is it coming from? How much must phosphorus inputs to the lake be reduced in order to attain the in-lake phosphorus criteria? How much will these reductions cost? Are there sufficient opportunities for phosphorus reduction in the basin to attain the in-lake criteria in all lake segments? The answers to these questions allowed managers to develop the quantitative load reduction targets by watershed that are included in the phosphorus agreement in *Opportunities for Action*. A number of additional questions affecting implementation of the phosphorus reduction strategy were researched separately or are the subject of studies currently underway. For example, phosphorus transport through the watershed was addressed by Hoffmann et al. [1996] and through a recent study [Hughes et al., 1998] of phosphorus movement between agricultural fields and streams. The results of these studies will be essential to future evaluations of progress toward phosphorus reduction goals.

Similarly, research on sediment toxicity has been equally critical to the Basin Program's toxics reduction strategy. An initial screening study [McIntosh, ed., 1994] that analyzed sediment throughout the lake for the presence of trace metals and toxic organic compounds identified the three sites of concern (Cumberland Bay, Inner Burlington Harbor, Outer Malletts Bay) that are now a key focus of management efforts. This and subsequent research in Cumberland Bay lead to a major clean-up effort (involving removal of PCB-laden sludge) currently underway at that site. Decisions on whether to pursue remediation at the other sites of concern are being driven by further studies [e.g., McIntosh et al., 1997; Diamond et al., 1998] of contaminant distribution and potential ecological effects at these sites. The multi-year investment in a comprehensive sediment assessment effort has been crucial to the development of a rational toxics management strategy.

Public involvement is essential for long-term commitment

The environmental management literature is replete with references to the need for public involvement in the management process [e.g., Born and Sonzogni, 1995; Mazmanian and Sabatier, 1983], and the Lake Champlain experience has borne this out. The history of water chestnut management in the Lake Champlain Basin is one example. Water chestnut is a nonnative invasive plant that forms dense mats of vegetation in lake bays and other shallow areas, choking out native species and restricting human access to the water. Mechanical harvesting and hand pulling can effectively control the plant, provided that infested areas are harvested prior to seeds dropping from the plants *and that harvesting is repeated for at least five consecutive years at each site.* When adequate funding was provided by federal and state agencies during the 1980s, management efforts successfully prevented the plant from spreading north in Lake Champlain. However, when program funding was significantly reduced in the 1990s, water chestnut plants spread northward in Lake Champlain to new areas never before infested, and to several other inland waterbodies in Vermont [Lake Champlain Basin Program, 1998]. Strong public concern over this issue for the last several years, channeled primarily through the Lake Champlain citizens advisory committees, played an important role in restoring funding for this program in 1997 and 1998. Government agencies and state legislatures make funding decisions on an annual basis and often do not support this type of management program consistently. While adequate funding for water chestnut control was provided this year, continued pressure from the public will likely be required to sustain current water chestnut funding levels through (and beyond) the five consecutive years required to make a real difference.

Many other plan recommendations require similar long-term commitments of funding or other resources. An informed and involved public can play a big part in keeping these commitments a priority for relevant agencies, legislative bodies, or private groups.

The LCBP's education and outreach program has been key to many of the Program's successes. The education and outreach program is responsible for delivering school and community presentations, developing educational literature that presents technical information in a readable and interesting format, coordinating media relations, providing educator training, organizing annual State of the Lake conferences, and hosting public meetings as necessary to receive input on the plan. LCBP education and outreach staff also organize the annual Celebrate the Lake week (a special series of educational activities hosted by a wide variety of basin organizations and individuals) and maintain the LCBP's world wide web site. While a number of other entities also conduct educational activities on Lake Champlain issues, the LCBP's particular strength is coordinating the provision of services throughout the basin and linking outreach efforts to plan priorities.

LOOKING TO THE FUTURE: NEW CHALLENGES

The last year has been an important time of transition, moving the LCBP from plan development into plan implementation. Historically, this type of redirection has proven to be difficult, not only for the Lake Champlain Basin, but also for similar programs elsewhere in the country. Anticipating this, the LCMC included recommendations in the plan for specific institutional arrangements needed for the implementation phase, and called for the plan to be updated every two years with continued citizen involvement. However, the plan could not ensure continued momentum, nor could it address changes in leadership or changes in staff which affect institutional memory. The LCBP must overcome a variety of challenges for the program to meet its goals.

How to maintain and broaden public and financial support

Making real progress and staying visible are probably the best ways to ensure continued support; each year the workplan will have to be developed with this intention. An annual State of the Lake report which documents accomplishments and shortfalls, and is presented to the public for comment will be important to maintain accountability. It will also provide a crucial mechanism for the program to be responsive to public concerns in annual work plans. The State of the Lake report will also be important to sustain public interest and to document program success to members of congress and other legislative bodies. Congressional support has been critical to the procurement of ongoing funds to implement the plan. Fiscal year 1999 will be the first year that earmarked funds for Lake Champlain will be within all relevant federal agency base budgets, including the EPA's. This arrangement, if maintained into the future, should provide a steady source of funds to continue core program activities.

While federal funding has been a primary support to the LCBP, substantial contributions have also come from the states of New York and Vermont and numerous local entities. As the program matures, a wider range of funding sources must be explored and opportunities for leveraging funds must be pursued. Support from a full spectrum of organizations including public agencies, private entities and individuals will be needed to implement the plan and sustain the LCBP into the future.

How to evaluate progress

Another challenge will be to develop ways to effectively evaluate progress in meeting the goals set forth in *Opportunities for Action* and to communicate this progress to the public through the State of the Lake report and other means. LCBP's 1998 workplan includes some initial work on the development of environmental indicators. However, this work cannot be done in isolation; the states of Vermont and New York and the EPA are all also (to varying degrees) developing indicators to document changes in the environment. Strong partnerships with these other agencies will help in developing a set of meaningful indicators for the Lake Champlain Basin. With indicators in place, the LCBP will be better able to determine the effectiveness of particular management strategies and adjust these strategies as necessary.

How to address emerging issues without undermining past accomplishments

The LCMC recognized that the plan needed to be a flexible document and addressed this in the plan's subtitle: "An Evolving Plan for the Future of the Lake Champlain Basin". Not only may refinements to management strategies be warranted, but new issues may emerge and deserve attention by managers. The plan addresses this need for change by requiring biennial plan updates. However, the potential to lose important ground as a result of these updates, after so many years of negotiation during the initial planning phase, is a concern. Guidelines will need to be established before the planning process begins to guard against this type of loss. In addition, care will have to be taken to ensure that the energy devoted to the planning process does not detract from implementation efforts.

Applied research, designed to increase understanding of priority issues and to inform the development of management strategies, will remain important. In 1998, for example, the LCBP is supporting research in Missisquoi Bay and the South Lake to assist in refining phosphorus reduction strategies for these important lake segments. Other research efforts, such as an evaluation of the effectiveness of best management practices, may be necessary to more accurately determine progress towards phosphorus goals. Continued coordination between the LCBP and the Lake Champlain Research Consortium will be necessary to update research priorities and respond to emerging needs.

CONCLUSION

The Special Designation Act, informed by lessons learned through almost a century of watershed management activities, created an innovative management program for the Lake Champlain Basin. During the past seven years, program partners have learned many additional lessons. Program participants now face a variety of challenges including: how to maintain and broaden support for implementation efforts, how to measure and communicate progress, and how to remain responsive to emerging issues. The accumulated knowledge gained during the first seven years should help the program meet these challenges and continue to be a model for successful watershed management.

This paper represents the views of the authors and does not necessarily represent the views of the Environmental Protection Agency.

APPENDIX: LCBP COMMITTEE DESCRIPTIONS

Lake Champlain Management Conference (LCMC)

The LCMC existed from 1991-1996 and was responsible for overseeing plan development. Membership was specified in the Lake Champlain Special Designation Act and included representation from the following organizations or interests:

- New York and Vermont Lake Champlain Citizens Advisory Committees
- New York State Department of Environmental Conservation
- Vermont Agency of Natural Resources
- US Environmental Protection Agency
- US Department of Agriculture

- US Fish & Wildlife Service
- National Park Service
- University of Vermont, School of Natural Resources
- State University of New York, Plattsburgh
- Lake Champlain Committee
- Clinton County Legislature
- Vermont State Senate
- Vermont House of Representatives
- New York State Senate (appointee)
- Town of Westport, NY
- City of Burlington, VT
- Winooski Valley Park District
- The farming community (one farmer from NY, one from VT)
- The business community (one marina owner, one development corporation owner, one retail business owner, one consultant)
- Plattsburgh - North Country Chamber of Commerce
- Lake Champlain Technical Advisory Committee
- Lake Champlain Research Consortium

Lake Champlain Steering Committee

The Lake Champlain Steering Committee (LCSC) was established in 1988 through a memorandum of understanding among New York, Vermont and Quebec designed to improve lake management coordination. The LCSC assumed responsibility for overseeing implementation of the plan *Opportunities for Action* following the dissolution of the Lake Champlain Management Conference in 1996. The membership of the LCSC was expanded in 1997, as recommended in the plan, to include the following agencies and organizations:

- New York, Vermont and Quebec Citizens Advisory Committees
- New York State Department of Environmental Conservation
- New York State Department of Economic Development
- New York State Department of Agriculture and Markets
- New York State Office of Parks, Recreation and Historic Preservation
- Vermont Agency of Natural Resources
- Vermont Department of Agriculture, Food and Markets
- Vermont Agency of Transportation
- Vermont Agency of Commerce & Community Development
- Quebec Ministry of the Environment
- Two additional Quebec agencies (as needed)
- Lake Champlain Technical Advisory Committee
- US Environmental Protection Agency
- US Department of Agriculture
- US Department of Interior
- Local Government (currently Burlington VT and Plattsburgh NY are represented)

Technical Advisory Committee (TAC)

The TAC was established in 1991 to advise the Lake Champlain Management Conference on technical and scientific issues related to plan development. The TAC was reorganized somewhat in 1997 and now provides similar guidance to the Lake Champlain Steering Committee on plan implementation. Current agencies, organizations or specialties represented on the TAC include:

- University of Vermont, School of Natural Resources
- Adirondack Aquatic Institute
- VT Department of Environmental Conservation, Water Quality Division
- VT Department of Fish & Wildlife
- NYS Department of Environmental Conservation, Water Bureau
- NYS Department of Environmental Conservation, Fish & Wildllife
- NYS Department of Environmental Conservation, Public Lands
- Lake Champlain Research Consortium
- VT Department of Agriculture, Food and Markets
- NYS Soil and Water Conservation Committee
- Essex County (NY) Planning Office
- US Geological Survey
- US Department of Agriculture
- US Fish & Wildlife Service
- Quebec Ministry of the Environment
- Clinton County Health Department
- Miner Institute, Chazy, NY
- Recreation specialist (currently vacant)
- Economist (currently vacant)
- Planning and policy specialist (currently vacant)

Citizens Advisory Committees

The States of New York and Vermont and the Province of Quebec all have Lake Champlain Citizens Advisory Committees (CACs) created by the Steering Committee in 1988. The New York CAC has fourteen members appointed by the Commissioner of NYSDEC; the Vermont CAC currently has twelve members appointed by the Governor and Legislature, and the Quebec CAC has eight members appointed by the Minister of the Environment. The CACs make recommendations on the management of Lake Champlain to the Steering Committee and serve as liaisons to the public. Each committee operates independently in accordance with state or provincial authorizations. The plan, *Opportunities for Action*, recommended that the CACs include representatives from environmental groups, agriculture, business/industry, sportspersons, and local government. Most of these interests are currently represented on each of the CACs. In addition, CAC membership includes a variety of concerned citizens.

REFERENCES

Born, S.M., and W.C. Sonzogni, Integrated Environmental Management: Strengthening the Conceptualization, *Environmental Management* 19:167-181, 1995.

Diamond, J., A.L. Richardson and C. Daley, Ecological effects of sediment-associated contaminants in inner Burlington Harbor, Lake Champlain, Lake Champlain Basin Program Technical Report (draft), Grand Isle, VT, 1998.

Hoffman, J.P., E.A. Cassell, J.C. Drake, S. Levine, D.W. Meals, and D. Wang, Understanding phosphorus cycling, transport and storage in stream ecosystems as a basis for phosphorus management, Lake Champlain Basin Program Technical Report No. 20, Grand Isle, VT, 1996.

Holmes, T.P. and A. Artuso, Preliminary economic analysis of the Draft Plan for the Lake Champlain Basin Program, Lake Champlain Basin Program Technical Report No. 12B, Grand Isle, VT, 1995a.

Holmes, T.P. and A. Artuso, Preliminary economic analysis of the Draft Plan for the Lake Champlain Basin Program- Part 2, Lake Champlain Basin Program Technical Report No.14, Grande Isle, VT, 1995b.

Holmes, T.P. and A. Artuso, An economic analysis of the Draft Final Plan for the Lake Champlain Management Conference, Lake Champlain Basin Program Technical Report No. 17B, Grand Isle, VT, 1996.

Hughes, J.W., W.E. Jokela, D. Wang and C. Borer, Determination and quantification of factors buffering pollutant transfer from agricultural land to streams in the Lake Champlain Basin, Lake Champlain Basin Program Technical Report (draft), Grand Isle, VT, 1998.

Lake Champlain Committee, Burlington toxic reduction project - final report on the joint project of the Burlington Board of Health and the Lake Champlain Committee, submitted to Lake Champlain Basin Program, Grand Isle, VT, 1998.

Lake Champlain Basin Program, Background technical information for *Opportunities for action - an evolving plan for the future of the Lake Champlain Basin,* Lake Champlain Basin Program Technical Report # 16, Grand Isle, VT, 1996.

Lake Champlain Basin Program, State of the lake: a report on progress toward implementation of opportunities for action - an evolving plan for the future of the Lake Champlain Basin, Lake Champlain Basin Program, Grand Isle, VT, 1998.

Lake Champlain Management Conference, Opportunities for action - an evolving plan for the future of the Lake Champlain Basin, Lake Champlain Basin Program, Grand Isle, VT, 1996.

Mazmanian, D.A., and P.A Sabatier, Implementation and public policy, Scott Foresman, Illinois, 1983.

McIntosh, A., ed., Lake Champlain sediment toxics assessment program: an assessment of sediment associated contaminants in Lake Champlain, phase I. Lake Champlain Basin Program Technical Report No. 5, Grand Isle, VT, 1994.

McIntosh, A., M. Watzin and E. Brown, eds., Lake Champlain sediment toxics assessment program: an assessment of sediment-associated contaminants in Lake Champlain - phase II. Lake Champlain Basin Program Technical Report No. 23B, Grand Isle, VT, 1997.

Retired Engineers And Professionals, Final report on pollution prevention assistance to the vehicle service industry, submitted to the Lake Champlain Basin Program, Grand Isle, VT, 1996.

Retired Engineers And Professionals, workplan for a municipal pollution prevention project, submitted to the Lake Champlain Basin Program, Grand Isle, VT, 1997.

Vermont Department of Environmental Conservation and New York State Department of Environmental Conservation, A phosphorus budget, model, and load reduction strategy for Lake Champlain, VT DEC, Waterbury, VT, 1997.

Watzin, M.C., A research and monitoring agenda for Lake Champlain: proceedings of a workshop, December 17-19, 1991, Burlington, VT. Lake Champlain Basin Program Technical Report No. 1, Grand Isle, VT, 1992.

Yellow Wood Associates, Inc., Report on institutional arrangements for watershed management of the Lake Champlain Basin, Lake Champlain Basin Program Technical Report No. 11B, Grand Isle, VT, 1995.

Phosphorus Management in Lake Champlain

Eric Smeltzer

ABSTRACT

Eutrophication management in Lake Champlain has expanded over the past several years into a comprehensive approach involving the analysis of lakewide response to multiple point and nonpoint sources of phosphorus. A user survey analysis of the relationship between phosphorus levels and recreational use impairment was used to derive in-lake total phosphorus criteria for 13 segments of Lake Champlain. Annual water, chloride, and phosphorus loadings to the lake were measured by a field sampling program and used to support the development of a whole-lake phosphorus mass balance model. The model used a minimum-cost optimization procedure to identify the load reductions needed in each sub-watershed to attain the in-lake phosphorus criteria. Watershed phosphorus loading targets established by the model provided the basis for a Lake Champlain phosphorus reduction agreement negotiated by the States of Vermont and New York and the U.S. Environmental Protection Agency in 1996.

INTRODUCTION

Lake Champlain contains a diversity of environments with respect to phosphorus levels and trophic state (Figure 1). Lake regions such as Malletts Bay and the Main Lake have phosphorus concentrations in the low-mesotrophic range of 0.009-0.012 mg·L^{-1}. Eutrophic conditions prevail in areas such as St. Albans Bay, Missisquoi Bay and the South Lake, where phosphorus levels are in the range of 0.024-0.058 mg·L^{-1}. Water quality problems are most acute in these eutrophic bay areas. However, public concerns have been expressed at times throughout the lake for symptoms of eutrophication including algae blooms, reduced water clarity, and shoreline periphyton growth.

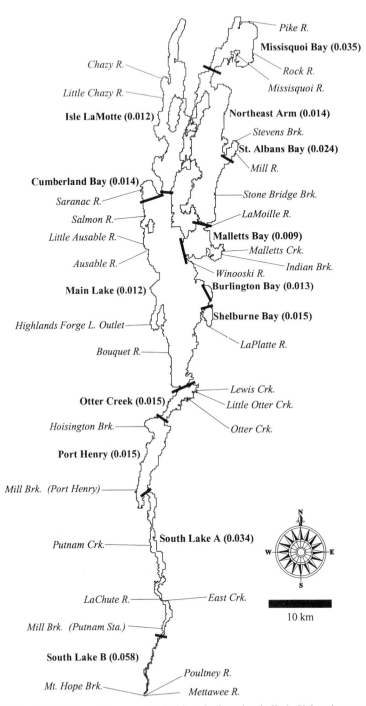

Figure 1. Map of Lake Champlain segments (bold) and tributaries (italics). Values in parentheses are 1991-1992 mean total phosphorus concentrations (mg·L^{-1}) in each lake segment [Vermont DEC and New York State DEC, 1997].

Phosphorus enrichment and eutrophication became a major water quality management issue in Lake Champlain during the 1970s. The first assessments of phosphorus sources to the lake [U.S. Environmental Protection Agency, 1974; Henson and Gruendling, 1977; Bogdan, 1978] estimated the total annual phosphorus load to the lake to be in the range of 536-804 t·yr^{-1}, with approximately half of the total derived from point source discharges within the lake basin.

A comprehensive planning effort conducted in the 1970s [Lake Champlain Basin Study, 1979] recommended several management actions to hold constant or reduce phosphorus inputs to the lake until 1990. These actions included the continuation of phosphorus detergent bans in Vermont, New York, and Canada, the construction of advanced wastewater treatment facilities for phosphorus removal at a number of Vermont municipal treatment plants located near the lake, and the implementation nonpoint source best management practices in priority watersheds.

Even as these recommendations were pursued during the 1980s, it became apparent that management policies, and the limnological data to support sound policies, were inadequate to protect Lake Champlain from the cumulative effects of phosphorus loading [Vermont Agency of Natural Resources and New York State Department of Environmental Conservation (DEC), 1988]. Lake Champlain receives phosphorus inputs from more than 90 point source discharges throughout its basin and nonpoint source runoff from 31 major tributaries. However, no single phosphorus source is dominant on a lakewide basis. Numeric water quality standards were needed to define acceptable levels of phosphorus in the lake. Updated loading estimates and a lake phosphorus mass balance model were needed to establish finite limits for the cumulative phosphorus loading to each segment of the lake. Point and nonpoint source management programs in Vermont, New York, and Quebec needed to be structured to ensure attainment of the allowable phosphorus loads for Lake Champlain.

The purpose of this paper is to describe work done during the 1990s to address these phosphorus management needs for Lake Champlain. As a result of this work, a phosphorus reduction agreement was reached between the States of Vermont and New York and the U.S. Environmental Protection Agency and incorporated into a comprehensive basin plan [Lake Champlain Management Conference, 1996].

NUMERIC PHOSPHORUS CRITERIA

The first step in developing a phosphorus reduction plan for Lake Champlain was the establishment of numeric, in-lake phosphorus concentration criteria consistently between the various management jurisdictions. The total phosphorus criteria for each lake segment listed in Table 1 were incorporated into Vermont's state water quality standards in 1991 and subsequently endorsed as management goals by Vermont, New York, and Quebec in a Lake Champlain Water Quality Agreement signed in 1993 [Lake Champlain Phosphorus Management Task Force, 1993].

The criteria were derived, in part, from a lake user survey and an analysis of the relationship between phosphorus levels and recreational impairment of Lake Champlain [Heiskary and Walker, 1988; Smeltzer and Heiskary, 1990; North American Lake Management Society, 1992]. A user survey form was used in Lake Champlain from 1987-1991 as part of a citizen volunteer water quality monitoring program [Picotte, 1998]. Survey

TABLE 1. Lake Champlain total phosphorus concentration criteria.

Lake Segment	Existing Phosphorus Concentration[a] ($mg \cdot L^{-1}$)	Total Phosphorus Criterion[b] ($mg \cdot L^{-1}$)	Basis for Criterion[c]
Malletts Bay	0.009	0.010	Oligotrophic condition
Main Lake	0.012	0.010	
Port Henry	0.015	0.014	
Otter Creek	0.015	0.014	
Shelburne Bay	0.015	0.014	One percent time frequency of nuisance algal condition (from user survey analysis)
Burlington Bay	0.013	0.014	
Cumberland Bay	0.014	0.014	
Northeast Arm	0.014	0.014	
Isle LaMotte	0.012	0.014	
St. Albans Bay	0024	0.017	Expected response to advanced wastewater treatment
Missisquoi Bay	0.035	0.025	
South Lake A	0.034	0.025	Moderate degree of eutrophication
South Lake B	0.058	0.025	

[a]1991-1992 mean value [Vermont DEC and New York State DEC, 1997]
[b][Lake Champlain Phosphorus Management Task Force, 1993]
[c][Vermont DEC, 1990; Lake Champlain Basin Program, 1996]

questions asked the observers to rate the physical condition of the lake water and the recreational suitability at the same time samples were obtained for total phosphorus analysis.

The user survey results from over 900 individual observations distributed among 28 stations in Lake Champlain are illustrated in Figure 2. These results were used to quantify the instantaneous phosphorus levels at which critical transitions in user perceptions of water quality and recreational enjoyment occur in Lake Champlain. User descriptions such as "crystal clear" or "a little algae," and "beautiful" or "very minor problems" predominate when total phosphorus concentrations are below 0.025 $mg \cdot L^{-1}$. Above 0.025 $mg \cdot L^{-1}$, perceptions of "definite algal greenness" or "high algae," and "enjoyment slightly impaired" or "enjoyment reduced or impossible" represent the majority of responses. These results suggested that an instantaneous total phosphorus concentration of 0.025 $mg \cdot L^{-1}$ could be used to derive eutrophication criteria values for Lake Champlain.

Lake eutrophication criteria are best expressed as season or annual mean values, rather than as instantaneous "not to exceed" values [Walker, 1985a; North American Lake Management Society, 1992]. Cumulative frequency distributions for total phosphorus at Lake Champlain stations were used to evaluate the relationship between the mean phosphorus value and the frequency of the 0.025 $mg \cdot L^{-1}$ instantaneous nuisance criterion value [Walker, 1985a]. As shown in Figure 2, a mean value of 0.014 $mg \cdot L^{-1}$ represents a phosphorus level at which the 0.025 $mg \cdot L^{-1}$ nuisance condition would be exceeded only 1% of the time during the summer. A nuisance frequency of 1% (i.e. about one bad day per

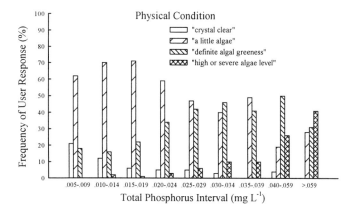
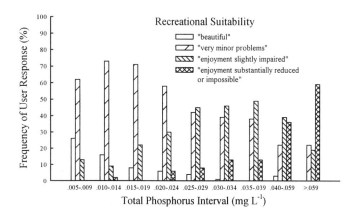
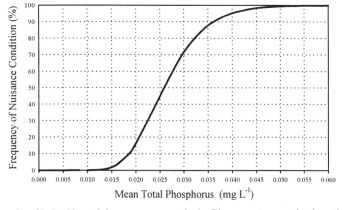

Figure 2. Results of Lake Champlain user survey analysis. The upper two graphs show the frequency of survey responses as a function of the instantaneous total phosphorus concentration measured at the time of observation. The lower graph shows the frequency (percent of time during the summer) of the 0.025 mg·L^{-1} nuisance algal condition as a function of the mean phosphorus value.

summer) was considered to be appropriately low for use in deriving nutrient water quality standards.

A mean total phosphorus criterion of 0.014 mg·L^{-1} was established for seven segments of Lake Champlain, as shown in Table 1. In other lake segments, higher or lower criteria values were established based on limitations of practical attainability, or to provide anti-degradation protection where existing phosphorus levels are below 0.014 mg·L^{-1} [Vermont DEC, 1990; Lake Champlain Basin Program, 1996]. The phosphorus criteria in Table 1 were used to guide the phosphorus modeling and load allocation studies described below.

WATER, CHLORIDE, AND PHOSPHORUS LOADS

A field study was conducted during 1990-1992 to quantify phosphorus loads to Lake Champlain from all major sources. The methods of sampling, data analysis, and results from the study are described in detail elsewhere [Vermont DEC and New York State DEC, 1997; Smeltzer and Quinn, 1996].

Hydrologic inputs were measured by a network of 31 tributary stream flow gages operated by the U.S. Geological Survey, representing 81% of the lake watershed area. Tributary samples (N = 36-115) were obtained near the mouth of each stream shown in Figure 1 and analyzed for total phosphorus, total dissolved phosphorus, and chloride. Sampling was conducted with an intentional bias toward high flow days to improve the precision of flow-stratified loading estimates [Verhoff et al., 1980]. Tributary annual mean mass loading rates were estimated using concentration vs. flow regression techniques provided by the FLUX program [Walker, 1987, 1990]. Other water and material budget terms were also estimated, including inputs from ungaged runoff, wastewater discharges, and direct precipitation, and losses by the Richelieu River outflow, water withdrawals, and evaporation [Vermont DEC and New York State DEC, 1997].

Mass loading rates are reported in this paper in units of metric tons per year (1 t·yr^{-1} = 1000 kg·yr^{-1}). Water flow rates are reported in units of cubic hectometers per year (1 hm^3·yr^{-1} = 10^6 m^3·yr^{-1}).

Water, chloride, and total phosphorus balances for Lake Champlain during the two-year sampling period are shown in Figure 3. Water and chloride inputs and losses balanced within statistical uncertainty limits. Residual budget errors were less than 2% for water and less than 4% for the conservative substance chloride. No attempt was made to account for groundwater inputs or losses, which were assumed to be minor. Net retention of phosphorus in the lake was estimated to be 80%.

Phosphorus loads from each tributary were estimated for calendar year 1991, which was defined as a hydrologic base year for phosphorus management purposes. Table 2 lists the base year flows and phosphorus loads from each tributary. Nonpoint source loading rates given in Table 2 were calculated by subtracting the loads discharged from all upstream point sources.

The total base year phosphorus loading rate to Lake Champlain was estimated to be 647 t·yr^{-1}, including inputs from monitored tributaries, unmonitored areas (estimated on a drainage area proportional basis relative to adjacent monitored tributaries), direct wastewater discharges to the lake, and precipitation direct to the lake surface. The 1991 value was within the range of 536-804 t·yr^{-1} estimated for the total phosphorus load to Lake Champlain during the 1970s [Bogdan, 1978], although the proportion of the total load derived from point sources (29%) was much lower in 1991 than in the 1970s.

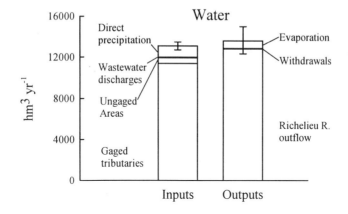
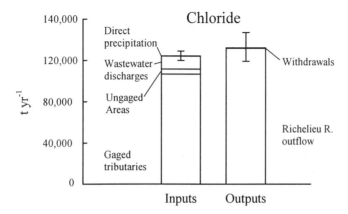
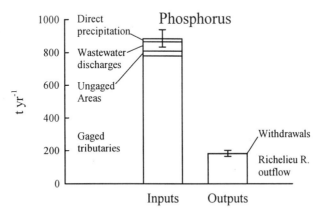

Figure 3. Water, chloride, and total phosphorus balances for Lake Champlain, March 1990 to February 1992. Error bars are 95% confidence intervals, calculated according to FLUX program procedures [Walker, 1987].

TABLE 2. Base year (1991) flows and phosphorus loading rates for Lake Champlain tributaries.

Tributary/Source	Drainage Area at Mouth (km^2)	Mean Flow (hm^3·yr^{-1})	Total Load (t·yr^{-1})	Point Source Load (t·yr^{-1})	Nonpoint Source Load (t·yr^{-1})
Otter	2,462	1,119	109.7	62.3	47.4
Winooski	2,828	1,543	83.8	24.3	59.5
Missisquoi	2,223	1,307	82.1	7.0	75.1
Pike	517	296	50.3	5.9	44.4
Mettawee/Barge Canal	1,098	487	37.1	3.4	33.7
Lamoille	1,909	1,100	29.6	3.1	26.5
Rock	152	69	28.9	0.0	28.9
Great Chazy	769	320	17.7	1.1	16.6
Poultney	692	273	17.1	3.0	14.1
Ausable	1,323	639	16.8	5.6	11.2
Saranac	1,575	776	16.4	8.7	7.7
Bouquet	712	281	13.5	0.0	13.5
LaPlatte	137	44	11.8	4.2	7.6
Little Otter	185	55	5.4	0.0	5.4
Lewis	209	90	5.2	0.0	5.2
Little Ausable	189	89	5.2	1.4	3.8
Mill	59	26	3.5	0.0	3.5
Stevens	59	14	3.4	0.0	3.4
Little Chazy	139	44	3.2	0.0	3.2
Malletts	76	31	1.7	0.0	1.7
Salmon	175	55	1.7	0.0	1.7
East	81	24	1.3	0.1	1.2
Putnam	160	67	1.3	0.0	1.3
LaChute	702	273	1.1	0.0	1.1
Indian	31	13	0.9	0.0	0.9
Stone Bridge	32	10	0.8	0.0	0.8
Mill (Port Henry)	73	25	0.6	0.0	0.6
Hoisington	28	10	0.5	0.0	0.5
Mill (Putnam Sta.)	27	9	0.4	0.0	0.4
Highlands Forge Lake Outlet	30	9	0.1	0.0	0.1
Mt. Hope	30	13	0.1	0.0	0.1
Ungaged Areas	1,028	424	21.7	0.0	21.7
Direct Wastewater Discharges		52	58.4	58.4	0.0
Direct Precipitation	1,130	915	16.0	0.0	16.0
TOTAL	20,840	10,503	647.3	188.5	458.8

MASS BALANCE MODEL

A steady-state phosphorus mass balance model was developed for Lake Champlain to simulate the response of total phosphorus concentrations in each segment of the lake to changes in phosphorus loading rates. A steady-state approach was chosen because the in-lake criteria values established to guide the analysis (Table 1) were expressed as seasonal or annual mean values. The model employed the linear branching network of 13 lake segments illustrated in Figure 4. Each segment was modeled as a mixed-reactor, since vertical water column phosphorus concentration gradients were small in comparison with concentration differences between lake segments [Vermont DEC and New York State DEC, 1997].

The steady-state mass balance equation for an individual lake segment is given in Equation 1 [Chapra and Sonzogni, 1979; Chapra and Reckhow, 1983].

$$V_i \, dc_i/dt = 0 = W_i + \Sigma_j \left\{Q_{ji}c_j - Q_{ij}c_i + E_{ij}(c_j - c_i)\right\} - k_i V_i c_i^2 \tag{1}$$

Where

V_i = volume of segment i (hm^3)
c_i = concentration in segment i (mg·L^{-1})
c_j = concentration in adjacent segment j (mg·L^{-1})
W_i = direct external mass loading to segment i (t·yr^{-1})
Σ_j = summation over all adjacent segments j
Q_{ji} = advective inflow to segment i from upstream segment j (hm^3·yr^{-1})
Q_{ij} = advective outflow from segment i to adjacent downstream segment j (hm^3·yr^{-1})
E_{ij} = bulk exchange flow between adjacent segments i and j (hm^3·yr^{-1})
k_i = second-order sedimentation coefficient for segment i (m^3·g^{-1}·yr^{-1})

The flow routing scheme shown in Figure 4 was used to define the connections between adjacent lake segments. The advective inflow and outflow terms (Q_{ji} and Q_{ij}) in Equation 1 include the accumulated net sum of water inputs from monitored tributaries, ungaged watershed areas, wastewater discharges, direct precipitation, evaporation, and water withdrawals. One lake segment (Malletts Bay) required special two-dimensional treatment in the model. Advective and exchange flows from Malletts Bay were partitioned between adjacent lake segments according to field observations [Myer, 1977; Myer and Gruendling, 1979], with 86% of the flow routed to the Main Lake segment and the remainder going to the Northeast Arm.

Bulk exchange flow terms (E_{ij}) were evaluated using chloride loads and in-lake concentrations measured during the field study [Vermont DEC and New York State DEC, 1997]. Sedimentation rates (k_i) were assumed to be zero for the conservative substance chloride. Equation 1 was solved for the exchange flows at each segment boundary by direct linear matrix inversion, using a modified version of the BATHTUB program [Walker, 1987, 1992]. The values used for the chloride model terms for each lake segment and the calculated exchange rates are given in Table 3.

The second-order formulation for the phosphorus sedimentation term (k_i) in Equation 1 was found to provide a better empirical fit of the BATHTUB model to a U.S. reservoir data set than the more common first-order models [Walker, 1985b]. Use of a second-order decay term is consistent with the concept of poorer nutrient recycling efficiency in eutrophic environments, compared with oligotrophic lakes.

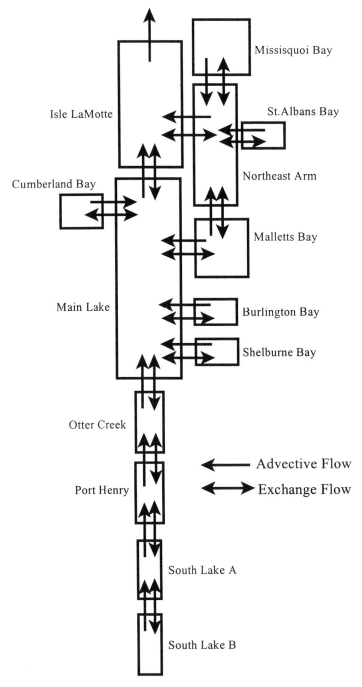

Figure 4. Lake Champlain mass balance model flow routing scheme.

TABLE 3. Chloride model terms for the period of March 1990 to February 1992.

Lake Segment	Segment Volume (hm^3)	Segment Area (hm^2)	External Inflow (hm^3·yr^{-1})	Chloride Load (t·yr^{-1})	Chloride Conc. (mg·L^{-1})	Exchange Flow[a] (hm^3·yr^{-1})
South Lake B	7.8	579	1,092	11,384	11.62	712
South Lake A	125	4,327	629	14,679	13.47	1,259
Port Henry	1,463	7,555	150	1,283	11.18	13,998
Otter Creek	955	2,849	1,648	15,769	10.72	49,427
Main Lake	16,787	41,414	3,530	38,250	10.61	8,861
Shelburne Bay	140	962	79	2,206	10.89	4,816
Burlington Bay	63	551	9	598	10.78	2,986
Cumberland Bay	63	1,075	950	5,941	10.18	8,672
Malletts Bay	722	5,506	1,529	14,098	9.43	272 (52[b])
Northeast Arm	3,380	24,825	130	997	9.29	1,968
St. Albans Bay	23	721	63	2,320	10.20	1,844
Missisquoi Bay	205	8,994	2,039	15,407	7.78	297
Isle LaMotte	1,892	18,559	514	4,931	10.33	

[a] At boundary with downstream lake segment.
[b] Value for the northern boundary of Malletts Bay with the Northeast Arm.

The optimal second-order sedimentation parameter estimate of 100 m^3·g^{-1}·yr^{-1}, derived independently from the U.S. reservoir data set [Walker, 1985b], was applied without modification to ten of the 13 Lake Champlain segments [Smeltzer and Quinn, 1996; Vermont DEC and New York State DEC, 1997]. The sedimentation rates for two lake segments (Malletts Bay and Missisquoi Bay) were increased to 400 m^3·g^{-1}·yr^{-1} to improve the fit of model-predicted lake phosphorus concentrations to the observed Lake Champlain data. Calibration of the model for the St. Albans Bay segment required setting the phosphorus sedimentation rate to zero and adding an internal load of 8.6 t·yr^{-1} to the direct mass inputs in order to obtain an acceptable model fit. The non-linear phosphorus mass balance equations were solved by an iterative procedure in the BATHTUB program [Walker, 1987].

The values used for the phosphorus model terms and the phosphorus concentrations in each lake segment predicted by the calibrated model are given in Table 4. Figure 5 shows that a good overall calibration fit (root mean squared error = 0.034, log$_{10}$ scale) was obtained between model-predicted phosphorus concentrations in each lake segment and the measured mean values. Uncertainty estimates for model input terms and model predictions (e.g. 95% confidence intervals, Figure 5) were derived from error analysis procedures in the FLUX and BATHTUB programs [Walker, 1987].

LOAD REDUCTION STRATEGY

The calibrated phosphorus mass balance model was used to evaluate alternative load reduction strategies to attain the in-lake criteria in each lake segment. A process for assigning phosphorus loading targets for point and nonpoint sources in each sub-watershed

TABLE 4. Phosphorus model terms for the 1991 base year.

Lake Segment	External Inflow ($hm^3 \cdot yr^{-1}$)	Phosphorus Load ($t \cdot yr^{-1}$)	Calibrated Sedimentation Coefficient ($m^3 \cdot g^{-1} \cdot yr^{-1}$)	Observed Phosphorus Conc. ($mg \cdot L^{-1}$)	Predicted Phosphorus Conc. ($mg \cdot L^{-1}$)
South Lake B	830	56.7	100	0.058	0.049
South Lake A	410	15.3	100	0.034	0.031
Port Henry	96	5.8	100	0.015	0.015
Otter Creek	1,283	121.3	100	0.015	0.014
Main Lake	2,708	133.0	100	0.012	0.012
Shelburne Bay	47	16.4	100	0.015	0.015
Burlington Bay	7	11.7	100	0.013	0.016
Cumberland Bay	828	38.1	100	0.014	0.015
Malletts Bay	1,176	33.0	400	0.009	0.009
Northeast Arm	66	6.5	100	0.014	0.014
St. Albans Bay	43	16.6[a]	0	0.024	0.022
Missisquoi Bay	1,720	167.9	100	0.035	0.035
Isle LaMotte	413	31.4	100	0.012	0.012

[a] Includes an internal load of 8.6 $t \cdot yr^{-1}$ added for model calibration.

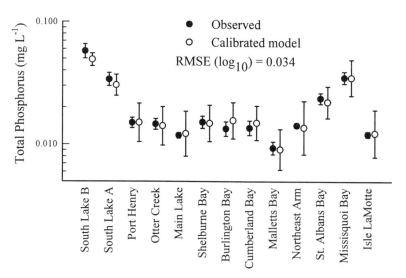

Figure 5. Phosphorus model calibration results, comparing the observed mean and model-predicted total phosphorus concentrations in each lake segment. Error bars are 95% confidence intervals.

TABLE 5. Base year (1991) phosphorus loads to Lake Champlain from Vermont and New York, and watershed target loads.

Lake Segment	1991 Load (t·yr⁻¹)			Target Load (t·yr⁻¹)		
	VT	NY	Total	VT	NY	Total
South Lake B	28.0	28.2	56.2	20.8	26.2	47.0
South Lake A	2.4	13.1	15.5	0.6	9.4	10.1
Port Henry	0.4	4.3	4.7	0.1	2.5	2.6
Otter Creek	121.7	0.1	121.8	56.1	0.0	56.2
Main Lake	88.0	38.9	126.9	76.6	35.0	111.6
Shelburne Bay	16.4	0.0	16.4	12.0	0.0	12.0
Burlington Bay	11.5	0.0	11.5	3.1	0.0	3.1
Cumberland Bay	0.0	38.0	38.0	0.0	25.5	25.5
Malletts Bay	32.9	0.0	32.9	28.6	0.0	28.6
Northeast Arm	3.2	0.0	3.2	1.2	0.0	1.2
St. Albans Bay	8.0	0.0	8.0	9.5	0.0	9.5
Missisquoi Bay[a]	167.3	0.0	167.3	109.7	0.0	109.7
Isle LaMotte	0.6	28.3	28.8	0.3	21.5	21.8
Direct Precipitation			16.0			16.0
TOTAL	480.4	150.9	647.3	318.6	120.2	454.9

[a]Includes loads from both Vermont and Quebec.

in each state was negotiated by the States of Vermont and New York and the U.S. Environmental Protection Agency [Lake Champlain Management Conference, 1996].

Point source loading targets were established for each sub-watershed before the phosphorus model was applied. Allowable point source loads (Table 5) were calculated by assuming an advanced treatment effluent total phosphorus concentration of 0.8 mg·L⁻¹ at all wastewater treatment facilities larger than 200,000 gal·day⁻¹ (757 m³·day⁻¹) in permitted flow, exempting facilities using aerated lagoon processes.

The remaining nonpoint source phosphorus load reductions needed in each lake segment watershed were identified by the model using a minimum-cost optimization procedure [Chapra et al., 1983; Holmes and Artuso, 1995; Vermont DEC and New York State DEC, 1997]. Information on the cost and effectiveness of agricultural and urban phosphorus control practices specific to each sub-watershed was obtained from the U.S. Natural Resources Conservation Service and from an economic analysis [Holmes and Artuso, 1995]. The calibrated mass balance model equations and cost-effectiveness data were transferred to a commercial spreadsheet program. Numeric solution techniques in the spreadsheet program were used to find the least-cost distribution of nonpoint source load reductions among the sub-watersheds. The optimization procedure was constrained by the need to attain the in-lake phosphorus criteria values without exceeding the maximum potential nonpoint source load reduction considered possible in each sub-watershed.

The nonpoint source loading targets resulting from the optimization procedure are shown in Table 5. Compliance with the 0.025 mg·L⁻¹ criterion for the South Lake B segment was

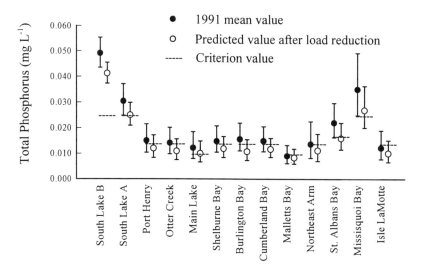

Figure 6. Phosphorus load reduction modeling results showing the observed 1991 mean total phosphorus concentration in each lake segment, and comparing the in-lake criteria values with the concentrations predicted after all watershed loading targets are achieved. Error bars are 95% confidence intervals for the observed mean and model-predicted values.

waived in the procedure because of the unrealistically large nonpoint source reductions needed to achieve that criterion. However, substantial reductions were still required in the South Lake B segment watershed (Table 5) in order to attain the 0.025 mg·L^{-1} criterion for the South Lake A segment downstream. The endpoint for the Missisquoi Bay segment was modified in the analysis from 0.025 mg·L^{-1} (Table 1) to 0.027 mg·L^{-1} for similar reasons. The load reduction analysis assumed that the 8.6 t·yr^{-1} internal load applied to the St. Albans Bay segment during the model calibration will decline to zero over time as excessive sediment phosphorus is gradually depleted in the bay [Martin et al., 1994].

The results of the load reduction analysis shown in Table 5 indicated that the total phosphorus load to Lake Champlain should be limited to 455 t·yr^{-1}, representing a net reduction of 192 t·yr^{-1} (30%) from the 1991 base year total load of 647 t·yr^{-1}. As shown in Figure 6, the model predicted that these load reductions will result in attainment of the in-lake criteria for most lake segments with a 50% or greater level of confidence.

The point and nonpoint source loading targets for each lake segment watershed derived from the modeling analysis (Table 5) were adopted as part of a comprehensive basin plan for Lake Champlain [Lake Champlain Management Conference, 1996]. The loading targets are to be achieved over a time period of 20 years.

DISCUSSION

Substantial progress was made during the 1990s in managing phosphorus in Lake Champlain. Agreements were reached between the various government jurisdictions on the desired in-lake water quality goals and on a division of responsibility for achieving the necessary phosphorus load reductions. Implementation of point and nonpoint source controls reduced phosphorus loading to the lake by 20% between 1991 and 1995 [Lake Champlain Management Conference, 1996]. The results of the analyses described in this paper contributed directly to the development of the water quality agreements and the formation of the comprehensive basin plan for Lake Champlain.

The relatively simple mass balance model used to derive the phosphorus loading targets for Lake Champlain has certain limitations that should be addressed by future research. By not accounting in detail for internal phosphorus cycling between the lake sediments and the water column, the model assumed a constant net sedimentation rate as external loadings change. However, studies in St. Albans Bay [Martin et al., 1994] and in other shallow lakes [Havens and James, 1997] have shown that sediment phosphorus storage may cause long response time lags that are not predicted by simple models.

Missisquoi Bay and the South Lake are two shallow, eutrophic regions of Lake Champlain where substantial loading reductions are planned. The sediment phosphorus release model applied to St. Albans Bay [Martin et al., 1994; Chapra and Canale, 1991] should be extended to these areas of Lake Champlain. Benthic phosphorus cycling research and modeling work in progress on Lake Champlain [HydroQual, Inc., 1998; Cornwell and Owens, 1998] should be used to refine predictions of the long-term water quality response of these regions to phosphorus management actions.

Acknowledgments. This work was conducted with the support and assistance of staff from the Vermont Department of Environmental Conservation and the New York State Department of Environmental Conservation. The study was supported by a grant from the U.S. Environmental Protection Agency Clean Lakes Program. Cooperative assistance was provided by the U.S. Geological Survey.

REFERENCES

Bogdan, K.G. 1978. Estimates of the annual loading of total phosphorus to Lake Champlain. Prep. for New England River Basins Commission. Burlington, VT.

Chapra, S.C. and K.H. Reckhow. 1983. *Engineering Approaches to Lake Management. Vol. 2: Mechanistic Modeling.* Butterworth. Boston.

Chapra, S.C. and W.C. Sonzogni. 1979. Great Lakes total phosphorus budget for the mid 1970s. *J. Water Poll. Contr. Fed.* 51:2524-2533.

Chapra, S.C., H.D. Wicke, and T.M. Heidtke. 1983. Effectiveness of treatment to meet phosphorus objectives in the Great Lakes. *J. Water Poll. Contr. Fed.* 55:81-91.

Chapra, S.C. and R.P. Canale. 1991. Long-term phenomenological model of phosphorus and oxygen in stratified lakes. *Water Research.* 25:707-715.

Cornwell, J.C. and M. Owens. 1998. Benthic phosphorus cycling in Lake Champlain. Draft report submitted to Lake Champlain Basin Program. Grand Isle, VT.

Havens, K.E. and R.T. James. 1997. A critical evaluation of phosphorus management goals for Lake Okeechobee, Florida, USA. *Lake and Reserv. Manage.* 13:292-301.

Heiskary, S.A. and W.W. Walker. 1988. Developing phosphorus criteria for Minnesota lakes. *Lake and Reserv. Manage.* 4:1-10.

Henson, E.B. and G.K. Gruendling. 1977. The trophic status and phosphorus loadings of Lake Champlain. U.S. Environ. Prot. Agency. EPA-600/3-77-106.

Holmes, T.P. and A. Artuso. 1995. Preliminary economic analysis of the draft plan for the Lake Champlain Basin Program. Prep. for Lake Champlain Management Conference. Lake Champlain Basin Program Tech. Rep. No. 12b. Grand Isle, VT.

HydroQual, Inc. 1998. Benthic phosphorus cycling in Lake Champlain: Results of an integrated field sampling/water quality modeling study. Draft report submitted to Lake Champlain Basin Program. Grand Isle, VT.

Lake Champlain Basin Program. 1996. Background technical information for: Opportunities for action - an evolving plan for the future of the Lake Champlain Basin. Tech. Rep. No. 16. Grand Isle, VT.

Lake Champlain Basin Study. 1979. Shaping the Future of Lake Champlain. Final Report. Prep. for New England River Basins Commission. Burlington, VT.

Lake Champlain Management Conference. 1996. Opportunities for action. An evolving plan for the future of the Lake Champlain Basin. Pollution prevention, control, and restoration plan. Lake Champlain Basin Program. Grand Isle, VT.

Lake Champlain Phosphorus Management Task Force. May 14, 1993 report prepared for the Lake Champlain Steering Committee. New York State Department of Environmental Conservation, Adirondack Park Agency, Quebec Ministry of the Environment, and Vermont Agency of Natural Resources.

Martin, S.C., R.J. Ciotola, P. Malla, N.G. Subramanyaraje Urs, and P.B. Kotwal. 1994. Assessment of sediment phosphorus distribution and long-term recycling in St. Albans Bay, Lake Champlain. Lake Champlain Basin Program Tech. Rep. No. 7c. Grand Isle, VT.

Myer, G.E. 1977. Currents of northern Lake Champlain. *In* J.C. Dawson (ed.). Proc. Lake Champlain Basin Environmental Conference. August 9-10, 1977. Chazy, NY.

Myer, G.E. and G.K. Gruendling. 1979. Limnology of Lake Champlain. Prep. for New England River Basins Commission. Burlington, VT.

North American Lake Management Society. 1992. Developing eutrophication standards for lakes and reservoirs. Report prepared by the Lake Standards Subcommittee. Alachua, FL. 51 pp.

Picotte, A. 1998. 1997 Lake Champlain Lay Monitoring Report. Vermont Department of Environmental Conservation. Waterbury, VT.

Smeltzer, E. and S.A. Heiskary. 1990. Analysis and applications of lake user survey data. *Lake and Reserv. Manage.* 6:109-118.

Smeltzer, E. and S. Quinn. 1996. A phosphorus budget, model, and load reduction strategy for Lake Champlain. *Lake and Reserv. Manage.* 12:381-393.

U.S. Environmental Protection Agency. 1974. Report on Lake Champlain New York and Vermont. National Eutrophication Survey Working Paper No. 154.

Verhoff, E.H., S.M. Yaksich, and D.A. Melfi. 1980. River nutrient and chemical transport estimation. *J. Environ. Eng. Div. A.S.C.E.* 106:591-608.

Vermont Agency of Natural Resources and New York State Department of Environmental Conservation. 1988. The Framework for the Vermont-New York Workplan on Lake Champlain. Waterbury, VT and Albany, NY.

Vermont Department of Environmental Conservation. 1990. A proposal for numeric phosphorus criteria in Vermont's Water Quality Standards applicable to Lake Champlain and Lake Memphremagog. Waterbury, VT.

Vermont Department of Environmental Conservation and New York State Department of Environmental Conservation. 1997. A phosphorus budget, model, and load reduction strategy for Lake Champlain. Lake Champlain diagnostic-feasibility study final report. Waterbury, VT and Albany, NY.

Walker, W.W. 1985a. Statistical bases for mean chlorophyll *a* criteria. pp. 57-62 *In* Lake and Reservoir Management: Practical Applications. Proceedings of the Fourth Annual Conference and International Symposium. October 16-19, 1984. McAfee, NJ. North American Lake Management Society.

Walker, W.W. 1985b. Empirical methods for predicting eutrophication in impoundments. Report 3. Phase II: Model Refinements.. Tech. Rep. E-81-9. Prep. for U.S. Army Corps Eng. Waterways Exp. Sta. Vicksburg, MS.

Walker, W.W. 1987. Empirical methods for predicting eutrophication in impoundments. Report 4. Applications Manual. Tech. Rep. E-81-9. Prep. for U.S. Army Corps Eng. Waterways Exp. Sta. Vicksburg, MS.

Walker, W.W. 1990. FLUX stream load computations. Version 4.4. Prep. for U.S. Army Corps Eng. Waterways Exp. Sta. Vicksburg, MS.

Walker, W.W. 1992. BATHTUB empirical modeling of lake and reservoir eutrophication. Version 5.1 draft. Updated December 1992. Prep. for U.S. Army Corps Eng. Waterways Exp. Sta. Vicksburg, MS.

List of Contributors

Denise Arruda
Saint Michael's College
Department of Biology
Winooski Park
Colchester, VT 05439

Anthony Artuso
Cornell University
Natural Resource Policy & Management
University of Charleston
Institute for Public Affairs & Policy Studies
66 George St.
Charleston, SC 29424

Richard S. Artz

Glenn A. Bauer
Assistant Professor of Biology
Saint Michael's College
Biology Department
Winooski Park
Colchester, VT 05439

Claude Bernard

Vicki Blazer
USGS
National Fish Health Research Lab. at Leetown
Box 700
Kearneysville, WV 25430

M. A. Borchardt
Marshfield Research Medical Foundation
Marshfield, Wisconsin 54449

M. Braner
Vermont Department of Health
108 Cherry
Burlington VT 05401

Susan Bulmer

E. Alan Cassell
School of Natural Resources
University of Vermont
Burlington, VT 05405

Richard J. Ciotola

Art Cohn
Director
Lake Champlain Maritime Museum
RR3 Box 4092
Vergennes, VT 05491

Denis Côté
Institut de recherche et de developpement en agroenvironement
Complexe Scientifique
2700 rue Einstein
Ste-Foy (Quebec)
G1P 3W8
CANADA

Ann Cousins

C. Daley
Tetra Tech, Inc.
10045 Red Run Blvd, Suite 110
Owings Mills, MD 21117

Jon C. Denner
U.S. Geological Survey
Montpelier, Vermont 05601

Jerome M. Diamond
Tetra Tech, Inc.
10045 Red Run Blvd, Suite 110
Owings Mills, MD 21117

Andrea F. Donlon
School of Natural Resources
Aiken Center
University of Vermont
Burlington, VT 05405

John C. Drake
Geology Department
303 Perkins Building
University of Vermont
Burlington, VT 05405

Alan Duchovnay

Diana Dunbar
Saint Michael's College
Department of Biology
Winooski Park
Colchester, VT 05439

Brian T. Dwyer
Harvard Medical School, Pathology
Thorn 630, BWM Hospital
20 Shattuck St
Boston, MA 02115

Steven D. Faccio
Vermont Institute of Natural Science
RR 2 Box 532
Churchill Road
Woodstock, VT 05091

Douglas E. Facey
Saint Michael's College
Department of Biology
Box 283
Winooski Park
Colchester, VT 05439

Phil Girton
VT Dept. of Forests, Parks & Recreation
111 West St.
Essex Jct., VT 05452-4695

James P. Hoffman
Botany Department
230 Marsh Life Science Building
University of Vermont
Burlington, VT 05405

Timothy P. Holmes
Holmes & Associates
PO Box 295
Saranac Lake, NY 12983

Thomas R. Hudspeth
UVM Environmental Program & School
of Natural Resources
153 S. Prospect
Burlington, VT 05401

Kenneth L. Hunkins
Rte. 9W
Lamont-Doherty Earth Observatory
Columbia University
Palisades, NY 10983

Tatsu Isaji

Gerald J. Keeler
Air Quality Laboratory
University of Michigan
Ann Arbor, Michigan

John King
Graduate School of Oceanography
University of Rhode Island

Mark S. LaBarr
Green Mountain Audubon Society
255 Sherman Hollow Road
Huntington, VT 05462

Cynthia Leclerc
Saint Michael's College
Department of Biology
Winooski Park
Colchester, VT 05439

Brendon J. Leddy

Suzanne N. Levine
School of Natural Resources
Aiken Center
University of Vermont
Burlington, VT 05405

Prashant Malla

Patricia L. Manley
Geology Department
Middlebury College
Middlebury, VT 05753

Thomas O. Manley
Geology Department
Middlebury College
Middlebury, VT 05753

Scott C. Martin
Civil & Environmental Engineering
Youngstown State University
Youngstown, OH 44555

Ryan P. Maynard

Alan W. McIntosh
332 Aiken Center
School of Natural Resources
University of Vermont
Burlington, VT 05401

Donald W. Meals, Jr.
School of Natural Resources
University of Vermont
Burlington, VT 05405

Daniel Mendelsohn

Aubert Michaud
Centre de recherche et d'experimentation en sols
Ministere del'agriculture, des pecheries et de l'alimentation du Quebec
2700 rueEinstein, Sainte-Foy
Quebec G1P-3W8
CANADA

Gerald S. Miller
National Oceanic and Atmospheric Administration
Great Lakes Environmental Research Laboratory
2205 Commonwealth Blvd.
Ann Arbor, Michigan 48105

Leah M. Moyer

Eric Perkins
Vermont State Unit
Office of Ecosystem Protection
EPA region 1
1 Congress St., Suite 1100
Boston, MA. 02114

Richard L. Poirot
VT Department of Environmental Conservation
Air Pollution Control Division
Bldg. 3 South, 3rd floor
103 South Main St.
Waterbury, VT 05671-0402

Lori Pyzocha
Saint Michael's College
Department of Biology
Winooski Park
Colchester, VT 05439

A. L. Richardson
Tetra Tech, Inc.
10045 Red Run Blvd, Suite 110
Owings Mills, MD 21117

Christopher C. Rimmer
Vermont Institute of Natural Science
RR 2 Box 532
Churchill Road
Woodstock, VT 05091

James H. Saylor
National Oceanic and Atmospheric Administration
Great Lakes Environmental Research Laboratory
2205 Commonwealth Blvd.
Ann Arbor, MI 48105

Timothy D. Scherbatskoy
School of Natural Resources
Aiken Center
University of Vermont
Burlington, VT 05405

Bret Schichtel
Washington University
Center for Air Pollution Impact and Trend Analysis (CAPITA)
Campus Box 1124
One Brookings Dr.
St. Louis, MO, 63130-4899

Angela D. Shambaugh
205 Aiken Center
School of Natural Resources
University of Vermont
Burlington, VT 05405-0086

James B. Shanley
U.S. Geological Survey
Montpelier, VT 05601

Eric Smeltzer
State Limnologist
VT Department of Environmental Conservation
103 South Main Street, Building 10 N
Waterbury, VT 05671-0408

Lee Steppacher
Environmental Protection Agency
Vermont State Program Unit
JFK Federal Building
Boston, MA. 02203

Patricia Straughan
Shelburne Farms
Shelburne, VT. 05482

Barbara J. B. Stunder

Douglas R. Thomas
Economic & Policy Resources, Inc.
PO Box 1660
Williston VT 05495-1660

Thi Sen Tran)

Subramanyaraje N. G. Urs

Deane Wang
School of Natural Resources
University of Vermont
Burlington, VT 05482

Mary Watzin
332 Aiken Center
School of Natural Resources
University of Vermont
Burlington, VT 05401

Paul Wishinski
Air Quality Planning Chief
VT Dept. of Environmental Conservation
103 South Main St.
Waterbury, VT 05671-0402